Emerging Trends in Plant Pathology

Emerging Trends in Plant Pathology

Edited by Edward Parker

MURPHY & MOORE
www.murphy-moorepublishing.com

Published by Murphy & Moore Publishing,
1 Rockefeller Plaza,
New York City, NY 10020, USA
www.murphy-moorepublishing.com

Emerging Trends in Plant Pathology
Edited by Edward Parker

International Standard Book Number: 978-1-63987-190-2 (Hardback)

Cataloging-in-Publication Data

Emerging trends in plant pathology / edited by Edward Parker.
p. cm.
Includes bibliographical references and index.
ISBN 978-1-63987-190-2
1. Plant diseases. 2. Diseased plants. 3. Plants--Wounds and injuries.
4. Plant pathologists. I. Parker, Edward.
SB601 .E44 2022
632--dc23

TABLE OF CONTENTS

PREFACE

Plant pathology can be defined as the scientific study of diseases in plants caused by pathogens and environmental conditions. It comprises the study of pathogen identification, disease etiology, disease cycles, economic impact, plant disease epidemiology, management of plant diseases, and how they affect humans and animals. Although all plants are subject to diseases, some plant varieties are more vulnerable to outbreaks of diseases, whereas others are more resistant to them. The study of plant pathology aims at preserving the environment, growing healthy crops, ensuring adequate food supply, and discovering new ways of controlling plant diseases. This book explains the emerging trends and innovative models around prospective developments with respect to plant pathology. Some of the diverse topics covered herein address the varied branches that fall under this category. This book will serve as a valuable source of reference for aspiring plant pathologists, students, and researchers.

This book has been the outcome of endless efforts put in by authors and researchers on various issues and topics within the field. The book is a comprehensive collection of significant researches that are addressed in a variety of chapters. It will surely enhance the knowledge of the field among readers across the globe.

It gives us an immense pleasure to thank our researchers and authors for their efforts to submit their piece of writing before the deadlines. Finally in the end, I would like to thank my family and colleagues who have been a great source of inspiration and support.

Editor

1

Grapevine, esca complex, and environment: the disease triangle

Michael FISCHER[1,*], Samira PEIGHAMI ASHNAEI[2]

[1] *JKI-Julius Kühn-Institute, Federal Research Centre for Cultivated Plants, Plant Protection in Fruit Crops and Viticulture, D-76833 Siebeldingen, Germany*
[2] *Plant Disease Research Department, Iranian Research Institute of Plant Protection (IRIPP), Agricultural Research, Education and Extension Organization (AREEO), Tehran, Iran*
*Corresponding author: michael.fischer@julius-kuehn.de

Competing Interests: The Author(s) declare(s) no conflict of interest.

Editor: Laura Mugnai, University of Florence, Italy.

Abstract. This review compiles the available knowledge on the triple impact of host-pathogens-environment in the progress of the esca disease complex of grapevine. The perennial crop grapevine encounters different biotic and abiotic factors responsible for numerous changes at the various growth stages. This review provides increased understanding of the esca disease complex, with emphasis on (1) the nature of esca-associated fungi as endophytes or pathogens in grapevine, (2) the importance of grapevine genotype and age in relation to resistance or susceptibility to the pathogens, (3) the significant effects of climatic changes, especially drought, on pathogen development and symptomatology, and (4) the physio-biochemical changes in the grapevines arising from the biotic and abiotic interactions. Drought often provides conditions favouring disease development in plants. Physiological and biochemical changes in plants play critical roles in this topic. The constantly increasing economic impacts of esca disease in many grape-producing countries, and the broad lack of knowledge so far, require precise studies on the transcriptional responses to biotic and abiotic factors in grapevines, as effects of "climate change" develop. On the viticultural side, improved management of water and adjusted nutrition balance in vineyards may become useful strategies to mitigate the widespread damage caused by grapevine wood pathogens.

Keywords. Climate change, esca disease complex, grapevine, water stress.

INTRODUCTION

Viticulture, like other fields of agriculture, is deeply affected by a wide variety of biotic and abiotic factors. For the past three decades, the economic burden of trunk diseases such as esca has become a limiting factor for grapevine production in many countries (Larignon and Dubos, 1997; Mugnai *et al.*, 1999; Dubos *et al.*, 2002; Bertsch *et al.*, 2009; Úrbez-Torres, 2011; Bruez *et al.*, 2013; Mohammadi *et al.*, 2013; De la Fuente *et al.*, 2016). Advances in control of grapevine leaf strip disease (GLSD), the main disease in esca complex, have only recently been made with the introduction of *Trichoderma* as a biological treatment for wound protection, and with the possibility to reduce symptom expression by applications of mixtures of foliar fertilizers (Di Marco *et al.*, 2004; Calzarano *et al.*, 2014; Calzarano and Di Marco, 2018).

Esca is a disease complex, generally characterized by the development of typical inner necrosis in grapevine wood tissues and external symptoms known as "tiger-striped" leaves or black measles on the berries, assigned to infection by pathogenic fungi that invade the perennial plants and their vascular systems (Larignon and Dubos, 1997; Mugnai *et al.*, 1999; Graniti *et al.*, 2000). Petri disease, as part of the esca complex (Surico, 2001; Gramaje and Armengol, 2011) is caused by several fungi in the genera *Phaeomoniella* and *Phaeoacremonium* (Kubátová *et al.*, 2004; Mostert *et al.*, 2006; Essakhi *et al.*, 2008; Gramaje *et al.*, 2009; Gramaje and Armengol, 2011). Mature vineyards (10 years or more) are also affected by white rot linked with Basidiomycete taxa such as *Fomitiporia mediterranea* and related species (Fischer, 2002; Fischer and Kassemeyer, 2003; Fischer and Binder, 2004; Fischer *et al.*, 2005; Fischer, 2006; Péros *et al.*, 2008; Luque *et al.*, 2009; Cloete, 2015; Fischer and González García, 2015).

In recent years there has been progressive evolution in the concept of esca and the diseases that are included within this complex (for example see Mugnai *et al.*, 1999; Calzarano *et al.*, 2001, 2014; Edwards *et al.*, 2001; Larignon *et al.*, 2001, 2009; Surico *et al.*, 2008; Surico, 2009; Lecomte *et al.*, 2012; Bertsch *et al.* 2013; Fontaine *et al.*, 2016). Following the suggestion of Surico (2009), GLSD (previously known as "young esca") is considered the most important and widespread disease within the esca complex. Other diseases that are discussed within esca include brown wood streaking, Petri disease (black goo or slow dieback), and esca proper (including white rot). Co-existence of GLSD and esca proper is a common feature in mature vineyards. However, all of the above diseases may overlap, as they also do with canker agents such as those caused by members of the Botryosphaeriaceae or Diatrypaceae (Rolshausen *et al.*, 2010; Gramaje *et al.*, 2018; Moyo *et al.*, 2018).

The above picture is not adopted throughout the community of pathologists and others with interests in esca, and differing views are taken, for instance, by Lecomte *et al.* (2012) or Larignon and collaborators (Larignon *et al.*, 2001; 2009). Other, and less specific, designations have also been commonly used, including "manifest and hidden esca" (Marchi *et al.*, 2006), "grapevine dieback disease" (Amponsah *et al.*, 2011), "esca decline" (Lecomte *et al.*, 2008, 2011) or "esca dieback foliar expression" (Guérin-Dubrana *et al.*, 2012).

The diseases included in the esca complex are often cryptic and symptoms usually take several years to develop, in some of the diseases discontinuously (Surico *et al.*, 2000, 2006; Christen *et al.*, 2007; Calzarano *et al.*, 2018). A potentially important group of pathogens are endophytic fungi (Petrini, 1986), which may be pathogenic or mutualistic, depending on the circumstances (Schulz and Boyle, 2005; Alvarez-Loayza *et al.*, 2011). In the endophytic phase they asymptomatically colonize their hosts to form latent infections, but may modify their behaviour and become invasive when the plant gets stressed (Verhoeff, 1974; Gubler *et al.*, 2005; Aroca *et al.*, 2006; Surico *et al.*, 2006; Slippers and Wingfield, 2007; Sakalidis *et al.*, 2011; Hofstetter *et al.*, 2012). There has been considerable debate during the past two decades whether, and to what extent, environmental factors influence fungal development in vineyards and the expression of symptoms (Calzarano *et al.*, 2018). Several biotic and abiotic factors (Cramer *et al.*, 2007; McDowell *et al.*, 2008; Deluc *et al.*, 2009; Andreini *et al.*, 2014; Kovacs *et al.*, 2017) naturally affect the physio-biochemical processes of grapevines, and subsequently their responses to pathogens, as in the case of GLSD phytoalexin patterns during leaf symptom development (Calzarano *et al.*, 2016; 2017a; 2017b).

It is well known that exposure of plants to different stresses generally induces the expression of various genes. As a result, various enzymes and plant hormones (Vanholme *et al.*, 2008; Vogt, 2010) with multiple biological functions are produced in response to different environmental stimuli. In this way, the variable virulence of potential pathogens, the host defense responses, and environmental conditions constitute the disease triangle (Agrios, 2005).

This eventually leads to an even more diffused picture of esca and the related diseases, making it challenging to clearly separate between diseases and fungal endophytes/pathogens and stress related symptomatology. In the case of grapevine and esca, whenever possible we refer to the specific designations of the diseases.

In this review, we refer to a large amount of literature, most of which is original research papers. In addition, we cite several review articles, which provide useful summaries of certain aspects of the "plant and stress" topic. A considerable proportion of the references refers to plants other than grapevine. While we particularly emphasize grapevine and the esca complex, numerous entries in Tables 1 and 2 also refer to other groups of plants. First, this underlines the overall significance of the subject, i.e. interrelations between stress factors and host plants. Secondly, data generated for non-grapevine hosts may assist understanding of esca and related disease phenomena.

ESCA-ASSOCIATED FUNGI: ENDOPHYTES OR PATHOGENS?

Some vascular-inhabiting fungi isolated from declining vines (Ferreira *et al.*, 1999) or other hosts have been considered as latent opportunistic pathogens, and cause diseases when their hosts are subjected to abiotic stresses (Pearce, 1996; Slippers and Wingfield, 2007; Jactel *et al.*, 2012).

One of the main features of GLSD is the variability in external symptom expression, whereby symptoms on leaves and berries may be obvious one year but are not apparent in another. The significant increase in foliar symptoms in 20-year-old potted vines induced by excessive watering (Surico *et al.*, 2010), and the relationship between growing season rainfall (particularly in July) and leaf symptoms expression (Calzarano *et al.*, 2018) have confirmed the role of rainy seasons and water in the soil in GLSD symptom development (Surico *et al.*, 2000; Marchi *et al.*, 2006). Factors extraneous to the plant-pathogen interactions (e.g. light, soil type, nutrients, and water availability) have been hypothesized to play roles in the erratic nature of the foliar symptoms of esca (Mugnai *et al.*, 1999; Calzarano *et al.*, 2007; 2009; 2014). The observed rates of mortality of vines are not necessarily, or not always, in accordance with the incidence of external symptoms (Andreini *et al.*, 2014). However, Calzarano and collaborators (2018) observed greater mortality of GLSD affected vines that showed severe leaf symptoms at the first appearance, compared to vines with low symptom severity.

Although several studies have dealt with various aspects of esca related pathogens, the causes of foliar symptom development still remain elusive (Surico *et al.*, 2006; Larignon *et al.*, 2009; Camps *et al.*, 2010). It was hypothesized that toxic metabolites produced by fungi in colonized wood reach leaves through the xylem vessels, triggering defense responses that result in the development of leaf symptoms (Evidente *et al.*, 2000; Tabacchi *et al.*, 2000; Andolfi *et al.*, 2011; Bertsch *et al.*, 2013). These plant responses include the formation of necrotic lesions on leaves as a hypersensitive reaction. In GLSD symptomatic vines, phytoalexins increased with increasing severity of leaf symptoms, confirming that these substances are synthesized as a reaction to lesions occurring on leaves, usually followed by formation of antimicrobial compounds such as stilbene derivatives (Heath, 2000; Calzarano *et al.*, 2016; 2017a; 2017b).

No strict correlation has been found between the occurrence of symptoms and the extent of pathogen colonization or wood necrosis (Calzarano and Di Marco, 2007; Liminana *et al.*, 2009; Fischer and Kassemeyer, 2012). In contrast, Lecomte *et al.* (2008) showed that vines exhibiting leaf symptoms had, on average, greater proportions of necrotic wood than asymptomatic plants. The degree of wood necrosis may also be positively correlated with the mortality of vines (Liminana *et al.*, 2009).

No significant difference was observed between the fungal communities that inhabit apparently healthy and visibly diseased individual plants (Hofstetter *et al.*, 2012). Both *Phaeomoniella chlamydospora* and *Phaeoacremonium* spp. have been isolated from asymptomatic tissue (Mugnai *et al.*, 1999; Bruez *et al.*, 2014; Elena *et al.*, 2018), supporting the concept of latent phases for these pathogens. Furthermore, the time delay between infection and symptom expression (Di Marco and Osti, 2008) may reinforce the role of other factors in esca disease development. A possible explanation for this situation may be the different environmental factors prevailing. The varying interactions between such factors and symptom expression caused by vascular pathogens has been extensively reviewed by Mundy and Manning (2011). Characterizing the impacts of abiotic factors in grapevine responses may be a key step towards obtaining complete knowledge of physicochemical changes in plants that lead to disease development and the appearance of symptoms.

GRAPEVINE GENOTYPE AND AGE

Genotype

Genotype relationships with possible tolerance of *Vitis* spp. to biotic and abiotic stresses has been the focus of many studies. Scion cultivars and the rootstocks may differ in their responses to stress. Rootstocks provide benefits, such as drought stress tolerance (Alsina *et al.*, 2007; Koundouras *et al.*, 2008; Gambetta *et al.*, 2012). However, information is scarce on the genetic backgrounds of these benefits, and specific responses to environmental factors.

Grapevine rootstocks have been shown to play important roles in adaptation to water deficit (Soar *et al.*, 2006; Marguerit *et al.*, 2012; Parker *et al.*, 2013), while scion transpiration rates and their acclimation to water deficit are, through different genetic architecture, controlled genetically by rootstocks (Marguerit *et al.*, 2012). Alsina *et al.* (2011) showed that scion stomatal conductance was more strongly down-regulated with drought-sensitive rootstocks (*V. riparia* × *V. rupestris* cv. 101-14Mgt) than with drought-tolerant ones (*V. berlandieri* × *V. rupestris* cv. Paulsen 1103).

Rootstock impacts may be mediated by chemical (Soar *et al.*, 2006; Alsina *et al.*, 2011), hormonal (particularly abscisic acid: ABA) and hydraulic signaling in responses to water deficit (Soar *et al.*, 2006; Vandeleur *et al.*, 2009; Marguerit *et al.*, 2012). At the hydraulic level in roots, radial water movement mediated by aquaporins was shown to have a significant potential contribution to drought stress adaption (Vandeleur *et al.*, 2009; Lovisolo *et al.*, 2010). Aquaporins also affect root hydraulic conductance (Perrone *et al.*, 2012), water uptake by fine roots (Koundouras *et al.*, 2008, 2009; Gambetta *et al.*, 2012) and, consequentially, vine growth (Pouzoulet *et al.*, 2014). Selection of appropriate rootstocks may increase tolerance to extended drought stress periods, and even resistance to pathogens.

Rootstock genotype may affect the frequency of external symptoms (Marchi, 2001; Murolo and Romanazzi, 2014). However, leaf symptoms are not necessarily related to alterations observed in wood (Eskalen *et al.*, 2001; Feliciano *et al.*, 2004; Calzarano and di Marco, 2007; Romanazzi *et al.*, 2009; Andreini *et al.*, 2013; Murolo and Romanazzi, 2014).

The comparative transcriptome analysis in Pinot Noir grapevine highlighted rootstock-dependent differences in the response of genes involved in the jasmonic acid (JA) metabolism and pathogenesis-related (PR) proteins (Berdeja *et al.*, 2014). JA is a well known regulator of wound responses against pathogens, and plays a role in water stress perception and regulation and, subsequently, in expression of several related genes (Bell and Mullet, 1991; Zhang and Memelink, 2009; see Table 1 for more information on gene expression induced by environmental factors in different plants). In grapevine, production of PR proteins is the most frequent defense reaction to fungal infection (Derckel *et al.*, 1999); they accumulate in leaves and berries after pathogen infection and contribute to grapevine resistance (Giannakis *et al.*, 1998). Berdeja *et al.* (2014) indicated that, under drought stress, the transcript abundance of some specific PR genes, e.g. PR1 and PR4, differed between rootstock cultivars. This indicates a link between JAs and ABA in water-stressed grapevines (Suhita *et al.*, 2004).

As for scion cultivars, *V. vinifera* Chardonnay and Merlot have been classified as less susceptible to esca than Cabernet Sauvignon, which is considered as susceptible due to the high proportions of symptomatic vines observed in different climatic conditions (Christen *et al.*, 2007; Andreini *et al.*, 2009; Bruez *et al.*, 2013; Murolo and Romanazzi, 2014). Varieties with small xylem vessels are considered less likely to express foliar symptoms because of a limited drought-induced xylem cavitation (Pouzoulet *et al.*, 2014). Furthermore, own rooted vines sometimes show lower incidence of esca disease than grafted vines (Andreini *et al.*, 2014), possibly because they avoid contamination by fungal spores on wounds at graft unions during propagation processes (Halleen *et al.*, 2003; Hofstetter *et al.*, 2012). In contrast, vines grafted on SO4, with low drought resistance, and those grafted on 1103P, conferring high drought resistance, had very similar amounts of apoplexy, while the rate in own-rooted vines, also with high drought resistance, was much higher (Marchi, 2001).

With other combinations of cultivar and rootstock, the rootstocks may be less important than other factors, both internal and external. This suggests that the susceptibility of vine genotypes to esca depends on environment and genotype factors (Marchi, 2001). The complexity of esca, and the effects of several environmental factors in appearance of symptoms, make evaluation of cultivar susceptibility very difficult. In-depth and broad genetic studies of varieties in response to biotic and abiotic stresses are necessary.

Age

Age of plants is a key factor governing the interaction between responses to the biotic and abiotic stresses, and whether plants are tolerant or susceptible to stress factors; the overall effect of stress combination on plants largely depends on their age (Pandey *et al.*, 2015). Age of vines also affects disease incidence, resulting in greater incidence of diseased plants in old vineyards. Older plants experience more infection cycles than young plants, through wounds they have received (McCutcheon *et al.*, 1993; Mugnai *et al.*, 1999; Zabalgogeazcoa, 2008; Amponsah *et al.*, 2011; Kovacs *et al.*, 2017). As a result, there are positive correlations between vine age and trunk disease incidence, and, on a global scale, the disease remains the greatest limitation to maintaining optimum yields (Brown *et al.*, 2016).

CLIMATIC AND ENVIRONMENTAL CHANGES AS ABIOTIC FACTORS

Climatic and edaphic factors have been reported to influence the incidence of esca and other grapevine trunk diseases (Graniti *et al.*, 2000; van Niekerk *et al.*, 2011; Sosnowski *et al.*, 2011; Calzarano *et al.*, 2018). This indicates that the pathogens are not the only agents responsible for the disease symptoms. Changes in environmental conditions are known to exacerbate disease symptoms in plants (Boyer, 1995). In latent vine infections, the presence of pathogens does not necessar-

Table 1. Gene expression in response to genotype and various environmental factors and the putative action in different plants.

Gene Expression	Plant species (as cited in respective research papers)	Environmental factors	Putative action	Reference
Pathogenesis-related proteins	*Vitis* ssp. *V. vinifera* *V. vinifera*	Pathogen infection; Genotype	Defense genes against pathogens	Giannakis *et al.*, 1998 Derckel *et al.*, 1999 Letousey *et al.*, 2010
Phenylalanine ammonia lyase (PAL)	*Vitis* sp. *V. vinifera* *V. vinifera* *V. vinifera*	Pathogen infection; Water stress	Related to synthesis of the phytoalexin in grapevine and general responses against biotic and abiotic stresses	Melchior and Kindl, 1991 Grimplet *et al.*, 2007 Deluc *et al.*, 2009 Letousey *et al.*, 2010
Abscisic acid (ABA)	*Zea mays* *Glycine max* *Z. mays* *V. vinifera* *Arabidopsis thaliana* *V. vinifera* *A. thaliana* *Solanum lycopersicum* *V. vinifera* *Vitis* spp. *Pinus taeda*	Water stress; Rootstock genotype	Stomatal closure; Regulation of transpiration and root hydraulic traits; Affecting SA-, JA and ET-related defense genes; Suppressing PAL activity and ROS production	Zhang and Davies, 1990 McDonald and Cahill, 1999 Kizis and Peges, 2002 Soar *et al.*, 2004 Kariola *et al.*, 2006 Soar *et al.*, 2006 Adie *et al.*, 2007 Asselbergh *et al.*, 2008 Vandeleur *et al.*, 2009 Lovisolo *et al.*, 2010 Lorenz *et al.*, 2011
Cytokinins	*S. lycopersicum* *Z. mays*	Water stress	Influencing shoot responses	Kudoyarova *et al.*, 2007 Alvarez *et al.*, 2008
Sugars	*V. vinifera* *Prunus persica* *V. vinifera* *V. vinifera* *V. vinifera* *P. persica* *P. persica*	Water stress; Genotype	Sugar accumulation in fruit	Freeman and Kliewer, 1983 Golding *et al.*, 2006 Castellarin *et al.*, 2007 Deluc *et al.*, 2009 Koundouras *et al.*, 2009 Lopresti *et al.*, 2014 Cirilli *et al.*, 2016
Amino acids	*Ocimum* sp. *V. vinifera* *V. vinifera*	Water stress; Rootstock genotype	Adjusting the osmotic potential in cytoplasm; Affecting berry composition	Khalid, 2006 Deluc *et al.*, 2009 Berdeja *et al.*, 2014
Polyamines (PAs)	*Oryza sativa* *Z. mays*	Saline stress; Pathogen infection	Involved in many physiological processes, such as cell growth and response to stress tolerance	Krishnamurthy and Bhagwat, 1989 Rodriguez-Kessler *et al.*, 2008
Anthocyanins	*Pisum sativum*	UV-B radiation; Water stress	Developing a light red color in leaves and playing a primary role in photosynthesis;	Noguees *et al.*, 1998
	V. vinifera		Playing a major role in wine quality, including colour, flavor and bitterness	Berdeja *et al.*, 2014
Lignin	*Z. mays*	Water stress	Important in plant defense	Alvarez *et al.*, 2008
Stilbenes	*Sorghum bicolor* *V. vinifera*	Pathogen infection; Water stress	Protection against environmental challenges	Yu *et al.*, 2005 Grimplet *et al.*, 2007
Flavonoids	*P. sativum* *Hordeum vulgare*	UV-B radiation; Water stress; Saline stress	Decreasing the auxin level in plant tissues and causing growth retardation	Noguees *et al.*, 1998 Ali and Abbas, 2003

ily result in the immediate appearance of symptoms (Di Marco and Osti, 2008), as abiotic factors affect the appearance and the severity of the disease (Lecomte *et al.*, 2011). As a result, incidence and symptom profiles of diseases associated with the different esca pathogens may vary between different climatic areas (van Niekerk *et al.*, 2011). Climate changes will not occur uniformly in different regions, particularly for host and pathogen exposure to drought (Schultz, 2016).

Environmental changes may affect the nature of host responses to the pathogens. Firstly, plants adjust to environmental challenges by tightly and differentially regulating their transcriptomes (Baker *et al.*, 1997; Chen *et al.*, 2002; Yamaguchi and Shinozaki, 2006). These alterations, often found associated with duration and severity of stresses, determine a plant's ability to respond to internal and external signals, and to adjust to changing conditions (Eastburn *et al.*, 2011; Pandey *et al.*, 2015). Plant responses to environmental changes, e.g. development of thicker wax layers on leaves or changes in stomatal densities, could impact the processes of infection and expression of symptoms (Campbell and Madden, 1990). Abiotic stresses have impacts on the defense-growth trade-off that plants face whenever there is a pathogen attack (Huang *et al.* 2008; Leakey *et al.* 2009; Kontunen-Soppela *et al.*, 2010), causing resistance or susceptibility to the pathogens (Kuldau and Yates, 2000; Amtmann *et al.*, 2008; Mittler and Blumwald, 2010). Environmental changes can also have direct effects on the pathogens. With esca, as indicated above, climatic conditions influence not only the distribution of pathogens but also the disease symptoms they cause (Surico *et al.*, 2000; Marchi *et al.*, 2006; van Niekerk *et al.*, 2011; Calzarano *et al.*, 2018).

Climatic changes have the potential to alter the incidence and severity of plant disease epidemics, as well as to reshape the co-evolutionary relationships between pathogens and host plants (Burdon *et al.*, 2006; Ziska and Runion, 2007; Crowl *et al.*, 2008). Drought and temperature, as the most influential factors, affect pathogens by altering growth rates, propagule germination, and the rates of inoculum production (Huber and Gillespie, 1992). Eventually, increased colonization of plants by pathogens may occur in the presence of abiotic stresses (Koga *et al.*, 2004).

Drought

With the background of "climate change", the frequencies and intensities of drought periods are increasing worldwide. Increased drought leads to enhanced plant respiration, resulting in carbon (C) losses at the plant level (Schultz, 2000; Moriondo *et al.*, 2011), and plant death due to C starvation (Martinez-Vilalta *et al.*, 2002; Breda *et al.*, 2006; McDowell *et al.*, 2008; Adams *et al.*, 2013; see Table 2 for additional information on the effect of water stress on different plants).

Low soil water content and resulting water deficit have been considered as causes of stress on grapevines (Lovisolo and Schubert, 1998; Lovisolo *et al.*, 2010; Lanari and Silvestroni, 2015). Among different stress combinations that occur in vineyards, the influential interaction between esca and drought has often been considered (Surico *et al.*, 2000, 2006, 2010; Edwards *et al.*, 2007b, c; Luque *et al.*, 2010; Fischer and Kassemeyer, 2012; Bostock *et al.*, 2014; Ramegowda and Senthil-Kumar, 2015).

Analysis of *V. vinifera* plants subjected to individual drought stress and/or pathogen infection revealed down-regulation of transcripts involved in photosynthesis, nutrient assimilation, and cellular homeostasis (Choi *et al.*, 2013). Two mechanisms have been suggested to illustrate how water deficit increases the susceptibility of grapevine to pathogen attack: first, limitation of photosynthesis (Flexas *et al.*, 1999; Escalona *et al.*, 1999) eliminates the plant's ability to produce defensive compounds, and, second, plant growth is reduced in the presence of the pathogen, allowing further progression of the pathogen and eventually expression of symptoms.

Drought-induced plant death under pathogen attack is influenced by the type of interaction between pathogen and host (Oliva *et al.*, 2014). Studies reviewed by Boyer (1995) showed that predisposition to disease is often observed in host plants experiencing soil water deficits, and, among other cases, this has been illustrated in increases of bacterial leaf scorch symptoms (in *Parthenocissus*: McElrone *et al.*, 2001,), and esca-related wood symptoms (caused by *P. chlamydospora* in *Vitis*: Fischer and Kassemeyer, 2012). Conversely, it was shown that resistance is typically restored when water stress is remediated (in *Pinus*: Johnson *et al.*, 1997).

The pathogen may act: i) simultaneously with drought, as an opportunistic agent taking advantage of the effects of water stress on the host; or ii) prior to drought events (Oliva *et al.*, 2014). Drought occurring during infection may have greater impact than previous drought episodes (Croise *et al.*, 2001). For example, vascular wilt pathogens can accelerate drought-induced mortality by damaging the xylem vascular system, causing phloem impairment and foliage wilting (Oliva *et al.*, 2014). These conditions tend to increase evapotranspiration, producing favourable conditions for the development of xylem embolism, as is also observed in esca (Pouzoulet *et al.*, 2014).

Table 2. Different plant species and their adaptive changes in response to water stress.

Plant species (as cited in respective research papers)	Physio-biochemical changes	Reference
Acer platanoides, *Populus tremula*, etc.	Stomatal closure	Aasamaa *et al.*, 2001
Juglans regiaxxnigra		Cochard *et al.*, 2002
Betula alleghaniensis, *B. davurica*, etc.		Gu and Rom, 2007
Vitis vinifera		Letousey *et al.*, 2010
Zea mays	Changes in photosynthetic rate and Carbon reserves	Westgate and Boyer, 1985
V. vinifera		Christen *et al.*, 2007
Arabidopsis thaliana		Hummel *et al.*, 2010
Pinus edulis		Sevanto *et al.*, 2014
V. vinifera	Enhanced respiration	Schultz, 2000
A. saccharum, *Thuja occidentalis*, etc.	Lower shoot hydraulic conductance and leaf specific conductivity	Tyree and Sperry, 1988
V. vinifera	Lower transectional areas in xylem vessels	Lovisolo and Schubert, 1998
Eucalyptus globulus	Impaired function of phloem	Pate and Arthur, 1998
Vitis sp.	Aquaporin gene expression	Galmes *et al.*, 2007
V. vinifera	Growth reduction	Shellie and Brown, 2012
Melissa officinalis	Chlorophyll loss	Munne-Bosch and Alegre, 1999
Z. mays	Alteration in root structure and function	Zhang *et al.*, 1995
V. berlandieri×V. rupestris		Dry *et al.*, 2000
Lycopersicom esculentum		Mingo *et al.*, 2004
Glycine max	ABA-responsive signaling pathway, e.g. the activation of JA-related defense genes, alteration in PAL activity, etc.	Ward *et al.*, 1989
Z. mays		Zhang and Davies, 1990
G. max		McDonald and Cahill, 1999
L. esculentum		Audenaert *et al.*, 2002
A. thaliana		Kariola *et al.*, 2006
A. thaliana		Adie *et al.*, 2007
V. vinifera		Grimplet *et al.*, 2007
Solanum lycopersicum		Asselbergh *et al.*, 2008
V. vinifera		Deluc *et al.*, 2009
Vitis sp.		Lovisolo *et al.*, 2010
S. lycopersicum	Cytokinin production	Kudoyarova *et al.*, 2007
Z. mays		Alvarez *et al.*, 2008
V. vinifera	Sugar accumulation	Castellarin *et al.*, 2007
V. vinifera		Deluc *et al.*, 2009
V. vinifera		Koundouras *et al.*, 2009
Ocimum sp.	Accumulation of amino acids, *e.g.* proline	Khalid, 2006
V. vinifera		Deluc *et al.*, 2009
V. vinifera		Berdeja *et al.*, 2014
G. max	Reduction of isoflavone content	Gutierrez-Gonzalez *et al.*, 2010
Z. mays	Changes in lignin content	Alvarez *et al.*, 2008

Evidence of esca complex and drought interactions in grapevine

The effects of esca pathogens and water stress on grapevines have been intensively studied. Water stress exacerbated decline symptoms associated with *P. chlamydospora* in young plantations (Ferreira *et al.*, 1999). In glasshouse experiments, leaf water potentials in vines subjected to water stress were reduced when the vines were infected by *P. chlamydospora*, indicating that infection altered host plant responses to water stress, making it difficult for vines to get water to their leaves (Edwards *et al.*, 2007b; c). Pasquier *et al.* (2013) showed that foliar symptoms of esca proper may impact stress-related pathways in grapevines, e.g. sHSP (small heatshock protein) which is induced during water, salt, and oxidative stresses (Löw *et al.*, 2000; Scharf *et al.*, 2001). Christen *et al.* (2007), however, showed that foliar symptoms (in the case of GLSD) was not simply a water transport-deficit disease, but that xylem dysfunction due to pathogen spread partly explained the appearance of the foliar symptoms.

Intensity and timing of water deficit

Water supply plays an important role in plants under stress. Several studies have shown that water availability induces modifications of vessel diameter in different plants, including *V. vinifera* (Lovisolo and Schubert, 1998; Fichot *et al.*, 2009; Bauerle *et al.*, 2011). Annual changes in esca symptom expression may be partly due to differences in the size of new vessels formed under different water regimes (Pouzoulet *et al.*, 2014).

Root systems that are able to maintain water uptake under low water availability may increase drought resistance (Passioura, 2002; Comas *et al.*, 2013; Barrios-Masias and Jackson, 2014). Hydraulic conductivity and suberization of grapevine roots can change dramatically when they are subjected to water deficit, so irrigation management could be used to modify the water uptake capacity of root systems (Barrios-Masias *et al.*, 2015). Moderate water deficit increases root growth of grapevine, as this is required for water uptake from deeper layers of soil, and maintains stomatal conductance under declining soil moisture (Alsina *et al.*, 2007, 2011). Water use efficiency is consequentially optimized (Shellie and Brown, 2012).

Intensity and timing of water deficit may play critical roles in grapevine health and productivity, particularly in semi-arid regions. Schreiner and Lee (2014) suggested that there will be only small negative impacts of late-season water deficit in Pinot noir grapevines, from evidence with plants grown in pot-in-pot microplots.

Wine grapes are often intentionally grown under water deficit regimes to meet wine quality goals (Castellarin *et al.*, 2007). In contrast, increased amounts of rainfall during summer and soils with high water reserves were shown to increase severity of GLSD and other diseases in the esca complex (Calzarano *et al.*, 2016; 2017a, 2017b; Guérin-Dubrana *et al.*, 2005, 2012; Marchi *et al.*, 2006). Apoplectic symptoms are often correlated with excess of soil water combined with hot weather, leading to dramatic imbalance between foliar transpiration and root absorption (Surico *et al.*, 2006).

OTHER ABIOTIC FACTORS

Adding plant nutrients (fertilizers) to soils, to promote plant growth, was reported to enhance esca disease expression. For esca proper, high availability of nutrients in a growing season increased the proportion of diseased vines with symptoms, and decreased the proportion of infected but symptomless vines (Calzarano and Di Marco, 2007; Calzarano *et al.*, 2009).

Soil salinity also is important. High salinity causes damage to plants from water deficit due to osmotic stress and by ion toxicity from excessive sodium ions (Zhu, 2003; Munns and Tester, 2008; Hasegawa, 2013), and this may provide favourable conditions for esca-related pathogens. However, grapevine is adapted to semi-arid environments, where drought and salinity are prevalent growth restraints, and is considered moderately tolerant to salinity stress (Walker *et al.*, 2002; Gil *et al.*, 2013). Combined drought and salt stresses led to enhanced sodium accumulation in roots and shoots (Ahmed *et al.*, 2013). Ma *et al.* (2015) characterized physiological and molecular responses of grapevine to short-term osmotic and sodium chloride stresses; a close relationship between drought and salinity may exist, leading to eventual impacts on plant physiological responses.

The duration and intensity of light may affect the life cycles of fungal pathogens, including those causing esca. Mittler *et al.* (1997) and Alvarez-Loayza *et al.* (2011) demonstrated that low light intensity favoured endosymbiotic development, while high light intensity triggered pathogenicity of the fungi. These authors also showed that mycelial melanin, correlated with increased production of reactive oxygen species (ROS) in pathogens and often associated with increased virulence to hosts, increased with light exposure, for instance caused by natural radiation. The rates of spore survival are also influenced by light conditions, mainly due to the UV-B portion of the spectrum, and to increased surface temperatures caused by radiation (Rotem *et al.*, 1985; Stevenson and Pennypacker, 1988; Braga *et al.*, 2015). However, no such studies, for instance with respect to colonization of pruning wounds, have been conducted with esca pathogens.

Pandey *et al.* (2015) reviewed enhanced damage caused by heat and drought stress combinations in plants. Both stress types share a number of physiological traits; their overall effects on plants are additive and aggravate both conditions. Sosnowski *et al.* (2007) reported that temperature and rainfall were related to the symptom development of Eutypiosis. Cool, rainy summers favoured GLSD or esca proper (chronic esca), and hot dry summers favoured apoplexy (acute esca) (Surico *et al.*, 2000). Marchi *et al.* (2006) reported on the relationship between annual incidence of manifest esca (diseased plants with foliar symptoms), hidden esca (asymptomatic plants through a growing season) and rainfall. Other data show that rainfall and temperature in (Northern Hemisphere) May-July or only in July were, respectively, directly and inversely related with esca leaf symptoms (Calzarano *et al.*, 2018). Incidence of *P. chlamydospora* and *Phaeoacremonium* spp. infections were greater in winter rainfall regions than in marginal and summer rainfall regions, and this was attributed to the climatic preferences of these fungi (van Niekerk *et al.*, 2011). Expression of PR-10 protein was found to be dependent on climatic conditions; therefore, it is possible that yearly temperature differences could influence the appearance of foliar symptoms in esca proper (Pasquier *et al.*, 2013).

PHYSIO-BIOCHEMICAL CHANGES IN GRAPEVINE IN ADAPTATION TO ENVIRONMENTAL CHANGES

Acclimation of plants to environmental factors leads to adaptive changes in root and shoot growth and physio-biochemical processes, which subsequently can alter molecular host-pathogen interactions. Beckman (1964) noted that physiological changes in plants, such as increased respiration and changes in water balance, result in wilting due to lack of water rather than toxins produced by vascular pathogens, such as *Phaeomoniella* or *Phaeoacremonium*.

Physiological changes

Advanced soil drying decreases the amount of nitrate taken up by plant roots, and this raises the xylem sap pH, increasing the sensitivity of stomata to the hormone ABA (Wilkinson and Davies, 1997; see also Woodall and Ward, 2002; Benjamin and Nielsen, 2006). The effects of stress on xylem sap were investigated by Agüero *et al.* (2008), who reported that sap from water stressed grapevines enhanced the *in vitro* growth of *P. chlamydospora* and *Phaeoacremonium* spp.

Xylem morphology

During developmental stages, xylem is responsive to environmental signals, and biotic and abiotic stresses may threaten xylem function (Pouzoulet *et al.*, 2014). Grapevine xylem is very efficient for water transport and by the formation of tyloses and gels is particularly vulnerable to cavitation (Alsina *et al.*, 2007). This results in possible dysfunction due to water stresses (Hacke *et al.*, 2000; McDowell *et al.*, 2008; Choat *et al.*, 2012; Vilagrosa *et al.*, 2012), and the presence of pathogens (Kuroda, 1991; Edwards *et al.*, 2007a; Raimondo *et al.*, 2010). Resistance to drought-induced cavitation is correlated with xylem vessel dimensions (Hacke *et al.*, 2001; Jacobsen *et al.*, 2005; Sperry *et al.*, 2006), which are the function of the plant's genetic makeup (Fichot *et al.*, 2009). Pouzoulet *et al.* (2014) documented that the grape varieties Cabernet Sauvignon and Thompson Seedless, both considered susceptible to esca, had wider and longer vessels than other varieties. Occlusion processes taking place in vessels of wide diameter, with loose clusters of tyloses, compared to smaller diameter vessels with more compact clusters, would create favourable environments for *P. chlamydospora* (Pouzoulet *et al.*, 2017).

Several studies in different plant species have observed increases of vessel grouping (Tyree *et al.*, 1994; Lopez *et al.*, 2005; Robert *et al.*, 2009; Lens *et al.*, 2011; Carlquist, 2012), and fewer solitary vessels (Arx *et al.* 2013), with increasing water limitation. Vessel grouping improves hydraulic redundancy and reduces the potential loss of water transport capacity associated with cavitation (Pouzoulet *et al.*, 2014). However, it was also hypothesized that the mechanism of tolerance towards esca is similar to that displayed by elms (*Ulmus* spp.) against Dutch elm disease, and therefore may not be correlated with vessel morphology (Venturas *et al.*, 2013). Thorough evaluation of xylem structure in different grapevine varieties may offer increased understanding of xylem vulnerability to drought stress, and susceptibility to vascular pathogens.

BIOCHEMICAL CHANGES

Plant biochemical responses to biotic and abiotic factors are complex and based on a number of molecular factors.

Photosynthetic responses

Esca-affected and drought-stressed grapevines provided information on the distinct functional behaviour patterns of photosynthetic response for these two stress types. In a study by Christen *et al.* (2007), esca-infected plants showed greater fluorescence intensity than drought stressed plants. However, the pool size of electron carriers, the electron transport per cross section and the electron transport per active reaction centre increased in the drought stress plants, whereas they decreased in the esca-infected plants.

Defense-related metabolism

In response to biotic and abiotic factors, a "trade-off" exists between growth and defense-related metabolism in plants (Treutter, 2005). When defense compounds are produced, plant resources are used, including amino acids, carbohydrates and other nutrients. Plants with symptoms reduce their carbohydrate reserves during winter dormancy (Petit *et al.*, 2006), and the lower pool of reserves may contribute to decreased plant development and vigour during the subsequent year (for grapevine, shown by Fontaine *et al.*, 2016).

Vines infected with *P. chlamydospora* showed reduced carbohydrate reserves compared to the control vines during winter dormancy, and were characterized by an overall loss of plant vigour (Petit *et al.*, 2006). Rudelle *et al.* (2005) suggested that high metabolic activity associated with secretory defense responses results from reduced starch storage. This associates with the first appearance of GLSD symptoms in a growing season in Trebbiano d'Abruzzo vineyards, at pre-bunch closure, and is when carbohydrate reserves in the wood are at their least. This condition could stimulate release of toxins by the pathogens in the vine wood, causing host leaf responses (Sumarah *et al.*, 2005; Eaton *et al.*, 2015). The reconstitution of these reserves between pre-bunch closure and veraison (Lebon *et al.*, 2008) was accompanied by reduced leaf symptom expression (Calzarano *et al.*, 2016; 2017a; 2017b). In contrast, if a vine has already experienced water stress, and has allocated amino acids to proline production in order to maintain osmotic pressure (Keller, 2005; Deluc *et al.*, 2009), less raw material will be available to produce plant defense compounds.

CONCLUSIONS

Considering the fact that grapevine culture and wine production are highly dependent on climatic conditions (van Leeuwen *et al.*, 2004; Ollat *et al.*, 2016), this review tried to provide new insights regarding biotic and abiotic factors in esca disease incidence.

Grapevines are economically important woody perennial fruit crops, with approx. 7.12 million ha cultivated and 74.5 million t of fruit harvested in 2014 (Gramaje *et al.*, 2018). Viticulture is facing important environmental challenges that need to be addressed through coordinated research. To achieve this goal, it will be important to assess the effects of biotic and abiotic stressors on grapevine performance, since plant disease responses may change significantly under the influence of environmental parameters. Evidence for global climate change is generally accepted (Solomon *et al.*, 2007; Bradley *et al.*, 2012; Matyssek *et al.*, 2012), and concomitant changes can lead to the disappearance or emergence of diseases in particular regions.

Water stress and the esca disease complex have been identified as the most important economic threats for viticulture (Luque *et al.*, 2010; Sosnowski *et al.*, 2011; van Niekerk *et al.*, 2011). The worldwide economic cost for the replacement of dead grapevines is estimated to be in excess of 1.5 billion dollars per year (Hofstetter *et al.*, 2012; see also De la Fuente *et al.*, 2016). Since plant responses to environmental stressors are known to affect responses to pathogens, deciphering how grapevines adapt to drought conditions is an essential step to improving water use efficiency, and to adequately managing the linked pathogens. Understanding the role of xylem plasticity among grapevine varieties, and its consequences for xylem vulnerability to drought stress and susceptibility to vascular pathogens, are essential. Further research is needed to determine the physiological and molecular mechanisms underlying the plant-environment-pathogen triangle. Bio-informatic analyses may be useful for comparing the expression of various sets of biotic and abiotic stress-related genes involved in general plant responses to pathogen infection (Fontaine *et al.*, 2016), and these analyses could improve understanding of the regulatory networks that control the plant responses to various stresses. To date, not enough information has been gathered on the effects of environmental stresses on the transcriptional responses of grapevine to the esca disease complex.

Long-term application of controlled conditions should be a priority, to increase understanding of the interactions between grapevine and the esca-related pathogens. The goal of this research should be to manipulate field growth conditions to favour vine hosts. Sustainable esca disease management will consist of a combination of appropriate grapevine cultivars and water regimes to mitigate the economic losses. Due to the extended periods of drought expected in the coming decades, the targeted use of protective endophytes, such as arbuscular mycorrhiza fungi, to improve plant nutrient uptake and osmotic stress tolerance, could provide improved drought tolerance (Schreiner, 2003; Schreiner and Linderman, 2005; Schreiner and Mihara, 2009; Trouvelot *et al.*, 2015).

ACKNOWLEDGMENTS

The authors thank Alan Phillips and Francois Halleen, who provided valuable critique of the manuscript of this paper.

LITERATURE CITED

Aasamaa K., Sober A., Rahi M., 2001. Leaf anatomical characteristics associated with shoot hydraulic conductance, stomatal conductance and stomatal sensitivity to changes of leaf water status in temperate deciduous trees. *Australian Journal of Plant Physiology* 28: 765–774.

Adams H.D., Germino M.J., Breshears D.D., Barron-Gafford G.A., Guardiola-Claramonte M., … Huxman T.E., 2013. Nonstructural leaf carbohydrate dynamics of *Pinus edulis* during drought-induced tree mortality reveal role for carbon metabolism in mortality mechanism. *New Phytologist* 197: 1142–1151.

Adie B.A., Perez-Perez J., Perez-Perez M.M., Godoy M., Sanchez-Serrano J.J., … Solano R., 2007. ABA is an essential signal for plant resistance to pathogens affecting JA biosynthesis and the activation of defenses in *Arabidopsis*. *The Plant Cell* 19: 1665–81.

Agrios G.N., 2005. *Plant Pathology*. 5th ed. Elsevier Academic Press, USA and UK, 952 pp.

Agüero C.B., Thorne E.T., Ibanez A.M., Gubler W.D., Dandekar A.M., 2008. Xylem sap proteins from *Vitis vinifera* L. Chardonnay. *American Journal of Enology and Viticulture* 59: 306–311.

Ahmed I.M., Cao F., Zhang M., Chen X., Zhang G., Wu F., 2013. Difference in yield and physiological features in response to drought and salinity combined stress during anthesis in Tibetan wild and cultivated barleys. *PLoS ONE* 8(10): e77869.

Ali R.M., Abbas H.M., 2003. Response of salt stressed barley seedlings to phenylurea. *Plant, Soil and Environment* 49: 158–162.

Alsina M.M., de Herralde F., Aranda X., Save R., Biel C., 2007. Water relations and vulnerability to embolism are not related: experiments with eight grapevine cultivars. *Vitis* 46: 1–6.

Alsina M.M., Smart D.R., Bauerle T., de Herralde F., Biel C., ... Save R., 2011. Seasonal changes of whole root system conductance by a drought-tolerant grape root system. *Journal of Experimental Botany* 62: 99–109.

Alvarez S., Marsh E.L., Schroeder S.G., Schachtman D.P., 2008. Metabolomic and proteomic changes in the xylem sap of maize under drought. *Plant, Cell & Environment* 31: 325–340.

Alvarez-Loayza P., White J.F., Torres M.S., Balslev H., Kristiansen T., … Gil N., 2011. Light converts endosymbiotic fungus to pathogen, influencing seedling survival and niche-space filling of a common tropical tree, *Iriartea deltoidea*. *PLoS ONE* 6: e16386.

Amponsah N.T., Jones E.F., Ridgway H.J., Jaspers M.V., 2011. Identification, potential inoculum sources and pathogenicity of botryosphaeriaceous species associated with grapevine dieback disease in New Zealand. *European Journal of Plant Pathology* 131: 467–482.

Amtmann A., Troufflard S., Armengaud P., 2008. The effect of potassium nutrition on pest and disease resistance in plants. *Physiologia Plantarum*133: 682–691.

Andolfi L., Mugnai L., Luque J., Surico G., Cimmino A., Evidente A., 2011. Phytotoxins produced by fungi associated with grapevine trunk diseases. *Toxins* 3: 1569–1605

Andreini L., Caruso G., Bertolla C., Scalabrelli G., Viti R., Gucci R., 2009. Gas exchange, stem water potential and xylem flux of some grapevine cultivars affected by esca disease. *South African Journal for Enology and Viticulture* 30: 142–147.

Andreini L., Guarino L., Viti R., Scalabrelli G., 2013. Evaluation of the effect of esca disease on bud break in *Vitis vinifera* L.: Possible relationship between cultivars and rootstocks. *Vitis* 52: 33–40.

Andreini L., Cardelli R., Bartolini S., Scalabrelli G., Viti R., 2014. Esca symptoms appearance in *Vitis vinifera* L.: influence of climate, pedo-climatic conditions and rootstock/cultivar combination. *Vitis* 53: 33–38.

Aroca A., García-Figueres F., Bracamonte L., Luque J., Raposo R., 2006. A survey of trunk disease pathogens within rootstocks of grapevines in Spain. *European Journal of Plant Pathology* 115: 195–202.

Arx G., Kueffer C., Fonti P., 2013. Quantifying plasticity in vessel grouping - added value from the image analysis tool Roxas. *IAWA Journal* 34: 433–445.

Asselbergh B., Achuo E.A., Höfte M., Van Gijsegem F., 2008. Abscisic acid-deficiency leads to rapid activation of tomato defense responses upon infection with *Erwinia chrysanthemi*. *Molecular Plant Pathology* 9: 11–24.

Audenaert K., de Meyer G.B., Hofte M.M., 2002. Abscisic acid determines basal susceptibility of tomato to *Botrytis cinerea* and suppresses salicylic acid-dependent signaling mechanisms. *Plant Physiology* 128: 491–501.

Baker B., Zambryski P., Staskawicz B., Dinesh-Kumar S.P., 1997. Signaling in plant-microbe interactions. *Science* 276: 726–733.

Barrios-Masias F.H., Jackson L.E., 2014. California processing tomatoes: morphological, physiological and phenological traits associated with crop improvement during the last 80 years. *European Journal of Agronomy* 53: 45–55.

Barrios-Masias F.H., Knipfer T., McElrone A.J., 2015. Differential responses of grapevine rootstocks to water stress are associated with adjustments in fine root hydraulic physiology and suberization. *Journal Experimental Botany* 19: 6069–6078.

Bauerle T.L., Centinari M., Bauerle W.L., 2011. Shifts in xylem vessel diameter and embolisms in grafted apple trees of differing rootstock growth potential in response to drought. *Planta* 234: 1045–1054.

Beckman C.H., 1964. Host responses to vascular infection. *Annual Review of Phytopathology* 2: 231–252.

Bell E., Mullet J.E., 1991. Lipoxygenase gene expression is modulated in plants by water deficit, wounding and methyl jasmonate. *Molecular Genetics and Genomics* 230: 456–462.

Benjamin J.G., Nielsen D.C., 2006. Water deficit effects on root distribution of soybean, field pea and chickpea. *Field Crops Research* 97: 248–253.

Berdeja M., Hilbert G., Dai Z., Lafontaine M., Stoll M., ... Delrot S., 2014. Effects of water stress and rootstock genotype on Pinot Noir berry composition. *Australian Journal of Grape and Wine Research* 20: 409–421.

Bertsch C., Larignon P., Farine S., Clément C., Fontaine F., 2009. The spread of grapevine trunk disease. *Science* 324: 721.

Bertsch C., Ramirez-Suero M., Magnin-Robert M., Larignon P., Chong J., … Fontaine F., 2013. Grapevine trunk diseases: complex and still poorly understood. *Plant Pathology* 62: 243–265.

Bostock R.M., Pye M.F., Roubtsova T.V., 2014. Predisposition in plant disease: exploiting the nexus in abiotic and biotic stress perception and response. *Annual Review of Phytopathology* 52: 517–549.

Boyer J.S., 1995. Biochemical and biophysical aspects of water deficits and the predisposition to disease. *Annual Review of Phytopathology* 33: 251–74.

Bradley B.A., Blumenthal D.M., Early R., Grosholz E.D., Lawler J.J., … Olden J.D., 2012. Global change, global trade, and the next wave of plant invasions. *Frontiers in Ecology and the Environment* 10: 20–28.

Braga G.U.L., Rangel D.E.N., Fernandes E.K.K., Flint S.D., Roberts D.W., 2015. Molecular and physiological effects of environmental UV radiation on fungal conidia. *Current Genomics* 61: 405–425.

Breda N., Huc R., Granier A., Dreyer E., 2006. Temperate forest trees and stands under severe drought: a review of ecophysiological responses, adaptation processes and long-term consequences. *Annals of Forest Science* 63: 625–644.

Brown A.A., 2016. *Trunk Disease of Grapevines in Texas.* M.S. Thesis, Texas A & M University, Texas, USA, 74 pp.

Bruez E., Lecomte P., Grosman J., Doublet B., Bertsch C., … Guérin-Dubrana L., 2013. Overview of grapevine trunk diseases in France in the 2000s. *Phytopathologia Mediterranea* 52: 262–275.

Bruez E., Vallance J., Gerbore J., Lecomte P., Da Costa J.P., ... Rey P., 2014. Analyses of the temporal dynamics of fungal communities colonizing the healthy wood tissues of esca leaf-symptomatic and asymptomatic vines. *PloS One* 9: 15.

Burdon J.J., Thrall P.H., Ericson A.L., 2006. The current and future dynamics of disease in plant communities. *Annual Review of Phytopathology* 44: 19–39.

Calzarano F., Di Marco S., 2007. Wood discoloration and decay in grapevines with esca proper and their relationship with foliar symptoms. *Phytopathologia Mediterranea* 46: 96–101.

Calzarano F., Di Marco S., 2018. Further evidence that calcium, magnesium and seaweed mixtures reduce grapevine leaf stripe symptoms and increase grape yield. *Phytopathologia Mediterranea* 57: 57, 459–471.

Calzarano F., Cichelli A., Odoardi M., 2001. Preliminary evaluation of variations in composition induced by esca on cv. Trebbiano D'Abruzzo grapes and wines. *Phytopathologia Mediterranea* 40: S443-S448.

Calzarano F., Amalfitano C., Seghetti L., Cozzolino V., 2009. Nutritional status of vines affected with esca proper. *Phytopathologia Mediterranea* 48: 20–31.

Calzarano F., Di Marco S., D Agostino V., Schiff S., Mugnai L., 2014. Grapevine leaf stripe disease symptoms (esca complex) are reduced by a nutrients and seaweed mixture. *Phytopathologia Mediterranea* 53: 543–558.

Calzarano F., D'Agostino V., Pepe A., Osti F., Della Pelle F., ... Di Marco S., 2016. Patterns of phytoalexins in the GLSD (esca complex)/grapevine pathosystem. *Phytopathologia mediterranea* 55: 410–426.

Calzarano F., Osti F., D'Agostino V., Pepe A., Della Pelle F., ... Di Marco S., 2017a. Levels of phytoalexins in vine leaves with different degrees of grapevine leaf stripe disease symptoms (esca complex of diseases). *Phytopathologia mediterranea* 56: 494–501.

Calzarano F., Osti F., D'Agostino V., Pepe A., Di Marco S., 2017b. Mixture of calcium, magnesium and seaweed affects leaf phytoalexin contents and grape ripening on vines with grapevine leaf stripe disease. *Phytopathologia Mediterranea* 56: 445–457.

Calzarano F., Osti F., Baránek M., Di Marco S., 2018. Rainfall and temperature influence expression of foliar symptoms of grapevine leaf stripe disease (esca complex) in vineyards. *Phytopathologia Mediterranea* 57: 488–505.

Campbell C.L. Madden L.V., 1990. *Introduction to Plant Disease Epidemiology.* John Wiley & Sons, Inc, New York, USA, 532 pp.

Camps C., Kappel C., Lecomte P., Leon C., Gomes E., … Derlot S., 2010. A transcriptomic study of grapevine (*Vitis vinifera*cv. Cabernet-Sauvignon) interaction with the vascular ascomycete fungus *Eutypa lata*. *Journal of Experimental Botany* 61: 1719–1737.

Carlquist S., 2012. How wood evolves: a new synthesis. *Botany* 90: 901–940.

Castellarin S.D., Matthews M.A., Di Gaspero G., Gambetta G.A., 2007. Water deficits accelerate ripening and induce changes in gene expression regulating flavonoid biosynthesis in grape berries. *Planta* 227: 101–112.

Chen W., Provart N.J., Glazebrook J., Kataqiri F., Chang H.S., ... Zhu T., 2002. Expression profile matrix of *Arabidopsis* transcription factor genes suggests their putative functions in response to environmental stresses. *The Plant Cell* 14: 559–574.

Choat B., Jansen S., Brodribb T.J., Cochard H., Delzon S., ... Zanne A.E., 2012. Global convergence in the vulnerability of forests to drought. *Nature* 491: 752–755.

Choi H.K., Iandolino A., da Silva F.G., Cook D.R., 2013. Water deficit modulates the response of *Vitis vinifera* to the Pierce's disease pathogen *Xylella fastidiosa*. *Molecular Plant-Microbe Interactions Journal* 26: 643–657.

Christen D., Schonmann S., Jermini M., Strasser R.J., Defago G., 2007. Characterization and early detection of grapevine (*Vitis vinifera*) stress responses to esca disease by *in situ* chlorophyll fluorescence and comparison with drought stress. *Environmental and Experimental Botany* 60: 504–14.

Cirilli M., Bassi D., Ciacciulli A., 2016. Sugars in peach fruit: a breeding perspective. *Horticulture Research* 3: 15067.

Cloete M., 2015. *The Characterization of Basidiomycetes Associated with Esca Disease in South Africa Grapevines*. PhD thesis. Stellenbosch University, Stellenbosch, South Africa, 128 pp.

Cochard H., Coll L., Le Roux X., Améglio T., 2002. Unraveling the effects of plant hydraulics on stomatal closure during water stress in walnut. *Plant Physiology* 128: 282–290.

Comas L., Becker S., Cruz V.M.V., Byrne P.F., Dierig D.A., 2013. Root traits contributing to plant productivity under drought. *Frontiers in Plant Science* 4: 442.

Cramer G.R., Ergül A., Grimplet J., Tillett R.L., Tattersall E.A., ... Cushman J.C., 2007. Water and salinity stress in grapevines: early and late changes in transcript and metabolite profiles. *Functional & Integrative Genomics* 7: 111–134.

Croise L., Lieutier F., Cochard H., Dreyer E., 2001. Effects of drought stress and high density stem inoculations with *Leptographium wingfieldii* on hydraulic properties of young Scots pine trees. *Tree Physiology* 21: 427–436.

Crowl T.A., Crist T.O., Parmenter R.R., Belovsky G., Lugo A.E., 2008. The spread of invasive species and infectious disease as drivers of ecosystem change. *Frontiers in Ecology and the Environment* 6: 238–46.

De la Fuente M., Fontaine F., Gramaje D., Armengol J., Smart R., ... Corio-Costet M.F., 2016. *Grapevine Trunk Diseases. A Review*. International Organization of Vine and Wine (*OIV*), 24 pp.

Deluc L.G., Quilici D.R., Decendit A., Grimplet J., Wheatley M.D., ... Cramer G.R., 2009. Water deficit alters differentially metabolic pathways affecting important flavor and quality traits in grape berries of Cabernet Sauvignon and Chardonnay. *BMC Genomics* 10: 212.

Derckel J.P., Baillieul F., Manteau S., Audran J.C., Haye B., ... Legendre L., 1999. Differential induction of grapevine defenses by two strains of *Botrytis cinerea*. *Phytopathology* 89: 197–203.

Di Marco S., Osti F., 2008. Foliar symptom expression of wood decay in *Actinidia deliciosa* in relation to environmental factors. *Plant Disease* 92: 1150–1157.

Di Marco S., Osti F., Cesari A., 2004. Experiments on the control of esca by *Trichoderma*. *Phytopathologia Mediterranea* 43: 108-115.

Dry P.R., Loveys B.R., During H., 2000. Partial drying of the rootzone of grape. II. Changes in the pattern of root development. *Vitis* 39: 9–12.

Dubos B., Larignon P., Lecomte P., Magnien C., Panon M.L., ... Leguay M., 2002: Les maladies du bois en viticulture. *ITV France* 113: 32–35.

Eastburn D.M., McElrone A.J., Bilgin D.D., 2011. Influence of atmospheric and climatic change on plant-pathogen interactions. *Plant Pathology* 60: 54–69.

Eaton C.J., Dupont P.Y., Solomon P., Clyton W., Scot B., Cox M.P., 2015. A core gene set describes the molecular basis of mutualism and antagonism in *Epichloë* spp. *Molecular Plant-Microbe Interactions*. 28(3): 218–231.

Edwards J., Marchi G., Pascoe I.G., 2001. Young esca in Australia. *Phytopathologia Mediterranea* 40: 303–310.

Edwards J., Pascoe I., Salib S., 2007a. Impairment of grapevine xylem function by *Phaeomoniella chlamydospora* infection is due to more than physical blockage of vessels with "goo". *Phytopathologia Mediterranea* 46: 87–90.

Edwards J., Salib S., Thomson F., Pascoe I.G., 2007b. The impact of *Phaeomoniella chlamydospora* infection on the grapevines physiological response to water stress-part 1: Zinfandel. *Phytopathologia Mediterranea* 46: 26–37.

Edwards J., Salib S., Thomson F., Pascoe I.G., 2007c. The impact of *Phaeomoniella chlamydospora* infection on the grapevines physiological response to water stress-part 2: Cabernet Sauvignon and Chardonnay. *Phytopathologia Mediterranea* 46: 38–49.

Elena G., Bruez E., Rey P., Luque J., 2018. Microbiota of grapevine woody tissues with or without esca-foliar symptoms in northeast Spain. *Phytopathologia Mediterranea* 57: 425–438.

Escalona J.M., Flexas J., Medrano H., 1999. Stomatal and non-stomatal limitations of photosynthesis under water stress in field-grown grapevines. *Australian Journal of Plant Physiology* 26: 421–433.

Eskalen A., Gubler W.D., Khan A., 2001. Rootstock susceptibility to *Phaeomoniella chlamydospora* and *Phaeoacremonium* spp. *Phytopathologia Mediterranea* 40: S433–S438.

Essakhi S., Mugnai L., Crous P.W., Groenewald J.Z., Surico G., 2008. Molecular and phenotypic characterization of novel *Phaeoacremonium* species isolated from esca diseased grapevines. *Persoonia* 21: 119–134.

Evidente A., Sparapano L., Andolfi A., Bruno G., 2000. Two naphthalenone pentaketides from liquid cultures of *Phaeoacremonium aleophilum*, a fungus associated with esca of grapevine. *Phytopathologia Mediterranea* 39: 162–8.

Feliciano A.J., Eskalen A., Gubler W.D., 2004. Differential susceptibility of three grapevine cultivars to *Phaeoacremonium aleophilum* and *Phaeomoniella chlamydospora* in California. *Phytopathologia Mediterranea* 43: 66–69.

Ferreira J.H.S., van Wyk P.S., Calitz F.J., 1999. Slow dieback of grapevine in South Africa: stress-related predisposition of young vines for infection by *Phaeoacremonium chlamydosporum. South African Journal of Enology and Viticulture* 20: 43–46.

Fichot R., Laurans F., Monclus R., Moreau A., Pilate G., Brignolas F., 2009. Xylem anatomy correlates with gas exchange, water-use efficiency and growth performance under contrasting water regimes: evidence from *Populus deltoides* x *Populus nigra* hybrids. *Tree Physiology* 29: 1537–1549.

Fischer M., 2002. A new wood-decaying basidiomycete species associated with esca of grapevine: *Fomitiporia mediterranea* (Hymenochaetales). *Mycological Progress* 1: 314–324.

Fischer M., 2006. Biodiversity and geographic distribution of basidiomycetes causing esca-associated white rot in grapevine: a worldwide perspective. *Phytopathologia Mediterranea* 45: S30–S42.

Fischer M., Binder M., 2004. Species recognition, geographic distribution and host–pathogen relationships: a case study in a group of lignicolous basidiomycetes, *Phellinus* s.l. *Mycologia* 96: 799–811.

Fischer M., Kassemeyer H.H., 2003. Fungi associated with esca disease of grapevine in Germany. *Vitis* 42: 109–116.

Fischer M., Kassemeyer H.H., 2012. Water regime and its possible impact on expression of esca symptoms in *Vitis vinifera*: growth characters and symptoms in the greenhouse after artificial infection with *Phaeomoniella chlamydospora*. *Vitis* 51: 129–135.

Fischer M., González García V., 2015. An annotated checklist of European basidiomycetes related to white rot of grapevine (*Vitis vinifera*). *Phytopathologia Mediterranea* 54: 281–298.

Fischer M., Edwards J., Cunnington J., Pascoe I., 2005. Basidiomycetous pathogens on grapevine: a new species from Australia – *Fomitiporia australiensis. Mycotaxon* 92: 85–96.

Flexas J., Escalona J.M., Medraro H., 1999. Water stress induces different levels of photosynthesis and electron transport rate regulation in grapevines. *Plant, Cell and Environment* 22: 39–48.

Fontaine F., Pinto C., Vallet J., Clement C., Gomes A.C., Spagnolo A., 2016. The effects of grapevine trunk diseases (GTDs) on vine physiology. *European Journal of Plant Pathology* 144: 707–721.

Freeman B.M., Kliewer W.M., 1983. Effect of irrigation, crop level and potassium fertilization on Carignane vines. II. Grape and wine quality. *American Journal of Enology and Viticulture* 34: 197–207.

Galmes J., Pou A., Alsina M.M., Tomas M., Medrano H., Flexas J., 2007. Aquaporin expression in response to different water stress intensities and recovery in Richter-110 (*Vitis* sp.): relationship with ecophysiological status. *Planta* 226: 671–681.

Gambetta G.A., Manuck C.M., Drucker S.T., Shaghasi T., Fort K., … McElrone A.J., 2012. The relationship between root hydraulics and scion vigour across *Vitis* rootstocks: what role do root aquaporins play? *Journal of Experimental Botany* 63: 6445–6455.

Giannakis C., Bucheli C.S., Skene K.G.M., Robinson S.P., Scott N.S., 1998. Chitinase and b-1, 3-glucanase in grapevine leaves: a possible defense against powdery mildew infection. *Australian Journal of Grape and Wine Research* 4: 14–22.

Gil M., Esteruelas M., Gonzalez E., Kontoudakis N., Jimenez J., ... Zamora F., 2013. Effect of two different treatments for reducing grape yield in *Vitis vinifera* cv. Syrah on wine composition and quality: berry thinning versus cluster thinning. *Journal of Agriculture and Food Chemistry* 61: 4968–4978.

Golding J.B., Satyan S., Liebenberg C., Walsh K., McGlasson W.B., 2006. Application of portable NIR for measuring sugar concentrations in peaches. *Acta Horticulturae* 713: 461–464.

Gramaje D., Armengol J., 2011. Fungal trunk pathogens in the grapevine propagation process: potential inoculum sources, detection, identification, and management strategies. *Plant Disease* 95: 1040–1055.

Gramaje D., Armengol J., Mohammadi H., Banihashemi Z., Mostert L., 2009. Novel *Phaeoacremonium* species associated with Petri disease and esca of grapevine in Iran and Spain. *Mycologia* 101: 920–929.

Gramaje D., Úrbez-Torres J.R., Sosnowski M.R., 2018. Managing grapevine trunk diseases with respect to etiology and epidemiology: current strategies and future propects. *Plant Disease* 102: 12–39.

Graniti A., Surico G., Mugnai L., 2000. Esca of grapevine: a disease complex or a complex of diseases? *Phytopathologia Mediterranea* 39: 16–20.

Grimplet J., Deluc L.G., Tillett R.L., Wheatley M.D., Schlauch K.A., … Cushman J.C., 2007. Tissue-specific mRNA expression profiling in grape berry tissues. *BMC Genomics* 8: 187.

Gu M.M., Rom C.R., Robbins J.A., Oosterhuis D.M., 2007. Effect of water deficit on gas exchange, osmotic solutes, leaf abscission, and growth of four birch genotypes (*Betula* L.) under a controlled environment. *Hort science* 42: 1383–1391.

Gubler W.D., Rolshausen P.E., Trouillas F.P., Úrbez-Torres J.R., Voegel T., … Weber E.A., 2005. *Grapevine trunk diseases in California*. Practical Winery and Vineyard (Jan./Feb.), 6–25.

Guérin-Dubrana L., Destrac-Irvine A., Goutouly J.P., Letouze A., Gaudillére J.P., 2005. Relationship between incidence of esca and black dead arm foliar symptom expression in the vineyard, ecophysiological indicators and cultural practices. *Phytopathologia Mediterranea* 44: 110 (abstract).

Guérin-Dubrana L., Labenne A., Labrousse J.C., Bastien S., Rey P., Gegout-Petit A., 2012. Statistical analysis of grapevine mortality associated with esca or eutypa dieback foliar expression. *Phytopathologia Mediterranea* 52: 276–288.

Gutierrez-Gonzalez J.J., Guttikonda S.K., Tran L.S.P., Aldrich D.L., Zhong R., … Sleper D.A., 2010. Differential expression of isoflavone biosynthetic genes in soybean during water deficits. *Plant and Cell Physiology* 51: 936–948.

Hacke U.G., Sperry J.S., Pittermann J., 2000. Drought experience and cavitation resistance in six shrubs from the Great Basin, Utah. *Basic and Applied Ecology* 1: 31–41.

Hacke U.G., Sperry J.S., Pockman W.T., Davis S.D., McCulloch K.A., 2001. Trends in wood density and structure are linked to prevention of xylem implosion by negative pressure. *Oecologia* 126: 457–461.

Halleen F., Crous P.W., Petrini O., 2003. Fungi associated with healthy grapevine cuttings in nurseries with special reference to pathogens involved in the decline of young vines. *Australasian Plant Pathology* 32: 47–52.

Hasegawa P.M., 2013. Sodium (Na^+) homeostasis and salt tolerance of plants. *Environmental and Experimental Botany* 92: 19–31.

Heath M.C., 2000. Hypersensitive response-related death. *Plant Molecular Biology* 44: 321–334.

Hofstetter V., Buyck B., Croll D., Viret O., Couloux A., Gindro K., 2012. What if esca disease of grapevine were not a fungal disease? *Fungal Diversity* 54: 51–67.

Huang D., Wu W., Abrams S.R., Cutler A.J., 2008. The relationship of drought-related gene expression in *Arabidopsis thaliana* to hormonal and environmental factors. *Journal of Experimental Botany* 59: 2991–3007.

Huber L., Gillespie T.J., 1992. Modeling leaf wetness in relation to plant-disease epidemiology. *Annual Review of Phytopathology* 30: 553–577.

Hummel I., Pantin F., Sulpice R., Piques M., Rolland G., … Muller B., 2010. *Arabidopsis* plants acclimate to water deficit at low cost through changes of carbon usage: an integrated perspective using growth, metabolite, enzyme, and gene expression analysis. *Plant Physiology* 154: 357–372.

Jacobsen A.L., Ewers F.W., Pratt R.B., Paddock W.A., Davis S.D., 2005. Do xylem fibers affect vessel cavitation resistance? *Plant Physiology* 139: 546–556.

Jactel H., Petit J., Desprez-Loustau M.L., Delzon S., Piou D., … Koricheva J., 2012. Drought effects on damage by forest insects and pathogens: a meta-analysis. *Global Change Biology* 18: 267–276.

Johnson J.W., Gleason M.L., Parker S.K., Provin E.B., Iles J.K., 1997. Duration of water stress affects development of *Sphaeropsis* canker on Scots pine. *Journal of Arboriculture* 23: 73–76.

Kariola T., Brader G., Helenius E., Li J., Heino P., Palva E.T., 2006. Early responsive to dehydration 15, a negative regulator of abscisic acid responses in *Arabidopsis*. *Plant Physiology* 142: 1559–73.

Keller M., 2005. Deficit irrigation and vine mineral status. *American Journal of Enology and Viticulture* 56: 267–283.

Khalid K.A., 2006. Influence of water stress on growth, essential oil and chemical composition of herbs (*Ocimum* sp.). *International Agrophysics* 20: 289–296.

Kizis D., Pages M., 2002. Maize DRE-binding proteins DBF1 and DBF2 are involved in rab17 regulation through the drought-responsive element in an ABA-dependent pathway. *The Plant Journal* 30: 679–689.

Koga H., Dohi K., Mori M., 2004. Abscisic acid and low temperature suppress the whole plant-specific resistance reaction of rice plants to the infection of *Magnaporthe grisea*. *Physiological and Molecular Plant Pathology* 65: 3-9.

Kontunen-Soppela S., Parviainen J., Ruhanen H., Brosche M., Keinanen M., ... Vapaavuori E., 2010. Gene expression responses of paper birch (*Betula papyrifera*) to elevated CO_2 and O_3 during leaf maturation and senescence. *Environmental Pollution* 158: 959–968.

Koundouras S., Hatzidimitriou E., Karamolegkou M., Dimopoulou E., Kallithraka S., ... Kotseridis Y., 2009. Irrigation and rootstock effects on the phenolic concentration and aroma potential of *Vitis vinifera* L. cv. Cabernet Sauvignon grapes. *Journal of Agricultural and Food Chemistry* 57: 7805–7813.

Koundouras S., Tsialtas I.T., Zioziou E., Nikolaou N., 2008. Rootstock effects on the adaptive strategies of grapevine (*Vitis vinifera* L. cv. Cabernet Sauvignon) under contrasting water status: leaf physiological and structural responses. *Agriculture, Ecosystems & Environment* 128: 86–96.

Kovacs C., Balling P., Bihari Z., Nagy A., Sandor E., 2017. Incidence of grapevine trunk diseases is influenced by soil, topology and vineyard age, but not by *Diplodia seriata* infection rate in the Tokaj Wine Region, Hungary. *Phytoparasitica* 45: 21–32.

Krishnamurthy R., Bhagwat K.A.M., 1989. Polyamines as modulators of salt tolerance in rice cultivars. *Plant Physiology* 91: 500–504.

Kubátová A., Kolařrík M., Pažoutová S., 2004. *Phaeoacremonium rubrigenum* - hyphomycete associated with bark beetles found in Czechia. *Folia Microbiologica* 49: 99–104.

Kudoyarova G., Vysotskaya L.B., Cherkozyanova A., Dodd I.C., 2007. Effect of partial root zone drying on the concentration of zeatin-type cytokinins in tomato (*Solanum lycopersicum* L.) xylem sap and leaves. *Journal of Experimental Botany* 58: 161–168.

Kuldau G.A., Yates I.E., 2000. Evidence for *Fusarium* endophytes in cultivated and wild plants. In: *Microbial Endophytes* (B.W. Bacon, J.F. White, eds.), Marcel Dekker, New York and Basel, 85–120.

Kuroda K., 1991. Mechanism of cavitation development in the pine wilt disease. *European Journal of Forest Pathology* 21: 82–89.

Lanari V., Silvestroni O., 2015. Plant and leaf physiological responses to water stress in potted Vignoles grapevine. *Hort Science* 50: 1492–1497.

Larignon P., Dubos B., 1997. Fungi associated with esca disease in grapevine. *European Journal of Plant Pathology* 103: 147–157.

Larignon P., Fulchic R., Ceré L., Dubos B., 2001. Observation on black dead arm in French vineyards. *Phytopathologia Mediterranea* 40: 336–342.

Larignon P., Fontaine F., Farine S., Clement C., Bertsch C., 2009. Esca and Black Dead Arm: two major actors of grapevine trunk diseases. *Comptes Rendus Biologies* 332: 765–783.

Leakey A.D., Xu F., Gillespie K.M., McGrath J.M., Ainsworth E.A., Ort D.R., 2009. Genomic basis for stimulated respiration by plants growing under elevated carbon dioxide. *Proceedings of the National Academy of Sciences*, USA, 106: 3597–3602.

Lebon G., Lebon G., Wojnarowiez G., Holzapfel B., Fontaine F., ... Clément C., 2008. Sugars and flowering in the grapevine (*Vitis vinifera* L.). *Journal of Experimental Botany* 59(10): 2565–2578.

Lecomte P., Darrieutort G., Laveau C., Blancard D., Louvet G., ... Guérin-Dubrana L., 2011. Impact of biotic and abiotic factors on the development of esca decline. Integrated protection and production in viticulture, *IOBC-WPRS Bulletin* 67: 171–180.

Lecomte P., Darrieutort G., Liminana J.M., Comont G., Muruamendiaraz A., ... Fermaud M., 2012. New insights into esca of grapevine: the development of foliar symptoms and their association with xylem discoloration. *Plant Disease* 96: 924–934.

Lecomte P., Darrieutort G., Liminana J.M., Louvet G., Tandonnet J.P., ... Blancard D., 2008. Eutypiose et esca II. Vers une gestion raisonnée des maladies de dépérissement. *Phytoma* 616: 37–41.

Lens F., Sperry J.S., Christman M.A., Choat B., Rabaey D., Jansen S., 2011. Testing hypotheses that link wood anatomy to cavitation resistance and hydraulic conductivity in the genus *Acer*. *New Phytologist* 190: 709–723.

Letousey P., Baillieul F., Perrot G., Rabenoelina F., Boulay M., ... Fontaine F., 2010. Early events prior to visual symptoms in the apoplectic form of grapevine esca disease. *Phytopathology* 100: 424–431.

Liminana J.M., Pacreau G., Boureau F., Menard E., David S., ... Dumot V., 2009. Inner necrosis in grapevine rootstock mother plants in the Cognac area (Charentes, France). *Phytopathologia Mediterranea* 48: 92–100.

Lopez B.C., Sabatae S., Gracia C.A., Rodriguez R., 2005. Wood anatomy, description of annual rings, and responses to ENSO events of *Prosopis pallida* HBK, a wide-spread woody plant of arid and semi-arid lands of Latin America. *Journal of Arid Environments* 61: 541–554.

Lopresti J., Goodwin I., McGlasson B., Holford P., Golding J., 2014. Variability in size and soluble solids concentration in peaches and nectarines. In: *Horticultural Reviews* Vol. 42 (J. Janick, ed.) Wiley-Blackwell, New Jersey, USA, 253–311.

Lorenz W.W., Alba R., Yu Y.S., Bordeaux J.M., Simoes M., Dean J.F.D., 2011. Microarray analysis and scale-free gene networks identify candidate regulators in drought-stressed roots of loblolly pine (*P. taeda* L.). *BMC Genomics* 12: 264.

Lovisolo C., Schubert A., 1998. Effects of water stress on vessel size and xylem hydraulic conductivity in *Vitis vinifera* L. *Journal of Experimental Botany* 49: 693–700.

Lovisolo C., Perrone I., Carra A., Ferrandino A., Flexas J., ... Schubert A., 2010. Drought-induced changes in development and function of grapevine (*Vitis* spp.) organs and in their hydraulic and non-hydraulic interactions at the whole-plant level: a physiological and molecular update. *Functional Plant Biology* 37: 98–116.

Löw D., Brändle K., Nover L., Forreiter C., 2000. Cytosolic heat-stress proteins Hsp17.7 class I and Hsp17.3 class II of tomato act as molecular chaperones *in vivo*. *Planta* 211: 575–582.

Luque J., Martos S., Aroca A., Raposo R., García-Figueres F., 2009. Symptoms and fungi associated with declining mature grapevine plants in Northeast Spain. *Journal of Plant Pathology* 91: 381–390.

Luque J., Martos S., Garcia-Figueres F., 2010. Effects of water stress and inoculation with *Eutypa lata* and *Neofusicoccum parvum* on young grapevine plants. In: *Abstracts, 7th Int. Workshop on Grapevine Trunk Diseases*, 17-21-January, 2010, Santa Cruz, Chile (abstract).

Ma Y., Wang J., Zhong Y., Cramer G.R., Cheng Z.M., 2015. Genome-wide analysis of the cation/ proton antiporter (CPA) super family genes in grapevine (*Vitis vinifera* L.). *Plant Omics* 8: 300–311.

Marchi G., 2001. Susceptibility to esca of various grapevine (*Vitis vinifera*) cultivars grafted on different rootstocks in a vineyard in the province of Siena (Italy). *Phytopathologia Mediterranea* 40: 27–36.

Marchi G., Peduto F., Mugnai L., Di Marco S., Calzarano F., Surico G., 2006. Some observations on the relationship of manifest and hidden esca to rainfall. *Phytopathologia Mediterranea* 45: S117–S126.

Marguerit E., Brendel O., Lebon E., Van Leeuwen C., Ollat N., 2012. Rootstock control of scion transpiration and its acclimation to water deficit are controlled by different genes. *New Phytologist* 194: 416–429.

Martinez-Vilalta J., Prat E., Oliveras I., Pinol J., 2002. Xylem hydraulic properties of roots and stems of nine mediterranean woody species. *Oecologia* 133: 19–29.

Matyssek R., Wieser G., Calfapietra C., de Vries W., Dizengremel P., ... Paoletti E., 2012. Forests under climate change and air pollution: gaps in understanding and future directions for research. *Environmental Pollution* 160: 57–65.

McCutcheon T.L., Carrol G.C., Schwab S., 1993. Genotypic diversity in populations of a fungal endophyte from Douglas fir. *Mycologia* 85: 180–186.

McDonald K.L., Cahill D.M., 1999. Evidence for a transmissible factor that causes rapid stomatal closure in soybean at sites adjacent to and remote from hypersensitive cell death induced by *Phytophthora sojae*. *Physiological and Molecular Plant Pathology* 55: 197–203.

McDowell N.G., Pockman W.T., Allen C.D., Breshears D.D., Cobb N., ... Yepez E.A., 2008. Mechanisms of plant survival and mortality during drought: why do some plants survive while others succumb to drought? *New Phytologist* 178: 719–739.

McElrone A.J., Sherald J.L., Forseth I.N., 2001. Effects of water stress on symptomatology and growth of *Parthenocissus quinquefolia* infected by *Xylella fastidiosa*. *Plant Disease* 85: 1160–1164.

Melchior F., Kindl H., 1991. Coordinate- and Elicitor-dependent expression of stilbene synthase and phenylalanine ammonia-lyase genes in *Vitis* cv. Optima. *Archives of Biochemistry and Biophysics* 288: 552–557.

Mingo D.M., Theobald J.C., Bacon M.A., Davies W.J., Dodd I.C., 2004. Biomass allocation in tomato (*Lycopersicon esculentum*) plants grown under partial rootzone drying: enhancement of root growth. *Functional Plant Biology* 31: 971–978.

Mittler R., Blumwald E., 2010. Genetic engineering for modern agriculture: challenges and perspectives. *Annual Review of Plant Biology* 61: 443–462.

Mittler R., Simon L., Lam E., 1997. Pathogen-induced programmed cell death. *Journal of Cell Science* 110: 1333–1344.

Mohammadi H., Banihashemi Z., Gramaje D., Armengol J., 2013. Fungal pathogens associated with grapevine trunk diseases in Iran. *Journal of Agricultural Science and Technology* 15: 137–150.

Moriondo M., Giannakopoulos C., Bindi M., 2011. Climate change impact assessment: The role of climate extremes in crop yield simulation. *Climatic Change* 104: 679–701.

Mostert L., Halleen F., Fourie P., Crous PW., 2006. A review of *Phaeoacremonium* species involved in Petri disease and esca of grapevines. *Phytopathologia Mediterranea* 45: 12–29.

Moyo P., Damm U., Mostert L., Halleen F., 2018. *Eutypa*, *Eutypella*, and *Cryptovalsa* species (Diatrypaceae) associated with *Prunus* species in South Africa. *Plant Disease* 102: 7.

Mugnai L., Graniti A., Surico G., 1999. Esca (Black Measles) and brown wood-streaking: two old and elusive diseases of grapevines. *Plant Disease* 83: 404–418.

Mundy D.C., Manning M.A., 2011. Physiological response of grapevines to vascular pathogens: a review. *New Zealand Plant Protection* 64: 7–16.

Munne-Bosch S., Alegre L., 1999. Role of dew on the recovery of water-stressed *Melissa officinalis* L. plants. *Journal of Plant Physiology* 154: 759–766.

Munns R., Tester M., 2008. Mechanisms of salinity tolerance. *Annual Review of Plant Biology* 59: 651–681.

Murolo S., Romanazzi G., 2014. Effects of grapevine cultivar, rootstock and clone on esca disease. *Australasian Plant Pathology* 43: 215–221.

Noguees S., Allen D.J., Morison J.I.L., Baker N.R., 1998. Ultraviolet-B radiation effects on water relations, leaf development and photosynthesis in droughted pea plants. *Plant Physiology* 117: 173–81.

Oliva J., Stenlid J., Martinez-Vilatta J., 2014. The effect of fungal pathogens on the water and carbon economy of trees: implications for drought-induced mortality. *New Phytologist* 203: 1028–1035.

Ollat N., Touzard J.M., van Leeuwen C., 2016. Climate change impacts and adaptations: New challenges for the wine industry. *Journal of Wine Economics* 11: 139–149.

Pandey P., Ramegowda V., Senthil-Kumar M., 2015. Shared and unique responses of plants to multiple individual stresses and stress combinations: physiological and molecular mechanisms. *Frontiers in Plant Science* 6: 723.

Parker A., de Cortázar-Atauri I.G., Chuine I., Barbeau G., Bois B., ... van Leeuwen C., 2013. Classification of varieties for their timing of flowering and veraison using a modelling approach: A case study for the grapevine species *Vitis vinifera* L. *Agricultural and Forest Meteorology* 180: 249–264.

Pasquier G., Lapaillerie D., Vilain S., Dupuy J.W., Lomenech A.M., ... Donèche B., 2013. Impact of foliar symptoms of "esca proper" on proteins related to defense and oxidative stress of grape skins during ripening. *Proteomics* 13: 108–118.

Passioura J.B., 2002. Environmental biology and crop improvement. *Functional Plant Biology* 29: 537–546.

Pate J. Arthur D., 1998. σ^{13}C analysis of phloem sap carbon: novel means of evaluating seasonal water stress and interpreting carbon isotope signatures of foliage and trunk wood of *Eucalyptus globulus*. *Oecologia* 117: 301–311.

Pearce R.B., 1996. Antimicrobial defences in the wood of living trees. *New Phytologist* 132, 203–233

Péros J.P., Berger G., Jamaux-Despréaux I., 2008. Symptoms, wood lesions and fungi associated with esca in organic vineyards in Languedoc-Roussillon (France). *Journal of Phytopathology* 156: 297–303.

Perrone I., Gambino G., Chitarra W., Vitali M., Pagliarani C., ... Lovisolo C., 2012. The grapevine root-specific aquaporin VvPIP2; 4N controls root hydraulic conductance and leaf gas exchange under well-watered conditions but not under water stress. *Plant Physiology* 160: 965–977.

Petit A.N., Vaillant N., Boulay M., Clement C., Fontaine F., 2006. Alteration of photosynthesis in grapevines affected by esca. *Phytopathology* 96: 1060–1066.

Petrini O., 1986. Taxonomy of endophytic fungi of aerial plant tissues. In: *Microbiology of the Phyllosphere* (N.J. Fokkema, J. van den Heuve, eds), Cambridge University Press Cambridge, UK, 175–187.

Pouzoulet J., Pivovaroff A.L., Santiago L.S., Rolshausen P.E., 2014. Can vessel dimension explain tolerance towards fungal vascular wilt diseases in woody plants? Lessons from Dutch elm disease and esca disease in grapevine. *Frontiers in Plant Science* 5: 253.

Pouzoulet J., Scudiera E., Schiavon M., Rolshausen P.E., 2017. Xylem vessel diameter affects the compartmentalization of the vascular pathogen *Phaeomoniella chlamydospora* in grapevine. *Frontiers in Plant Science* 8: 1442.

Raimondo F., Nardini A., Salleo S., Cacciola S.O., Lo Gullo M.A., 2010. A tracheomycosis as a tool for studying the impact of stem xylem dysfunction on leaf water status and gas exchange in *Citrus aurantium* L. trees. *Structure and Function* 24: 327–33.

Ramegowda V. Senthil-Kumar M., 2015. The interactive effects of simultaneous biotic and abiotic stresses on plants: mechanistic understanding from drought and pathogen combination. *Journal of Plant Physiology* 176: 47–54.

Robert E.M.R., Koedam N., Beeckman H., Schmitz N., 2009. A safe hydraulic architecture as wood anatomical explanation for the difference in distribution of the mangroves *Avicennia* and *Rhizophora*. *Functional Ecology* 23: 649–657.

Rodrıguez-Kessler M., Ruiz O.A., Maiale S., Ruiz Herrera J., Jimenez-Bremont J.F., 2008. Polyamine metabolism in maize tumors induced by *Ustilago maydis*. *Plant Physiology and Biochemistry* 46: 805–814.

Rolshausen P.E., Úrbez-Torres J.R., Rooney-Latham S., Eskalen A., Smith R.J., Gubler W.D., 2010. Evaluation of pruning wound susceptibility and protection against fungi associated with grapevine trunk diseases. *American Journal of Enology and Viticulture* 61: 113–119.

Romanazzi G., Murolo S., Pizzichini L., Nardi S., 2009. Esca in young and mature vineyards, and molecular diagnosis of the associated fungi. *European Journal of Plant Pathology* 125: 277–290.

Rotem J., Wooding B., Aylor D.E., 1985. The role of solar radiation, especially ultraviolet, in the mortality of fungal spores. *Phytopathology* 75: 510–514.

Rudelle J., Octave S., Kaid-Harche M., Roblin G., Fleurat-Lessard P., 2005. Structural modifications induced by *Eutypa lata* in the xylem of trunk and canes of *Vitis vinifera*. *Functional Plant Biology* 32: 537–547.

Sakalidis M.L., Hardy G.E.S., Burgess T.I., 2011. Endophytes as potential pathogens of the baobab species *Adansonia gregorii*: a focus on the Botryosphaeriaceae. *Fungal Ecology* 4: 1–14.

Scharf K.D., Siddique M., Vierling E., 2001. The expanding family of *Arabidopsis thaliana* small heat stress proteins and a new family of proteins containing alpha-crystalline domains (Acd proteins). *Cell Stress Chaperones* 6: 225–237.

Schreiner R.P., 2003. Mycorrhizal colonization of grapevine rootstocks under field conditions. *American Journal of Enology and Viticulture* 54: 143–149.

Schreiner R.P., Lee J., 2014. Effects of post-veraison water deficit on "Pinot noir" yield and nutrient status in leaves, clusters, and musts. *Hort Science* 49: 1335–1340.

Schreiner R.P., Linderman R.G., 2005. Mycorrhizal colonization in dryland vineyards of the Willamette Valley, Oregon. *Small Fruits Review* 4: 41–55.

Schreiner R.P., Mihara K.L., 2009. The diversity of arbuscular mycorrhizal fungi amplified from grapevine roots (*Vitis vinifera* L.) in Oregon vineyards is seasonally stable and influenced by soil and vine age. *Mycologia* 101: 599–611.

Schultz H.R., 2000. Climate change and viticulture: A European perspective on climatology, carbon dioxide and UV-B effects. *Australian Journal of Grape and Wine Research* 6: 2–12.

Schultz H.R., 2016. Global climate change, sustainability, and some challenges for grape and wine production. *Journal of Wine Economics* 11: 181–200.

Schulz B., Boyle C., 2005. The endophytic continuum. *Mycological Research* 109: 661–686.

Sevanto S., McDowell N.G., Dickman L.T., Pangle R., Pockman W.T., 2014. How do trees die? A test of the hydraulic failure and carbon starvation hypotheses. *Plant, Cell & Environment* 37: 153–161.

Shellie K., Brown B., 2012. Influence of deficit irrigation on nutrient indices in wine grape (*Vitis vinifera* L.). *Agricultural Sciences* 3: 268–273.

Slippers B., Wingfield M.J., 2007. Botryosphaeriaceae as endophytes and latent pathogens of woody plants: diversity, ecology and impact. *Fungal Biology Reviews* 21: 90–106.

Soar C.J., Spiers J., Maffei S.M., Loveys B.R., 2004. Gradients of stomatal conductance, xylem sap ABA and bulk leaf ABA along canes of *Vitis vinifera* cv. Shiraz: molecular and physiological studies investigating the source. *Functional Plant Biology* 31: 659–669.

Soar C.J., Dry P.R., Loveys B.R., 2006. Scion photosynthesis and leaf gas exchange in *Vitis vinifera* L. cv. Shiraz: mediation of rootstock effects via xylem sap ABA. *Australian Journal of Agricultural Research* 12: 82–96.

Solomon S., Qin D., Manning M., Chen Z., Marquis M., ... Miller H.L., 2007. *Climate Change*. The Physical Science Basis, Contribution of Working Group I to the Fourth Assessment Report of the Intergovernmental Panel on Climate Change, Cambridge University Press. Agenda, Cambridge, UK and USA, 996 pp.

Sosnowski M.R., Shtienberg D., Creaser M.L., Wicks T.J., Lardner R., Scott E.S., 2007. The influence of climate on foliar symptoms of Eutypa dieback in grapevines. *Phytopathology* 97: 1284–1289.

Sosnowski M., Luque J., Loschiavo A., Martos S., Garcia-Figueres F., ... Scott E., 2011. Studies on the effect of water and temperature stress in grapevines inoculated with *Eutypa lata*. *Phytopathologia Mediterranea* 50: 127–138.

Sperry J.S., Hacke U.G., Pittermann J., 2006. Size and function in conifer tracheids and angiosperm vessels. *American Journal of Botany* 93: 1490–1500.

Stevenson R.E. Pennypacker S.P., 1988. Effect of radiation, temperature, and moisture on conidial germination of *Alternaria solani*. *Phytopathology* 78: 926–930.

Suhita D., Raghavendra A.S., Kwak J.M., Vavasseur A., 2004. Cytoplasmic alkalization precedes reactive oxygen species production during methyl jasmonate and abscisic acid induced stomatal closure. *Plant Physiology* 134: 1536–1545.

Sumarah M.W., Miller J.D., Blackwell B.A., 2005. Isolation and metabolite production by *Penicillium roqueforti*, *P. paneum* and *P. crustosum* isolated in Canada. *Mycopathologia*, 159(4): 571–577.

Surico G., 2001. Towards commonly agreed answers to some basic questions on esca. *Phytopathologia Mediterranea* 40: S487–490.

Surico G., 2009. Towards a redefinition of the diseases within the esca complex of grapevine. *Phytopathologia. Mediterranea* 48: 5–10.

Surico G., Marchi G., Braccini P., Mugnai L., 2000. Epidemiology of esca in some vineyards in Tuscany (Italy). *Phytopathologia Mediterranea* 39: 190–205.

Surico G., Mugnai L., Marchi G., 2006. Older and more recent observations on esca: a critical overview. *Phy-*

topathologia Mediterranea 45: S68–S86.

Surico G., Mugnai L., Marchi G., 2008. The esca disease complex. In: *Integrated management of disesases caused by fungi, phytoplasma and bacteria* (A. Ciancio, K.G. Mukerji, eds), Springer, Heidelberg, New York, 119–136.

Surico G., Marchi G., Mugnai L., Peduto F., 2010. Espressione dei sintomi fogliari dell'esca in funzione della disponibilità di acqua nel suolo e germogliamento delle gemme. In: *Il Mal dell'Esca della Vite: Interventi di ricerca e Sperimentazione per il Contenimento della Malattia. Progetto MesVit.* (Surico G., Mugnai L., ed.), Arsia Regione Toscana, Firenze, Italy: 276-291.

Tabacchi R., Fkeyrat A., Poliart C., Dubin G.M., 2000. Phytotoxins from fungi of esca of grapevine. *Phytopathologia Mediterranea* 39: 156–61.

Treutter D., 2005. Significance of flavonoids in plant resistance and enhancement of their biosynthesis. *Plant Biology* 7: 581–591.

Trouvelot S., Bonneau L., Redecker D., van Tuinen D., Adrian M., Wipf D., 2015. Arbuscular mycorrhiza symbiosis in viticulture: a review. *Agronomy for Sustainable development* 35: 1449–1467.

Tyree M.T. Sperry J.S., 1988. Do woody plants operate near the point of catastrophic xylem dysfunction caused by dynamic water stress? *Plant Physiology* 88: 574–580.

Tyree M.T., Davis S.D., Cochard H., 1994. Biophysical perspectives of xylem evolution: is there a tradeoff of hydraulic efficiency for vulnerability to dysfunction? *IAWA Journal* 15: 335–360.

Úrbez-Torres J.R., 2011. The status of Botryosphaeriaceae species infecting grapevines. *Phytopathologia Mediterranea* 50: S5–S45.

van Leeuwen C., Friant P., Choné X., Tregoat O., Koundouras S., Dubourdieu D., 2004. Influence of climate, soil, and cultivar on terroir. *American Journal of Enology and Viticulture* 55: 207–217.

van Niekerk J.M., Bester W., Halleen F., Crous P.W., Fourie P.H., 2011. The distribution and symptomatology of grapevine trunk disease pathogens are influenced by climate. *Phytopathologia Mediterranea* 50: S98–S111.

Vandeleur R.K., Mayo G., Shelden M.C., Gilliham M., Kaiser B.N., Tyerman S.D., 2009. The role of plasma membrane intrinsic protein aquaporins in water transport through roots: diurnal and drought stress responses reveal different strategies between isohydric and anisohydric cultivars of grapevine. *Plant Physiology* 149: 445–460.

Vanholme R., Morreel K., Ralph J., Boerjan W., 2008. Lignin engineering. Current opinion in *Plant Biology* 11: 1–8.

Venturas M., Lopeza R., Martina J.A., Gascob A., Gil L., 2013. Heritability of *Ulmus minor* resistance to Dutch elm disease and its relationship to vessel size, but not to xylem vulnerability to drought. *Plant Pathology* 63: 500–509.

Verhoeff K., 1974. Latent infections by fungi. *Annual Review of Phytopathology* 12: 99–110.

Vilagrosa A., Chirino E., Peguero-Pina J.J., Barigah T.S., Cochard H., Gil-Pelegrin E., 2012. Xylem cavitation and embolism in plants living in water-limited ecosystems. In: *Plant Responses to Drought Stress* (R. Aroca, ed.), Springer, Berlin, Germany, 63–109.

Vogt T., 2010. Phenylpropanoid biosynthesis. *Molecular Plant* 3: 2–20.

Walker R.R., Blackmore D.H., Clingeleffer P.R., Correll R., 2002. Rootstock effects on salt tolerance of irrigated field-grown grapevines (*Vitis vinifera* L.cv. Sultana). 1. Yield and vigour inter-relationships. *Australian Journal of Grape and Wine Research* 8: 3–14.

Ward E.W., Cahill D.M., Bhattacharyya M.K., 1989. Abscisic acid suppression of phenylalanine ammonia-lyase activity and mRNA, and resistance of soybeans to *Phytophthora megasperma* f.sp. *glycinea*. *Plant Physiology* 91: 23–7.

Westgate M.E., Boyer J.S., 1985. Carbohydrate reserves and reproductive development at low water potentials in maize. *Crop Science* 25: 762–769.

Wilkinson S., Davies W.J., 1997. Xylem sap pH increase: A drought signal received at the apoplastic face of the guard cell that involves the suppression of saturable abscisic acid uptake by the epidermal symplast. *Plant Physiolology* 113: 559–573.

Woodall G.S., Ward B.H., 2002. Soil water relations, crop production and root pruning of a belt of trees. *Agricultural Water Management* 53: 153–169.

Yamaguchi-Shinozaki K., Shinozaki K., 2006. Transcriptional regulatory networks in cellular responses and tolerance to dehydration and cold stresses. *Annual Review of Plant Biology* 57: 781–803.

Yu C.K.Y., Springob K., Jurger S., Nicholson R.L., Chu I.K., ... Lo C., 2005. A stilbene synthase gene (*SdSTS1*) is involved in host and non host defense responses in *Sorghum*. *Plant Physiology* 138(1): 393–401.

Zabalgogeazcoa I., 2008. Fungal endophytes and their interaction with plant pathogens. *Spanish Journal of Agricultural Research* 6: 138–146.

Zhang J., Davies W.J., 1990. Changes in the concentration of ABA in xylem sap as a function of changing soil water status will account for changes in leaf conductance. *Plant, Cell and Environment* 13: 277–285.

Zhang H., Memelink J., 2009. Regulation of secondary metabolism by jasmonate hormones. In: *Plant-derived Natural Products: Synthesis, Function and Application* (A.E. Osbourn, V. Lanzotti, eds.), Springer, Berlin Heidelberg, Germany, 181–194.

Zhang J., Zhang X.P., Liang J.S., 1995. Exudation rate and hydraulic conductivity of maize roots are enhanced by soil drying and abscisic acid treatment. *New Phytologist* 131: 329–336.

Zhu J.K., 2003. Regulation of ion homeostasis under salt stress. *Current Opinion in Plant Biology* 6: 441–445.

Ziska L.H., Runion G.B., 2007. Future weed, pest, and disease problems for plants. In: *Agroecosystems in a Changing Climate* (P.C.D. Newton, R.A. Carran, G.R. Edwards, P.A. Niklaus, eds), CRC Press, Boca Raton, FL, USA, 261–287.

2

New records of *Penicillium* and *Aspergillus* from withered grapes in Italy, and description of *Penicillium fructuariae-cellae* sp. nov.

Marilinda LORENZINI[1,*], Maria Stella CAPPELLO[2], Giancarlo PERRONE[3], Antonio LOGRIECO[3], Giacomo ZAPPAROLI[1]

[1] *Università degli Studi di Verona, Dipartimento di Biotecnologie, 37134 Verona, Italy*
[2] *Istituto di Scienze delle Produzioni Alimentari, CNR, 73100 Lecce, Italy*
[3] *Istituto di Scienze delle Produzioni Alimentari, CNR, 70126 Bari, Italy*
*Corresponding authors: lorenzini.marilinda@gmail.com

Summary. *Penicillium* and *Aspergillus* are common pathogenic fungi of grapes, that occur frequently on withered berries used in the Italian passito wine production. Members of these genera isolated from withered grapes were identified using molecular and morphological approaches. The isolates were examined by amplification of internal transcribed spacer region, β-tubulin, calmodulin and RNA polymerase II second largest subunit. *Penicillium bilaiae*, *Aspergillus pallidofulvus* and *A. puulaauensis* are reported for the first time from *Vitis vinifera*. Two *Penicillium* isolates showed a distinct phylogenetic position and different morphological characteristics from *P. bissettii* and *P. vasconiae*, the two most closely related species. These isolates are assigned to the new species *Penicillium fructuariae-cellae*, that is here described. An *in vitro* pathogenicity assay was carried out to evaluate the infectivity to grape berries by *Penicillium* and *Aspergillus* isolates recovered in this study. All examined isolates colonized the berries when artificially inoculated, but to a lesser extent than *Botrytis cinerea*. This suggests that these fungi may contribute, with other pathogenic species, to the onset of post-harvest diseases of grapes.

Competing Interests: The Author(s) declare(s) no conflict of interest.

Editor: Dimitrios I. Tsitsigiannis, Agricultural University of Athens, Greece.

Keywords. Saprophytic/pathogenic fungi, grapes, phylogeny, taxonomy, post-harvest diseases.

INTRODUCTION

Fungal contamination of grapes causes severe and economically important losses for world food and beverage industries. Table, wine and raisin grapes can be infected by several species of fungi, on grapevines and/or during post-harvest processes. Grapes for Italian passito wine production are particularly vulnerable to fungal infections, during withering carried out in fruit drying rooms (*fruttaio*) (off-vine withering) (Mencarelli and Tonutti, 2013).

Penicillium and *Aspergillus* spp. are among the most frequent saprophytic fungal pathogens on withered grapes (Torelli *et al.*, 2006; Lorenzini *et al.*, 2016; Stefanini *et al.*, 2017). Their presence is very important since they are causal agents of bunch rot and can be mycotoxin producers (Torelli *et al.*, 2006; Somma *et al.*, 2012). In addition, grape contamination by these fungi can lower the quality of the resulting wines. The detrimental effects of withered grapes infected by *P. expansum* and *P. crustosum* on the quality of Amarone wine, a dry red passito wine, has recently been documented (Zapparoli *et al.*, 2018).

During previous surveys on fungi associated with withered grapes (Lorenzini *et al.*, 2016, 2018), eight species of *Penicillium* (*P. adametzoides*, *P. expansum*, *P. crocicola*, *P. crustosum*, *P. glabrum*, *P. griseofulvum*, *P. oxalicum* and *P. ubiquetum*) and five species of *Aspergillus* (*A. flavus*, *A. sydowii*, *A. tubingensis*, *A. uvarum* and *A. welwitschiae*) were identified by phylogenetic analyses. The placements of two isolates (*Penicillium* sp. P3 and *Aspergillus* sp. AS100) were not clear enough for reliable species delimitations.

In the present study, isolates P3 and AS100 were phylogenetically and morphologically analyzed to clarify their taxonomic positions. Three isolates of *Penicillium* and *Aspergillus*, recovered from withered grapes during the current survey, were also identified. These fungi belonged to species that are reported for the first time from *Vitis vinifera*. Two isolates were assigned to a new species of *Penicillium*. The pathogenicity of fungi isolated from grape berries was also assayed.

MATERIALS AND METHODS

On the basis of our previous studies (Lorenzini *et al.*, 2016; 2018) carried out on grape berries of the Garganega and Corvina varieties, collected from fruit-drying rooms located in two Northern Italian winemaking areas (Soave and Valpolicella), five representative strains of *Aspergillus* and *Penicillium* were isolated and identified. Three isolates recovered from withered grapes in this study (designated Pdb1, Pls8 and ASls13) were isolated according to Lorenzini *et al.* (2016). The other two isolates (P3 and AS100) were obtained during a previous sampling (Lorenzini *et al.*, 2016). These isolates are deposited at the Westerdijk Fungal Biodiversity Institute (CBS, Utrecht, the Netherlands) and ITEM Agro-Food Microbial Culture Collection of the Institute of Science and Food Production (CNR-ISPA, Bari, Italy) (Table 1).

DNA was extracted from pure culture of each isolate as previously described (Lorenzini and Zapparoli, 2014). Each DNA extract was used to amplify the internal transcribed spacer (ITS) region, using primers ITS1/ITS4 (White *et al.*, 1990), partial β-tubulin gene (*benA*), using primers Bt2a/Bt2b (Glass and Donaldson, 1995), partial calmodulin gene (*CaM*), using primers cmd5/cmd6 (Hong *et al.*, 2006), and parts of the second largest subunit of RNA polymerase II (*rpb2*), using primers fRPB2-5F2/fRPB2-7C (Liu *et al.*, 1999). The amplified products were purified using the NucleoSpin Gel and PCR Clean-up Kit (Macherey-Nagel), and were sequenced in both directions using the same primers applied for amplification (Eurofins Genomics, Edersberg, Germany). The generated sequences were deposited at GenBank (Table 2).

Combined and individual analyses were conducted using the partial DNA sequences of five isolates recovered from withered grapes, and other reference taxa belonging to the genus *Penicillium* (sections *Lanata-divaricata* and *Sclerotiora*) and *Aspergillus* (*A. versicolor* clade and section *Circumdati*), retrieved from GenBank (Table 2). Maximum Likelihood (ML) analysis of the combined data sets was performed using MEGA7 v. 7.0.25 software. The combined data sets were analysed as three or four distinct partitions. For each individual data set, the most optimal substitution model was calculated in MEGA7 (Kumar *et al.*, 2016) using the Akaike Information Criterion (AIC). Maximum Likelihood analyses of the individual data sets were also conducted using MEGA7, and robustness of the trees was evaluated by 1,000 bootstrap (BS) replicates. A second measure for statistical support was performed using Bayesian Evolutionary Analysis Sampling Trees (BEAST) Version v1.10.1, 2002-2018 (Drummond and Rambaut, 2007), and the previously obtained most optimal substitution model was used in the analyses. The Markov Chain Monte Carlo (MCMC) analysis used four chains and started from a random tree topology. Burn-in was set to 25%

Table 1. Fungus isolates recovered from withered grape berries in this study.

Specie	Isolate designation	CBS number	ISPA number	Grape variety
Penicillium fructuariae-cellae	P3	CBS 145110^T	ITEM 18276^T	Corvina
Penicillium fructuariae-cellae	Pdb1	CBS 145111	ITEM 18277	Garganega
Penicillium bilaiae	Pls8	CBS 145112	ITEM 18278	Garganega
Aspergillus pallidofulvus	ASls13	CBS 145108	ITEM 18279	Garganega
Aspergillus puulaauensis	AS100	CBS 143103	ITEM 18280	Corvina

Table 2. Isolates used in phylogenetic analyses.

Species	Strain identification	Source	Locality	GenBank numbers			
				ITS	β-tubulin	Calmodulin	*rpb2*
Aspergillus section *Circundati*							
A. affinis	ATCC MYA-4773^{T}	Leaf litter	Italy	GU721090	GU721092	GU721091	-
A. auricomus	NRRL 391^{T}	ni	ni	EF661411	EF661320	EF661379	-
A. bridgeri	NRRL 13000^{T}	ni	ni	EF661404	EF661335	EF661358	-
A. cretensis	NRRL 35672^{T}	ni	ni	FJ491572	AY819977	FJ491534	-
A. elegans	NRRL 4850^{T}	ni	ni	EF661414	EF661349	EF661390	-
A. fresenii	NRRL 4077^{T}	Soil	India	EF661409	EF661341	EF661382	-
A. insulicola	NRRL 6138^{T}	Soil	Venezuela	EF661430	EF661353	EF661396	-
A. melleus	NRRL 5103^{T}	Soil	India	EF661425	EF661326	EF661391	-
A. muricatus	NRRL 35674^{T}	Grassland soil	Philippines	EF661434	EF661356	EF661377	-
A. neobridgeri	NRRL 13078^{T}	Soil	USA	EF661410	EF661345	EF661359	-
A. occultus	CBS 137330^{T}	Air sample	Netherlands	KJ775443	KJ775061	KJ775239	-
A. ochraceopetaliformis	NRRL 4752^{T}	Scalp lesion	Brazil	EF661429	EF661350	EF661388	-
A. ochraceus	NRRL 398^{T}	ni	ni	EF661419	EF661322	EF661381	-
A. ostianus	NRRL 420^{T}	ni	ni	EF661421	EF661324	EF661385	-
A. pallidofulvus	NRRL 4789^{T}	ni	ni	EF661423	EF661328	EF661389	-
*A. pallidofulvus**	ITEM 18279 = CBS 145108	Vitis vinifera	Italy	MK039437	MK045335	MK045340	-
A. persii	NRRL 35669^{T}	Toenail	Italy	FJ491580	AY819988	FJ491559	-
A. pseudoelegans	CBS 112796^{T}	Soil	Costa Rica	FJ491590	AY819962	FJ491552	-
A. pseudosclerotiorum	UTHSCSA DI15-13^{T}	Lung biopsy	-	LT574713	LT574748	LT574783	-
A. pulvericola	CBS 137327^{T}	Indoor hause	Micronesia	KJ775440	KJ775055	KJ775236	-
A. robustus	NRRL 6362^{T} (outgroup)	Soil	Kenya	EF661176	EU014101	EF661357	-
A. roseoglobulosus	NRRL 4565^{T}	Decaing leaves	Bahamas	FJ491583	AY819984	FJ491555	-
A. salwaensis	DTO 297B3^{T}	Soil	Qatar	KJ775447	KJ775056	KJ775244	-
A. sclerotiorum	NRRL 415^{T}	Apple	USA	EF661400	EF661337	EF661384	-
A. sesamicola	CBS 137324^{T}	Sesame sead	Denmark	KJ775437	KJ775063	KJ775233	-
A. steynii	NRRL 35675^{T}	Grean bean coffee	India	EF661416	EF661347	EF661378	-
A. subramanianii	NRRL 6161^{T}	Shelled brazil nuts	Canada	EF661403	EF661339	EF661397	-
A. tanneri	NRRL 62425^{T} (outgroup)	Human lung	USA	JN853798	JN896582	JN896583	-
A. westerdijkiae	NRRL 3174^{T} = CBS 112803	*Andropogum sorghum*	South Africa	EF661427	EF661329	EF661360	-
A. westlandensis	CBS 137321^{T}	Air sample	Netherlands	KJ775434	KJ775066	KJ775230	-
A. versicolor clade							
A. amoenus	NRRL 4838^{T}	ni	ni	NR_137462	EF652304	EF652392	-
A. austroafricanus	NRRL 233^{T}	ni	South Africa	NR_135443	JN853963	JN854025	-
A. creber	NRRL 58592^{T}	Indoor air sample	USA	NR_135442	JN853980	JN854043	-

(Continued)

Table 2. (Continued).

Species	Strain identification	Source	Locality	GenBank numbers			
				ITS	β-tubulin	Calmodulin	*rpb2*
A. cvjetkovicii	NRRL 227^{T}	Soil	USA	EF652440	EF652264	EF652352	-
A. fructus	NRRL 239^{T}	Date fruit	USA	EF652449	EF652273	EF652361	-
A. griseoaurantiacus	DTO_267D8^{T}	Indoor air dust	Micronesia	KJ775553	KJ775086	KJ775357	-
A. hongkongensis	HKU49^{T}	Human nail	Hong Kong	AB987907	LC000552	LC000565	-
A. jensenii	NRRL 58600^{T}	Indoor air sample	USA	NR_135444	JN854007	JN854046	-
A. multicolor	NRLL 4775^{T} (outgroup)	ni	ni	EF652477	EF652301	EF652389	-
A. protuberus	NRRL 3505^{T}	Brined meat	UK	NR_135353	EF652284	EF652372	-
A. puulaauensis	NRRL 35641^{T}	Dead hardwood branch	USA	NR_135445	JN853979	JN854034	-
A. puulaauensis	**ITEM 18280 = CBS 145103**	**Fruit of *Vitis vinifera***	Italy	MK039438	MK045336	KU554606	-
A. sydowii	NRRL 250^{T}	ni	n	EF652450	EF652274	EF652362	-
A. tabacinus	NRRL 4791^{T}	ni	ni	NR_135361	EF652302	EF652390	-
A. tennesseensis	NRRL 13150^{T}	Ex toxic dairy feed	USA	NR_135447	JN853976	JN854017	-
A. venenatus	NRRL 13147^{T}	Toxic dairy feed	USA	NR_135448	JN854003	JN854014	-
A. versicolor	NRRL 238^{T}	ni	USA	EF652442	EF652266	EF652354	-
A.subversicolor	NRRL 58999^{T}	Coffee berry	India	NR_135446	JN853970	JN854010	-
Penicillium section *Lanata–divaricata*							
P. abidjanum	CBS 246.67^{T}	Soil	Ivory Coast	NR_111502	GU981650	KF296383	JN121469
P. amphipolaria	DAOMC 250551^{T} = KAS 2555 = CBS 140997	Leaves	Panama	KT887872	KT887833	KT887794	na
P. annulatum	CV 37 = CBS 135126^{T}	Soil	South Africa	NR_138303	JX091514	JX141545	KF296410
P. araracuarense	CBS 113149^{T}	Leaf litter	Colombia	GU981597	GU981642	KF296373	KF296414
P. austrosinenese	CGMCC 3.18797^{T} = CBS 144505	Acidic soil	China	KY495003	KY495112	KY494943	KY495061
P. bissettii	DAOMC 167011^{T} = KAS1951 = CBS 140972	Soil	Canada	KT887845	KT887806	KT887767	KY495055
P. brasilianum	CBS 253.55^{T}	Herbarium specimen	Brazil	AF178512	GU981629	AB667857	KF296420
P. brefeldianum	CBS 235.81^{T}	Human alimentary tract	ni	GU981580	GU981623	AB667857	KF296421
P. camponotum	KAS2177 = DAOMC 250557^{T} = CBS 140982	Ants	Canada	KT887855	KT887816	KT887777	na
P. caperatum	CBS 443.75^{T}	Soil	Australia	KC411761	GU981660	KF296392	KF296422
P. cataractum	CBS 140974 = KAS2145 = DAOMC 250534^{T}	Nuts of *Carya cordiformis*	Canada	KT887847	KT887808	KT887769	na
P. fructuariae-cellae	**ITEM 18276^{T} = CBS 145110^{T}**	**Fruit of *Vitis vinifera***	**Italy**	**MK039434**	KU554679	MK045337	MK520927
P. fructuariae-cellae	**ITEM 18277 = CBS 145111**	**Fruit of *Vitis vinifera***	**Italy**	**MK039435**	**MK045333**	**MK045338**	**MK520928**
P. cluniae	CBS 326.89^{T}	Soil	Spain	KF296406	KF296471	KF296402	KF296424
P. coeruleum	CBS 141.45^{T}	ni	ni	NR_138293	GU981655	KF296393	KF296425
P. cremeogriseum	CBS 223.66^{T}	Soil	Ukraine	NR_111505	GU981624	KF296403	KF296426
P. curticaule	CV2842 = CBS 135127^{T}	Soil	South Africa	FJ231021	JX091526	JX141536	KF296417
P. daleae	CBS 211.28^{T}	Soil	Poland	NR_111503	GU981649	KF296385	KF296427
P. echinulonalgiovense	CBS 328.59^{T}	Soil	Japan	GU981587	GU98163	KX961269	KX961301

(Continued)

Table 2. (Continued).

Species	Strain identification	Source	Locality	GenBank numbers			
				ITS	β-tubulin	Calmodulin	*rpb2*
P. ehrlichii	CBS 324.48^T	ni	Poland	GU981578	KF296464	KF296395	KF296428
P. elleniae	CBS 118135^T	Leaf litter	Colombia	GU981612	GU981663	KF296389	KF296429
P. excelsum	HF-2015 = CCT 7772^T	Nut shell	Brazil	KR815341	KP691061	KR815342	na
P. flaviroseum	CGMCC 3.18805^T = CBS 144479	Acidic soil	China	KY495032	KY495141	KY494972	KY495083
P. glabrum	CBS 125543^T (outgroup)			GU981567	GU981619	KM089152	JF417447
P. glaucoroseum	NRRL 908^T = CBS 138908	Soil	USA	KF296407	KF296469	KF296400	KF296430
P. globosum	CGMCC 3.18800^T = CBS 144639	Acidic soil	China	KY495014	KY495123	KY494954	KY495067
P. griseoflavum	CGMCC 3.18799^T = CBS 144525	Acidic soil	China	KY495011	KY495120	KY494951	KY495064
P. griseopurpureum	CBS 406.65^T	Soil	England	KF296408	KF296467	KF296384	KF296431
P. guangxiense	CGMCC 3.18793^T = CBS 144526	Soil	China	KY494986	KY495095	KY494926	KY495045
P. hainanense	CGMCC 3.18798^T = CBS 144527	Acidic soil	China	KY495009	KY495118	KY494949	KY495062
P. infrabuccalum	KAS2181^T= DAOMC 250537	Ants	Canada	KT887817	KT887817	KT887778	na
P. janthinellum	CBS 340.48^T	Soil	Nicaragua	NR_111504	GU981625	KF296401	JN121497
P. javanicum	CBS 341.48^T	Rut of *Cammelia sinensis*	Indonesia	NR_111511	GU981657	KF296387	JN121498
P. jianfenglingense	CGMCC 3.18802^T = CBS 144640	Soil	China	KY495016	KY495125	KY494956	KY495069
P. koreense	KACC 47721^T	Soil	Korea	KJ801939	KM000846	na	na
P. laevigatum	CGMCC 3.18801^T = CBS 144481	Acidic soil	China	KY495015	KY495124	KY494955	KY495068
P. levitum	CBS 345.48^T	Modeling clay	USA	NR_111510	GU981654	KF296394	KF296432
P. limosum	CBS 339.97^T	Marine sediment	Japan	NR_111496	GU981621	KF296398	KF296433
P. lineolatum	CBS 188.77^T	Soil	Japan	NR_111500	GU981620	KF296397	KF296434
P. ludwigii	CBS 417.68^T	Polished rice	Japan	NR_138339	KF296468	KF296404	KF296435
P. malacosphaerulum	CV2855 = CBS 135120^T	Soil	South Africa	FJ231026	JX091524	JX141542	KF296438
P. mariae-crucis	CBS 271.83^T	*Secale cereale*	Spain	NR_111506	GU981630	KF296374	KF296439
P. meloforme	CBS 445.74^T	Soil	Papua New Guinea	NR_153203	GU981656	KF296396	KF296440
P. ochrochloron	CBS 357.48^T	Copper sulphate solution	USA	NR_111509	GU981672	KF296378	KF296445
P. onobense	CBS 174.81^T	Soil	Spain	NR_111497	GU981627	KF296371	KF296447
P. ortum	CV 102 = CBS 135669^T	Soil	South Africa	NR_138304	JX091520	JX141551	KF296443
P. oxalicum	CBS 219.30^T (outgroup)	Soil	USA	MH855125	KF296462	KF296367	JN121456
P. panissanguineum	DAOMC 250562^T = KAS 2209 = CBS 140989	Termite mounds	Tanzania	KT887862	KT887823	KT887784	na
P. paraherquei	ATCC 22354 = CBS 338.59^T	Soil	Japan	AF178511	KF296465	KF296372	KF296449
P. pedernalense	CBS 140770^T	Waste compost	Ecuador	KU255398	KU255396	KY494968	KY495079
P. penarojense	CBS 113178^T	Leaf litter	Colombia	NR_138289	GU981646	KF296381	KF296450
P. piscarium	CBS 362.48^T	Cod-liver oil emulsion	Norway	NR_111507	GU981668	KF296379	KF296451
P. pulvillorum	CBS 280.39^T	Acidic soil	UK	NR_138292	GU981670	KF296377	KF296452
P. raperi	NRRL 2674 = CBS 281.58^T	Soil	UK	AF033433	GU981622	KF296399	KF296453

(Continued)

Table 2. (Continued).

Species	Strain identification	Source	Locality	GenBank numbers			
				ITS	β-tubulin	Calmodulin	*rpb2*
P. reticulisporum	NRRL 3447 = CBS 122.68^T	Soil	Japan	AF033437	GU981665	KF296391	KF296454
P. rolfsii	CBS 368.48^T	Pineapple	USA	JN617705	JU981667	KF296375	KF296455
P. rubriannulatum	CGMCC 3.18804^T = CBS 144641	Acidic soil	China	KY495029	KY495138	KY494969	KY495080
P. simplicissimum	CBS372.48^T	Flannel bag	South Africa	NR_138290	GU981632	KF296368	JN121507
P. singorense	DTO 133C6 = CBS 138214^T	House dust	Thailand	KJ775674	KJ775167	KJ775403	na
P. skrjabinii	CBS 439.75^T	Soil	Russia	NR_111498	GU981626	KF296370	EU427252
P. soliforme	CGMCC 3.18806^T = CBS 144482	Acidic soil	China	KY495038	KY495147	KY494978	KY495047
P. stolkiae	NRRL 5816 = CBS 315.67^T (outgroup)	ni	ni	NR_121233	JN617717	AF481135	JN121488
P. spiliferum	CGMCC 3.18807^T = CBS 14483	Acidic soil	China	KY495040	KY495149	KY494980	KY495090
P. subrubescens	DTO188-D6 = CBS 132785^T	Soil	Finland	KC346350	KC346327	KC346330	KC346306
P. svalbardense	CBS 122416^T	Glacial ice	Greenland	GU981603	KC346325	KC346338	KF296457
P. tanzanicum	DAOMC 250514^T = KAS 1946 = CBS 140968	ni	Tanzania	KT887841	KT887802	KT887763	KY495066
P. terrarumae	CBS 131811^T	Soil	China	KC346349	KC346326	KC346339	KC346316
P. vanderhammenii	CBS126216^T	Leaf litter	Colombia	NR_137577	GU981647	KF296382	KF296458
P. vasconiae	CBS 339.79^T	Soil	Spain	NR_138291	GU981653	KF296386	KF296459
P. viridissimum	CGMCC 3.18796^T = CBS 14484	Acidic soil	China	KY495004	KY495113	KY494944	KY495059
P. wotroi	CBS 118171^T	Leaf litter	Colombia	GU981591	GU981637	KF296369	KF296460
P. yunnanense	CGMCC 3.18794^T = CBS 14485	Acidic soil	China	KY494990	KY495099	KY494930	KY495048
P. zonatum	CBS 992.72^T	Coastal marsh soil	USA	NR_111501	GU981651	KF296380	KF296461
Penicillium section *Sclerotiora*							
P. adametzii	CBS 209.28^T	Soil	Canada	NR_103661	JN625957	KC773796	-
P. adametzioides	CBS 313.59^T	Soil	Japan	JN686433	JN799642	JN686387	-
P. alexiae	DTO 118H8^T	Soil	Tunisia	KC790400	KC773778	KC773803	-
P. amaliae	CV 1875 = CBS 134209^T	*Protea repens*	South Africa	JX091443	JX091563	JX141557	-
P. angulare	CBS 130293^T	Polypore on dead conifer	USA	KC773828	KC773779	KC773804	-
P. arianeae	DTO 20B8^T	Soil	Netherlands	KC773833	KC773784	KC773811	-
P. austrosinicum	HMAS 248734^T = CGMCC 3.18410	ni	ni	NR_153272	KX885041	KX885051	-
P. bilaiae	NRRL 3391^T	Soil	Ukraine	NR_111679	JN625966	JN626009	-
P. bilaiae	**ITEM 18278 = CBS 145112**	**Fruit of *Vitis vinifera***	**Italy**	**MK039436**	**MK045334**	**MK045339**	-
P. brocae	CBS 116113^T	Coffee berry	Mexico	NR_111868	KC773787	KC773814	-
P. cainii	DAOM 239914^T	Nuts of *Juglans nigra*	Canada	NR_120000	JN686366	JN686389	-
P. choerospondiatis	HMAS 248813^T = CGMCC 3.18411	ni	ni	NR_153274	KX885043	KX885053	-
P. daejeonium	CNU 100097 = KACC 46609^T	Grape fruit	South Korea	JX436489	JX436493	JX436491	-
P. exsudans	HMAS 248735^T = CGMCC 3.18412	ni	ni	NR_153273	KX885042	KX885052	-
P. guanacastense	DAOM 239912^T	Gut of caterpillar	Costa Rica	NR_111673	JN625967	JN626010	-

(Continued)

Table 2. (Continued).

Species	Strain identification	Source	Locality	GenBank numbers			
				ITS	β-tubulin	Calmodulin	*rpb2*
P. herquei	CBS 336.48^T	Leaf of *Agauria pirifolia*	France	JN626101	JN625970	JN626013	-
P. hirayamae	CBS 229.60^T	Milled rice	Thailand	JN626095	JN625955	JN626003	-
P. jacksonii	DAOM 239937^T	Forest soil	Australia	NR_111675	JN686368	JN686391	-
P. johnkrugii	DAOM 239943^T	Forest soil	Malaysia	NR_111676	JN686378	JN686401	-
P. jugoslavicum	CBS 192.87^T	Seed of *Helianthus annuus*	Ex Yugoslavia	NR_120269	KC773789	KC773815	-
P. lilacinoechinulatum	CBS 454.93^T	Soil	Japan	KC773837	KC773790	KC773816	-
P. malachiteum	CBS 647.95^T	Soil	Japan	NR_120271	KC773794	KC773820	-
P. mallochii	DAOM 239917^T	Caterpillar	Costa Rica	NR_111674	JN625973	JN626016	-
P. maximae	NRRL 2060^T	Cellophane	USA	NR_121343	KC773795	KC773821	-
P. restingae	MS-2014 = URM 7075^T	Soil	ni	KF803354	KF803349	KF803352	-
P. sanshaense	HMAS 248820^T = CGMCC 3.18413	ni	ni	NR_153276	KX885050	KX885060	-
P. sclerotiorum	NRRL 2074^T	Air	Indonesia	JN626132	JN626001	JN626044	-
P. vanoranjei	DTO 99H6^T	Soil	Tunisia	KC695696	KC695686	KC695691	-
P. verrucisporum	HMAS 248819^T = CGMCC 3.18415	ni	ni	KX885069	KX885049	KX885059	-
P. viticola	FKI-4410 = JCM 17636^T	Grape	Japan	AB606414	AB540174	AB540173	-
P. levitum	CBS 345.48^T (outgroup)	ni	ni	NR_111510	GU981654	KF296394	-

* Strains studied in this paper are in bold type;
- Sequences not used in this study;
^{T} = Ex-type strain; ni = no information about the source and locality; na = not available
ATCC, American Type Culture Collection (Manassas, VA); CBS, Westerdkjik Fungal Biodiversity Institute (Utrecht, the Netherlands); DTO, Applied and Industrial Mycology Department Collection, Westerdkjik Fungal Biodiversity Institute (Utrecht, Netherlands); NRRL, Agriculture Research Service Culture Collection (Peoria, NY); UTHSCSA, University of Texas Health Science Center (San Antonio, TX); HKU, University of Hong Kong, (Hong Kong, China). ITEM, Agro-Food Microbial Culture Collection of Institute of Science and Food Production (ISPA) (Bari, Italy); JCM, Japan Collection of Microorganism (Saitama, Japan); CGMCC. Chinese General Microbiological Culture Collection Center (Beijing, China); URM, University Recife Mycology Culture Collection (Recife, Brazil); KACC, Korean Agricultural Culture Collection (Suwon, South Korea); CCT, Coleção de Cultura Tropical (Campinas, Brazil); DAOMC, Canadian Collection of Fungal Cultures (Ottawa, Canada); DAOM, Canadian National Mycological Herbarium (Ottawa, Canada).

and Tracer v. 1.5.0 (Rambaut and Drummond, 2009) was used to confirm the convergence of chains. The phylograms obtained through the ML analyses were used for presenting the data. Phylograms were redrawn from the tree files using FigTree v1.4 2006-2012. Bootstrap values less than 70% and posterior probability (pp) values less than 0.95 were removed from the phylograms. All ambiguous positions were removed for each sequence pair. Evolutionary analyses were conducted in MEGA7. All isolates of *Aspergillus* and *Penicillium* (Pdb1, P3, Pls8, ASls13 and AS100) were inoculated onto Czapek yeast extract agar (CYA), yeast extract sucrose agar (YES), creatine sucrose agar (CREA) (Visagie *et al.* 2014) and malt extract agar (MEA, 2% w v^{-1} malt extract, 0.1% w v^{-1} peptone, 2% w v^{-1} dextrose, 1.5% w v^{-1} agar) at 25°C in the darkness. The colony diameters (mm) were measured daily for 7 d (three replicate plates for each isolate) and the experiments were performed twice. The colony diameters were also measured on CYA at 15, 30 and 37°C. The cultural characteristics and micro-morphology of each strain was examined on CYA, YES and MEA after 7, 14 or 30 d at 25°C. Colony morphology was also examined on CREA after 7 d incubation.

Growth tests for the fungal isolates were carried out under acidic, neutral and alkaline conditions, as reported by Diao *et al.* (2018).

Preparations for microscopy were made from colonies grown on CYA, YES and MEA after 7, 14 or 30 d. Lactic acid (60% v v^{-1}) was used as mounting fluid and excess conidia were washed away with ethanol (70% v v^{-1}). ethanol. Characters were recorded and analyzed using stereomicroscopy (Leica EZ4D, Leica Microsystems). Measurements of fungal components were carried out using light microscopy (Leica DM750) equipped with camera module (Leica ICC50W). Lengths and widths were determined for 20 conidiophores, metulae and phialides, 50 conidia, 20 cleistothecia and 30 ascospores (when present) from each isolate.

The five isolates were tested for their ability to cause disease on grape berries (white fresh table and red withered wine grapes). The berries were surface sterilized by immersion for 5 min in 0.5% NaOCl solution, then rinsed twice with sterile distilled water and placed in compartmentalized square culture dishes. Suspensions of conidia were prepared, adjusted to 10^4 conidia mL^{-1} and then inoculated by berry piercing, as reported by Lorenzini and Zapparoli (2014). Mock inoculation (controls) consisted of berries wounded and inoculated with sterile water, and a positive control consisted of berries wounded and inoculated with *Botrytis cinerea* ITEM 17200. The experiment was performed twice, each with three replicates, which each consisted of 25 berries. After 7 and 14 d at 22°C, the disease index (DI) was assessed on a scale of 0 to 4, as previously described (Lorenzini and Zapparoli, 2014). Variance analysis (ANOVA) was used for the DI data to evaluate isolate differences in pathogenicity. Tukey's multiple comparison test (Tukey, HSD) was applied to determine statistically significant differences.

RESULTS

Phylogenetic analysis

Using the BLASTn tool in GenBank, the *benA*, *CaM*, ITS and *rpb2* gene sequences of Pdb1 (ITEM 18277) showed 99% similarity to *Penicillium* sp. P3 (ITEM 18276[T]) (Lorenzini *et al.* 2016) for *benA*, greater than 89% similarity to *P. bissettii* and *P. annulatum* for *CaM*, greater than 98% similarity to different *Penicillium* species (e.g. *P. janthinellum, P. reticulisporum, P. ochrochloron, P. bissettii and P. javanicum*) for ITS, and 95% similarity to *Penicillium* sp. for *rpb2*. The alignment results of the *benA*, *CaM*, ITS and *rpb2* gene sequences of P3 were similar to those of Pdb1. Based on these data, the phylogenetic position of both isolates (Pdb1 and P3) was evaluated using members of *Penicillium* section *Lanata-divaricata*, according to Diao *et al.* (2018). The ML combined phylogenetic tree (*benA*+*CaM*+ITS+*rpb2*) with the greatest log likelihood (-21693.88) is shown in Figure 1a. The Pdb1 and P3 isolates were grouped together (BS/pp = 100/1), and were distantly related to *P. bissettii* DOAMC 167011 and *P. vasconiae* CBS 339.79. Data from phylogenetic analyses using *benA*, *CaM*, ITS and *rpb2* individually (data not shown) were in concordance with those based on the combine dataset. However, analyses of the combined dataset provided greater support than the individual datasets. The molecular identification of all isolates recovered in this study is based, therefore, on phylogeny from the combined dataset of gene sequences.

The comparative analysis by GenBank database of *benA*, *CaM* and ITS gene sequences of isolate Pls8 (ITEM 18278) showed 99% similarity to different strains of *Penicillium bilaiae* (section *Sclerotiora*). The phylogenetic position of Pls8 was therefore evaluated using members of *Penicillium* section *Sclerotiora*, reported by Wang *et al.* (2017). The combined phylogenetic tree (*benA*+*CaM*+ITS) with the greatest log likelihood (-11066.6111) confirmed the identification (Figure 1b), as the Pls8 (ITEM 18278) was grouped with *P. bilaiae* NRRL3391 (BS/pp = 100/1).

Comparative analysis using GenBank of *benA*, *CaM* and ITS gene sequences of ASls13 (ITEM 18279) showed 99% similarity to *A. melleus, A. pallidofulvus, A. sulphureus* and *A. petrakii* for *benA*, 99% similarity to *A.*

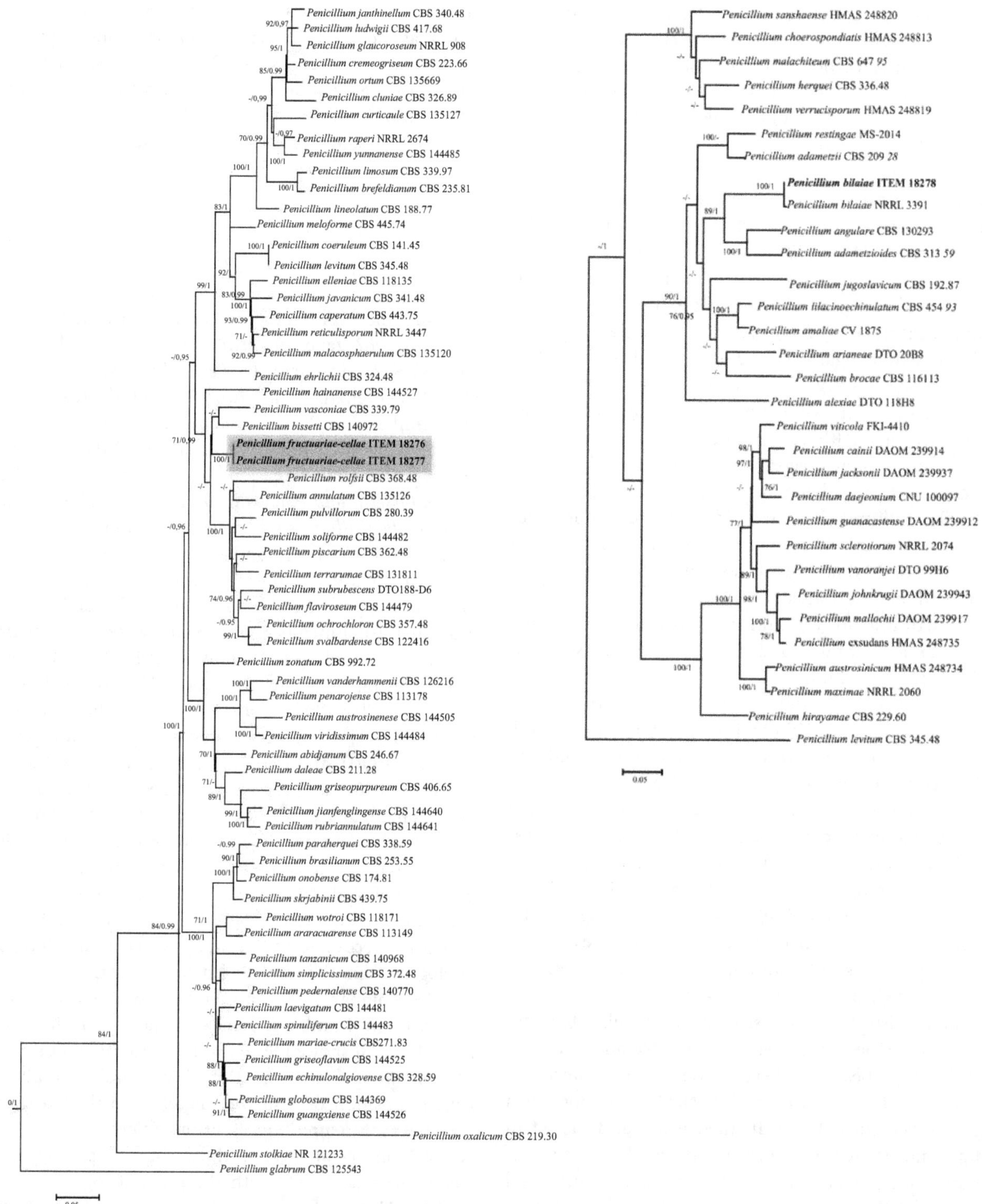

Figure 1. Phylogenetic tree of combined sequences of *Penicillium* spp. (a) Maximum Likelihood combined (*benA+CaM+ITS+rpb2*) tree of *P. fructuariae-cellae* ITEM 18277 (Pdb1) and ITEM 18276^{T} (P3) and representative taxa of *Penicillium* section *Lanata-divaricata*. *Penicillium stolkiae* CBS 315.67, *P. oxalicum* CBS 219.30 and *P. glabrum* CBS 125543 are outgroups. (b) Maximum Likelihood combined (*benA+CaM+ITS+rpb2*) tree of *P. bilaiae* ITEM 18278 (Pls8) and representative taxa of *Penicillium* section *Sclerotiora*. *Penicillium levitum* CBS 345.48 is outgroup. The BI posterior probabilities values and bootstrap percentages of the ML analysis are indicated above the nodes (BS/pp). Values less than 70% bootstrap support in the ML analysis and less than 0.95 posterior probability in the Bayesian analysis are indicated with a hyphen. Branches with high support are thickened.

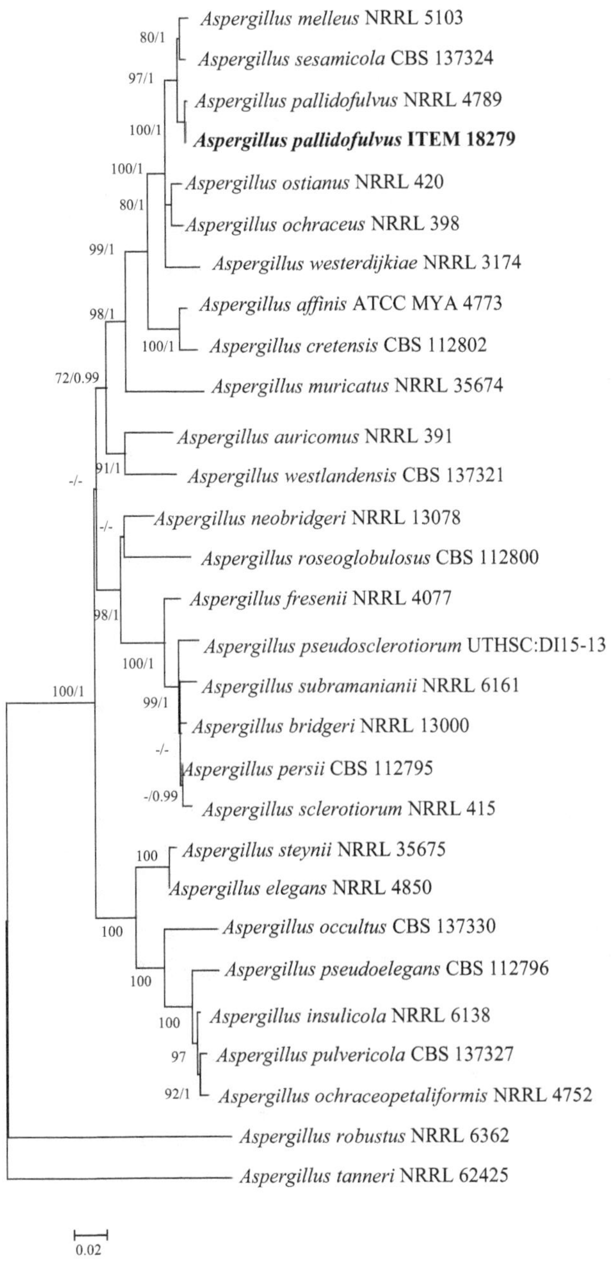

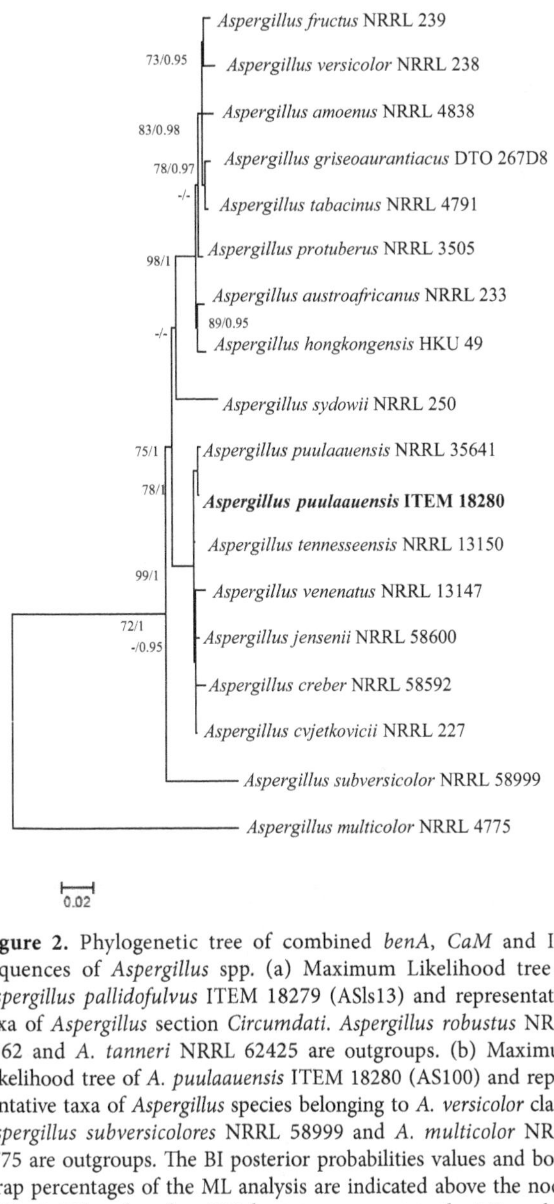

Figure 2. Phylogenetic tree of combined *benA*, *CaM* and ITS sequences of *Aspergillus* spp. (a) Maximum Likelihood tree of *Aspergillus pallidofulvus* ITEM 18279 (ASls13) and representative taxa of *Aspergillus* section *Circumdati*. *Aspergillus robustus* NRRL 6362 and *A. tanneri* NRRL 62425 are outgroups. (b) Maximum Likelihood tree of *A. puulaauensis* ITEM 18280 (AS100) and representative taxa of *Aspergillus* species belonging to *A. versicolor* clade. *Aspergillus subversicolores* NRRL 58999 and *A. multicolor* NRRL 4775 are outgroups. The BI posterior probabilities values and bootstrap percentages of the ML analysis are indicated above the nodes (BS/pp). Values less than 70% bootstrap support in the ML analysis and less than 0.95 posterior probability in the Bayesian analysis are indicated with a hyphen. Branches with high support are thickened.

pallidofulvus and different strains of *Aspergillus* sp. for *CaM*, and 99% similarity to different *Aspergillus* species (e.g. *A. ochraceus, A. melleus* and *A. pallidofulvus*) for ITS. Based on these results, the phylogenetic analysis was performed using the *Aspergillus* taxa of section *Circumdati*, according to Siqueira *et al.* (2017). The ML combined phylogenetic tree (*benA*+*CaM*+ITS) with the greatest log likelihood (-3612.040) showed that ASls13 strongly belongs to *A. pallidofulvus,* as it grouped with the relevant strain, NRRL4789 (BS/pp = 100/1) (Figure 2a).

The GenBank comparative analysis of *benA*, *CaM* and ITS sequences of AS100 (ITEM 18280) showed 99% similarity to *A. puulaauensis, A. versicolor* and *A. jensenii* for *benA*, and 99% similarity to different species of *Aspergillus* (e.g. *A. puulaauensis, A. cvjetkovicii, A. tennesseensis, A. versicolor, A. jensenii* and *A. creber*) for *CaM* and ITS. In the ML combined tree (*benA*+*CaM*+ITS) with the greatest log likelihood (-3938.2941), the clustering of AS100 with *A. puulauensis* NRRL35641 was highly supported (BS/pp = 78/1), as

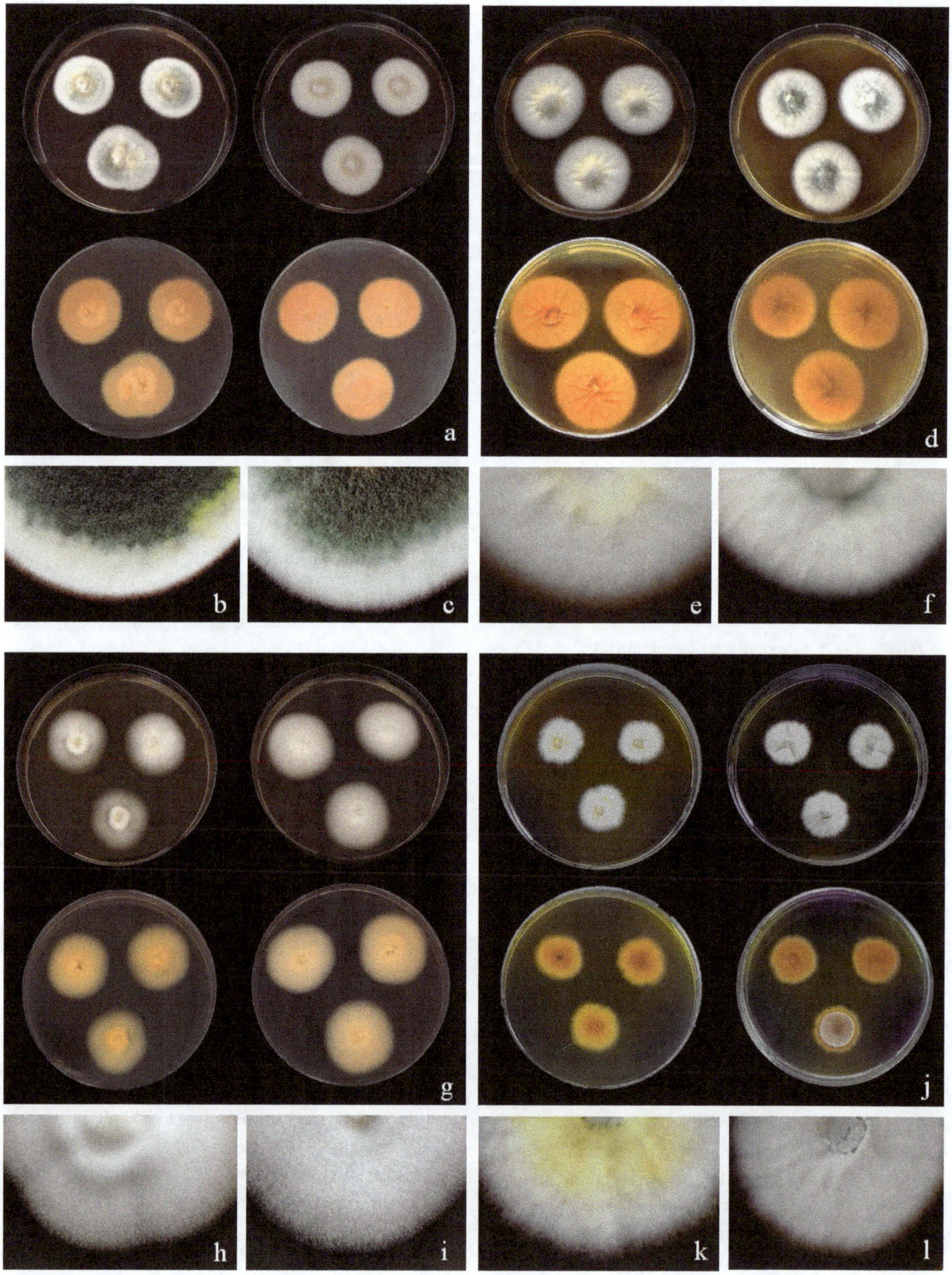

Figure 3. Macro-morphology of *Penicillium fructuariae-cellae* on culture media. Colony upper surfaces (top row) of ITEM 18277 (isolate Pdb1, left) and ITEM 18276^T (isolate P3, right), and reverse sides (bottom row) on CYA (a), YES (d), MEA (g) and CREA (j). Colony texture of ITEM 18277 (Pdb1) (b, e, h, k) and ITEM 18276^T (P3) (c, f, i, l) on, respectively, CYA, YES, MEA or CREA.

Table 3. Micromorphological characteristics of *Penicillium* and *Aspergillus* isolates after 7 d growth on different agar media. All dimensions are in μm. Numbers of metulae or phialides are indicated in parentheses..

	Media	*P. fructuariae-cellae* ITEM 18276^{T} (P3)	*P. fructuariae-cellae* ITEM 18277 (Pdb1)	*P. bilaiae* ITEM 18278 (Pls8)	*A. pallidofulvus* ITEM 18279 (ASls13)	*A. puulaauensis* ITEM 18280 (AS100)
Cleistothecia*	CYA	121-221 × 80-194	74-120 × 67-100	-	-	-
	YES	88-192 × 88-178	126-277 × 120-231	-	-	-
	MEA	124-191 × 77-133	35-87 × 25-56	-	-	-
Ascospore	CYA	2.5–4.5	2.5–4	-	-	-
Conidiophores/Stipes	CYA	56-200 × 2-3	55-313 × 2-3.5	34-110 × 2-3.5	155-760 × 4.5-10	115-360 × 4.5-7.5
	YES	137-413 × 2.5-4	68-227 × 2.5-5	26-97 × 2-3.5	288-630 × 4.5-10.5	168-325 × 3.5-9.5
	MEA	92-129 × 2.5-$3.5^{\S}$	86-120 × $2\text{-}3^{\S}$	31-123 × 2-3.5	200-890 × 5-9	190-810 × 4-7
Vescicles	CYA	-	-	3.5-5	12-30	11-19.5
	YES	-	-	2.5-4.5	20-51	15-22
	MEA	-	-	3-4	13-31	9-19
Metulae/Branches	CYA	9-16 × 2-3 (2-4)	10-47 × 2-3.5 (2-4)	-	5.5-8 × 2.5-4.5	4.5-6.5 × 2-3
	YES	13-24 × 2 -4 (2-3)	8.5-18 × 2-4.5 (2-4)	-	6-12 × 3-6	3.5-7 × 2.5-4
	MEA	12-19 × 2.5-4 (2-4) **	10-26 × 2-4 (2-3) **	-	7.5-10 × 3-6	4-8 × 2-3.5
Phialides	CYA	3.5-8 × 2-3 (2-10)	4.5-10 × 2-3 (3-10)	6-9 × 2.5-3.5 (3-8)	5-8 × 2-3.5	4.5-7.5 × 2-3.5
	YES	7.5-15 × 2-4 (2-7)	5.5-11 × 2-3.5 (3-9)	5-10 × 2-3 (2-6)	6-9 × 2.5-4	4-7 × 2-3.5
	MEA	6-9.5 × 2-3 (3-5) **	5-9.5 × 2-3.5 (3-5) **	4.5-8.5 × 2-3.5 (2-4)	6.5-8 × 2.5-4	4.5-7 × 2-3.5
Conidia	CYA	2-3.5	2-3.5	2-3.5	2.5-4	2-3.5
	YES	2-4	2-3	2-3	2-3.5	2-3.5
	MEA	2-3.5 **	2-3.5 **	2-3.5	2.5-4.5	2-3.5

- = structures absent; § n < 4; * measurements based on 14 d old cultures; ** measurements based on 30 d old cultures.

shown in Figure 2b. Phylogenetic analysis based on the ITS dataset placed AS100 in a group containing most of the *Aspergillus* taxa of the *A. versicolor* clade (data not shown).

Culture and morphological characteristics

Penicillium isolates Pdb1 (ITEM 18277) and P3 (ITEM 18276^{T}) had similar colony morphology on the different media (Figure 3). On CYA, colonies were compact, velvety, with entire margins, and were white; initially yellowish (for Pdb1) or cream (for P3) then becoming gray-green due to abundant sporulation. Spherical or suboval cleistothecia were observed, often covered with a network of hyphae. The cleistothecia matured after three or more weeks, containing evanescent asci and hyaline ascospores, which were smooth-walled and globose to subglobose (Table 3). The colonies were surrounded by diffused soluble pigment in the agar developing as a red-brown colony halos. Hyaline exudates were also observed. Reverse sides of the colonies were light brown and pale cream shades (Figures 3a, b and c).

The growth test revealed little variability among the isolates. The colony diameters at 25°C were 36 to 41 mm for isolate Pdb1 and 43 to 48 mm for P3, at 15°C were 16 to 20 mm for Pdb1 and 17 to 20 mm for P3 mm, and at 30°C were 46 to 49 mm for Pdb1 and 47 to 49 mm for P3. The colony diameters at 37°C were 8 to 10 mm for Pdb1 and 10 to 13 mm for P3. On YES, the colonies were moderately deep and radially sulcate, with regular margins. The mycelium was white and pale yellow in the centre for Pdb1, or white and greenish to grayish in the centre for P3. Sporulation was moderate and pale yellow for Pdb1 and pale gray-green for P3. The cleistothecia were spherical or suboval. The colonies were surrounded by diffused soluble pigment in the agar, as a faint yellow zones surrounded by faint purpuric-brown halos. The colony textures were pubescent and exudates were not observed. The reverse sides of the colonies were orange to brown in colour (Figures 3b, e and f). The colony diameters were 43 to 50 mm for isolate Pdb1 and 42 to 46 mm for P3. On MEA, the colonies were compact, velvety, sometimes radially wrinkled, with entire and plane margins. The mycelia were whitish and non-sporulating mycelium. The cleistothecia were spherical or suboval, and hyaline exudates sometimes were observed. The reverse sides of colonies were yellow and white for Pdb1 and white to pale cream for P3 (Figures 3g, h and i). The colony diameters were 38 to 40 mm for isolate Pdb1 and

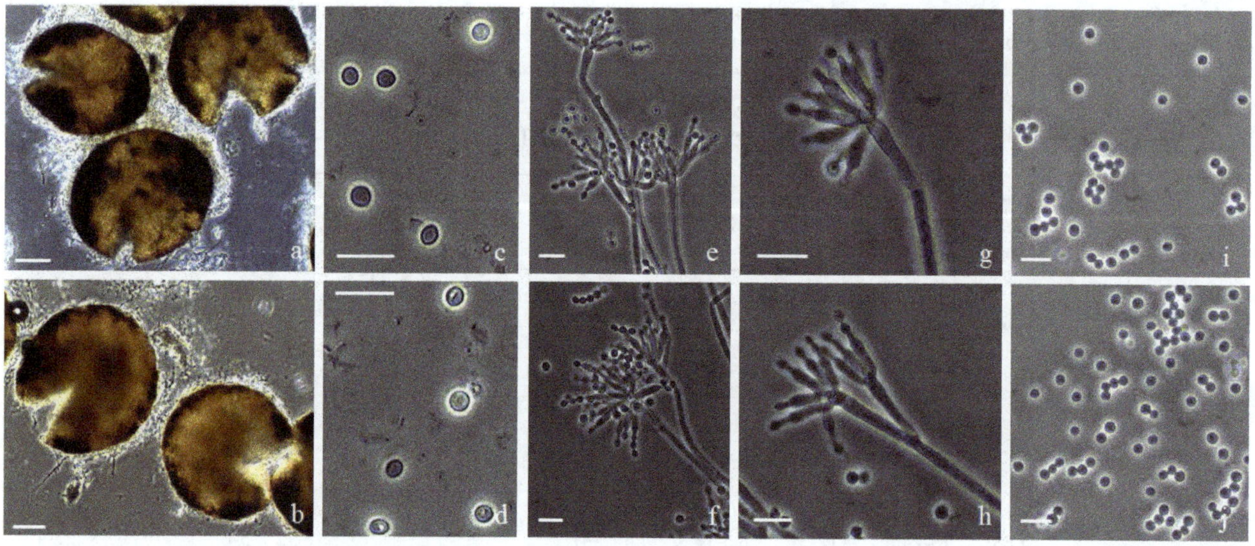

Figure 4. Micro-morphology of *Penicillium fructuariae-cellae* on CYA. Cleistothecia (a and b), ascospores (c and d), conidiophores (e to h), and conidia (i and j) of, respectively, ITEM 18277 (isolate Pdb1, top row) and ITEM 18276^T (isolate P3, bottom row). Bars = 50 μm (a and b), or 10 μm (c to j).

43 to 45 mm for P3. The colony diameters at 25°C after 7 d on PDA pH 4 were 37 to 40 mm for Pdbl and 39 to 41 for P3, on ¼ strength PDA pH 7 were 35 to 38 mm for Pdbl and 38 to 42 mm for P3, and on Horikoshi agar pH 10 were 29 to 30 mm for Pdbl and 32 to 33 mm for P3. On CREA, the isolates grew well and produced acid (Figures 3j, k, l). On CYA and YES, conidiophores were monoverticillate or biverticillate, and a minor proportion were divaricate. Metulae were oblong and divergent, phialides were ampuliform, and conidia were smooth walled and globose to subglobose (Figure 4; Table 3). On MEA, conidiophores were rarely observed.

On CYA, colonies of *P. bilaiae* Pls8 were convex, with concentric folds, conidia were abundantly produced, and aerial mycelium was lanose and floccose and grey-green to white. The colony margins were margin entire and white and sporulation was heavy. Exudates were dark, superficial or embedded, and the colonies were surrounded by diffused soluble pigment into agar as faint purpuric-brown halos. The colony reverse sides were orange-yellow. On YES, colonies were heavily wrinkled, white and pubescent at the margins, green-gray in colour, and felty in the centres. The colonies sporulated heavily. The margins entire and exudates were not observed. The colonies were surrounded by soluble pigment diffused into the agar as faint yellow zones surrounded by faint purpuric-brown halos. The colony reverse sides were orange to brown. On MEA, the colonies were velutinous, floccose and raised in the centre, with villose white aerial mycelium and entire margins which were entire, low, plane and white. The colonies were; sporulating heavily, without exudates, and were surrounded by diffused soluble pigment into the agar as faint yellow zones. The reverse sides of the colonies were yellow to pale orange. Colony diameters were 22 to 26 mm on CYA, 20 to 22 mm on YES and 16 to 20 mm on MEA. On CREA, the fungal growth was weak with good acid production. On CYA, YES, and MEA the conidiophores were monoverticillate, the stipes were smooth walled and mostly globose vesiculate or subglobose to ellipsoidal. The phialides were ampulliforms and wide at the bases. Conidia were globose to subglobose (Table 3).

On CYA, colonies of *A. pallidofulvus* ASls13 was velutinous, the mycelium was white, and sporulation was pale yellow to cream, without exudates. Reverse sides of colonies were light brown to brown. On YES, the colonies were moderately powdery to velutinous, the mycelium was white without exudates, and sporulation was light yellow Colony reverse sides were pale brown to yellow. On MEA, the colonies were velutinous, felty and floccose and the mycelium was white with light yellow sporulation, without exudates. Colony reverse sides were cream to pale brown. Colony diameters were 56 to 60 mm on CYA, 70 to 75 mm on YES and 55 to 61 mm on MEA. On CREA, growth was weak with no acid production. On CYA, YES and MEA, the conidiophores were biseriate, stipes were hyaline to pale brown, vesicles were globose, metulae were oblong covering entire vesicles, and phialides were ampulliform. Conidia were globose, subglobose to ovoid and smooth (Table 3).

Table 4. Disease index (%) on grape berries after 14 d, for three isolates of *Penicillium* (P3, Pb1b and Pls8) and two of *Aspergillus* (ASls13 and AS100), all recovered from withered grapes, and for *B. cinerea* ITEM 17200 as positive control.

Species	Isolate designation	Disease index (%) on grape berries	
		White fresh table	Red withered wine
Penicillium fructuariae-cellae	P3	24 ± 0 a	44 ± 1 a
Penicillium fructuariae-cellae	Pdb1	23 ± 1 a[1]	44 ± 1 a
Penicillium bilaiae	Pls8	24 ± 2 a	44 ± 2 a
Aspergillus pallidofulvus	ASls13	22 ± 1 a	47 ± 2 a
Aspergillus puulaauensis	AS100	23 ± 1 a	43 ± 1 a
Botrytis cinerea	ITEM 17200	100 ± 0 b	98 ± 1 b

[1]Values (mean of three independent measurements ± standard deviation) with different letters are significantly different (ANOVA, Tukey HSD, $P < 0.05$)

On CYA, colonies of *A. puulaauensis* AS100 was sulcate and raised in the centre. Sporulation green to gray. Colony margins were regular, plane and white without exudates. Colony reverse sides were yellow to light orange. On YES, colonies were moderately powdery to felty, sulcate, raised in the centre, and white with green to grey and light pink shadows. Sporulation was green to gray. Colony margins were regular and no exudates were produced. Colony reverse sides were yellow to light brown. On MEA, colonies were sulcate, and raised in the centre. Mycelium was white and sporulation was light yellow. Colony margins were regular, plane and white, without exudates. Colony reverse sides were yellow to light orange. The colony diameters were 18 to 22 mm on CYA, 24 to 25 mm on YES and 18 to 20 mm on MEA. On CREA, growth was moderate growth without acid production. On CYA, YES and MEA, the conidiophores were biseriate, and stipes were smooth, and hyaline to light yellow. Vesicles were spatulate or subspherical, and metulae were oblong covering entire vesicles. Phialides were oblong from which globose or subglobose conidia developed (Table 3).

Pathogenicity assay

All *Penicillium* and *Aspergillus* isolates obtained in this study displayed ability to infect grape berries, although infection was much less than for *B. cinerea* (Table 4). Wounded inoculated berries initially showed mycelium around the inoculation sites and subsequently necrotic areas became visible, particularly in white fresh table berries. Red withered berries inoculated by *Penicillium* isolates Pdb1, P3 and Pls8 caused typical symptoms of *Penicillium* infection characterized by tufts of white and green mycelium erupting from the berry skins (Figure 5). Abundant sporulation was observed on berries infected by *A. pallidofulvus* isolate ASls13. The pathogens were re-isolated from inoculated berries, fulfilling Koch's postulates.

Taxonomy

The phylogenetic analysis based on four different loci and the morphological analysis showed that two isolates of *Penicillium* (Pdb1 and P3) recovered from withered grapes were distinct from any known species within the *Penicillium* section *Lanata-divaricata*. Therefore, these isolates are here described as members of a new *Penicillium* species.

Figure 5. Details of the pathogenicity assay of ITEM 18277 (isolate Pdb1) (a and c) and ITEM 18276^T (isolate P3) (b and d) on white fresh table grapes (left) and red withered wine grapes (right).

Penicillium fructuariae-cellae Lorenzini, Zapparoli & Perrone **sp. nov.**

MycoBank: MB 831228 – Figures 3 and 4

In: subgenus *Aspergilloides*, section *Lanata–divaricata*.

ITS barcode: MK039434. Alternative markers: *benA* =KU554679; *CaM* = MK045337; *rpb2* = MK520927.

Etymology. Latin, *fructuariae-cellae*, meaning fruit-drying room for grape withering, the place where two representative strains were isolated.

Type specimen. ITALY, Verona, Marano di Valpolicella, on Corvina withered grapes stored in fruit-drying room, Dec. 2013, *coll.* M. Lorenzini and G. Zapparoli, *isol.* M. Lorenzini and G. Zapparoli, P3 (holotype CBS 145110^T; ex-type strain ITEM 18276^T).

Colony morphology. Colony diameters (mm), 7 d: CYA 43–48; YES 42–46; MEA 43–45; CREA 35–37; CYA 15°C 17–20; CYA 30°C 47–49; CYA 37°C 10–13.

Colonies on CYA after 7 d at 25°C were compact, velvety; margins entire and white; sporulation abundant, conidia *en masse* pale gray-green; cleistothecia spherical or suboval covered with networks of hyphae; asci evanescent, ascospores hyaline, smooth-walled, globose to subglobose; soluble red-brown pigments and hyaline exudates produced; colony reverse sides pale brown and pale cream. Colonies on YES after 7 d at 25°C were moderately deep, radially sulcate, with regular margins; mycelia white and green to gray; sporulation moderate, conidia *en masse* pale gray-green; cleistothecia observed; soluble yellow and faint purpuric-brown pigments produced from colonies; exudates absent; colony reverse sides orange to brown. Colonies on MEA after 7 d at 25°C were compact, sometimes radially wrinkled, with entire and plain margins with velvety texture; mycelia white; sporulation poor; cleistothecia observed; hyaline exudates sometimes observed; colony reverse sides white yellow-pale cream. Colonies grew well on CREA after 7 d at 25°C, with good acid production.

Conidiophores (on CYA) monoverticillate, biverticillate, with a minor proportion divaricate; stipes smooth, 55–313 × 2–3.5 µm, *metulae* divergent, 2–4 per stipe or branch, 9–47 × 2–3.5 µm; *phialides* ampulliform, 2–10 per metula, 3.5–10 × 2–3 µm; *conidia* smooth walled, globose to subglobose, 2–3.5 µm (mean = 2.5 µm ± 0.4 µm n = 50); *cleistothecia* covered with a network of hyphae, 74–221 × 67–194 µm (n = 20); asci evanescent; *ascospores* hyaline, smooth-walled, globose to subglobose 2.5–4.5 µm (mean = 3.0 µm ± 0.4 µm, n = 30).

Other strains examined. ITALY, Verona, Montecchia di Crosara, Garganega withered grapes stored in a fruit-drying room, Nov. 2017, *coll.* M. Lorenzini and G. Zapparoli, *isol.* M. Lorenzini and G. Zapparoli, Pdb1, ITEM 18277 = CBS 145111, ITS barcode: MK039435. Alternative markers: *benA* = MK045333; *CaM* = MK045338; *rpb2* = MK520928.

Notes. *Penicillium fructuariae-cellae* is classified in section *Lanata–divaricata*, and is distantly related to other *Penicillium* species. The multi-locus phylogeny placed it closed to *P. bissettii* KAS1951 and *P. vasconiae* CBS 339.79 (Figure 1a). *Penicillium fructuariae-cellae* produces red-brown pigments in CYA compared with *P. bissettii* and *P. vasconiae* that do not produce pigments. *Penicillium fructuariae-cellae* mainly differs from *P. vasconiae* and *P. bissettii* in conidiophore structure and size as it has longer conidiophores than *P. vasconiae*, and shorter conidiophores than *P. bissetti*. *Penicillium fructuariae-cellae* also differs from *P. bissettii* by having smooth stipes. *Penicillium fructuariae-cellae* differs from *P. vasconiae* in phialide shape (*P. vasconiae* has long tapped neck phialides) and having smooth and small conidia.

DISCUSSION

Phylogenetic analysis and morphological observations of the isolates recovered in this study from withered grapes compile the first report of *P. bilaiae*, *A. pallidofulvus* and *A. puulaauensis* from *Vitis vinifera*.

The species identification of Pls8 (ITEM 18278) as *P. bilaiae* was taxonomically clear due to its congruence with phylogenetic and morphological data from the *P. bilaiae* holotype NRRL 3391. However, isolate Pls8 showed slower growth in agar media and some micro-morphological differences (i.e. longer stipes, wider and more numerous phialides) than holotype PB-50 described by Pitt (1979) and Savard *et al.* (1994). Prior to the present study, *P. bilaiae* was detected in Portuguese grapes through morphological observations (Serra *et al.*, 2005), a method that does not provide sufficient data for reliable identification at the species level. *Penicillium bilaiae* is morphologically similar to both *P. alexiae* and *P. adametzioides* (Visagie *et al.*, 2013). Hence, the present study provides the first taxonomic evidence of the occurrence of *P. bilaiae* on *Vitis vinifera*.

The assignment of isolate ASls13 (ITEM 18279) to *A. pallidofulvus* was confirmed by its genealogy and macromorphology, according to the description of the holotype *A. pallidofulvus* NRRL 4749 (Visagie *et al.*, 2014). However, micro-morphological observations of ASls13 showed differences in the size of its stipes and phialides (respectively smaller and shorter for ASls13 than the holotype), and with no production of sclerotia by ASls13, in contrast with the *A. pallidofulvus* holotype. This species, recently introduced into section *Circumdati*, has also been isolated from green coffee beans in India and clinical samples (Visagie *et al.* 2014; Masih *et al.*, 2016). The recovery of this species from grapes and the results of pathogenicity assays show that *A. pallidofulvus* may exhibit pathogenic behaviour in grapevine. Nevertheless, further investigation is required to determine occurrence for this fungus on withered grapes and its infectivity under post-harvest environmental conditions.

The use of multi-locus phylogenetic analysis (*CaM*, *benA* and ITS) resolved the taxonomic position of isolate AS100 (ITEM 18280), identifying it as *A. puulaauensis*. Assignment of the isolate to this species has previously proved impossible using only the *CaM* gene sequence (Lorenzini *et al.*, 2016). Moreover, AS100 and the holotype *A. puulaauensis* NRRL 35641 (Jurjevic *et al.*, 2012) both showed identical colony macro- and micro-morphological characters. The recovery of *A. puulaauensis* from grape berries further supports its worldwide and cosmopolitan distribution, since this species has previously been reported from disparate environments including Hawaiian plants, Atlantic sponges, Italian cheese, air samples in North America and clinical samples (Jurjevic *et al.*, 2012; Siqueira *et al.*, 2017; Bovio *et al.*, 2018; Decontardi *et al.*, 2018).

Based on multi-locus phylogenetic analyses, isolates Pdb1 (ITEM 18277) and P3 (ITEM 18276^T) represented a distinct species, herein named *P. fructuariae-cellae*. These two isolates form a phylogenetic cluster based on *benA*+*CaM*+ITS+*rpb2* combined gene genealogies, distinct from any currently described species in section *Lanata–divaricata*, which was recently updated with 13 new *Penicillium* species collected from Chinese acidic soils (Diao *et al.*, 2018). *Penicillium fructuariae-cellae* is acid-preferential, like most of the new species described by Diao *et al.* (2018). This is a physiological characteristic congruent with the acidic habitat (grapes) from which it was isolated. Isolates of P3 and Pdb1 are distantly related to *P. vasconiae* CBS 339.79 (Ramírez and Martínez, 1980) and the recently described species *P. bissettii* (Visagie *et al.*, 2016). Moreover, *P. fructuariae-cellae* displays some differences in macro-morphological characteristics (e.g. colony colour, texture and pigment production) compared with related species. The pigment production by these two isolates distinguishes them from *P. vasconiae* and *P. bissettii* that produce no pigments. *Penicillium fructuariae-cellae* produces monoverticillate, sometimes biverticillate conidiophores, whereas *P. vasconiae* is strictly monoverticillate and *P. bissettii* is biverticillate/terverticillate. The conidiophores of *P. fructuariae-cellae* are longer than those of *P. vasconiae* ($<$ 50 μm; Ramírez and Martínez, 1980) and shorter than conidiophores of *P. bissettii* (190–670 μm; Visagie *et al.*, 2016). Its stipes are smooth, while in *P. bissettii* they are rough. Its phialides are also shorter than those of *P. vasconiae* (9–13 μm; Ramírez and Martínez, 1980). As well, of *P. fructuariae-cellae* produced smaller conidia than *P. vasconiae* (4–4.5 μm; Ramírez and Martínez, 1980). The conidial surfaces of *P. fructuariae-cellae* were smooth, while those of *P. vasconiae* are conspicuously echinulate (Ramírez and Martínez, 1980). These morphological differences together with phylogenetic information, support the uniqueness of *P. fructuariae-cellae*.

Based on the results of pathogenicity assays, the detrimental effects of these fungi on withered grapes was quite significant. Although these *Penicillium* and *Aspergillus* isolates were less pathogenic than *B. cinerea*, they could make major contributions to berry rotting. Decay of berry surfaces, like that observed on infected berries caused by each isolate due to mycelial growth and necrosis, is important for susceptibility to subsequent fungal infections by the same and other pathogens (Padgett and Morrison, 1990). Incidence and symptoms of grape diseases caused by these fungi under fruit-drying room conditions require further investigation.

In conclusion, this study describes isolates belonging to species of *Penicillium* and *Aspergillus* from withered grape berries that have not been previously reported from *V. vinifera*. A new species of *Penicillium* from this host, *P. fructuariae-cellae*, is herein described. The recovery of these species highlights the complexity of fungal species affecting withered grapes (Lorenzini *et al.*, 2016; 2018). According to pathogenicity assays, *P. fructuariae-cellae*, *P. bilaiae*, *A. pallidofulvus* and *A. puulaauensis* are able to infect grapes but with much lower infectivity than *B. cinerea*, which is the most important pathogen occurring on withered grapes. It is likely that the species identified in this study are less pathogenic than other *Penicillium* and *Aspergillus* species frequently reported on withered grapes (e.g. *P. expansum*, *P. crustosum*, *A. tubingensis* and *A. uvarum*) (Lorenzini *et al.*, 2016). Further investigations are necessary to ascertain the pathogenic role of *P. fructuariae-cellae*, *P. bilaiae*, *A. pallidofulvus* and *A. puulaauensis* under withering conditions, as well as their interaction with other causal fungal agents causing rots of grape berries.

LITERATURE CITED

Bovio E., Garzoli L., Poli A., Prigione V., Firsova D., ... Varese G.C., 2018. The culturable mycobiota associated with three Atlantic sponges, including two new species: *Thelebolus balausformis* and *T. spongiae*. *Fungal Systematics and Evolution* 1: 141–167.

Decontardi S., Soares C., Lima N., Battilani P., 2018. Polyphasic identification of Penicillia and Aspergilli isolated from Italian grana cheese. *Food Microbiology* 73: 137–149.

Diao Y-Z., Chen Q., Jiang X.-Z., Houbraken J., Barbosa R.N., … Wu W.-P., 2018. *Penicillium* section *Lanata-divaricata* from acidic soil. *Cladistics*, 0: 1–36.

Drummond A.J., Rambaut A., 2007. BEAST: Bayesian evolutionary analysis by sampling trees. *BMC Evolutionary Biology*, 7: 214.

Glass N.L., Donaldson G.C., 1995. Development of primer sets designed for use with the PCR to amplify conserved genes from filamentous ascomycetes. *Applied and Environmental Microbiology,* 61: 1323–1330.

Hong S.B., Cho H.S., Shin H.D., Frisvad J.C., Samson R.A., 2006. Novel *Neosartorya* species isolated from soil in Korea. *International Journal of Systematics and Evolutionary Microbiology* 56: 477–486.

Jurjevic Z., Peterson S.W., Horn B.W., 2012. *Aspergillus* section *Versicolores*: nine new species and multilocus DNA sequence based phylogeny. *IMA Fungus* 3: 59–79.

Kumar S., Stecher G., Tamura K., 2016. MEGA7: Molecular Evolutionary Genetics Analysis version 7.0 for bigger datasets. *Molecular Biology and Evolution* 33: 1870–1874.

Liu Y., Whelen S., Hall B.D., 1999. Phylogenetic relationships among ascomycetes: evidence from an RNA polymerase II subunit. *Molecular Biology and Evolution* 16: 1799–1808.

Lorenzini M., Zapparoli G., 2014. Characterization and pathogenicity of *Alternaria* spp. strains associated with grape bunch rot during post-harvest withering. *International Journal of Food Microbiology* 186: 1–5.

Lorenzini M., Cappello M.S., Logrieco A., Zapparoli G., 2016. Polymorphism and phylogenetic species delimitation in filamentous fungi from predominant mycobiota in withered grapes. *International Journal of Food Microbiology* 238: 56–62.

Lorenzini M., Simonato B., Favati F., Bernardi P., Sbarbati A., Zapparoli G., 2018. Filamentous fungi associated with natural infection of noble rot on withered grapes. *International Journal of Food Microbiology* 272: 83–86.

Masih A, Singh P.K., Kathuria S., Agarwal K., Meis J.F., Chowdhary A., 2016. Identification by molecular methods and matrix-assisted laser desorption ionization–time of flight mass spectrometry and antifungal susceptibility profiles of clinically significant rare *Aspergillus* species in a referral chest hospital in Delhi, India. *Journal of Clinical Microbiology* 54: 2354–2364.

Mencarelli F., Tonutti P., 2013. *Sweet, Reinforced and Fortified Wines: Grape Biochemistry, Technology and Vinification.* John Wiley & Sons, Ltd, Chichester, UK.

Padgett M., Morrison J.C., 1990. Changes in grape berry exudates during fruit development and their effect on mycelial growth of *Botrytis cinerea. Journal of the American Society for Horticultural Science* 115: 269–273.

Pitt J.I., 1979. *The genus* Penicillium *and its teleomorphic states* Eupenicillium *and* Talalvmyces. London, Academic Press, UK.

Rambaut A., Drummond A.J., 2009. Tracer v. 1.5. Available from: http://tree.bio.ed.ac.uk/software/tracer/.

Ramìrez C., Martìnez A.T., 1980. Some new species of *Penicillium* recovered from the atmosphere in Madrid and from other substrata. *Mycopathologia* 72: 181–191.

Savard M.E., Miller J.D., Blais L.A., Seifert K.A., Samson R.A., 1994. Secondary metabolites of *Penicillium bilaii* strain PB-50. *Mycopathologia* 127: 19–27.

Serra R., Braga A., Venâncio A., 2005. Mycotoxin-producing and other fungi isolated from grapes for wine production, with particular emphasis on ochratoxin A. *Research in Microbiology* 156: 515–521.

Siqueira J.P.Z., Sutton D.A., Gené J., García D., Wiederhold, N., … Guarro J., 2017. Multilocus phylogeny and antifungal susceptibility of *Aspergillus* section *Circumdati* from clinical samples and description of *A. pseudosclerotiorum* sp. nov. *Journal of Clinical Microbiology* 55: 947–958.

Somma S., Perrone G., Logrieco A., 2012. Diversity of black Aspergilli and mycotoxin risks in grape, wine and dried vine fruits. *Phytopathologia Mediterranea* 51: 131–147.

Stefanini I., Carlin S., Tocci N., Albanese D., Donati C., … Cavalieri D., 2017. Core microbiota and metabolome of *Vitis vinifera* L. cv. Corvina grapes and musts. *Frontiers in Microbiology* 8: 457.

Torelli E., Firrao G., Locci R., Gobbi E., 2006. Ochratoxin A-producing strains of *Penicillium* spp. isolated from grapes used for the production of "passito" wines. *International Journal of Food Microbiology* 106: 307–312.

Visagie C.M., Houbraken J., Rodriques C., Silva Pereira C., Dijksterhuis J., … Samson R.A., 2013. Five new *Penicillium* species in section *Sclerotiora*: a tribute to the Dutch Royal family. *Persoonia* 31, 42–62.

Visagie C.M., Varga J., Houbraken J., Meijer M., Kocsube S., … Samson S.A., 2014. Ochratoxin production and taxonomy of the yellow aspergilli (*Aspergillus* section *Circumdati*). *Studies in Mycology* 78: 1–61.

Visagie C.M., Renaud J.B., Burgess K.M.N., Malloch D.W., Clark D., … Seifert K.A., 2016. Fifteen new species of *Penicillium. Persoonia* 36: 247–280.

Wang X.C., Chen K., Zeng Z.Q., Zhuang W.Y., 2017. Phylogeny and morphological analyses of *Penicillium* section *Sclerotiora* (Fungi) lead to the discovery of five new species. *Scientific Reports* 7: 8233.

White T.J., Bruns T., Lee S., Taylor J., 1990. Amplification and direct sequencing of fungal ribosomal RNA genes for phylogenetics. In: *PCR Protocols: a Guide to Methods and Applications* (M.A. Innis, D.H. Gelfand, J.J. Sninsky and T.J. White eds.), Academic Press San Diego, CA, USA, 315–322.

Zapparoli G., Lorenzini M., Tosi E., Azzolini M., Slaghenaufi D., ... Simonato B., 2018. Changes in chemical and sensory properties of Amarone wine produced by *Penicillium* infected grapes. *Food Chemistry* 263: 42–50.

3

Phytotoxins produced by *Lasiodiplodia laeliocattleyae* involved in Botryosphaeria dieback of grapevines in Brazil

Pierluigi REVEGLIA[1,2,*], Marco MASI[2], Alessio CIMMINO[2], Sami MICHEREFF[3], Tamara CINELLI[3], Laura MUGNAI[3], Antonio EVIDENTE[2]

[1] *National Wine and Grape Industry Centre, School of Agricultural and Wine Sciences, Charles Sturt University, Locked Bag 588 Wagga Wagga NSW 2678, Australia*
[2] *University of Naples Federico II, Department of Chemical Sciences Complesso Universitario Montesant'Angelo, Via Cintia 4, 80126, Naples, Italy*
[3] *Dipartimento di Scienze delle Produzioni Agroalimentari e dell'Ambiente, Sez. Patologia vegetale ed entomologia, Università di Firenze, Piazzale delle Cascine 28, 50144 Firenze, Italy*
Corresponding author: pierluigi.reveglia@unina.it

Abstract. Botryosphaeria dieback (BD) is an important trunk disease affecting grapevines. Several *Lasiodiplodia* species have been shown to be involved in BD affecting the perennial organs of grapevine, mainly causing cankers. (*R*)-(-)-mellein and tyrosol, two well-known fungal phytotoxins, were isolated from the organic extract of culture filtrate of *Lasiodiplodia laeliocattleyae* (syn. *egyptiacae*), which had been isolated from grapevines affected by BD in Brazil. This increases knowledge of the secondary metabolites produced by *Lasiodiplodia* species, confirming that (*R*)-(-)-mellein is a toxin typically produced by Botryosphaericeae species.

Keywords. Grapevine wood disease, phytotoxins, (*R*)-(-)-mellein, tyrosol.

Competing Interests: The Author(s) declare(s) no conflict of interest.

Editor: E. Paplomatas, Agricultural University of Athens, Greece.

INTRODUCTION

Botryosphaeria dieback (BD) is a grapevine trunk disease that causes serious problems for grape production, including emerging wine-producing countries such as Brazil. The economic impacts of grapevine trunk diseases result from significant yield reductions from diseased vines and increased production costs for application of control measures. Many efforts have been made to find new and effective management practices for the disease (Mondello *et al.*, 2018). BD also affects many fruit tree crops, including mango, olive, walnut and almond (de Oliveira Costa *et al.*, 2010; Olmo *et al.*, 2016; Rodríguez-Gálvez *et al.*, 2017).

BD of grapevine is caused by several Botryosphaeriaceae species, associated with decline symptoms including dieback, wood canker and spur dieback (Úrbez-Torres, 2011; Billones-Baijens and Savocchia 2018; Gramaje *et al.*, 2018; Mondello *et al.*, 2018).

Like other trunk disease pathogens, Botryosphaeriaceous species can produce toxic metabolites belonging to different compound classes including aromatic compounds, isocoumarins, jasmonates, naphthalenones, polyketides, and phenols (Martos *et al.*, 2008; Masi *et al.*, 2018). In particular, *Lasiodiplodia* species were investigated for production of phytotoxic metabolites and other substances (including jasmonic acid, mellein, lasiodiplodin, theobroxide, butyrolactones, botryosphaeran, botryorodines, lasiodiplodan) are produced *in vitro* by different isolates of *L. theobromae* and other *Lasiodiplodia* spp. such as *L. mediterranea*. These toxic metabolites have also been purified from *in vitro* cultures of strains isolated from host pants other than grapevine, and tested for their toxic activity (Aldridge *et al.* 1971, Husan *et al.* 1993, Matsuura *et al.*, 1998, He *et al.*, 2004, Miranda *et al.*, 2008, Kitoaka *et al.*, 2009, Andolfi *et al.*, 2014).

In a recent study, phytotoxic metabolites produced in liquid cultures by six species of *Lasiodiplodia* isolated in Brazil, and causing Botryosphaeria dieback of grapevine, were chemically identified. As determined by LC-MS, *L. brasiliense, L. crassispora., L. iraniensis, L. pseudotheobromae* produced jasmonic acid, while *L. brasiliense* synthesized jasmonic acid and (3*R*,4*S*)-4-hydroxymellein. *Lasiodiplodia euphorbiaceicola* and *L. hormozganensis* produced some low molecular weight lipophilic toxins, that were isolated and identified (Cimmino *et al.*, 2017). In particular, from culture filtrate of *L. euphorbiaceicola*, (*R*)-(-)-mellein, (3*R*,4*R*)-(-)- and (3*R*,4*S*)-(-)-4-hydroxymellein, and tyrosol were isolated, and identified. Tyrosol and *p*-hydroxybenzoic acid were also isolated from culture filtrates of *L. hormozganensis* (Cimmino *et al.*, 2017).

Knowledge of the non-specific phytotoxins produced by Botryosphaeriaceous species is increasing (Masi *et al.*, 2018), and it is increasingly important to determine the full spectrum of these metabolites to understand their roles in disease and symptom development (Meh *et al.*, 2013). This note reports the isolation of phytotoxins produced *in vitro* from *L. laeliocattleyae* (syn. *egyptiacae*) (Jayawardena *et al.*, 2018), a pathogen of grapevine (Correia *et al.*, 2016) and agent of mango dieback and fruit rot (Rosado *et al.*, 2016). To our knowledge this is the first report of phytotoxic metabolites isolated from *L. laeliocattleyae.*

MATERIALS AND METHODS

Fungal isolates and culture conditions

The strain of *L. laeliocattleyae* (CMM0206) used in this study was obtained from the collection of Universidade Federal Rural de Pernambuco, Recife, Brazil. It was inoculated and grown in stationary culture, as reported for other strains of *Lasiodiplodia* (Cimmino *et al.*, 2017), in modified Difco Czapek Dox medium (Benton), with 0.5% yeast and 0.5% malt extract (Difco). The cultures were grown for 21 d at 25 °C in the dark. The mycelium was removed and the liquid cultures were lyophilized prior to the extraction procedure.

Extraction of low molecular weight phytotoxic metabolites

The lyophilized residues of the culture filtrates (2.85 L) were dissolved in 300 mL of water and extracted with EtOAc (3 × 300 mL) at the same pH as the original culture (pH 8). The organic extracts were then combined, dried (Na_2SO_4), filtered, and evaporated under low pressure. The organic residue (264.0 mg) was purified by silica gel column chromatography using $CHCl_3$-*i*-PrOH (95:5, v/v), and six fractions of homogenous groups were collected. The residue (21.9 mg) of fraction #1 was purified by preparative TLC on silica gel using $CHCl_3$ as eluent. This yielded a white solid, which was identified as *R*(-)- mellein (**1**, Figure 1, 13.7 mg). The residue of fraction #3 was purified on preparative TLC on silica gel, using $CHCl_3$-i-PrOH (97:3, v/v) as eluent, yielding a white solid, which was identified as tyrosol (**2**, Figure 1, 12.4 mg).

Chemical analyses and characterization

Optical rotations were measured in MeOH on a Jasco P-1010 digital polarimeter (Jasco). ^1H NMR spectra were recorded at 400 or 500 MHz in $CDCl_3$ on Bruker and Varian instruments, with MeOH as an internal standard. ESI MS and LC/MS analyses were performed using the LC/MS TOF system AGILENT (Agilent Technologies) 6230B, HPLC 1260 Infinity. Analytical and preparative TLC was carried out on silica gel plates (Kieselgel 60, F254, 0.25 mm and 0.5 mm) (Merck). TLC spots were visualized by exposure to UV radiation, or by spraying first with 10% H_2SO_4 in MeOH, and then with 5% phosphomolybdic acid in EtOH, followed by heating at 110°C for 10 min. Column chromatography was performed using silica gel (Kieselgel 60, 0.063-0.200 mm) (Merck). Standard sample of (*R*)-(-)-mellein was obtained from the culture filtrates of *Sardiniella urbana* (Cimmino *et al.*, 2018), and of tyrosol from *Lasiodiplodia euphorbiaceicola* (Cimmino *et al.*, 2017).

Phytotoxicity bioassay

The phytotoxic activity of chromatographic organic extract fractions was assayed on lemon fruit, using previously reported protocol (Andolfi *et al.*, 2014b; Cimmino *et al.*, 2017).

RESULTS

The culture filtrates of *L. laeliocattleyae* were exhaustively extracted with ethyl acetate. The organic extract and the resulting aqueous phases were assayed for phytotoxic activity, and both showed some phytotoxicity in the bioassay conditions. The phytotoxicity results were essentially the same as previously outlined (Cimmino et al, 2017). The organic extract was purified by column chromatography. When assayed on lemon fruit, the residue of fractions #1 and #3 produced intense necrotic spots. These were further purified by preparative TLC, obtaining two pure metabolites. The purified compounds were identified as (*R*)-(-)-mellein and tyrosol (**1** and **2**, Figure 1) by comparison with standard samples (Cimmino *et al.*, 2017; 2018). The identity of metabolite **1** was confirmed comparing the $[a]^{25}_{D}$, ^{1}H NMR and ESIMS(+) data with those reported in previous studies (Djoukeng *et al.*, 2009; Evidente *et al.*, 2010). Metabolite **2** was identified by comparing its ^{1}H NMR and ESIMS data with those reported in literature (Kimura *et al.*, 1973; Evidente *et al.*, 2010).

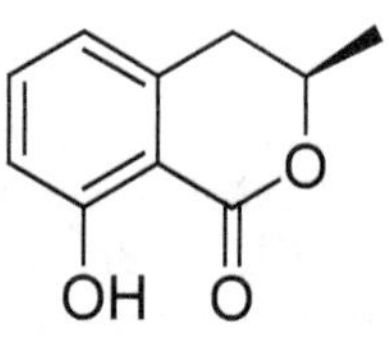

1, (*R*)-(-)-Mellein

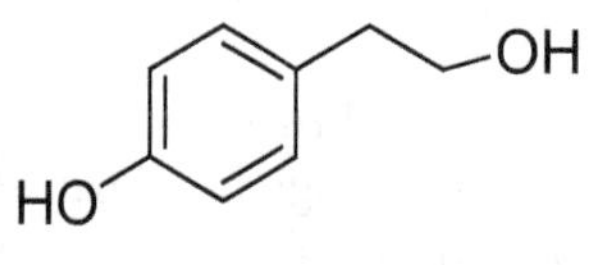

2, Tyrosol

Figure 1. Structures of *R*(-)-mellein **1** and tyrosol **2** isolated from *Lasiodiplodia laeliocattleyae* H141a.

DISCUSSION

Melleins are metabolites produced by many fungi in different genera which are involved in numerous plant diseases. These compounds give different phytotoxic, zootoxic and moderate antifungal effects. (*R*)-(-)-mellein (**1**), produced by different Botryosphaeriaceae, produces toxic effects on grapevine leaves and grapevine calli (Vankatasubbaiah *et al.*, 1991; Djoukeng *et al.*, 2009,; Evidente *et al.*, 2010). Furthermore, this compound was detected in infected wood samples and in green shoots of grapevines affected with Botryosphaeria dieback (Abou-Mansour *et al.*, 2015). *R*-(-)-mellein was produced *in vitro* by several different species of Botryosphaeriaceae such us *Diplodia mutila, Neofusicoccum parvum, Neofusiccoum australe, Neofusicoccum luteum* isolated from grapevine in different grape-growing areas in the world, posing questions about its involvement in the virulence of Botryosphaeria dieback pathogens (Reveglia *et al.*, 2018).

Melleins are isocoumarins, and together with jasmonic acid, its esters, dihydrofuranones and closely related compounds, these are specifically related to Botryosphaeriaceous pathogens. Eutypine and analogues are only produced by *Eutypa* species, the cause of Eutypa dieback (Masi *et al.*, 2018).

Tyrosol (**2**), is a ubiquitous phytotoxic secondary metabolite that has been isolated from *N. parvum* (Evidente *et al.*, 2010) and *N. australe*, both of which are well-known Botryosphaeria dieback agents (Andolfi *et al.*, 2012). Furthermore, metabolite **2** was produced by some *Lasiodiplodia* species such as *L. euphorbiaceicola* and *L. hormozganensis* (Cimmino *et al.*, 2017). More recently, it was also isolated from *Diplodia seriata, N. luteum* and *D. mutila* associated with grapevine wood infections (Reveglia *et al.*, 2018). The phytotoxic activity of tyrosol was shown on tomato cuttings (Evidente *et al.*, 2010), but other results point to activity of tyrosol as an active quorum sensing compound in *Candida albicans* (Chen *et al.*, 2004), controlling of this organism. Tyrosol was also shown to have a synergistic inhibitory effect on radish and grain sorghum when tested with vanillic acid (Einhelling *et al.*, 1978,;Yu *et al.*, 1994; Evidente *et al.*, 2010).

This first report on phytotoxic metabolites produced by *L. laeliocattleyae*, one of the pathogens associated with cankers and diebacks on grapevine in Brazil, adds further information on the complex interactions between virulence factors produced by the internationally common pathogens involved in Botryosphaeria dieback of grapevine and the diseases they cause.

ACKNOWLEDGEMENT

This research was partly supported by academic grants from the Dipartimento di Scienze Chimiche, Università di Napoli Federico II, Italy. Antonio Evidente is associated with "Istituto di Chimica Biomolecolare del CNR", Pozzuoli, Italy.

LITERATURE CITED

Aldridge D.C., Galt S., Giles, D., Turner W.B., 1971. Metabolites of *Lasiodiplodia theobromae*. *Journal Chemical. Society C: Organic* 1623–1627.

Abou-Mansour E., Débieux M., Ramírez-Suero M., Bénard-Gellon M., Magnin-Robert A.,... Larignon P., 2015. Phytotoxic metabolites from *Neofusicoccum parvum*, a pathogen of Botryosphaeria dieback of grapevine. *Phytochemistry* 115: 207–215.

Andolfi A., Maddau L., Cimmino A., Linaldeddu B.T.,Franceschini A.,... Evidente A., 2012. Cyclobotryoxide, a phytotoxic metabolite produced by the plurivorous pathogen *Neofusicoccum australe*. *Journal of Natural Products* 75: 1785–1791.

Andolfi, A., Maddau L., Cimmino A., Linaldeddu B.T., Basso S.,... Evidente A, 2014. Lasiojasmonates A-C, three jasmonic acid esters produced by Lasiodiplodia sp., a grapevine pathogen. *Phytochemistry* 103: 145–153.

Billones-Baaijens R., Savocchia S., 2018. A review of Botryosphaeriaceae species associated with grapevine trunk diseases in Australia and New Zealand. *Australasian Plant Pathology* Doi.org/10.1007/s13313-018-0585-5.

Chen H., Fujita M., Feng Q., Clardy J. G., Fink R., 2004. Tyrosol is a quorum-sensing molecule in *Candida albicans*. *Proceedings of the National Academy of Sciences*, 101: 5048–5052.

Cimmino A., Cinelli T., Masi M., Reveglia P., da Silva M.A.,... Evidente A., 2017. Phytotoxic lipophilic metabolites produced by grapevine strains of *Lasiodiplodia* species in Brazil. *Journal of Agricultural and Food Chemistry* 65: 1102–1107.

Cimmino A., Maddau L., Masi M., Linaldeddu B.T., Evidente A., 2018. Secondary metabolites produced by *Sardiniella urbana*, a new emerging pathogen on European hackberry. Natural product research. doi.org/10.1080/14786419.2018.1477154.

Correia K. C., Silva M. A., de Morais Jr M. A., Armengol, J., Phillips A.J., Michereff S.J. 2016. Phylogeny, distribution and pathogenicity of *Lasiodiplodia* species associated with dieback of table grape in the main Brazilian exporting region. *Plant Pathology*, 65: 92–103.

de Oliveira Costa V. S., Michereff S. J., Martins R. B., Gava C. A. T., Mizubuti, E. S. G., Câmara M.P. S. 2010. Species of Botryosphaeriaceae associated on mango in Brazil. *European Journal of Plant Pathology*, 127: 509–519.

Djoukeng, J.D., Polli S., Larignon P., Abou-Mansour E., 2009. Identification of phytotoxins from *Botryosphaeria obtusa*, a pathogen of black dead arm disease of grapevine. *European Journal of Plant Pathology* 124:303–308.

Einhelling, F.A., Rasmussen J.A, 1978. Synergistic inhibitory effects of vanillic and p-hydroxybenzoic acids on radish and grain sorghum. *Journal of Chemical Ecology* 4: 425–436.

Evidente A., Punzo B., Andolfi A., Cimmino A., Melck D., Luque J., 2010. Lipophilic phytotoxins produced by *Neofusicoccum parvum*, a grapevine canker agent. *Phytopathologia Mediterranea* 49: 74–79 (and references therein cited).

Gramaje, D., Úrbez-Torres J.R, Sosnowski M.R., 2018. Managing grapevine trunk diseases with respect to etiology and epidemiology: current strategies and future prospects. *Plant Disease* 102, 12–39.

He G., Matsuura H., Yoshihara T., 2004. Isolation of an α-methylene-γ-butyrolactone derivative, a toxin from the plant pathogen *Lasiodiplodia theobromae*. *Phytochemistry* 65: 2803–2807.

Husain A., Ahmad A., Agrawal K.P., 1993. (-)-Jasmonic acid, a phytotoxic substance from *Botryodiplodia theobromae*: characterization by NMR spectroscopic methods. *Journal of Natural Products* 56: 2008–2011.

Jayawardena R. S., Purahong W., Zhang W., Wubet T., Li X., Liu M., ... & Yan J., 2018. Biodiversity of fungi on *Vitis vinifera* L. revealed by traditional and high-resolution culture-independent approaches. *Fungal Diversity* 90: 1-84.

Kitaoka N., Nabeta K., Matsuura H., 2009. Isolation and structural elucidation of a new cyclohexenone compound from *Lasiodiplodia theobromae*. *Biosciences, Biotechnology & Biochemistry* 73: 1890–1892.

Kimura Y., Tamura S., 1973. Isolation of l-β-phenyllactic acid and tyrosol as plant-growth regulators from *Gloeosporium laeticolor*. *Agricultural and Bioogical Chemistry* 37: 2925–2925.

Martos S., Andolfi A., Luque J., Mugnai L., Surico G., Evidente A., 2008. Production of phytotoxic metabolites by five species of Botryosphaeriaceae causing decline on grapevines, with special interest in the species *Neofusicoccum luteum* and *N. parvum*. *European Journal of Plant Pathology* 121: 451–461.

Masi M., Cimmino A., Reveglia P., Mugnai L., Surico G., Evidente A., 2018. Advances on fungal phytotoxins and their role in grapevine trunk diseases. *Journal of Agricultural and Food Chemistry*, 66: 5948–5958 (and references therein cited).

Matsuura H., Obara N., Chisaka N., Ichihara A., Yoshihara T., 1998. Novel cyclohexene compound from *Lasiodiplodia theobromae* IFO 31059. *Bioscience, Biotechnology & Biochemistry* 62: 2460–2462.

Mehl, J. W., Slippers B., Roux J., Wingfield M. J., 2013. 14 Cankers and other diseases caused by the Botryosphaeriaceae. *Infectious Forest Diseases*, 298.

Miranda C.C.B.O., Dekker R.F.H., Serpeloni, J. M., Fonseca E.A.I, Cólus I.M.S., Barbosa A.M. 2008. Anticlastogenic activity exhibited by botryosphaeran, a new exopolysaccharide produced by *Botryosphaeria rhodina* MAMB-05. *International Journal of Biology and Macromolecules* 42: 172–177.

Mondello V., Songy S., Battiston E., Pinto C., Coppin C.,Trotel-Aziz P.,...Fontaine F., 2018. Grapevine Trunk Diseases: A Review of Fifteen Years of Trials for Their Control with Chemicals and Biocontrol Agents. *Plant Disease* 102: 1189–1217.

Olmo D., Armengol J., León M., Gramaje D., 2016. Characterization and pathogenicity of Botryosphaeriaceae species isolated from almond trees on the island of Mallorca (Spain). *Plant Disease*, 100: 2483–2491.

Reis P., Magnin-Robert M., Nascimento T., Spagnolo A., Abou-Mansour E., Fontaine F., 2016. Reproducing Botryosphaeria dieback foliar symptoms in simple model system. *Plant Disease* 100: 1071–1079.

Reveglia P., Savocchia S., Billones-Baaijens R., Masi M., Cimmino A., Evidente A., 2018. Phytotoxic metabolites by nine species of Botryosphaeriaceae involved in grapevine dieback in Australia and identification of those produced by *Diplodia mutila, Diplodia seriata, Neofusicoccum australe* and *Neofusicoccum luteum. Natural Product Research.* in press.

Rodríguez-Gálvez E., Guerrero P., Barradas C., Crous, P. W., Alves A., 2017. Phylogeny and pathogenicity of *Lasiodiplodia* species associated with dieback of mango in Peru. *Fungal Biology* 121: 452–465.

Rosado A., Machado, W.C., Freire A. R., F. D. C. O., Pereira O.L. 2016. Phylogeny, identification, and pathogenicity of *Lasiodiplodia* associated with postharvest stem-end rot of coconut in Brazil. *Plant Disease* 100: 561–568.

Úrbez-Torres J.R. 2011. The status of Botryosphaeriaceae species infecting grapevine. *Phytopathologia Mediterrranea* 50: 5–45.

Venkatasubbaiah P., Sutton T.B., Chilton W.S., 1991. Effect of phytotoxins produced by *Botryosphaeria obtusa*, the cause of black rot of apple fruit and frog-eye leaf spot. *Phytopathology* 81: 243–247.

Yu J.Q., Matsui Y, 1994. Phytotoxic substances in root exudates of cucumber (*Cucumis sativus* L.). *Journal of Chemical Ecology* 20: 21–31.

4

Severity of Diplodia shoot blight (caused by *Diplodia sapinea*) was greatest on *Pinus sylvestris* and *Pinus nigra* in a plantation containing five pine species

Ayşe Gülden ADAY KAYA[1,*], Şule YELTEKIN[2], Tuğba DOĞMUŞ LEHTI-JÄRVI[2], Asko LEHTIJÄRVI[3], Steve WOODWARD[4]

[1] *Isparta Applied Sciences University, Yenişarbademli Vocational School, Isparta, Turkey*
[2] *Isparta Applied Sciences Faculty of Forestry, 32260 Isparta, Turkey*
[3] *Bursa Technical University, Faculty of Forestry, 16310 Bursa, Turkey*
[4] *University of Aberdeen, Institute of Biological and Environmental Sciences, Aberdeen AB24 3UU, Scotland, United Kingdom*
*Corresponding author: guldenaday@isparta.edu.tr

Competing Interests: The Author(s) declare(s) no conflict of interest.

Editor: Matteo Garbelotto, University of California, Berkeley CA, USA.

Summary. The severity of Diplodia shoot blight in *Pinus nigra* and *P. sylvestris* seed orchards, and in *P. taeda*, *P. pinaster* and *P. radiata* industrial plantations was examined in two 20 × 20 m plots of each tree species. All 159 trees assessed in the plots showed symptoms of *Diplodia sapinea* infections. Average disease severity in the tree crowns varied from 42 to 68% (mean = 53.8%) over all tree stands. Fungus isolates obtained from affected trees were identified as *D. sapinea*, based on morphological characteristics of cultures and conidia grown on agar plates. The identifications were confirmed by sequence analysis of the ITS rDNA of a subsample of isolates. Multilocus genotyping with RAMS primers showed that 13 *D. sapinea* genets were present in the plots. *Diplodia sapinea* had serious impacts on *P. sylvestris* and *P. nigra* in the seed orchards Marmara Region of Turkey, and is likely to have been the main causal agent of shoot blight and reductions in seed production at this location.

Keywords. Disease severity, RAMS, pathogenicity, pine plantations.

INTRODUCTION

Diplodia sapinea (Desmaz.) J. Kickx [syn. *Sphaeropsis sapinea* (*Diplodia pinea*) (Fr.: Fr.) Dyko & Sutton] causes a range of disease symptoms on conifer trees, including stunting, browning of needles, shoot blight, twig and branch dieback, crown wilt and bark cankers, root disease, and damping-off and root rot of seedlings. In addition, the fungus causes blue stain of sapwood in fallen or freshly cut timber, in which it is considered a saprotroph (Brookhouser and Peterson, 1971; Peterson, 1977). The fungus has wide distribution in temperate and tropical regions. *Pinus nigra* Arnold, *Pinus radiata* Don. and *Pinus sylvestris* L. are particularly susceptible to infection (Wingfield and Know-Davies, 1980; Palmer and Nicholls, 1985; Swart *et al.*, 1987; Chou and Mackenzie, 1988; Rees and Webber, 1988; Nicholls and Ostry, 1990; Stanosz and Cummings, 1996; Stanosz *et al.*, 2001). Predisposing factors such as drought stress or physical wounding through forestry practices may have important roles in increasing tree susceptibility to *D. sapinea* infections (Swart and Wingfield, 1991; Blodgett and Stanosz, 1997; Adams *et al.*, 2002; Blodgett and Bonello, 2003).

Although *D. sapinea* was first reported in Turkey in 1993 on *Pinus pinea* L. and *Pinus pinaster* Ait. (Ünligil and Ertaş, 1993), little is known about the incidence and severity of Diplodia diseases, or genetic variation in the pathogen populations, in Turkish forests. Some of the most severe damage from Diplodia shoot blight has occurred in the southern part of Turkey. The pathogen was first noted on dead twigs or canker samples of *Pinus brutia* var. *eldarica* (Medw.) Silba and *Pinus brutia* Ten. in Kahramanmaraş, Turkey, in 2000 (Sümer, 2000). *Diplodia sapinea* was later shown to be the main causal agent of shoot blight of *P. brutia* in the Isparta region, in the western Taurus Mountains (Doğmuş Lehtijärvi *et al.* 2007). In 2013, similar symptoms were reported in *Pseudotsuga menziesii* (Mirb.) Franco plantations in İzmit province (Aday Kaya *et al.*, 2014), and *D. sapinea* was suggested as the likely causal agent.

Genetic variation in *D. sapinea* populations has previously been determined based on morphological characteristics and virulence of the isolates (De Wet *et al.*, 2002). In recent years, however, molecular genetic analyses of several populations worldwide have been performed. The results of these studies indicate high variation in the genetic diversity among *D. sapinea* populations (Burgess *et al.*, 2004; Bihon *et al.*, 2012 a, b). The pathogen is predominantly asexual and wound-associated die-back appears to be caused by clones of the pathogen occurring in narrow time frames (Bihon *et al.*, 2012b). However, endophytic infections show high levels of genetic diversity (Bihon *et al.*, 2011, Bihon *et al.*, 2012 a, b), probably resulting from a cryptic, heterothallic sexual cycle (Bihon *et al.*, 2014).

To date, genetic differentiation and pathogenicity in Turkish *D. sapinea* isolates has not been examined. Random amplified microsatellites (RAMS) and internal simple sequence repeat (ISSR) are reliable tools for analysis of genetic variation in several important forest pathogens (Hantula *et al.*, 1996). Using RAMS, Doğmuş Lehtijärvi *et al.* (2014) demonstrated that 60 isolates of *D. sapinea* in *P. nigra* and *P. sylvestris* seed orchards were genetically identical.

The objectives of the study reported here were: i) to estimate the disease severity caused by *D. sapinea* in *P. nigra* and *P. sylvestris* seed orchards, and in *P. taeda*, *P. pinaster* and *P. radiata* industrial plantations in Turkey, ii) to characterize isolates of the fungus using a combination of morphological and molecular techniques, iii) to investigate pathogenicity, host specialization and virulence of isolates, and iv) to determine genetic variation within the *Diplodia* isolates.

MATERIALS AND METHODS

Disease severity

Surveys were conducted in October 2014 in two seed orchards planted with *Pinus nigra* and *P. sylvestris*, and in industrial plantations of *P. taeda*, *P. pinaster* and *P. radiata* in the Kerpe Research Forest, İzmit province, Marmara Region of Turkey. These pine plots (sections) were located 130 to 870 m apart from each other (Table 1). Soil types of the stands were brown forest soil, light hydromorphic grey brown podzolic brown soil, and andesite brownstone soil on the main rock. For each tree species, two 20 × 20 m plots were surveyed. Diameter at breast height (DBH), tree height and disease severity caused by *D. sapinea* were recorded for all trees in the plots, and mature, open cones remaining on shoots were collected from each tree. Severity of Diplodia shoot blight was scored using a 1-5 scale (Table 2), and calculated using the Townsend-Heuberger formula (Townsend-Heuberger, 1943):

$$\text{Disease Severity (\%)} = [\Sigma(n.V)/Z.N] \times 100$$

Where

n = number of samples that are in different disease scales,
V = scale value,
Z = greatest scale value,
N = total number of samples

Estimation of inoculum densities in cones

A total of 477 cones were collected from sampled trees in the investigated plots. Three cones were taken from each tree to estimate *D. sapinea* inoculum densities. Cone samples from each tree were pooled in a 100 mL plastic beaker containing 80 mL of deionized water with two drops of Tween®80 (Sigma-Aldrich, Inc.), which was then shaken for 3 h at 24°C at 110 rpm. After shaking, the cones were removed from the beakers and the

Table 1. Locations of pine plantings assessed for Diplodia shoot dieback.

Tree species	Coordinates	Section ID	Type of stand	Size of section (ha)	Origin of tree species	Soil
Pinus nigra	41° 9' 5.83" N 30 °12'19.39" E 41° 8' 59.03" N 30°12'20.47" E	10c	Plantation	8.4	Turkey- Dursunbey	Brown forest soil
P. sylvestris	41° 9' 26.06" N 30°13'3.77" E 41° 9' 23.21" N 30°13'3.02" E	18e1, 19b	Seed orchard	34	Turkey- Sarıkamış	Brown forest soil
P. taeda	41° 9' 5.75" N 30°12'27.06" E 41° 8' 58.88" N 30°12'27.91" E	15r	Plantation	28.1	USA	Light hydromorphic gray brown podzolic brown soil
P. pinaster	41° 9' 11.73" N 30°12' 21.62" E 41° 9' 9.05" N 30°12' 25.03" E	9g	Plantation	19.1	France	Light hydromorphic gray brown podzolic brown soil
P. radiata	41° 9' 27.19" N 30°12' 29.55" E 41° 9' 23.50" N 30°12' 28.66" E	9a	Plantation	9.0	Australia	Andesite brownstone land on the main rock

Table 2. Disease severity scores used for assessments in pine stands.

Score	Damage class	Symptoms used for scoring
1	No damage	-
2	Light damage	≤five main branches with advanced defoliation
3	Moderate damage	Several (≥five) main branches or a whole crowns, groups of branches seriously damaged.
4	Severe damage	Most of the main branches seriously damaged.
5	Dead	-

volume of water in each beaker adjusted to 100 mL with deionized water. The numbers of conidia in the suspensions were enumerated with five replicate counts using a compound microscope and a haemocytometer.

Identification and assessment of genetic variation of the pathogen

Cones taken from pines were examined for the presence of pycnidia typical of *D. pinea* under a dissecting microscope (Table 3). Pycnidia were transferred to potato dextrose agar (PDA Merck 1.10130) and incubated at 24°C for 5 d. In total, 106 isolates were obtained. Morphological features of the conidia collected from pynidiaon cones were examined under a binocular microscope, and cultural characteristics of the isolates were examined using pure cultures grown on fresh PDA.

Isolates were sub-cultured to cellophane membranes covering PDA and incubated at 24°C for 7 d. The mycelium was then scraped off the membrane surfaces and ground in liquid nitrogen using a mortar and pestle. DNA was extracted using DNeasy Plant Mini Kits (Qiagen) following the manufacturer's instructions. PCR amplification of the internal transcribed spacer (ITS) region of the rDNA gene was performed using the primer set ITS1 (5'-TCCGTAGGTGAACCTGCGG-3') and ITS4 (5'-TCCTCCGCTTATTGATATGC-3') (White *et al.* 1990) in 50 μL reactions. Each reaction containing 50 ng genomic DNA, 250 nM of each primer, 200 μM of each dNTP, 25 mM $MgCl_2$, 1U *Taq* polymerase, 1 × Q solution, and 1 × PCR buffer (Promega Corporation). PCR was conducted in a Biorad MJ Mini Personal Thermal Cycler. PCR conditions were: denaturing at 95°C for 10 min, followed by 30 cycles of amplification (20 s denaturation at 94°C, 25 s annealing at 55°C, and 2 min extension at 72°C). PCR products were separated by electrophoresis in 1% (w/v) agarose.

Genetic variation among the isolates was determined using multilocus genotyping with the M13 marker (Zamponi *et al.* 2007) and RAMS-primers VDH $(TCG)_5$, $DDB(CCA)_5$ and $DDB(CGA)_5$, where B = G/T/C, D = G/A/T, H = A/T/C and V = A/C/G (Hantula *et al.* 1996).

Table 3. Diplodia shoot dieback severity assessments.

Stand No.	Tree species	No. of trees	Damage scale and disease severity (%)[1)]									
			No damage (1)		Light damage (2)		Moderate damage (3)		Severe damage (4)		Dead (5)	
			Tree No.	%	Tree No.	%	Tree No.	%	Tree No.	%	Tree No.	%
1	*Pinus sylvestris* 1	11	0	0	5	18.7	6	32.7	0	0	0	0
2	*Pinus sylvestris* 2	10	0	0	3	12	6	26.0	1	8	0	0
		21		0		15.1		34.3		4		0
3	*Pinus nigra* 1	10	0	0	0	0	3	18.0	6	48.0	1	10.0
4	*Pinus nigra* 2	9	0	0	0	0	3	20.0	6	53.3	0	0.0
		19		0		0		19.0		50.6		5.0
5	*Pinus pinaster* 1	26	0	0	0	0	12	27.7	14	43.0	0	0
6	*Pinus pinaster* 2	17	0	0	8	18.8	9	31.7	0	0.0	0	0
		43		0		9.4		29.7		21.5		0
7	*Pinus radiata* 1	21	0	0	6	11.4	8	22.8	6	22.8	1	4.8
8	*Pinus radiata* 2	12	0	0	0	0	4	20.0	8	53.3	0	0
		33		0		5.7		21.4		38.0		2.4
9	*Pinus taeda* 1	22	0	0	7	12.7	10	27.2	5	18.0	0	0
10	*Pinus taeda* 2	21	0	0	0	0	21	80.0	0	0	0	0
		43		0		6.3		53.6		9.0		0

Primers were synthesized by IonTek. A total of 50 isolates was analysed, with ten isolates from each filed plot examined. DNA was extracted from each isolate as described above and subjected to PCR. DNA was denatured at 95°C for 10 min, after which 37 cycles of amplification were carried out (30 s denaturation at 95°C, 45 s annealing at the primer dependent temperature and 2 min extension at 72°C), followed by 7 min at 72°C (Maresi *et al.* 2007).

Amplification products were separated by electrophoresis (6 V cm^{-1} for 180 min) in 1.0% agarose gels (Biobasic Inc.) and 1.0% SynerGel gels (Diversified Biotech) in TAE buffer, and lengths of the products was estimated using DNA molecular size markers with 100 bp repeats (Thermo Fisher Scientific).

The presence or absence of amplification products was scored, with only clear and reproducible markers considered for the analysis. To estimate genetic variation, analysis of molecular variance (AMOVA) was calculated based on RAMS markers. Variation was assessed among and within isolate groupings based on the host species.

Genetic variation between *D. sapinea* isolates was computed using Jaccard's coefficient of similarity in Arlequin V 6.1 of NTSYSpc 2.1 software (Exeter Software Co.) and GenAlEx 6.5 (Peakall and Smouse, 2006).

Pathogenicity, host specialization and virulence of isolates

Seedling inoculations

Pathogenicity of the *D. sapinea* isolates was assessed using potted seedlings of *P. sylvestris*, *P. nigra*, *P. radiata*, *P. taeda* and *P. pinaster*, which were obtained from local forest nurseries. Three-year-old seedlings growing in a mixture of 60% clay, 20% sand and 20% humus were placed in a growth chamber and incubated at 20°C for 28 d prior to inoculation and during incubation. Seedlings were irrigated with tap water at 48 h intervals. Twenty-five isolates of *D. pinea* grown on PDA were used for the inoculations. Seedlings were inoculated 2 cm above the root collar. The inoculation point was cleaned with 70% (v:v) ethanol. A circular 3 mm diam. wound was made in the stem of each seedling using a sterilized cork borer to remove the bark. An equal-sized agar plug colonized by *D. sapinea* was inserted into each wound and wrapped with Parafilm M®. Experimental controls were mock-inoculated with non-colonized agar plugs. Five seedlings were inoculated for each isolate-host combination, and a randomized complete block experimental design used in the trial. Inoculated seedlings were incubated in the growth chamber for 4 weeks, after which lesion lengths were measured. Random re-isolations were made onto fresh PDA to confirm *Diplodia* sp. as the causal agent of the lesions.

Cross inoculations on twigs

Host specialization and virulence of isolates were also tested on 30 cm long twigs of the five host species. All twigs were obtained from İzmit-Kerpe Research forests. The same 25 isolates as in the seedling inoculation assay were grown for 7 d on PDA and then used for twig inoculations. Twigs were under bark-inoculated 8 cm above the base of each twig using a method similar to that described above for seedling inoculations. Five twigs were used for each fungal isolate-host species combination.

In addition, five control twigs were inoculated with sterile PDA discs. Inoculated twigs were kept at 20°C in a growth chamber for 1 month before recording lesion lengths using the methods described above.

Analysis of variance (ANOVA) was performed on lesion length data using the SPSS MGLM procedure (SPSS Inc.), and differences among mean values were assessed using Duncan's multiple range test ($P < 0.05$).

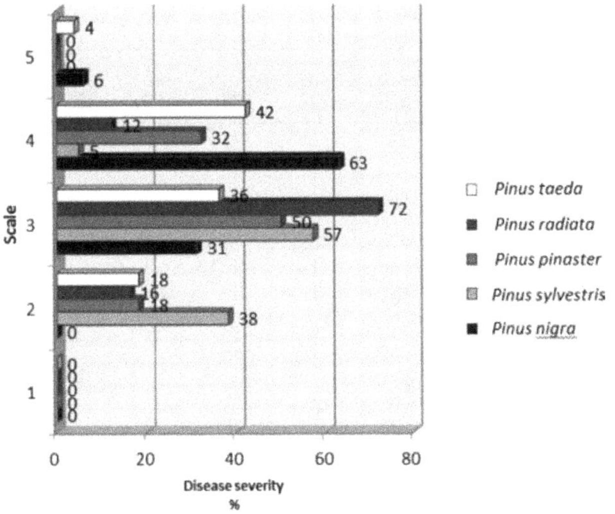

Figure 1. Percentage disease severity in each disease scale group for different *Pinus* spp.

RESULTS

Disease severity

All 159 trees assessed in the plots had symptoms of *Diplodia* infections. Needles on recently killed shoots were reddish brown. Needles that were dead for longer periods were greyish brown or dark grey.

The average disease severity varied from 42 to 68% (mean = 53.8%) in the stands (Table 2). The predominant disease severity class was moderate, whereas high severity was less common (Figure 1). Comparisons of the damage classes between host species indicated that light damage (score = 2) occurred more frequently in *P. sylvestris* stands, whereas moderate damage (score = 3) was greater in *P. taeda* stands. Frequencies of severely damaged and dead individuals were greatest in the *P. nigra* stands (Table 3).

Estimation of inoculum densities in cone samples

Cones from all sampled stands yielded conidia characteristic of *D. sapinea*. Pycnidia were abundant on the cones and shoots of the specimens examined. The number of conidia extracted from cones varied between tree species. Numbers of conidia were greater from cones of *P. nigra* and *P. sylvestris*, particularly when compared with those obtained from cones of *P. radiata* or *P. taeda*. There was no significant difference in the numbers of conidia (overall mean = 1.3×10^6) obtained from *P. sylvestris* and *P. pinaster* cones ($P \leq 0.05$) (Table 4).

Identification and genetic diversity in the pathogen populations

A total of 106 isolates was obtained from shoot and cone samples, and identified as *D. sapinea*. Conidia were brown to dark brown, thick-walled, with a mean width of 18.0 μm (SD ± 2.6) (range 9 to 22 μm) and mean length of 35.0 μm (SD ± 5.1) (range 20–41 μm) (n = 100).

Morphological identifications of the isolates were confirmed by sequence analysis of the ITS rDNA of representative isolates. The sequences showed homologies > 99% with GenBank accessions of *D. sapinea*.

Amplification of genomic DNA of 50 isolates of *D. sapinea* using the M13 and three RAMS markers produced 78 fragments of which 13 were polymorphic. All fragments were clear and reproducible, with sizes ranging from 350 to 870 base pairs (Table 5). The variation was analysed according to the presence or absence of the markers.

Table 4. Occurrence of *Diplodia sapinea* in cones of pine species and recovery of conidia in suspensions.

Host species	No. of cones examined	*Diplodia sapinea* positive (%)	Conidia per cone ($\times 10^6$)
Pinus sylvestris	54	81	1.3
P. nigra	51	88	1.1
P. pinaster	129	76	1.3
P. radiate	129	79	0.5
P. taeda	99	54	0.6
	Total = 462	Mean = 75.6	Mean = 0.96

Table 5. Presence/absence (1/0) vector of *Diplodia sapinea* isolates tested with RAMS and M13 primers.

Sample ID	Population name	CCA-870	CCA-800	CGA-750	M13-650	M13-600	CCA-570	CGA-550	CGA-510	CCA-480	TCG-440	TCG-420	CCA-400	CCA-370
D1	Pn	1	1	0	1	1	0	0	0	1	0	1	1	1
D2	Pr	0	1	0	1	0	0	0	1	1	1	1	0	1
D3	Pt	0	1	0	0	0	1	0	0	0	0	0	0	0
D4	Pp	1	1	1	1	0	1	0	0	0	1	1	0	1
D5	Ps	0	1	0	0	0	0	0	0	1	0	1	1	1
D6	Ps	0	1	0	0	0	0	0	0	1	0	1	1	1
D7	Ps	0	1	0	0	1	0	0	0	1	0	1	1	1
D8	Ps	0	1	0	0	1	0	0	0	1	0	1	1	1
D9	Ps	0	1	0	0	1	0	0	0	1	0	1	1	1
D10	Ps	0	1	0	0	1	0	0	0	1	0	1	1	1
D11	Ps	0	1	0	0	1	0	0	0	1	0	1	1	1
D12	Ps	0	1	0	0	1	0	0	0	1	0	1	1	1
D13	Ps	0	1	0	0	1	0	0	0	1	0	1	1	1
D14	Ps	0	1	0	0	1	0	0	0	1	0	1	1	1
D15	Pn	0	1	0	1	1	0	1	0	1	0	1	1	1
D16	Pn	0	1	0	0	0	0	1	0	1	1	0	1	1
D17	Pn	0	1	0	0	0	0	1	0	1	1	0	1	1
D18	Pn	0	1	0	0	0	0	1	0	1	1	0	1	1
D19	Pn	0	1	0	0	0	0	1	0	1	1	0	1	1
D20	Pn	0	1	0	1	0	0	1	0	1	1	0	1	1
D21	Pn	0	1	0	1	0	0	1	0	1	1	0	1	1
D22	Pn	0	1	0	0	0	0	1	0	1	1	0	1	1
D23	Pn	0	1	0	1	0	0	1	0	1	1	0	1	1
D24	Pp	0	1	0	1	0	0	1	0	0	0	0	0	0
D25	Pp	0	1	0	1	0	0	1	1	0	1	1	0	1
D26	Pp	0	1	0	1	0	0	1	1	0	1	1	0	1
D27	Pp	0	1	0	1	0	0	1	1	0	1	1	0	1
D28	Pp	0	1	0	1	0	0	1	1	0	1	1	0	1
D29	Pp	0	1	0	1	0	0	1	1	0	1	1	0	1
D30	Pp	0	1	0	1	0	0	1	1	0	1	1	0	1
D31	Pp	0	1	0	1	0	0	1	1	0	1	1	0	1
D32	Pp	0	1	0	1	0	0	1	1	0	1	1	0	1
D33	Pt	0	1	0	1	1	1	1	1	1	1	1	0	1
D34	Pt	0	1	0	1	1	1	1	1	1	1	1	0	1
D35	Pt	0	1	0	1	1	1	1	1	1	1	1	0	1
D36	Pt	0	1	0	1	1	1	1	1	1	1	1	0	1
D37	Pt	0	1	0	1	1	1	1	1	1	1	1	0	1
D38	Pt	0	1	0	1	1	1	1	1	1	1	1	0	1
D39	Pt	0	1	0	1	1	1	1	1	1	1	1	0	1
D40	Pt	0	1	0	1	1	1	1	1	1	1	1	0	1
D41	Pt	0	1	0	1	1	1	1	1	1	1	1	0	1
D42	Pr	0	0	0	1	0	0	0	1	1	1	1	0	1
D43	Pr	0	0	0	1	0	0	0	1	1	1	1	0	1
D44	Pr	0	0	0	1	0	0	0	1	1	1	1	0	1
D45	Pr	0	0	0	1	0	0	0	1	1	1	1	0	1
D46	Pr	0	0	0	1	0	0	0	1	1	1	1	0	1
D47	Pr	0	0	0	1	0	0	0	1	1	1	1	0	1
D48	Pr	0	0	0	1	0	0	0	1	1	1	1	0	1
D49	Pr	0	0	0	1	0	0	0	1	1	1	1	0	1
D50	Pr	0	0	0	1	0	0	0	1	1	1	1	0	1

ANOVA showed that variation among and within *D. sapinea* isolates obtained from different pine species amounted to, respectively, 44 and 56% of the total variation. Population divergence measured with F_{ST} also indicated differentiation in *D. sapinea* among *P. nigra*, *P. taeda* and *P. pinaster* (F_{ST} = 2.432) (Table 6).

Thirteen *D. sapinea* genotypes were present in all host species (Table 7). Genotypes of *D. sapinea* infecting *Pinus sylvestris* and *P. taeda* were unique to those hosts, whereas *P. nigra*, *P. pinaster* and *P. radiata* shared nine genets.

Genotype size was also estimated based on the RAMS analysis, and ranged from a single tree to approx. 115 m^2 (15 trees). Two genotypes were shared within two stand replicates. The largest genotype was located in a *P. radiata* stand.

Table 6. Genetic variation within *Diplodia sapinea* populations.

Variation source	df	SS	MS	Est. Var.	%
Between populations	4	58.640	14.660	1.359	56
Among populations	45	48.300	1.073	1.073	44
Total	49	106.940		2.432	100

Table 7. Genet size of *Diplodia sapinea* populations.

Tree species	Stand number	Genet code	Number of isolates/per genet	Estimated genet size m^2
Pinus nigra	1	1.genet	2	20
		2.genet	3	5
	2	3.genet	3	38
		4. genet	2	16
P. sylvestris	1	5.genet	2	6
		6.genet	3	18
	2	6.genet	5	100
P. pinaster	1	7.genet	1	1
		8.genet	1	1
		9.genet	3	55
	2	9.genet	5	68
P. taeda	1	10.genet	1	1
		11.genet	4	18
	2	11.genet	5	106
P. radiata	1	12.genet	1	1
		13.genet	4	114
	2	13.genet	5	93
Total		13	50	661

Pathogenicity, host specialization and virulence of isolates

Seedling inoculations

Diplodia sapinea inoculations caused dark brown to black discolouration around the inoculation points in seedlings of all pine species tested. Infection frequency in the inoculated seedlings was 100%. The pathogen was successfully re-isolated from symptomatic stem tissues, thus fulfilling Koch's postulates. Susceptibility of the tree species to *D. sapinea*, and the virulence of the *D. sapinea* isolates among those host species, differed significantly. Compared to the other *Pinus* species tested, *P. radiata* was highly susceptible to *D. sapinea*. There were also significant differences in the extents of the lesions produced by *D. sapinea* on the host species tested ($P < 0.05$) (Table 8). The longest lesions were on *P. radiata* and *P. sylvestris*. Lesion lengths caused by isolates were greater on *P. radiata*, *P. sylvestris* than *P. nigra* seedlings than on ($P < 0.05$).

Cross inoculations on twigs

Most of the isolates used in the inoculations caused lesions on the twigs of the five host species. Differences between the isolates obtained from the sampled pine species, in terms of ability to induce lesions on all tested twigs were significant ($P < 0.05$). In general, isolates from *P. taeda* caused longer lesions on *P. radiata* twigs (Table 8) than on the other host species. *Pinus radiata*, *P. sylvestris* and *P. nigra* twigs were susceptible to all isolates. MLGM analyses indicated that there was no host specificity from the cross inoculations (Table 9).

DISCUSSION

The studies reported in this paper have confirmed the presence of *D. sapinea* on *Pinus nigra*, *P. sylvestris*,

Table 8. Mean lesion lengths caused by *Diplodia sapinea* on seedlings of different *Pinus* spp. hosts.

Host	Mean lesion length (mm) (± S.D.)
Pinus radiata	18.1±0.580 A
P. sylvestris	16.5± 0.564 B
P. nigra	12.5± 0.564 B
P. taeda	5.4± 0.591 C
P. pinaster	5.4± 0.593 C

Means within a column followed by different letter are significantly different ($P < 0.05$) based on Duncan's Multiple Range test.

Table 9. Results of *Diplodia sapinea* cross inoculation of twigs.

Pinus. radiata isolates		*P. taeda* isolates		*P. pinaster* isolates		*P. sylvestris* isolates		*P. nigra* isolates	
P. sylvestris	23.8a	*P. radiata*	30.0a	*P. nigra*	13.0a	*P. nigra*	17.6a	*P. sylvestris*	15.5a
P. radiata	21.8a	*P. sylvestris*	19.3b	*P. radiata*	12.7a	*P. radiata*	14.8a	*P. nigra*	10.8b
P. nigra	10.8b	*P. nigra*	10.3bc	*P. sylvestris*	11.5ab	*P. sylvestris*	12.9ab	*P. radiata*	9.4bc
P. taeda	5.9c	*P. taeda*	5.8c	*P. pinaster*	5.3b	*P. pinaster*	6.0b	*P. pinaster*	6.3c
P. pinaster	5.3c	*P. pinaster*	4.4c	*P. taeda*	5.1b	*P. taeda*	4.5c	*P. teada*	5.7c

Means within a column followed by different letter are significantly different ($P < 0.05$) based on Duncan's Multiple Range test.

P. taeda, *P. pinaster* and *P. radiata* in the Kerpe Research Forest, in İzmit province of Turkey. On the basis of morphological characteristics of the pathogen in culture, ITS sequencing of isolates obtained from sampled trees with the symptoms and the pathogenicity tests performed on excised twigs of same host species, it was clear that *D. sapinea* was responsible for shoot blight and reductions in seed production in *Pinus* spp. trees at this location.

Typical Diplodia shoot blight symptoms (Maresi *et al.*, 2007) were observed, scattered throughout the crowns of the trees. The most serious damage occurred on the middle and upper crowns. It is not clear why there was such high incidence of Diplodia shoot dieback in the Kerpe Research Forest, but there were no records to suggest that this was connected with predisposing factors such as wounding or drought. However, the trees were planted in unsuitable climatic conditions.

The design of the Kerpe research forest was not optimal for critical comparison of the tree species for disease incidence and severity, as the trees were not planted in blocks homogenous for site factors. *Pinus sylvestris* and *P. nigra* were growing on brown forest soil, *P. pinaster* and *P. taeda* on light hydromorphic grey brown podzolic soil and *P. radiata* on andesite brownstone soil on the main rock (Table 1). For tree species pairs growing on the same soil type, the disease was more severe on *P. nigra* and *P. pinaster* than on *P. sylvestris* and *P. taeda*. (Table 3).

Colony morphology of the isolates obtained here was identical to that of the *D. sapinea* described in De Wet *et al.* (2002). The presence of abundant pycnidia on branches, stems and cones of infected trees illustrated the potential of this fungus to cause epidemics when conditions are favourable for infection and disease development (Nicholls and Ostry, 1990).

In previous research, cones collected from tree canopies yielded conidia more frequently than cones collected from the ground (Doğmuş Lehtijärvi *et al.*, 2014). In the present study, therefore, we sampled only from the tree canopies. It is very likely that cones in the canopies are more important for spread of the fungus than cones on the ground, and tree cones should be used for estimation of inoculum potential at particular sites (Munck and Stanosz, 2009).

Diplodia sapinea has been detected in the seeds of many *Pinus* species, including *P. radiata*, *P. nigra*, *P. sylvestris*, *P. resinosa*, as well as *P. patula*, *P. wallichiana*, *P. elliottii*, *P. taeda*, *P. oocarpa*, *P. caribaea*, *P. michoacana* and other tropical pines (Bihon *et al.*, 2011). However, despite the probable critical importance of seed infection in the epidemiology of disease caused by this fungus, and implications for dissemination via plant material (e.g. from seed orchards or through international trade), seed and plant disemination is likely to be very important (Bihon *et al.*, 2012a).

Diplodia sapinea causes degeneration of edible seeds on *P. pinea*, with direct economic consequences (Vagniluca *et al.*, 1995; Santini *et al.*, 2008). Moreover, the frequent occurrence of *Diplodia* symptoms on pine trees planted in urban landscapes and recreational areas has raised awareness of this pathogen among public administrations.

Multilocus genotyping with RAMS primers indicated that the trees in the sampled stands were colonized by thirteen *D. sapinea* genets. Genets found in *P. taeda* and *P. sylvestris* occurred only on these hosts, while the remaining genets were shared among other host species. Previous work in 2012 with *D. sapinea* in the same seed orchard showed that one genet occurred within *P. nigra* and *P. sylvestris* stands (Doğmuş Lehtijarvi *et al.*, 2014). This indicates that the disease in the stand had possibly started with wound infections by one or a few clones of the pathogen (cf. Bihon *et al.*, 2012b). In the present study, there were six distinguishable genets on the same hosts. The difference in number of genets could be due to spread of *Leptoglossus occidentalis* Heidemann to the study area. This alien invasive insect is a new vector of the fungus in Europe (Luchi *et al.*, 2012), and could possibly have introduced more genets to the site. *Leptoglossus occidentalis* was first detected in Turkey in 2009 in Istanbul (Arslangündoğdu and Hızal, 2010), approx. 100 km from the Kerpe Research Forest. Owing to the rapid increase in the known distribution of this insect around the Marmara Region over 2 to 3 years (Öztemiz and Doğanlar, 2015), it is possible that arrival and establishment of *L. occidentalis* in the area of the present study has affected the *D. sapinea* population, as the 13 genotypes were all closely related.

Bihon *et al.* (2012a), in examination of genetic differences between single *D. sapinea* populations from South Africa, Ethiopia, Argentina and Australia using microsatellite markers, found genetic variation between continents and within Australia.

Similarities between isolates of *D. pinea* were high when analysed with RAPD markers and isoenzyme analysis (Stanosz *et al.*, 1999). Burgess *et al.* (2004) studied six different populations of *D. pinea* isolated from *P. radiata* in New Zealand using SSR markers, and found small variations and diversity amongst populations. In the present study, the greatest numbers of polymorphic loci were in the pathogen population from *P. nigra*, with the lowest numbers of loci in the *D. sapinea* isolates from *P. radiata*.

The genetic similarities between *D. sapinea* genotypes and populations across continents suggests that the fungus has had a long asexual history of dispersal. Clones generally accumulate unique alleles in geographically isolated populations (Taylor *et al.*, 2000). In addition, plant genetic material such as seeds or seedlings for breeding programmes, has been exchanged widely between Turkey and other countries. Because *D. sapinea* is a latent pathogen, genotypes of the fungus have probably been spread with these materials (Burgess *et al.*, 2001; Flowers *et al.*, 2001).

The present study used three RAMS and M13 markers to detect genetic variation between *D. sapinea* isolates obtained from five different host pine species, although the isolate sampling was made at one site. Further analyses using SSR markers or other microsatellite markers are required to confirm these results and to determine if there is host specialization in *D. sapinea*. In the results presented here, cross inoculations suggested that there was no host specialization of this pathogen for the isolates tested.

Pathogenicity of *D. sapinea* did not relate to isolate origin or tested host in this study. This is similar to the results of Chou (1976), who found no differences in pathogenicity amongst New Zealand isolates of *D. pinea*. Doğmuş Lehtijärvi *et al.* (2009) previously reported the ability of a single isolate of *D. sapinea* to infect a range of coniferous species in Turkey, including *Cedrus libani* and *P. nigra*. In the present study, inoculations with 25 *D. sapinea* isolates resulted lesions on the tested coniferous tree species, with the greatest virulence on *P. nigra* and *P. sylvestris* and least on *P. taeda* and *P. pinaster*.

Outbreaks of disease caused by *D. sapinea* have occurred periodically wherever pines are grown. No direct and effective control measures are available for this pathogen in plantations and seed orchards. However, irrigation of seed orchards may prevent drought stress that can precede outbreaks of Diplodia shoot blight. Wounding of trees should also be minimized during forestry management to reduce infections of *D. sapinea*. Most importantly, foresters need to consider the risk of planting *P. sylvestris* and *P. nigra* in regions with unsuitable climate and location, such as low altitude, which may subject trees to stresses that predispose them to *Diplodia* infections.

ACKNOWLEDGEMENTS

The authors thank the Ministry of Forestry for providing the seedlings used in this research. Financial support was provided by SDÜ BAP 3826-YL1-13 and TÜBİTAK - TOVAG (Project No: 114O138).

LITERATURE CITED

Adams G.C., Wu N.T., Eisenberg B.E., 2002. Virulence and double stranded RNA in *Sphaeropsis sapinea*. *Forest Pathology* 32: 309–329.

Aday Kaya A.G., Lehtijärvi A., Kaya Ö.D., Doğmuş Lehtijärvi H.T., 2014. First report of *Diplodia pinea* on *Pseudotsuga menziesii* in Turkey. *Plant Disease* 98: 689.

Arslangundogdu A., Hizal E., 2010. The western conifer seed bug, *Leptoglossus occidentalis* (Heidemann, 1910), recorded in Turkey (Heteroptera: Coreidae). *Zoology in the Middle East* 50: 138–139.

Bihon W., Slippers B., Burgess T., Wingfield M.J., Wingfield B.D., 2011. Sources of *Diplodia pinea* endophytic infections in *Pinus patula* and *P. radiata* seedlings in South Africa. *Forest Pathology* 41: 370–375.

Bihon W., Burgess T., Slippers B., Wingfield M.J., Wingfield B.D., 2012a. High levels of genetic diversity and cryptic recombination is widespread in introduced *Diplodia pinea* populations. *Australasian Plant Pathology* 41: 41–46.

Bihon W., Slippers B., Burgess T., Wingfield M.J., Wingfield B.D., 2012b. Diverse sources of infection and cryptic recombination revealed in South African *Diplodia pinea* populations. *Fungal Biology* 116: 112–120.

Bihon W., Wingfield M.J., Slippers B., Duong T.A., Wingfield B.D., 2014. MAT gene idiomorphs suggest a heterothallic sexual cycle in a predominantly asexual and important pine pathogen. *Fungal Genetics and Biology* 62: 55–61.

Blodgett J.T., Stanosz G.R., 1997. *Sphaeropsis sapinea* morphotypes differ in aggressiveness, but both infect on wounded red or jack pines. *Plant Disease* 81: 143–147.

Blodgett J.T., Bonello P., 2003. The aggressiveness of *Sphaeropsis sapinea* on Austrian pine varies with isolate group and site of infection. *Forest Pathology* 33: 15–19.

Brookhouser L.W., Peterson G.W., 1971. Infection of Austrian, Scots and Ponderosa pines by *Diplodia pinea*. *Phytopathology* 6: 439–444.

Burgess T.I., Wingfield M.J., Wingfield B.D., 2004. Global distribution of *Diplodia pinea* genotypes revealed using simple sequence repeat (SSR) markers. *Australasian Plant Pathology* 33: 513–519.

Burgess T., Wingfield M.J., Wingfield B.D., 2001. Simple sequence repeat markers distinguish between morphotypes of *Sphaeropsis sapinea*. *Applied Environmental Microbiology* 67: 354–362.

Chou C.K.S., 1976. A shoot dieback in *Pinus radiata* caused by *Diplodia pinea* II. Inoculation studies. *New Zealand Journal of Forestry Science* 6: 409–420.

Chou C.K.S., Mackenzie M., 1988. Effect of pruning intensity and season on *Diplodia pinea* infection of *Pinus radiata* stems through pruning wounds. *European Journal of Forest Pathology* 18: 437–444.

De Wet J., Wingfield M.J., Coutinho T., Wingfield B.D., 2002. Characterization of the 'C' morphotype of the pine pathogen *Sphaeropsis sapinea*. *Forest Ecology and Management* 161: 181–188.

Doğmuş Lehtijärvi H.T., Lehtijärvi A., Karaca G., Aday A.G., 2007. *Sphaeropsis sapinea* Dyko & Sutton associated with shoot blight on Brutian pine in Southwestern Turkey. *IUFRO Meeting- Acta Silvatica et Lignaria Hungarica, Special Edition*, Hungary, 95–99.

Doğmuş Lehtijärvi H.T., Lehtijärvi A., Karaca G., Aday A.G., Oskay F., 2009. Susceptibility of different coniferous seedlings inoculated with *Diplodia pinea*. Proceedings of the Conference of IUFRO Working Party 7.02.02. *SDÜ Orman Fakültesi Dergisi*, 48–57.

Doğmuş Lehtijärvi H.T., Aday Kaya A.G., Lehtijärvi A., Oskay F., Kaya Ö.D., 2014. Occurrence and genetic similarity of *Diplodia pinea* on shoots and cones in seed orchards of *Pinus* spp. in north-western Turkey. *Plant Protection Science* 50: 217–220.

Flowers J., Nuckles E., Hartman J., Vaillancourt L., 2001. Latent infection of Austrian and Scots pine tissues by *Sphaeropsis sapinea*. *Plant Disease* 85: 1107–1112.

Hantula J., Dusabenyagasani M., Hamelin R.C., 1996. Random amplified microsatellites (RAMS) - a novel method for characterizing genetic variation within fungi. *European Journal of Forest Pathology* 26: 159–166.

Luchi N., Mancini V., Feducci M., Santini A., Capretti P., 2012. *Leptoglossus occidentalis* and *Diplodia pinea*: a new insect-fungus association in Mediterranean forests. *Forest Pathology* 42: 246–251.

Maresi G., Luchi N., Pinzani P., Pazzagli M., Capretti P., 2007. Detection of *Diplodia pinea* in asymptomatic pine shoots and its relation to the normalized insolation index. *Forest Pathology* 37: 272–280.

Munck I.A., Stanosz G.R., 2009. Quantification of conidia of *Diplodia* spp. extracted from red and jack pine cones. *Plant Disease* 93: 81–86.

Nicholls T.H., Ostry M.E., 1990. *Sphaeropsis sapinea* cankers on stressed red and jack pines in Minnesota and Wisconsin. *Plant Disease* 74: 54–56.

Öztemiz S., Doğanlar M., 2015. Invasive plant pests (Insecta and Acarina) of Turkey. *Munis Entomology & Zoology* 10: 144–159.

Palmer M.A., Nicholls T.H., 1985. Shoot blight and collar rot of *Pinus resinosa* caused by *Sphaeropsis sapinea* in forest tree nurseries. *Plant Disease*, 69: 739–740.

Peakoll, R., Smouse, R.E., 2006. GenAlEx 6.5: genetic analysis in Excel. Population genetic software for teaching and research--an update. *Bioinformatics* 28: 2537–2539.

Peterson G.W., 1977. Infection, epidemiology and control of Diplodia blight of Austrian, Ponderosa and Scots pines. *Phytopathology* 67: 511–514.

Rees A.A., Webber J.F., 1988. Pathogenicity of *Sphaeropsis sapinea* to seed, seedlings and saplings of some Central American pines. *Transactions of the British Mycological Society* 91: 273–277.

Santini A., Pepori A., Ghelardini L., Capretti P., 2008. Persistance of some pine pathogens in coarse woody debris and cones in a *Pinus pinea* forest. *Forest Ecology and Management* 256: 502–506.

Stanosz G.R., Cummings-Carlson J., 1996. Association of mortality of recently planted seedlings and established saplings in red pine plantations with Sphaeropsis collar rot. *Plant Disease* 80: 750–753.

Stanosz G.R., Swart W.J., Smith D.R., 1999. RAPD marker and isozyme characterization of *Sphaeropsis sapinea* from diverse coniferous hosts and locations. *Mycological Research* 103: 1193–1202.

Stanosz G.R., Blodgett J.T., Smith D.R., Kruger E.L., 2001. Water stress and *Sphaeropsis* as a latent pathogen of red pine seedlings. *New Phytologist* 149: 531–538.

Sümer S., 2000. Orman Bakanlığı Makamına Kahramanmaraş Yöresi için Hazırlanan Rapor, İstanbul (in Turkish).

Swart W.J., Wingfield M.J., Knox-Davies P.S., 1987. Conidial dispersal of *Sphaeropsis sapinea* in three climatic regions of South Africa. *Plant Disease* 71: 1038–1040.

Swart W.J., Wingfield M.J., 1991. Biology and control of *Sphaeropsis sapinea* on *Pinus* species in South Africa. *Plant Disease* 75: 761–766.

Taylor J.W., Jacobson D.J., Kroken S., Kasuga T., Geiser D.M., Hibbett D.S., Fisher M.C., 2000. Phylogenetic speciesrecognition and species concepts in fungi. *Fungal Genetics and Biology* 31: 21–32.

Townsend G.R., Heuberger J.W., 1943. Methods for estimating losses caused by diseases in fungicide experi ments. *The Plant Disease Reporter* 27: 340–343.

Ünligil H., Ertaş A., 1993. Damage caused by *Sphaeropsis sapinea* to pine trees near İstanbul [İstanbul yakınlarındaki çam ağaçlarında *Sphaeropsis sapinea* (Fr.) Dyko & Sutton mantar hastalığı]. İstanbul Üniversitesi *Orman Fakültesi Dergisi* 43: 131–137 (in Turkish).

Vagniluca S., Goggioli V., Capretti P., 1995. Cankers and shoot blights of *Pinus pinea* in Italy. In: *Proceedings of Shoot and Foliage diseases in forest trees a Joint Meeting of the IUFRO Working Parties S20602 and S20604*, Italy: 284–286.

White T.J., Bruns T., Lee S., Taylor J., 1990. *Amplification and direct sequencing of fungal ribosomal RNA genes for phylogenetics.* In: *PCR Protocols: a Guide to Methods and Applications.* (Innis M.A., Gelfand D.H., Sninsky J.J., White T.J., ed.). New York, USA, 315–322.

Wingfield M.J., Know-Davies P.S., 1980. Observation on diseases in pine and eucalyptus plantations in South Africa. *Phytophylactica* 12: 57–63.

Zamponi L., Paffetti D., Tegli S., Lakomy, P., 2007. Genetic variation in *Heterobasidion abietinum* populations detected with the M13 minisatellite marker. *Forest Pathology* 37: 321–328.

5

Phytoprotection potential of *Fusarium proliferatum* for control of Botryosphaeria dieback pathogens in grapevine

Vincenzo MONDELLO[1], Alessandro SPAGNOLO[1], Philippe LARIGNON[2], Christophe CLÉMENT[1], Florence FONTAINE[1,*]

[1] *SFR Condorcet FR CNRS 3417, Université de Reims Champagne-Ardenne, Résistance Induite et Bioprotection des Plantes, RIBP EA 4707, BP 1039, Reims, Cedex 2 51687, France*

[2] *Institut Français de la Vigne et du Vin Pôle Rhône-Méditerranée, 7 Avenue Cazeaux, 30230 Rodilhan, France*

*Corresponding author: florence.fontaine@univ-reims.fr

Competing Interests: The Author(s) declare(s) no conflict of interest.

Editor: Hanns-Heinz Kassemeyer, Staatliches Weinbauinstitut Freiburg, Germany.

Summary. The economic impact of grapevine trunk diseases (GTDs) is increasing worldwide, due to the lack of efficient and simple control protocols for these disease complexes. Possible and efficient GTD management strategies must consider the complexity of host physiological alterations affecting metabolism and defense responses determined by GTD pathogens, and linked to disease expression. In this complexity, the use of biocontrol agents could give advantages in GTD control methods. The effect of the potential biocontrol agent (BCA) *Fusarium proliferatum* was evaluated using *in vitro* tests and in dual inoculation with the Botryosphaeria dieback agent *Neofusicoccum parvum in planta*. Artificial inoculations were performed in greenhouse and vineyard experiments at three key vine growth stages, the onset of G (separated clusters), I (flowering) or M (veraison) stages. The biocontrol potential was assessed using pathogenicity tests and transcriptomic analyses. Results showed that the *F. proliferatum* has potential for phytoprotection, with disease control efficiency related to host plant growth stage. Flowering was confirmed as the growth stage when disease control was least, and efficiency of activated defense responses against pathogen infection was minimum.

Keywords. Botryosphaeria dieback, *Diplodia seriata*, *Neofusicoccum parvum*, defense responses, biological control.

INTRODUCTION

Grapevine trunk diseases (GTDs) are recognized as one of the most destructive and important problem of grapevine worldwide (Bertsch *et al.*, 2013; Fontaine *et al.*, 2016a; Guerin-Dubrana *et al.*, 2019). GTDs are caused by several unrelated fungi which are able to live and exclusively colonize the wood of grapevine perennial organs, causing wood necrosis and/or discolouration, vascular infections and decays (Mugnai *et al.*, 1999; Bertsch *et al.*, 2013). Externally, diseased vines show general and progressive dieback, often associated with specific foliar symptoms affected by different diseases and host cultivars. These diseases can initially cause loss of productivity and eventually death of affected vines. Their potential is increased by long latent asymptomatic periods, and by the capability of pathogens to infect vines through pruning wounds (Bertsch *et al.*, 2013) which may remain susceptible to GTD infections up to 4 months (Eskalen *et al.*, 2007). Among GTDs, Botryosphaeria dieback, caused by species in the Botryosphaeriaceae (Chethana *et al.*, 2016) is one of the most widespread GTDs in grape growing regions (Úrbez-Torres, 2011; Spagnolo *et al.*, 2014a; Fontaine *et al.*, 2016a). Esca disease and Eutypa dieback are also important grapevine diseases (Bertsch *et al.*, 2013).

After the banning of some disease management active ingredients, such as sodium arsenite, benomyl and carbendazim-based products (Bertsch *et al.*, 2013; Fontaine *et al.*, 2016a; Gramaje *et al.*, 2018; Songy *et al.*, 2019), which gave good control of some important GTDs but may have negative effects on humans and environment, a range of fungicides, natural molecules and biological control agents (BCAs) has been tested to set up efficient control strategies towards the increasing incidence of GTDs (Mondello *et al.*, 2018). To date, beyond *Trichoderma* species, other fungal and bacterial agents have been reported to be effective against GTD pathogens, although some have only been tested either *in vitro* or in nurseries (Gramaje *et al.*, 2018; Mondello *et al.*, 2018). Among these agents, several studies have shown how they could efficiently contribute in GTD management strategies. Beside other specific advantages (broad-spectrum activity and long-lasting effect as wound protectants), an interesting potential of BCAs is their effects on host/pathogen relationships, such as systemic induced resistance (SIR) responses (Handlesman and Stabb, 1996; Pal *et al.*, 2006; Berg, 2009). In this way, BCAs could reduce disease incidence by improving host plant resistance to abiotic and biotic stresses, especially for those provoked by GTDs (Sosnowski *et al.*, 2011; van Niekerk *et al.*, 2011; Spagnolo *et al.*, 2014, 2017; Pinto *et al.*, 2018). BCAs could also have positive effects on GTDs through induced changes of host plant metabolism determined by plant defense responses, and disorders caused by fungal toxins (Abou-Mansour *et al.*, 2015; Christen *et al.*, 2015; Burruano *et al.*, 2016; Fontaine *et al.*, 2016a; Trotel-Aziz *et al.*, 2019).

Previous studies focused on physiological changes occurring in GTD-affected grapevines have highlighted that: i) the primary host metabolism is involved, with carbohydrate metabolism variations also related to the activation of host defense responses (Fontaine *et al.*, 2016b); ii) defense responses are activated at least at transcriptomic levels with different induction levels for the same gene according to the host growth stage, especially for Botryosphaeria dieback (Spagnolo *et al.*, 2014, 2017); and iii) these defense responses are probably not enough to avoid the pathogen colonization and disease development, especially during flowering. At flowering, grapevines showed the greatest gene induction levels but also the greatest weakness towards the two Botryosphaeria dieback pathogens, *Diplodia seriata* and *Neofusicoccum parvum* (Spagnolo *et al.*, 2014, 2017; Fontaine *et al.*, 2016b). Furthermore, these pathogens are able to disperse their inoculum at flowering, aggravating the possibility of infections at this stage (Kuntzmann *et al.*, 2009; Amponsah *et al.*, 2009; Valencia *et al.*, 2015).

The aim of the present study was to expand knowledge of physiological changes occurring in green stems of grapevines cv. Mourvèdre that were artificially infected with GTD pathogens at different grapevine growth stages. Furthermore, because of the close relationship between host plant physiology and GTD development, the effects were also assessed of a potential fungal BCA to limit Botryosphaeriaceae *in vitro* and *in planta* through dual inoculation tests. Assessed effects were of: i) artificial inoculation with the pathogen *N. parvum*; and ii) dual inoculation with *N. parvum* and the potential BCA *Fusarium proliferatum*, in greenhouse experiments 2 months post-inoculation and in the field at different grapevine growth stages. *In vitro* biocontrol evaluation tests, pathogenicity tests and transcriptomic analyses were carried out in this study.

MATERIALS AND METHODS

In vitro *evaluation of biocontrol ability of* Fusarium proliferatum

The *F. proliferatum* strain ("Fus") used in this study was frequently isolated epiphytically and endophytically by Larignon *et al.* (2013), from tissues of symptomless, greenhouse-trained young grapevines.

This fungus is a common pathogen of maize but can also colonize trees, vegetables and other cereals (Cendoya *et al.*, 2017). A preliminary *in vitro* evaluation of the biocontrol ability of the Fus strain towards Botryosphaeriaceae included the pathogens *Diplodia seriata* (two strains: 98-1 and 99-7, described in Reis *et al.*, 2016) and *Neofusicoccum parvum* (three strains: Bourgogne (Np Bour) (Ramirez-Suero *et al.*, 2014), Np Sainte Victoire (NpSV) (Larignon *et al.*, 2015) and Np bt67 (Trotel-Aziz *et al.*, 2019).

Two different assays were carried out. The first assessed direct antagonistic effects of the potential BCA against the pathogens, and the second was to verify any antibiotic activity of the BCA's secondary metabolites. The direct antagonistic effects of the potential BCA were evaluated by dual culture assays following the adapted protocol of Bézert *et al.* (1996). A 3 mm diam. agar culture plug of the potential BCA and a plug of the respective pathogen were placed along the same diameter in a Petri dish (9 cm diam.) containing potato dextrose agar (PDA). Controls consisted in Petri dishes each inoculated with the pathogen and with a sterile PDA plug. Each experimental treatment was replicated five times. The plates were then maintained in the dark at 24 ± 1°C. The direct biocontrol activity of *F. proliferatum* was evaluated by measuring the area of mycelial growth (expressed in cm^2), recorded each 24 h until the pathogen covered the entire Petri dish agar surface in the experimental controls. Colony area was determined by measuring the colony margins, and was calculated using the ImageJ software (http://imagej.nih.gov/ij).

The antibiotic activity of *F. proliferatum* secondary metabolites was evaluated by growing the different pathogens on PDA enriched with cultural filtrate of *F. proliferatum*. Three-d-old agar plugs of *F. proliferatum* were put into flasks (one plug per flask) containing 100 mL of sterile potato dextrose broth (PDB), and was then maintained in agitation at 100 rpm for 14 d. The cultural filtrate was collected eliminating the mycelium pellets by non-sterile filtration, and was then sterilized by filtration (0.45 and 0.22 μm pore size filters in sequence), and was then added to autoclaved liquid PDA (55°C) to obtain PDA medium with concentrations of 0 (control), 25 or 50% of culture filtrate. These plates were then each inoculated with a 3 mm plug of a pathogen in the centre of the plate. Each experimental treatment was replicated three times. The antibiotic activity of the *F. proliferatum* secondary metabolites was evaluated by daily measuring of the surface of the pathogen colony up to 9 d post-inoculation (dpi). The colony surface areas were calculated using the ImageJ free software (http://imagej.nih.gov/ij).

Greenhouse assays: plant material, fungus strains and inoculations

To evaluate the *F. proliferatum* control capability *in planta*, greenhouse co-inoculation tests with *F. proliferatum* and Botryosphaeriaceae fungi were carried out. *Neofusicoccum parvum*, due to its high aggressiveness towards *Vitis vinifera*, was preferred to *D. seriata*. Also because of the *in vitro* results, the Fus biocontrol activity was evaluated towards *N. parvum* strain NpSV, whose growth was reduced in dual cultural assay and greatly inhibited by Fus culture filtrate. Four-month-old potted plants of grapevine cv. Mourvèdre were each inoculated with *F. proliferatum* strain Fus using a PDA plug from a 7-d-old colony. One (FN+7d) and two (FN+14d) weeks post-inoculation with Fus, the plants were each co-inoculated at the same internode with the *N. parvum* strain NpSV, about 2 cm above the Fus inoculation point. Control plants were inoculated, respectively, with sterile PDA, Fus alone or NpSV alone. Five plants per experimental treatment were used. Two months after pathogen inoculation, the plants were cut, and the portion of each inoculated internode was observed and internal necrosis was measure. The presence of the potential BCA and pathogen were also verified (Figure 1). Collected data

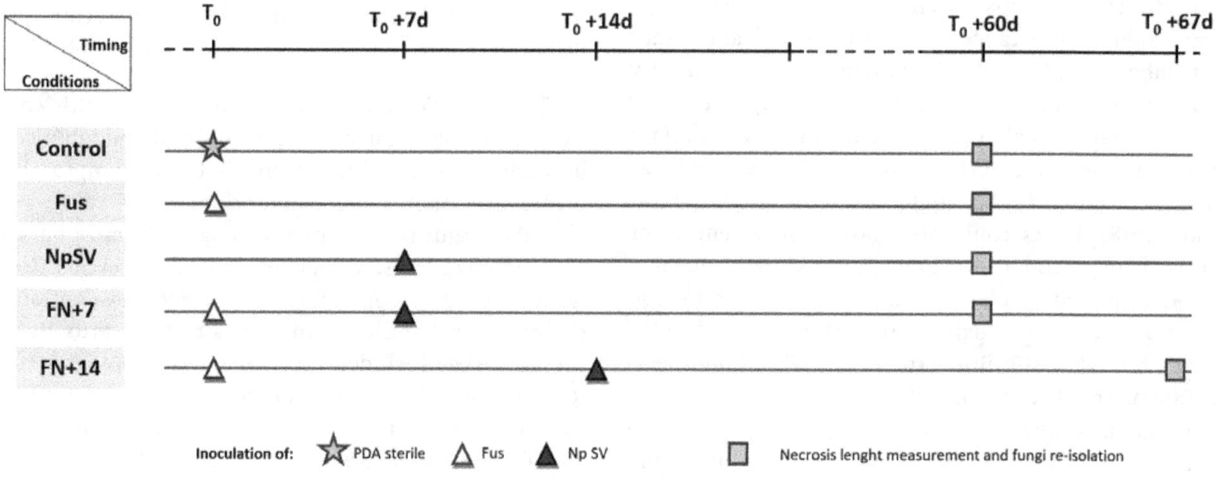

Figure 1. The co-inoculation test scheme, indicating the different conditions assayed and the timing of the inoculations BCA –pathogen, as performed in the greenhouse test.

were submitted to statistical analyses (Kruskal–Wallis and Dunn's multiple comparison tests).

Field tests: plant material, fungus strains and inoculations

To further evaluate the biocontrol activity of the Fus strain in natural conditions, and the role of host plant growth stage in disease development, a co-inoculation test was also performed in the field, as described in Reis *et al.* (2019). The experimental site was a vineyard of cv. Mourvèdre/3309, planted in 1997, and located at Rodilhan (Costières de Nîmes, France). The vineyard was owned by the Lycée agricole Marie-Durand of Rodilhan. Similar to the greenhouse essay, the Botryosphaeria dieback agent *N. parvum* strain NpSV and *F. proliferatum* strain Fus were used. These fungi were inoculated into green stems of standing vines, either alone (NpSV or Fus) or in combination (FN), using the method described by Spagnolo *et al.* (2017). In the case of the dual inoculations (FN), *N. parvum* was inoculated 7 d after Fus. The inoculations were performed at the onset of the grapevine Baggiolini phenological stages G (separated clusters), I (flowering) or M (veraison) (Baggiolini, 1952). Individual plants received the respective experimental treatments, and eight repetitions per treatment were applied. For each growth stage, non-wounded and non-inoculated green stems (C1), or stems inoculated with sterile malt agar (C2), were the experimental controls.

Observation of lesion development and re-isolation tests (for five of the 8 repetitions of each treatment) were performed at the end of vegetative season for samples from the three growth stages, as described by Larignon and Dubos (1997).

TRANSCRIPT ANALYSES

To study the effects of presence of the pathogens and the BCA on host plant physiology, inoculated stems in the vineyard were collected and submitted to transcriptomic analyses. Except for C1 stems, all the treatments were considered for transcript analysis. Samples for RNA extraction (three of 8 repetitions) were collected 20 dpi, and each consisted of the portion of the inoculated internode. For co-inoculated stems, each sample consisted of the portion of internode inoculated with both NpSV and Fus. Samples were collected, stored and processed to obtain a fine powder, according to the protocol outlined by Spagnolo *et al.* (2017).

RNA extraction

The Plant RNA Purification Reagent (Thermo Fischer Scientific Inc.) was used to extract total RNA from 2 x 50 mg of powdered green stem tissues from each assayed plant, and was DNase treated. The quality of RNA was checked by agarose gel electrophoresis, and the quantity was determined by measuring the absorbance at 260 nm.

Real-time RT-PCR analysis of gene expression

Reverse transcription was performed on 150 ng of total RNA, using the Verso cDNA synthesis kit (Thermo Fischer Scientific Inc.). Real-time PCR was performed with Absolute Blue QPCR SYBR Green (Thermo Fischer Scientific Inc.), using a CFX96 thermocycler system (Bio-Rad).

The thermal profile used was: 15s at 95°C (denaturation) and then 40 cycles each of 1 min at 60°C (annealing/extension). Melting curve assays were performed from 65–95°C at 0.5°C·s^{-1}. Melting peaks were visualized to check the specificity of each amplification. Results were expressed as the values of relative expression (ΔΔCt), corresponding to the mean of three independent experiments. The genes analyzed were considered significantly up- or down-regulated when changes in their expression were, respectively, >2-fold or <0.5-fold. The specific primers for the 16 targeted genes are listed in Table 1.

RESULTS

Evaluation of the biocontrol ability of Fusarium proliferatum

In dual culture tests, *F. proliferatum* strain Fus limited the growth of *D. seriata* and *N. parvum* strains, forming clear limits between the two touching colonies in each Petri plate after 4-7 d of co-culture, even where the differences with the controls were not statistically significant. The one exception was that of the strain Np bt67 (Figure 2, A to E). A slight decrease (7-20%) was recorded in growth of the *N. parvum* strains in dual culture with Fus, 1 d before the contact between both colonies in each plate. This effect was not recorded for the tested *D. seriata* strains (data not shown). The contact zones were often characterized by the contemporary presence of intertwining hyphae of Fus and the pathogens (Figure 2, A' and B'). Fus was also able to grow over most of the pathogens, starting at 10 dpi (Figure 2C).

Stronger Fus biocontrol activity was observed for the Fus culture filtrate (Fus CF), with differences related to

Table 1. Primers of genes analyzed by real-time reverse-transcription polymerase chain reaction.

Function	Gene	Primer Sequences	GenBank or TC TIGR* Accession Number
Housekeeping genes	*EF1* (EF1-α elongation factor)	5'-GAACTGGGTGCTTGATAGGC-3'	GU585871
		5'-AACCAAAAATATCCGGAGTAAAAGA-3'	
	60SRP (60S ribosomal protein L18)	5'-ATCTACCTCAAGCTCCTAGTC-3'	XM_002270599
		5'-CAATCTTGTCCTCCTTTCCT-3'	
Phenylpropanoid metabolism	*CHI* (Chalcone isomerase)	5'-GCAGAAGCCAAAGCCATTGA-3'	NM_001281104
		5'-GCCGATGATGGACTCCAGTAC-3'	
	PAL (Phenylalanine ammonia lyase)	5'-TCCTCCCGGAAAACAGCTG-3'	X75967
		5'-TCCTCCAAATGCCTCAAATCA-3'	
	POX4 (Peroxidase-like 4)	5'-AACATCCCCCCTCCCACTT-3'	XM_002269882
		5'-TGCATCTCGCTTGGCCTATT-3'	
	STS (Stilbene synthase)	5'-AGGAAGCAGCATTGAAGGCTC-3'	FJ851185
		5'-TGCACCAGGCATTTCTACACC-3'	
Defense protein	*CHV5* (Chitinase class V)	5'-CTACAACTATGGCGCTGCTG-3'	AF532966
		5'-CCAAAACCATAATGCGGTCT-3'	
	GLUC (β-1,3 glucanase)	5'-TCAATGGCTGCAATGGTGC-3'	DQ267748
		5'-CGGTCGATGTTGCGAGATTTA-3'	
	PPO (Polyphenol oxidase)	5'-TGGTCTTGCTGATAAGCCTAGTGA-3'	XM_002727606
		5'-TCCACATCCGATCGACATTG-3'	
	PR6 (Serine-protease inhibitor 6)	5'-AGGGAACAATCGTTACCCAAG-3'	AY156047
		5'-CCGATGGTAGGGACACTGAT-3'	
	SAMS (S-adenosylmethionine synthetase)	5'-CCTGAAATCAAAGTTCTCCTTCACA-3'	XM_002266322
		5'-CCGGGCCTGAAATCAAAGTT-3'	
	TL (Thaumatin-like)	5'-CCTAACACCTTAGCCGAATTCGC-3'	AF532965
		5'-GGCCATAGGCACATTAAATCCATC-3'	
Detoxification and Stress tolerance	*epoxH2* (Epoxide hydrolase 2)	5'-TCTGGATTCCGAACTGCATTG-3'	XM_002270484
		5'-ACCCATGATTAGCAGCATTGG-3'	
	GST5 (Glutathione s-transferase 5)	5'-GCAGAAGCTGCCAGTGAAATT-3'	XM_002277883
		5'-GGCAAGCCATGAAAGTGACA-3'	
	HSP (alpha crystalline heat shock protein)	5'-TCGGTGGAGGATGACTTGCT-3'	XM_002272382
		5'-CGTGTGCTGTACGAGCTGAAG-3'	
	SOD (Superoxide dismutase)	5'-GTGGACCTAATGCAGTGATTGGA-3'	AF056622
		5'-TGCCAGTGGTAAGGCTAAGTTCA-3'	
Primary metabolism	*PglyDH* (Phosphoglycerate dehydrogenase)	5'-CGTCGAAGATGCTCAATGATGA-3'	XM_002285322
		5'-CCCCCACGAGCAACATTAATT-3'	
Water stress	*TIP1* (Tonoplast intrinsic protein)	5'-ATCACCAACCTCATTCATATGC-3'	AF271661
		5'-GTTGTTGTCTCAACCCATTTCC-3'	

* see http://www.jcvi.org/cms/research/projects/tdb/overview/

the Botryosphaeriaceae species and the strain (Figure 3, A and B). *Diplodia seriata* strains showed similar reactions to the Fus CF, with a severe growth reduction even from the lowest aliquot of filtrate. Growth of the Ds99-7 strain was inhibited until 3 dpi on 25% Fus CF and 7 dpi on 50% filtrate. This strain then had abnormal and slow growth with sparse hyphae (Figure 4). Different from *D. seriata*, *N. parvum* growth in Fus CF-amended media varied according to the strain. Fus CF showed the greatest activity against NpSV growth compared to the control, with statistically significant reductions at both 25 and 50% filtrate. In contrast, there was no statisti-

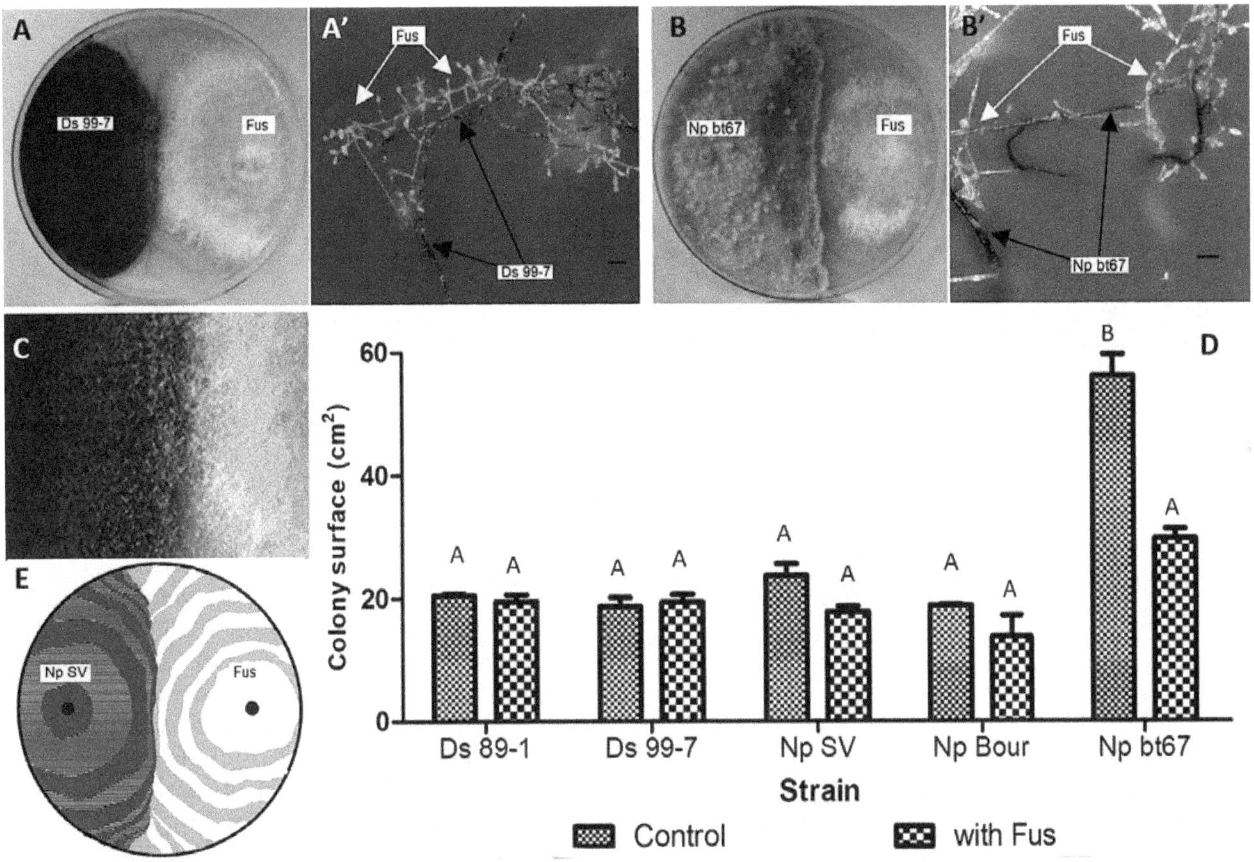

Figure 2. *In vitro* biocontrol activity of *Fusarium proliferatum* strain Fus on the tested Botryosphaeriaceae in dual culture tests. In A and B the limited growth of Ds99-7 and Np bt67 strains was noted after 4 - 7 d of dual inoculation with *F. proliferatum*. In A' and B', the respective zones of confrontation, observed with a microscope, often showed intertwining hyphae of the two strains (indicated by arrows). In C, the Fus strain are growing as white flocks over the Botryosphaeriaceae, as observed at 10 dpi. In D, the mean colony surface (in $cm^2 \pm SE$; two-way ANOVA + Bonferroli test), of the five strains compared to the controls and recorded on the day when the colonies touched, as calculated using a graphical representation of the co-cultures (E). Scale bars = 20μm. dpi = day post inoculation. Columns accompanied by the same letter are not significant different ($P \leq 0.05$).

cally significant difference the Np Bour growth values, although Fus CF growth was stimulated growth at 25% of filtrate but inhibited growth at 50%. Intermediated response was recorded for the Np Bt67 strain, which was significantly inhibited only when growing on plates with the 50% of Fus CF.

Greenhouse pathogenicity tests

The Fus and NpSV strains were re-isolated from the edges of internal green stem lesions at 60 dpi. No fungi were isolated from vines that received sterile PDA plugs. Both the strains induced lesions starting from the inoculation points, while no lesions occurred in the control plants. The presence of *F. proliferatum* strain Fus in FN+7d co-inoculations gave a significant reduction ($P \leq$ 0.05) in the lesion length caused by NpSV. In contract, NpSV gave increased of lesion lengths in co-inoculated vines when the pathogen was inoculated 14 d after inoculation with Fus (FN+14d), although the difference was statistically significant compared to control (Figure 5)

Field pathogenicity tests

The pathogen *N. parvum* and the potential BCA *F. proliferatum* were always re-isolated from the edges of the lesions associated with their inoculations, so the postulates associating disease and effects were fulfilled. No fungi were isolated from the lesions of control stems C2, indicating the lesions developed as a consequence of the wound in inoculated conditions. Mean lesion sizes associated with *F. proliferatum* in the single (Fus) and dual

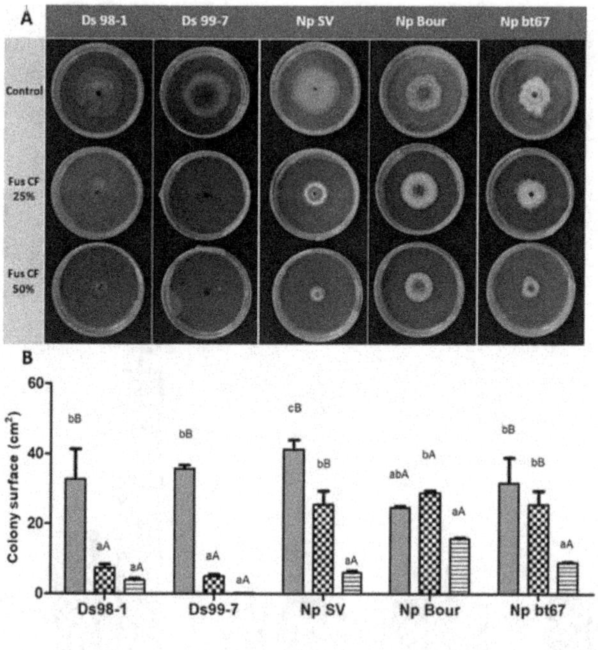

Figure 3. Biological control activity of the two different culture filtrate aliquots of *F. proliferatum* strain Fus (Fus CF) against growth of Botryosphaeriaceae fungi at 7 dpi. In B), the ANOVA results (two-way + Bonferroli test) on mean colony areas (cm^2) indicated a different Fus CF effect, for species and strain. Columns accompanied by the same letters are not significantly different (lowercase, $P \leq 0.05$, uppercase $P \leq 0.01$).

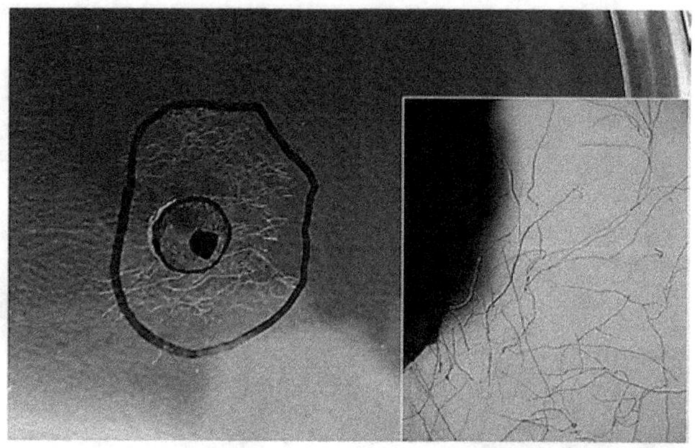

Figure 4. Detail of inhibited growth of strain Ds 99-7 on PDA plates amended with 50% of Fus cultural filtrate (Fus CF), observed with a microscope at 8 dpi. The mycelium rarely formed and had sparse hyphae.

inoculations with the pathogen (FN) were never statistically different from the controls (C2). Statistical differences were recorded for *N. parvum* in all the three vine phenological stages for single inoculation, and at the I and M stages for the in dual inoculations. At G stage, the longest lesions were recorded for Np condition (mean = 10.8 mm (± 1.3 mm)). At stages I (flowering) and M (veraison) longer lesions were associated with the pathogen from single (NpSV) and dual (FN) inoculations. At flowering, *N. parvum* produced longer lesions in the presence of Fus (mean = 24.0 mm (± 3.0 mm))

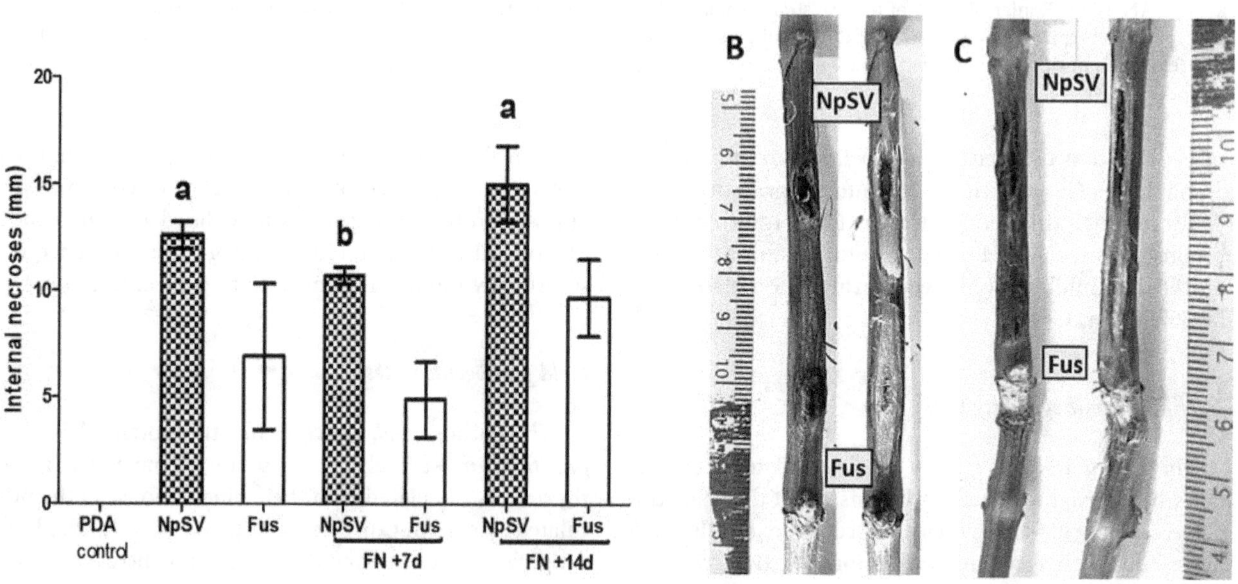

Figure 5. In A, mean lesion lengths (± SE) on rooted grapevine cv. Mourvèdre plants after inoculation with *N. parvum* strain NpSV and *F. proliferatum* strain Fus, at 60 dpi for single (NpSV, Fus) or dual inoculations (carried out in the greenhouse. Differences among the means were evaluated by the Mann-Whitney non-parametric test. Bars accompanied by the same letter are not statistically significant different ($P \leq 0.05$). In B and C, external and internal lesions produced by Fus and NpSV in FN+7d and in FN+14d co-inoculated plants.

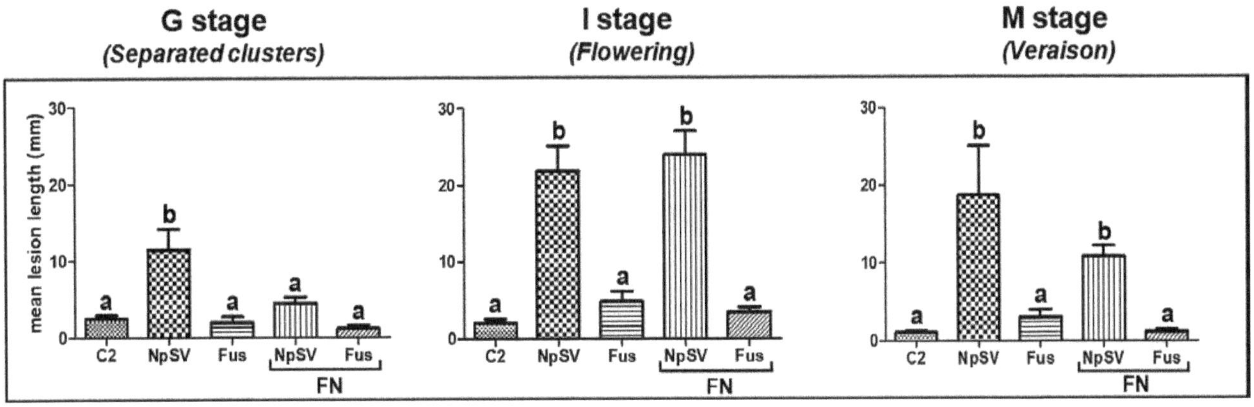

Figure 6. Mean lesion lengths (± SEs) on green grapevine stems after inoculation with *N. parvum* strain Saint Victoire and *F. proliferatum* strain Fus at the onset of the G, I and M stages, after single (NpSV, Fus) or dual (Inoculation. Control stems (C2) were wounded and inoculated with sterile malt agar. Differences among the means were evaluated by the Dunn's Multiple Comparison Test, following which the null hypothesis (equal means) was rejected in the Kruskal–Wallis test, assuming significance of $P \leq 0.05$. Columns accompanied by the same letters are not significantly different ($P \leq 0.05$).

than when inoculated alone (mean = 21.8 (± 3.3 mm)), although this difference was not statistically significant ($P > 0.05$). In contrast, at M stage (veraison), lesion lengths were greater from the *N. parvum* single inoculation (mean = 18.7 mm (± 3.4 mm)) than from the dual inoculations (mean = 10.8 mm (± 1.3 mm)) (Figure 6).

Characterization on targeted plant responses

The transcript analyses were used to evaluate the influence of the potential BCA *F. proliferatum* on host plant defense responses towards the Botryosphaeria dieback pathogen *N. parvum*. The genes were selected based on similar previous studies (Spagnolo *et al.*, 2014, 2017; Magnin-Robert *et al.*, 2011; 2014; 2016). A total of 16 genes encoding components of the phenylpropanoid pathway (genes *CHI, PAL, POX4, STS*), Pathogenesis-related (PR) and other defense proteins (*CHV5, GLUC, PPO, PR6, SAMS, TL*), proteins involved in detoxification processes (*epoxH2, GTS5, HSP, SOD*) and in primary metabolism (*PglyDH*) or water stress (*TIP1*), were chosen measure a profile of the grapevine response.

As already observed in previous studies (Spagnolo *et al.*, 2014; 2017), most of changes in the expression levels concerned the phenylpropanoid pathway and *PR*-genes, especially at the G and I vine growth stages. At these stages maximum levels of gene induction were detected in stems co-inoculated with *N. parvum* and *F. proliferatum* (FN) (Figure 7). For example, the relative expression of gene *STS* increased by up to 34-fold in FN at the G stage, and expression of *PR6* was increased almost 33-fold in FN at I stage. At flowering, a more homogenous background of upregulation was observed, determined by increased induction from both the Np and Fus inoculations, and a simultaneous decrease in some genes from the FN treatments. A general decrease in gene induction occurred at veraison (M stage), with slight upregulations occurring mostly in the co-inoculated stems (FN). Among the stress tolerance and detoxification genes, only *GST5* at the G stage and *HSP* at the I stage were slightly induced, especially in presence of the BCA either alone or in presence of the pathogen. No genes were down regulated by any of the treatments.

DISCUSSION

Antagonistic effects of Fusarium proliferatum *against Botryosphaeriaceae species*

Fusarium proliferatum strain Fus showed antagonistic effects towards *N. parvum* and *D. seriata* in the laboratory tests. The *in vitro* dual culture test highlighted the role of the potential BCA in limiting growth of the tested Botryosphaeriaceae pathogens, which were unable to overgrow the antagonist Fus when in contact with its mycelium. The slight inhibition recorded 1 d before confrontation with Fus, observed only for *N. parvum* due to the slower growth rate compared to *D. seriata*, suggested a role of antibiosis. Antibiosis was demonstrated from the results of culture filtrate test. *Fusarium proliferatum* metabolites limited the growth of the pathogens, with greater and more homogeneous inhibition of the *D. seriata* strains than of the *N. parvum* strains. The different strains of *N. parvum* showed different amounts of

Function	Gene	G stage (Separated clusters)			I stage (Flowering)			M stage (Veraison)		
		NpSV	Fus	FN	NpSV	Fus	FN	NpSV	Fus	FN
Phenyl-propanoid metabolism	CHI	1,16	0,79	1,69	2,04	2,21	2,24	1,05	0,92	1,26
	PAL	2,22	1,21	3,52	0,66	1,20	0,99	0,96	2,05	1,99
	POX4	1,81	1,07	3,45	2,68	1,48	2,02	1,11	1,37	1,53
	STS	5,71	3,08	34,07	7,13	7,50	9,27	1,06	0,95	2,19
Defense protein	CHV5	3,39	3,61	13,78	7,70	7,51	13,21	2,96	1,59	2,43
	GLU	4,69	0,96	23,30	4,70	6,80	14,29	1,93	1,16	2,71
	PPO	1,09	1,47	3,41	3,64	2,22	5,06	2,68	2,51	2,87
	PR6	3,57	2,40	7,35	12,63	25,75	32,73	0,88	0,62	1,06
	SAM	1,16	0,97	1,57	1,21	1,48	1,51	0,95	1,23	1,62
	TL	2,85	3,01	7,39	7,31	4,08	8,40	2,44	4,63	2,13
Detoxification and Stress tolerance	epoxH2	1,16	0,84	1,24	1,44	1,13	1,15	0,87	1,13	1,39
	GST5	1,93	1,18	3,60	0,85	1,47	1,26	0,80	1,01	1,24
	HSP	1,19	1,02	1,00	0,60	2,85	3,46	1,09	1,13	1,17
	SOD	1,07	0,88	1,23	1,23	1,31	1,24	0,81	0,98	1,01
Primary metabolism	PglyDH	1,56	1,05	1,36	0,62	1,13	1,47	0,82	1,05	1,40
Water stress	TIP1	0,94	0,90	1,01	3,04	1,40	1,81	0,73	0,96	0,99

Down-regulated <0,5 1 >2 Up-regulated

Figure 7. Expression levels of selected genes (determined with quantitative reverse-transcription polymerase chain reactions). The values (each the mean of three technical replicates) represent expression levels (ΔΔCt) of treatments relative to the controls (C2). The expression of each gene was considered up-regulated when the value was > 2-fold compared to the controls, or down-regulated when the value was < 0.5-fold compared to the controls.

growth inhibition, with NpSV as the most susceptible to Fus CF at both tested culture filtrate concentrations. *Fusarium proliferatum* was able to grow over the pathogens at 10 dpi, which also indicated direct antagonism effects, as reported previously for *F. proliferatum* and other *Fusarium* species. For instance, *F. proliferatum* was shown to be effective against the oomycete *Plasmopara viticola*, which causes grapevine downy mildew (Falk *et al.*, 1996). *Fusarium sporotrichioides* inhibited *in vitro* the growth of *Phoma negriana*, responsible of lesions and necrosis on grapevine shoots (Krol, 2008), while *F. lateritium* was shown to metabolize toxins produced by *Eutypa lata*, a cause of Eutypa dieback (Christen *et al.*, 2005). Some of the mechanisms involved in the *Fusarium* spp. biocontrol, such as mycoparasitism (Falk *et al.*,1996), extracellular β-glucosidase and endo-1,4-β-glucanase production (Bakshi *et al.*, 2001) and cyanide degradation (Christen *et al.*, 2005), were also found in other BCAs such as *Trichoderma* spp., *Ampelomyces quisqualis*, *Penicillium purpurogenum*, *Coniothyrium minitans* and *Pseudomonas fluorescens* (Haran *et al.*, 1996; Kubicek *et al.*, 1993; Rotem *et al.*, 1999; Whipps *et al.*, 2008; Winding *et al.*, 2004).

Fusarium proliferatum *reduced the necrosis produced by* N. parvum *strain NpSV* in planta

Under controlled conditions, *F. proliferatum* also showed biocontrol activity, reducing *N. parvum* strain NpSV aggressiveness when inoculated 7 d before the pathogen. In contrast, presence of *F. proliferatum* tended to increase the size of pathogen internal necroses when inoculated 14 d previously, although this effect was not statistically significant. In co-inoculations performed in the field, *F. proliferatum* showed biocontrol activity by reducing lesion sizes associated to *N. parvum* at the G (separated clusters) and, to a lesser extent, at M (veraison) grapevine stages. In contrast, the mean sizes of lesions associated with *N. parvum*, from single and dual inoculations, was greater at flowering, confirming the hypothesis that flowering is the period when the grapevine cv. Mourvèdre is more susceptible to Botryosphaeria dieback pathogens. As reported in previous studies (Spagnolo *et al.*, 2014; 2017), this susceptibility could be determined by high host plant metabolic activity focused towards inflorescence development. This metabolic re-orientation developing reproductive organs could have also determined the inefficient biocontrol of *F. proliferatum* at I growth stage, which was unexpected considering the relative transcriptomic data. Beside biocontrol capabilities towards *N. parvum*, *F. proliferatum* was also not pathogenic to grapevine cv. Mourvèdre, although in greenhouse inoculations the strain induced longer lesions than in the field. These differences may be explained by the different host types used (cuttings *vs* standing vines), and the less stressing environmental conditions of the greenhouse compared to the field, which may have modulated the *F. proliferatum* behaviour. Greenhouse conditions could have also favored aggressiveness *F. proliferatum* and *N. parvum* in unstressed plants which were less reactive to artificial infections.

Changes induced by Fusarium proliferatum *to the grapevine-*Neofusicoccum parvum *interaction*

This also focused on the physiological changes occurring in annual stems in the host/pathogen/BCA tritrophic interaction of grapevine/*N. parvum*/*F. proliferatum*, at three host growth stages.

Fusarium proliferatum strain Fus weakly induced genes at the G stage (genes *STS*, *CHV5*, *PR6*, *TL*) and the M stage (*PAL*, *PPO*, *TL*). This indicated host plant tolerance towards *F. proliferatum* and supported the observed in field non-pathogenicity of this strain to *V. vinifera* cv Mourvèdre. In contrast, at the I stage Fus induced genes at levels equivalent to those of the pathogen *N. parvum*. This indicated a possible switch at flowering, from "neutral" to "parasitic" perception of *F. proliferatum* by the host defense system (Kogel *et al.*, 2006). The greatest induction recorded for the most of genes in all FN treatments suggests the role of *F. proliferatum* strain Fus in priming the host defense system towards pathogen infections, which resulted in reduced necrosis from *N. parvum* strain NpSV observed at the G stage and to a lesser extent, at the M stage. This priming effect, elicited by several BCAs (*Trichoderma* sp., Perazzolli *et al.*, 2011; *Pythium oligandrum*, Yacoub *et al.*, 2016) was reported for *Aureobasidium pullulans* strain Fito_278 (Pinto *et al.*, 2018) and *Bacillus subtilis* PTA-271 (Aziz *et al.*, 2015; Trotel-Aziz *et al.*, 2019) towards pathogens associated Botryosphaeria dieback.

Changes in the grapevine-*N. parvum* interaction determined by the potential BCA were characterized by gene expression associated either with plant defense/stress responses (secondary metabolism) or with energy metabolism (glycolysis, Krebs cycle). In agreement with Spagnolo *et al.* (2014), most of the modifications detected in the present study were more related to the host phenological stage than to a particular inoculation tested. Fasoli *et al.* (2012) also highlighted the importance of grapevine growth stage for gene expression. Gene expression in different organs is probably based on the host developmental stage rather than on organ type.

The over-expression of the genes *CHI*, *PAL*, *POX4* and especially *STS* at both the G and I stages on the infected stems from the NpSV and FN treatments could be related to activation of the phenylpropanoid pathway linked to plant defense. Phenylpropanoids are related in plant defense responses, playing preformed or inducible functions, forming physical or chemical barriers to infection, or local and/or systemic signalling for the defense gene induction (Dixon *et al.*, 2002). Similarly, stilbenes may limit development of GTD fungi in wood. *In vitro* tests have shown ability of different phenolic compounds to limit growth of several GTD-associated pathogens in the Botryosphaeriaceae and *Phaeomoniella chlamydospora* (Fontaine *et al.*, 2016b; Lambert *et al.*, 2012; Lima and Dias, 2012). Furthermore, the stilbene resveratrol could also act as a signaling molecule in the activation of defense-related responses in *Vitis* cells (Chang *et al.*, 2011).

The induction of *GLUC* expression, recorded in single or dual infected stem treatments, especially at the G (NpSV and FN) and I (NpSV, Fus and FN) growth stages, could be related to the multiple role of the related proteins. β-1,3-glucanases are abundant in plants, and are involved in cell division, movement of materials through plasmodesmata and in plant resistance towards abiotic stresses. These proteins also protect plants against fungal pathogens, alone or in association with other antifungal proteins such as chitinases or isoenzymes (Balasubramanian *et al.*, 2012). The synergistic effects between β-1,3-glucanases and chitinases towards fungal infections is frequently observed in other crops (including pea, bean, tomato, tobacco, maize, soybean, potato, and wheat), and their presence led to increased disease resistance in plants (Balasubramanian *et al.*, 2012; Saboki *et al.*, 2011; Jach *et al.*, 1995). In the present study, *GLUC* induction was always coupled with increases in chitinase *CHV5* at the G growth stage (NpSv and FN treatments), the I stage (all treatments) and the M stage (FN treatment). At the transcriptomic level, this synergism of *Gluc/CHV5* could be also have occurred in green stems cv. Mourvèdre infected by *F. proliferatum* and Botryosphaeriaceous pathogens, either alone or in combination, as observed here and in the previous study Spagnolo *et al.* (2017). For defense protein genes, the observed high induction of *PR6* genes at flowering (the I stage) from all treatments (NpSV, Fus, and FN) has been previously reported in grapevines naturally affected by

GTD pathogens (Magnin-Robert *et al.*, 2014; Falk *et al.*, 1996; Magnin-Robert *et al.*, 2014; Valtaud *et al.*, 2009; Letousey *et al.*, 2010) and in artificially inoculated plants (Camps *et al.*, 2010), and has been described in grapevine cells in the presence of extracellular compounds produced by Botryosphaeria dieback pathogens (Ramirez-Suero *et al.*, 2014). The strong induction of defense genes (*PR6, PPO, TL, GLU, CHV5*) observed in FN-treated vines at the I growth stage, was either unrelated to the reduction of lesion sizes observed at G and M stages, or not great enough to limit lesion development at flowering. The implication of inflorescences reducing responsiveness to pathogen attack in green stems (Spagnolo *et al.*, 2014; 2017), supports the observation of little biological control effect from *F. proliferatum* at flowering.

CONCLUSIONS

Results from this study agree with previous observations, and confirm the importance of flowering in the defense of grapevine cv. Mourvèdre towards the development of *N. parvum* infections, even in the presence of the potential BCA *F. proliferatum*. This fungus limited the growth of pathogens both through antibiosis and direct antagonism *in vitro*, and development of pathogen-induced necrosis *in planta*, by priming effects enhancing some plant responses to infection. Based on these results, further studies should assess the physiological mechanisms that could influence host metabolism and responses to biotic stress during flowering.

The results observed in the *in planta* dual inoculation assays suggest further analysis of possible use of *F. proliferatum* as BCA is warranted, since the priming of plant defense response is an efficient and low cost way to improve resistance to abiotic and biotic stresses (van Hulten *et al.*, 2006). Toxicological studies should also be carried out when considering this fungus as a possible BCA against GTDs, as variable toxigenic potential was detected in *F. proliferatum* populations by Stepien *et al.* (2011). This would reduce the risk that toxic fungal metabolites would enter human or animal food chains. Furthermore, the role of *F. proliferatum* as opportunistic human pathogen should be also considered, since it and other *Fusarium* species are reported to be associated with infections of immunocompromised patient (Summerbell *et al.*, 1988; Tulin, 2018).

ACKNOWLEDGMENTS

The National Program FranceAgrimer financed this research. The authors thank Nancy Terrier (INRA-UMR Sciences of enology, Montpellier) for her help during vineyard sampling.

Conflicts of Interest: The authors declare no conflict of interest.

LITERATURE CITED

Abou-Mansour E., Débieux J.L., Ramírez-Suero M., Bénard-Gellon M., Magnin-Robert M., ... P. Larignon, 2015. Phytotoxic metabolites from *Neofusicoccum parvum*, a pathogen of Botryosphaeria dieback of grapevine. *Phytochemistry* 115: 207–215.

Amponsah N.T., Jones E.E., Ridgway H.J.M.V. Jaspers, 2009. Rainwater dispersal of *Botryosphaeria* conidia from infected grapevines. *New Zealand Plant Protection* 62: 228–233.

Aziz A., Verhagen B., Villaume S., Höfte M., Baillieul F., ... Trotel-Aziz P., 2015. Beneficial Bacteria Prime Local and Systemic Immunity against *Botrytis cinerea* in Grapevine. In: *Biocontrol of Major Grapevine Diseases*, 64–69.

Baggiolini M., 1952. Les stades repères dans le développement annuel de la vigne et leur utilisation pratique. *Revue romande d'agriculture, de viticulture et d'arboriculture* 8 : 4–6. (In French)

Bakshi S., Sztejnberg A., Yarden O., 2001. Isolation and characterization of a cold-tolerant strain of *Fusarium proliferatum*, a biocontrol agent of grape downy mildew. *Phytopathology* 91: 1062–1068.

Balasubramanian V., Vashisht D., Cletus J., Sakthivel N., 2012. Plant beta-1,3-glucanases: Their biological functions and transgenic expression against phytopathogenic fungi. *Biotechnology Letters*, 34: 1983–1990.

Berg, G., 2009. Plant-microbe interactions promoting plant growth and health: Perspectives for controlled use of microorganisms in agriculture. *Applied Microbiology and Biotechnology* 84: 11–18.

Bertsch C., Ramírez-Suero M., Magnin-Robert M., Larignon P., Chong J., ... Fontaine F., 2013. Grapevine trunk diseases: Complex and still poorly understood. *Plant Pathology* 62: 243–265.

Bézert G., Chappe P., Mourey A., Loubinoux B., 1996. Action de *Bacillus* et d'actinomycètes sur les champignons du bleuissement du bois. *Bulletin des Académie et Société Lorraines des Sciences* 35 : 177–190. (In French)

Burruano S., Giambra S., Mondello V., Della Greca M., Basso S., ... Andolfi A., 2016. Naphthalenone polyketides produced by *Neofusicoccum parvum*, a fungus associated with grapevine Botryosphaeria dieback. *Phytopathologia Mediterranea* 55: 197–206.

Camps C., Kappel C., Lecomte P., Leon C., Gomes E., ... Delrot S., 2010. A transcriptomic study of grapevine (*Vitis vinifera* cv. Cabernet-sauvignon) interaction with the vascular ascomycete fungus *Eutypa lata*. *Journal of Experimental Botany* 61: 1719–1737.

Cendoya E., Pinson-Gadais L., Farnochi M.C., Ramirez M.L., Chéreau S., ... Richard-Forget F., 2017. Abiotic conditions leading to FUM gene expression and fumonisin accumulation by *Fusarium proliferatum* strains grown on a wheat-based substrate. *International Journal of Food Microbiology* 253: 12–19.

Chang X. L., Heene E., Qiao F., Nick P., 2011. The phytoalexin resveratrol regulates the initiation of hypersensitive cell death in *Vitis* cell. *Plos One* 6 (10) e26405.

Chethana K.W.T., Li X.H., Zhang W., Hyde K.D., Yan J.Y., 2016. Trail of decryption of molecular research on *Botryosphaeriaceae* in woody plants. *Phytopathologia Mediterranea* 55: 147–171.

Christen D., Tharin M., Perrin-Cherioux S., Abou-Mansour E., Tabacchi R., Défago G., 2005. Transformation of Eutypa Dieback and Esca Disease Pathogen Toxins by Antagonistic Fungal Strains Reveals a Second Detoxification Pathway Not Present in *Vitis vinifera*. *Journal of Agricultural and Food Chemistry* 53(18): 7043–7051.

Dixon R.A., Achnine L., Kota P., Liu C.J., Reddy M.S.S., Wang L.J., 2002. The phenylpropanoid pathway and plant defence - a genomics perspective. *Molecular Plant Pathology* 3: 371–390.

Eskalen A., Feliciano A.J., Gubler W.D., 2007. Susceptibility of Grapevine Pruning Wounds and Symptom Development in Response to Infection *by Phaeoacremonium aleophilum* and *Phaeomoniella chlamydospora*. *Plant Disease* 91: 1100–1104.

Falk S.P., Pearson R.C., Gadoury D.M., Seem R.C., Sztejnberg A., 1996. *Fusarium proliferatum* as a biocontrol agent against grape downy mildew. *Phytopathology* 86: 1010–1017.

Fasoli M., Dal Santo S., Zenoni S., Tornielli G.B., Farina L., ... Pezzotti M., 2012. The grapevine expression atlas reveals a deep transcriptome shift driving the entire plant into a maturation program. *Plant Cell* 24: 3489–3505.

Fontaine F., Gramaje D., Armengol J., Smart R., Nagy Z.A., ... Corio-Costet M-F., 2016a. Grapevine trunk diseases. A review. ©OIV publications, 1st Edition (Paris, France).

Fontaine F., Pinto C., Vallet J., Clément C., Gomes A.C., Spagnolo A., 2016b. The effects of grapevine trunk diseases (GTDs) on vine physiology. *European Journal of Plant Pathology* 144: 707–721.

Gramaje D., Úrbez-Torres J.R., Sosnowski M.R., 2018. Managing Grapevine Trunk Diseases with respect to etiology and epidemiology: Current Strategies and Future Prospects. *Plant disease* 102: 1, 12–39.

Guerin-Dubrana L., Fontaine F., Mugnai L., 2019. Grapevine trunk disease in European and Mediterra- nean vineyards: occurrence, distribution and associated disease-affecting cultural factors. *Phytopathologia Mediterranea* 58(1): 49–71. doi: 10.13128/ Phytopathol_Mediterr-25153

Handelsman J., Stabb E., 1996. Biocontrol of Soilborne Plant Pathogens. *The Plant Cell*, 8: 1855–1869.

Haran S., H. Schickler and I. Chet, 1996. Molecular mechanisms of lytic enzymes involved in the biocontrol activity of *Trichoderma harzianum*. *Microbiology* 142: 2321–2331.

Jach G., Gornhardt B., Mundy J., Logemann J., Pinsdorf E., Leah R., Schell J., C. Maas, 1995. Enhanced quantitative resistance against fungal disease by combinatorial expression of different barley antifungal proteins in transgenic tobacco. *The Plant Journal* 8: 97–109.

Kogel K.H., Franken P., Hückelhoven R., 2006. Endophyte or parasite-what decides? *Current Opinion in Plant biology* 9: 358–363

Król E., Machowicz-Stefaniak Z., 2008. Biotic effect of fungal communities inhabiting grapevine phyllosphere on *Phoma negriana*. *Biologia* 63: 466–470.

Kubicek C.P., Messner R., Gruber F., Mach R.L., Kubicek-pranz E.M., 1993. The *Trichoderma* cellulase regulatory puzzle - from the interior life of a secretory fungus. *Enzyme and Microbial Technology* 15: 90–99.

Kuntzmann P., S. Villaume and C. Bertsch, 2009. Conidia dispersal of *Diplodia* species in a French vineyard. *Phytopathologia Mediterranea* 48: 150–154.

Lambert C., Bisson J., Waffo-Teguo P., Papastamoulis Y., Richard T., ... Cluzet S., 2012. Phenolics and their antifungal role in grapevine wood decay: Focus on the *Botryosphaeriaceae* family. *Journal of Agricultural and Food Chemistry* 60: 11859–11868.

Larignon P., Dubos B., 1997. Fungi associated with esca disease in grapevine. *European Journal of Plant Pathology* 103: 147–157.

Larignon P., Baptiste C., Mallet J.F., Granier J.P., Bloy P., 2013. Greffage en vert, premier test contre les maladies du bois. *Phytoma* 33–35. (In French)

Larignon P., Spagnolo A., Bertsch C., Fontaine F., 2015. First report of young grapevine decline caused by *Neofusicoccum parvum* in France. *Plant Disease* 99 : 1859.

Letousey P., Baillieul F., Perrot G., Rabenoelina F., Boulay M., ... Fontaine F., 2010. Early events prior to visual symptoms in the apoplectic form of grapevine esca disease. *Phytopathology* 100: 424–431.

Lima M.R.M. Dias A.C.P., 2012. *Phaeomoniella chlamydospora*-induced oxidative burst in *Vitis vinifera* cell suspensions: Role of NADPH oxidase and Ca^{2+}. *Journal of Phytopathology* 160: 129–134.

Magnin-Robert M., Letousey P., Spagnolo A., Rabenoelina F., Jacquens L., ... Fontaine F., 2011. Leaf strip form of esca induces alteration of photosynthesis and defence reactions in presymptomatic leaves. *Functional Plant Biology* 38: 856–86.

Magnin-Robert M., Spagnolo A., Alayi T.D., Cilindre C., Mercier L., Schaeffer-Reiss C., ... Fontaine F., 2014. Proteomic insights into changes in wood of *Vitis vinifera* L. in response to esca proper and apoplexy. *Phytopathologia Mediterranea* 53 : 173–192.

Magnin-Robert, M., Spagnolo, A., Boulanger, A., Joyeux, C., Clément, C., ... Fontaine, F., 2016. Changes in plant metabolism and accumulation of fungal metabolites in response to esca proper and apoplexy expression in the whole grapevine. *Phytopathology* 106: 541–553.

Mondello V., Songy A., Battiston E., Pinto C., Coppin C., ... Fontaine F., 2018. Grapevine trunk diseases: a review of fifteen years of trials for their control with chemicals and biocontrol agents. *Plant Disease* 102: 1189–1217.

Mugnai L., Graniti A., Surico G., 1999. Esca (Black Measles) and Brown Wood-Streaking: Two Old and Elusive Diseases of Grapevines. *Plant Disease* 83: 404–418.

Pal K. K., Mc Spadden Gardener B., 2006. Biological Control of Plant Pathogens. *The Plant Health Instructor* 1–25.

Perazzolli M., Roatti B., Bozza E. Pertot I., 2011.*Trichoderma harzianum* T39 induces resistance against downy mildew by priming for defense without costs for grapevine. *Biological Control* 58: 1, 74–82.

Pinto C., Custódio V., Nunes M., Songy A., Rabenoelina F., ... Fontaine F., 2018. Understand the Potential Role of *Aureobasidium pullulans*, a Resident Microorganism From Grapevine, to Prevent the Infection Caused by *Diplodia seriata. Frontiers in Microbiology* 9:3047.

Ramirez-Suero M., Benard-Gellon M., Chong J., Laloue H., Stempien E., ... Bertsch C., 2014. Extracellular compounds produced by fungi associated with Botryosphaeria dieback induce differential defence gene expression patterns and necrosis in *Vitis vinifera* cv. Chardonnay cells. *Protoplasma* 251: 1417–1426

Reis P., Magnin-Robert M., Nascimento T., Spagnolo A., Abou-Mansour E., Fioretti C., ... Fontaine F., 2016. Reproducing Botryosphaeria dieback foliar symptoms in a simple model system. *Plant Disease* 100: 1071–1079.

Reis P., Pierron R., Larignon P., Lecomte P., Abou-Mansour E., ... Fontaine, F., 2019. *Vitis* Methods to Understand and Develop Strategies for Diagnosis and Sustainable Control of Grapevine Trunk Diseases. *Phytopathology*, 109: 916–931

Rotem Y., Yarden O., Sztejnberg, A., 1999. The mycoparasite *Ampelomyces quisqualis* expresses exgA encoding an exo-beta-1,3-glucanase in culture and during mycoparasitism. *Phytopathology* 89: 631–8.

Saboki E., Usha K. Singh B., 2011. Pathogenesis Related (PR) Proteins in: *Plant Defense Mechanism Age-Related Pathogen Resistance. Current Research and Technological Advances*, A. Méndez-Vilas (Ed.), 1043–1054.

Songy A., Vallet J., Gantet M., Boos A., ... Fontaine, F., 2019. Sodium arsenite effect on *Vitis vinifera* L. Physiology. *Journal of Plant Physiology* 238: 72–79.

Sosnowski M.R., Luque J., Loschiavo A.P., Martos S., Garcia-Figueres F., ... Scott E.S., 2011. Studies on the effect of water and temperature stress on grapevines inoculated with *Eutypa lata. Phytopathologia Mediterranea* 50: S127–S138.

Spagnolo A., Larignon P., Magnin-Robert M., Hovasse A., Cilindre C., ... Fontaine F., 2014. Flowering as the most highly sensitive period of grapevine (Vitis vinifera L. cv Mourvèdre) to the Botryosphaeria dieback agents *Neofusicoccum parvum* and *Diplodia seriata* infection. *International Journal of Molecular Sciences* 15: 9644–9669.

Spagnolo A., Mondello V., Larignon P., Villaume S., Rabenoelina F., ... Fontaine F., 2017. Defense responses in grapevine (cv. Mourvèdre) after inoculation with the Botryosphaeria dieback pathogens *Neofusicoccum parvum* and *Diplodia seriata* and their relationship with flowering. *International Journal of Molecular Sciences* 18: 393.

Stepien L., Koczyk G., Waskiewicz A., 2011. Genetic and phenotypic variation of *Fusarium proliferatum* isolates from different host species. *Journal of Applied Genetics* 52: 487–496.

Summerbell R.C., Richardson S.E., Kane J., 1988. *Fusarium proliferatum* as an agent of disseminated infection in an immunosuppressed patient. *Journal of Clinical Microbiology* 26: 82–87.

Trotel-Aziz P., Abou-Mansour E., Courteaux B., Rabenoelina F., Clément C., ... Aziz A., 2019. *Bacillus subtilis* PTA-271 Counteracts Botryosphaeria Dieback in Grapevine, Triggering Immune Responses and Detoxification of Fungal Phytotoxins. *Frontiers in Plant Sciences* 10: 25.

Tulin A., 2018. *Fusarium* - Pathogenicity, Infections, Diseases, Mycotoxins and Management. In: Plant Diseases, Pathogen Diversity, Genetic Diversity, Resistance and Molecular Markers. Edited by Tulin A. https://www.intechopen.com/books/fusarium-plant-diseases-pathogen-diversity-genetic-diversity-resistance-and-molecular-markers

Úrbez-Torres J.R., 2011. The status of *Botryosphaeriaceae* species infecting grapevines. *Phytopathologia Mediterranea* 50: S5-S45.

Valencia D., Torres C., Camps R., López E., Celis-Diez J., Besoain X., 2015. Dissemination of *Botryosphaeriaceae* conidia in vineyards in the semiarid Mediterranean climate of the Valparaíso Region of Chile. *Phytopathologia Mediterranea* 54: 394–402.

Valtaud C., Foyer C.H., Fleurat-Lessard P., Bourbouloux A., 2009. Systemic effects on leaf glutathione metabolism and defence protein expression caused by esca infection in grapevines. *Funcional Plant Biology* 36: 260–279.

van Hulten M., Pelser M., van Loon L.C., Pieterse C.M.J., Ton J., 2006. Costs and benefits of priming for defense in *Arabidopsis*. *Proceedings of the National Academy of Sciences of the United States of America* 103: 5602–5607

Van Niekerk J.M., Strever A.E., Gerhard Du Toit P., Halleen F., Fourie P.H., 2011. Influence of water stress on *Botryosphaeriaceae* disease expression in grapevines. *Phytopathologia Mediterranea* 50: S151–S165.

Whipps J. M., Sreenivasaprasad S., Muthumeenakshi S., Rogers C.W. Challen M.P., 2008. Use of *Coniothyrium minitans* as a biocontrol agent and some molecular aspects of sclerotial mycoparasitism. *Sustainable Disease Management in a European Context*: 323–330.

Winding A., Binnerup S.J. Pritchard H., 2004. Non-target effects of bacterial biological control agents suppressing root pathogenic fungi. *FEMS Microbiology Ecology* 47: 129–141.

Yacoub A., Gerbore J., Magnin N., Chambon P., Dufour, … Rey P., 2016. Ability of *Pythium oligandrum* strains to protect Vitis vinifera L., by inducing plant resistance against Phaeomoniella chlamydospora, a pathogen involved in Esca, a grapevine trunk disease. *Biological Control* 92: 7–16.

6

Soil-borne diseases caused by *Fusarium* and *Neocosmospora* spp. on ornamental plants in Italy

Vladimiro GUARNACCIA[1,2], Dalia AIELLO[3,*], Giancarlo POLIZZI[3], Pedro W. CROUS[1,4], Marcelo SANDOVAL-DENIS[1,4]

[1] *Westerdijk Fungal Biodiversity Institute, Uppsalalaan 8, 3584 CT, Utrecht, The Netherlands*
[2] *DiSAFA, University of Torino, Largo Paolo Braccini, 2, 10095 Grugliasco, TO, Italy*
[3] *Dipartimento di Agricoltura, Alimentazione e Ambiente, Università degli Studi di Catania, Via S. Sofia 100, 95123 Catania, Italy*
[4] *Faculty of Natural and Agricultural Sciences, Department of Plant Sciences, University of the Free State, P.O. Box 339, Bloemfontein 9300, South Africa*
*Corresponding author: dalia.aiello@unict.it

Competing Interests: The Author(s) declare(s) no conflict of interest.

Editor: Lizel Mostert, Faculty of AgriSciences, Stellenbosch, South Africa.

Abstract. During surveys conducted in 2010–2014, several disease symptoms were observed on eight ornamental plant species in different nurseries located in Sicily (Southern Italy). Two *Neocosmospora* and 31 *Fusarium* isolates were recovered from symptomatic plants. Fungus identification was based on morphological characteristics and phylogenetic analyses of fragments of the intergenic spacer (IGS), internal transcribed spacer (ITS) and large subunit (LSU) regions of the rDNA; partial fragments of the beta-tubulin (*TUB*), RNA polymerase largest subunit (*RPB1*), RNA polymerase second largest subunit (*RPB2*) and translation elongation factor 1-alpha (*EF-1α*) genes. The *Fusarium* species recovered from ornamental plants were *F. agapanthi* (from *Agapanthus africanus*), *F. anthophilum* (*Dasylirion longissimum*), *F. fujikuroi* (*Trachycarpus princeps*), *F. oxysporum* (*Bougainvillea glabra*, *Cordyline australis* 'Purpurea', *Dasylirion longissimum*, *Eremophila laanii* and *Philoteca myoporoides*) and *F. proliferatum* (*T. princeps*), while *N. solani* was isolated from crown and root rot of *Ficus carica*. The pathogenicity of representative isolates collected from each host was tested on seedlings or cuttings grown in a growth chamber. All the *Fusarium* and *Neocosmospora* isolates tested were pathogenic and reproduced symptoms identical to those observed in the field, except for *F. fujikuroi* on *T. princeps* and *F. oxysporum* on *D. longissimum* that were non-pathogenic.

Keywords. Morphology, multigene phylogeny, pathogenicity, root rot, wilt.

INTRODUCTION

During the last decade, Italy has significantly increased production of ornamental plants in nurseries, and several new species and products have been introduced for cultivation in greenhouses and open fields. Movement of ornamental plants through the peninsula led to the spread of pathogens to new areas, and introduction of new pathogens from abroad (Gullino and Garibaldi, 2006; Polizzi *et al.*, 2012; Aiello *et al.*, 2017, 2018).

In Sicily (Southern Italy), production of ornamentals has increased in the eastern area, where it replaced lemon orchards due to decline in demand for these fruits. Plant growth in nurseries is compromised by several foliar and root diseases, and among these diseases those caused by species of Nectriaceae are exceptionally common (Polizzi *et al.*, 2007; Vitale *et al.*, 2009; Aiello *et al.*, 2014, 2015; Gullino *et al.*, 2015).

Fusarium Link *sensu lato* was recently segregated into several *Fusarium*-like genera (i.e., *Bisifusarium* L. Lombard, Crous & W. Gams [*Fusarium dimerum* species complex (SC)], *Neocosmospora* E.F. Sm. [*Fusarium solani* SC] and *Rectifusarium* L. Lombard, Crous & W. Gams [*Fusarium ventricosum* SC]). These taxa are among the most important human, animal or plant pathogens, affecting an extensive variety of hosts (O'Donnell *et al.*, 2008, 2010; Lombard *et al.*, 2015). *Fusarium* and *Fusarium*-like genera are well-known as responsible for diseases on ornamental plants, including flowering crops, herbaceous ornamentals such as begonia, carnation and chrysanthemum, woody ornamentals such as *Bougainvillea*, *Hebe*, *Hibiscus* and *Pyracantha* spp. (Horst and Nelson, 1997; Sinclair and Lyon, 2005; Polizzi *et al.*, 2010a, 2010b, 2011; Bertoldo *et al.*, 2015; Lupien *et al.*, 2017), and palms such as *Arecastrum*, *Phoenix* and *Washingtonia* spp. (Elliott *et al.*, 2004).

Considering the importance of diseases caused by *Fusarium*-like fungi, the high economic losses caused by these pathogens and the relevance of these crops, surveys were conducted over a 5-year period in ornamental nurseries located in the Catania province, eastern Sicily, Italy. During the surveys conducted from 2010 to 2014, large numbers of palms, perennial herbaceous shrubs, and young cuttings were detected showing symptoms of crown and root rots, damping-off, wilt and dieback. The aims of the present study were to identify the Fusaria obtained from these affected ornamentals, using morphological characteristics and DNA sequence analyses, and to evaluate the pathogenicity of representative isolates on the hosts from which they were isolated.

MATERIALS AND METHODS

Field sampling and pathogen isolations

During 2010–2014, surveys were performed in ornamental plant-producing regions located in eastern Sicily (Table 1). The disease incidence (DI) was recorded for each host, based on the number of symptomatic plants in the total of those present in five investigated nurseries. Additionally, approx. 20 plants per species per nursery showing wilt, crown or root rot or damping-off symptoms, were randomly collected for analysis. Fragments (each 5 × 5 mm) of symptomatic tissues were cut from the margins of lesions, surface-sterilised in a sodium hypochlorite solution (10%) for 20 s, followed by 70% ethanol for 30 s, and rinsed three times in sterilised water. Tissue fragments were dried in sterilised filter paper, placed on 2% potato dextrose agar (PDA) amended with 100 μg mL^{-1} penicillin and 100 μg mL^{-1} streptomycin (PDA-PS), and were incubated at 25°C until characteristic *Fusarium*-like colonies were observed. Pure cultures were obtained by transferring single conidia to fresh PDA, with the aid of a Nikon SMZ1000 dissecting microscope.

Fungal isolates and morphological characterization

The cultural and micromorphological features of all the isolates included in this study were evaluated fol-

Table 1. Hosts, locations, symptoms and incidence (%) of diseases caused by *Fusarium* and *Neocosmospora* in Sicily (Southern Italy).

Hosts	Locations	Geographical coordinates	Collection year	Symptoms	Incidence %[a]
Agapanthus africanus	Carruba, Nursery 1	37.698004, 15.193944	2014	Damping-off	50
Bougainvillea glabra	Carruba, Nursery 1	37.698004, 15.193944	2010	Wilt	30
Cordyline australis 'Purpurea'	Carruba, Nursery 2	37.699090, 15.197300	2014	Wilt	20
Dasylirion longissimum	Riposto, Nursery 3	37.733699, 15.194320	2014	Wilt	10
Eremophila laanii	Carruba, Nursery 1	37.698004, 15.193944	2010	Wilt	50
Ficus carica	Carruba, Nursery 1	37.698004, 15.193944	2013	Crown and Root rot	50
Phyloteca myoporoides	Milazzo, Nursery 4	38.201964, 15.239530	2013	Wilt	30
Trachycarpus princeps	Grotte, Nursery 5	37.681233, 15.180316	2013	Root rot	40

[a] Number of symptomatic plants on the total of those cultivated.

lowing the procedures of Aoki *et al.* (2003), with some modification as described previously (Sandoval-Denis *et al.*, 2018).

DNA extraction, PCR amplification and sequencing

Fungus isolates were grown on PDA for 4-7 d at room temperature, under a natural day/night photoperiod. Total genomic DNA was extracted from fresh mycelium scraped from each colony surface, using the Wizard® Genomic DNA purification Kit (Promega Corporation). Fragments of seven nuclear loci, including the translation elongation factor 1-alpha (*EF-1α*), the intergenic spacer region of the rDNA (*IGS*), the internal transcribed spacer region of the rDNA (*ITS*), the large subunit of the rDNA (*LSU*), the RNA polymerase largest subunit (*RPB1*), RNA polymerase second largest subunit (*RPB2*) and beta-tubulin (*TUB*), were PCR amplified as described previously (O'Donnell *et al.*, 2009; 2010, Sandoval-Denis *et al.*, 2018). The PCR products were sequenced using the following primer pairs: EF-1/EF-2 for *EF-1α* (O'Donnell *et al.*, 2008), iNL11/iCNS1 plus the internal sequencing primer pair NLa/CNSa for IGS (O'Donnell *et al.*, 2009), ITS4/ITS5 for ITS (White *et al.*, 1990), LR0R/LR5 for LSU (Vilgalys and Hester, 1990; Vilgalys and Sun, 1994), Fa/G2R for *RPB1* (O'Donnell *et al.*, 2010), 5f2/7cr and 7cf/11ar for *RPB2* (Liu *et al.*, 1999; Sung *et al.*, 2007), and 2Fd/4Rd for TUB (Woudenberg *et al.*, 2009). Sequences generated in this study were uploaded to the GenBank and the European Nucleotide Archive (ENA) databases.

Phylogenetic analyses and molecular identification

Sequence alignments were performed individually for each locus using MAFFT on the European Bioinformatics Institute (EMBL-EBI) portal (http://www.ebi.ac.uk/Tools/msa/mafft/). BLASTn searches on GenBank, and pairwise sequence alignments on the *Fusarium* MLST database of the Westerdijk Fungal Biodiversity Institute (http://www.westerdijkinstitute.nl/fusarium/), were performed using *EF-1α* and *RPB2* sequences. This was to assess the distribution of the Fusaria isolates among the different *Fusarium* species complexes or *Neocosmospora*. Following this preliminary identification, different loci combinations were selected for each of the *Fusarium* species complexes and *Neocosmospora* isolates, according to the phylogenetic informativeness for each locus as compiled in published literature. These combinations were as follows: *EF-1α*, ITS, *RPB1*, *RPB2* and *TUB* for the *F. fujikuroi* species complex (FFSC) (Edwards *et al.* 2016); *EF-1α* and IGS for the *F. oxysporum* species complex (FOSC), and collapsed to haplotypes according to O'Donnell *et al.* (2009); *EF-1α*, ITS, LSU and *RPB2* for the genus *Neocosmospora* (O'Donnell *et al.*, 2008).

The different gene datasets were analysed independently and combined, using Maximum likelihood (ML) and Bayesian methods (BI) as described previously (Sandoval-Denis *et al.*, 2018).

Pathogenicity tests

Pathogenicity tests were performed on potted healthy seedlings or cuttings of all symptomatic species recovered with a subset of 16 representative isolates (Table 2). Each experiment was conducted twice, obtaining similar results in both tests. For each experiment three replicates per isolate were used with 20 to 50 plants per replicate. All plants were inoculated by placing two colonised 1 cm^2 plugs (PDA from 9-d-old mycelium cultures, grown at 25 ± 1°C in the dark) at the base of each plant stem. Uninoculated plants for all the host species served as controls. After inoculation, plants were covered with a plastic bag for 48 h and maintained at 25 ± 1°C and 95% relative humidity (RH) under a 12 h fluorescent light/dark regime until the symptoms were observed. All plants were irrigated two to three times per week, and were examined each week for disease symptoms. Disease incidence (DI) was determined for each host species. Fungi were re-isolated from symptomatic tissues and identified, to fulfil Koch's postulates.

RESULTS

Field sampling and pathogen isolations

Symptoms referable to *Fusarium* spp. were detected on eight ornamental species in five nurseries investigated in Eastern Sicily, Italy (Figure 1). The diseases were observed on seedlings and unrooted and rooted cuttings (1 to 12-month-old) during propagation stages in the greenhouses. Disease incidence varied from 10 to 50%, according to the host species (Table 1). The symptoms observed on ornamental plants consisted of damping-off, crown and root rot, and wilt (Table 1).

Damping-off consisted of root rot and stem decay at soil level, and occurred on young seedlings. Rotted roots were dark brown or black, and were partially or completely destroyed. Crown rot sometimes occurred in association with root rot. As consequence of crown and root rot, basal leaves turned necrotic while infected

Table 2. Collection details and sequence accession numbers of isolates included in this study, as well as disease incidence from pathogenicity tests conducted with the isolates.

Species	Culture number[a]	Host	Incidence %[b]	GenBank/ENA accession number[c]						
				ITS	*EF-1α*	IGS	LSU	*RPB1*	*RPB2*	*TUB*
Fusarium agapanthi	CPC 27740#	*Agapanthus africanus*	100	LS422776	LS420058		LS422776	LS420106	LS420122	LS420041
	CPC 27741#	*Agapanthus africanus*	100	LS422777	LS420059		LS422777	LS420107	LS420123	LS420042
Fusarium anthophilum	CPC 27742#	*Dasylirion longissimum*	75	LS422778	LS420060		LS422778	LS420108	LS420124	LS420043
	CPC 27743	*Dasylirion longissimum*		LS422779	LS420061		LS422779	LS420109	LS420125	LS420044
	CPC 27744	*Dasylirion longissimum*		LS422780	LS420062		LS422780	LS420110	LS420126	LS420045
Fusarium fujikuroi	CPC 27719#	*Trachycarpus princeps*	0	LS422781	LS420063		LS422781	LS420111	LS420127	LS420046
Fusarium oxysporum	CPC 27729#	*Philoteca myoporoides*	100		LS420064	LS420138				
	CPC 27730#	*Philoteca myoporoides*	100		LS420065	LS420139				
	CPC 27731	*Philoteca myoporoides*			LS420066	LS420140				
	CPC 27732	*Philoteca myoporoides*			LS420067	LS420141				
	CPC 27733	*Philoteca myoporoides*			LS420068	LS420142				
	CPC 27734#	*Eremophila laanii*	100		LS420069	LS420143				
	CPC 27735#	*Eremophila laanii*	100		LS420070	LS420144				
	CPC 27738#	*Bougainvillea glabra*	100		LS420071	LS420145				
	CPC 27739#	*Bougainvillea glabra*	100		LS420072	LS420146				
	CPC 27745#	*Dasylirion longissimum*	0		LS420073	LS420147				
	CPC 27746#	*Cordyline australis* 'Purpurea'	100		LS420074	LS420148				
	CPC 27747#	*Cordyline australis* 'Purpurea'	100		LS420075	LS420149				
	CPC 27748	*Cordyline australis* 'Purpurea'			LS420076	LS420150				
	CPC 27749	*Cordyline australis* 'Purpurea'			LS420077	LS420151				
	CPC 27750	*Cordyline australis* 'Purpurea'			LS420078	LS420152				
	CPC 27751	*Cordyline australis* 'Purpurea'			LS420079	LS420153				
Fusarium proliferatum	CPC 27711#	*Trachycarpus princeps*	100	LS422782	LS420080		LS422782	LS420112	LS420128	LS420047
	CPC 27712	*Trachycarpus princeps*		LS422783	LS420081		LS422783	LS420113	LS420129	LS420048
	CPC 27713	*Trachycarpus princeps*		LS422784	LS420082		LS422784	LS420114	LS420130	LS420049
	CPC 27714	*Trachycarpus princeps*		LS422785	LS420083		LS422785	LS420115	LS420131	LS420050
	CPC 27715	*Trachycarpus princeps*		LS422786	LS420084		LS422786	LS420116	LS420132	LS420051
	CPC 27716	*Trachycarpus princeps*		LS422787	LS420085		LS422787	LS420117	LS420133	LS420052
	CPC 27717	*Trachycarpus princeps*		LS422788	LS420086		LS422788	LS420118	LS420134	LS420053
	CPC 27718	*Trachycarpus princeps*		LS422789	LS420087		LS422789	LS420119	LS420135	LS420054
	CPC 27720	*Trachycarpus princeps*		LS422791	LS420089		LS422791	LS420121	LS420137	LS420056
Neocosmospora solani	CPC 27736#	*Ficus carica*	100	LT991945	LT991907		LT991952		LT991914	
(= *Fusarium solani*)	CPC 27737#	*Ficus carica*	100	LT991946	LT991908		LT991953		LT991915	

[a] CPC: Culture collection of P.W. Crous, housed at Westerdijk Fungal Biodiversity Institute, Utrecht, The Netherlands; Strains used in the pathogenicity tests are indicated with #.
[b] Percentage calculated from the number of symptomatic plants relative to the total number inoculated.
[c] ENA: European Nucleotide Archive; *EF-1α*: Translation elongation factor 1-alpha; IGS: Intergenic spacer region of the rDNA; ITS: Internal transcribed spacer regions of the rDNA and 5.8S region; LSU: Partial large subunit of the rDNA; *RPB1*: RNA polymerase largest subunit; *RPB2*: RNA polymerase second largest subunit; *TUB*: Beta-tubulin.

Figure 1. Natural and artificial symptoms referable to *Fusarium* and *Neocosmospora* spp. **a, b.** Wilt of *Bougainvillea glabra* cuttings (a) and *Cordyline australis* seedlings (b). **c.** Root rot with subsequent leaf chlorosis of *Trachycarpus princeps*. **d.** Damping-off of *Agapanthus africanus*. **e.** Wilt of *Phyloteca myoporoides*. **F to i.** Vascular discolouration and wilt after *Fusarium oxysporum* inoculation of *Bougainvillea glabra* (f and g), *Eremophila laanii* (h) and *Phyloteca myoporoides* (i). **j.** Crown and root rot caused by *Neocosmospora solani* (= *Fusarium solani*) on *Ficus carica* (left) compared with control plants (right).

Table 3. Characteristics of the data partitions used for phylogenetic analyses in this study.

Species[a]	Data partition[b]	ML evolutionary model[c]	Number of characters[d]				
			Total	Conserved	Variable	Informative	BI unique sites
F. fujikuroi SC	ITS	SYM+I	459	420	39	32	50
	RPB1	SYM+I+G	1279	1038	241	148	195
	RPB2	GTR+I+G	1570	1251	319	216	332
	EF-1α	SYM+G	455	317	133	76	148
	TUB	SYM+I+G	507	389	117	65	140
F. oxysporum SC	*EF-1α*	GTR+I+G	591	448	143	101	67
	IGS	GTR+I+G	2190	1420	744	554	218
Neocosmospora	ITS	GTR+I+G	496	404	90	65	117
	LSU	GTR+I+G	481	450	31	14	25
	RPB2	GTR+I+G	1603	1235	365	243	275
	EF-1α	GTR+G	324	222	97	48	107

[a] SC: species complex.
[b] *EF-1α*: Translation elongation factor 1-alpha. IGS: Intergenic spacer region of the rDNA. ITS: Internal transcribed spacer regions of the rDNA and 5.8S region. *RPB1*: RNA polymerase largest subunit. *RPB2*: RNA polymerase second largest subunit. *TUB*: Beta-tubulin.
[c] G: Gamma distributed rate variation among sites. GTR: Generalised time-reversible. I: Proportion of invariable sites. ML: Maximum-likelihood; SYM: Symmetrical model.
[d] BI: Bayesian inference.

plants sometimes wilted and died. Wilted plants had conspicuous vascular brown discolourations from the crown to the canopy.

A total of 33 monosporic *Fusarium*-like isolates were collected (Table 2). Among these, two isolates were obtained from damping-off, 12 from root rot, and 19 were from wilted plants.

Phylogenetic analyses and species identification

Pairwise sequence alignments on the *Fusarium* MLST database and GenBank BLASTn searches demonstrated that 16 isolates belonged to the FOSC and 15 isolates to the FFSC, while two isolates were assigned to the genus *Neocosmospora* (*F. solani* species complex) (Table 2).

Subsequent more inclusive multilocus phylogenetic analyses identified a total of five *Fusarium* spp. and one *Neocosmospora* sp. The alignment characteristics and statistics are summarized in Table 3. The phylogenetic analyses of the 15 FFSC isolates from ornamentals revealed a total of four species (*F. agapanthi* O'Donnell, T. Aoki, J. Edwards & Summerell, *F. anthophilum* (A. Braun) Wollenw., *F. fujikuroi* Nirenberg and *F. proliferatum*) from different hosts (Figure 2). Isolates belonging to FOSC were studied based on a two-gene analysis using *EF-1α* and IGS sequences and incorporated in the original alignments previously published by O'Donnell *et al.* (2009) including representatives of 257 known FOSC haplotypes. The FOSC isolates from ornamentals belonged to 15 different haplotypes: isolates CPC 27748 and 22749, from *Cordyline australis* 'Purpurea' showed identical DNA sequences, and corresponded to haplotype 122 of FOSC; isolate CPC 27733 from *Philoteca myoporoides* belonged to haplotype 188, while each of the remaining isolates corresponded to a previously undescribed haplotype (Figure 3). The phylogeny of the genus *Neocosmospora* was based on *EF-1α*, ITS, LSU and *RPB2* sequences, and showed that two isolates from *Ficus carica* (CPC 27736 and 27737) belonged to *N. solani* (Martius) L. Lombard & Crous (= *F. solani*) (Figure 4).

Pathogenicity tests

Fourteen *Fusarium* and two *Neocosmospora* isolates tested were pathogenic to the different inoculated original hosts, and produced symptoms similar to those observed on diseased plants in nurseries (Figure 1). Two isolates were non-pathogenic. Damping-off occurred on *Agapanthus africanus*, crown and root rot on *F. carica* and root rot with subsequent leaf chlorosis appeared on *Trachycarpus princeps*. The remaining host plants showed vascular discolouration and wilted. The DI (%) caused by *Fusarium* and *Neocosmospora* species on different hosts ranged from 75 to 100%, after 15 d to 3 months (Table 2).

All *F. agapanthi*, *F. proliferatum*, and *N. solani* isolates were pathogenic, and caused 100% DI on *A. africanus*, *T. princeps* and *F. carica*, whereas *F. anthophilum*

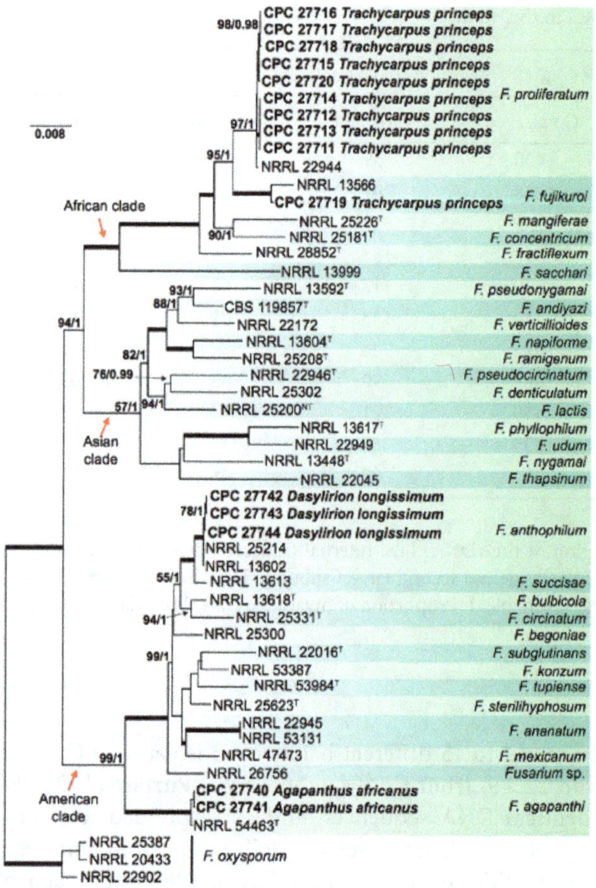

Figure 2. Maximum-likelihood (ML) phylogram of the *Fusarium fujikuroi* species complex obtained from combined ITS, *RPB1*, *RPB2*, *EF-1α* and *TUB* sequences. Branch lengths are proportional to distance. Numbers on the nodes are ML bootstrap values greater than 55%; and Bayesian posterior probability values greater than 0.95. Full supported branches and isolates obtained from ornamentals plants are indicated in bold. Ex-type strains are indicated with T and ex-neotype strains are indicated with NT.

caused disease with lower DI on *Dasylirion longissimum* (75%). *Fusarium oxysporum* isolates gave high DI (100%) on *Bougainvillea glabra*, *C. australis* 'Purpurea', *Eremophila laanii* and *P. myoporoides*, but was non-pathogenic on *D. longissimum*. Similarly, *F. fujikuroi* caused no symptoms on the original host *T. princeps*. The pathogens were re-isolated from the artificially inoculated plants, and were identified as previously described, fulfilling Koch's postulates. No symptoms were observed on control (uninoculated) plants.

DISCUSSION

The most important plant pathogenic *Fusarium* species is the soil-borne *F. oxysporum* Schltdl. (Gordon and

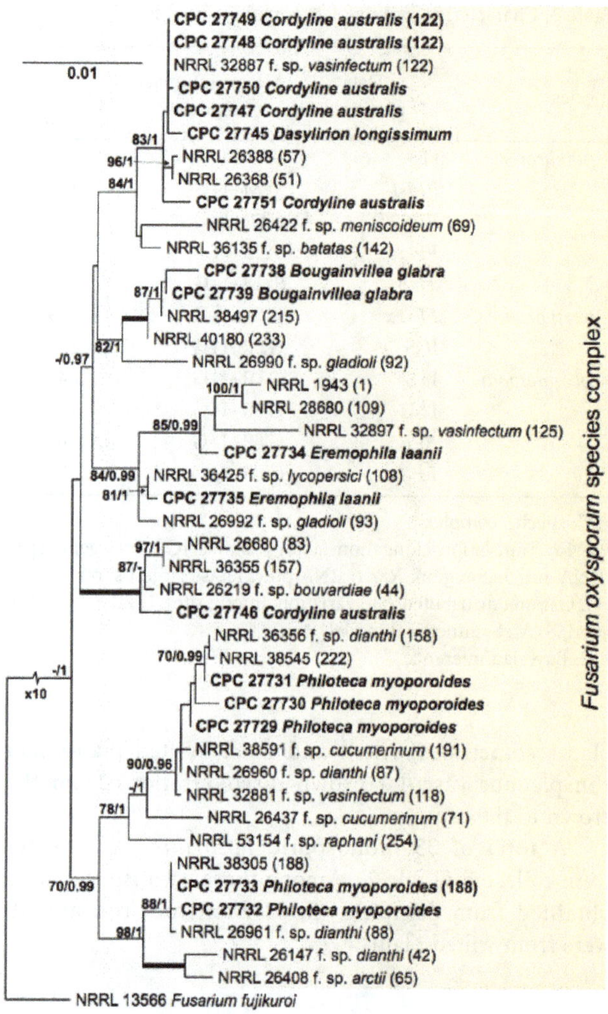

Figure 3. Maximum-likelihood (ML) phylogram of the *Fusarium oxysporum* species complex obtained from combined *EF-1α* and IGS sequences of isolates obtained in this study and representatives of the most closely related haplotypes. Branch lengths are proportional to distance. Numbers on the nodes are ML bootstrap values greater than 55%; and Bayesian posterior probability values greater than 0.95. Full supported branches and isolates obtained from ornamentals plants are indicated in bold. Numbers between parentheses indicate the corresponding haplotype.

Martyn, 1997; Gullino *et al.*, 2012), currently encompassing nearly 150 *formae speciales* (ff. spp.) and races. The broad host plant range of this fungus includes valuable ornamental plants such as *Chrysanthemum*, *Dianthus*, *Gerbera*, *Gladiolus*, and *Lilium* spp. (Engelhard and Woltz, 1971; Linderman, 1981; Farr and Rossman, 2018), on which it causes symptoms ranging from vascular wilt to crown and root rot (Engelhard and Woltz, 1971; Linderman, 1981).

Fusarium proliferatum (Matsush.) Nirenberg ex Gerlach & Nirenberg is another important species, which has

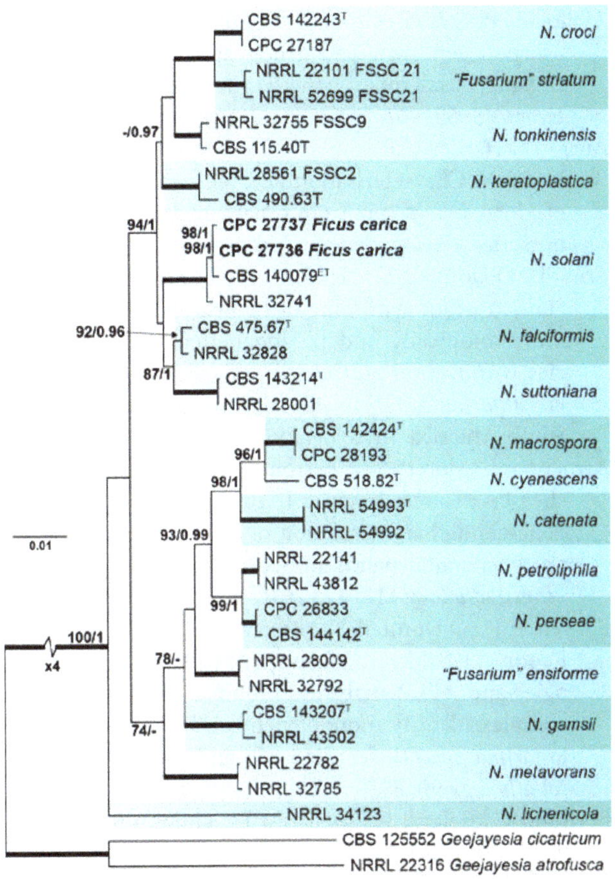

Figure 4. Maximum-likelihood (ML) phylogram of *Neocosmospora* (= *Fusarium solani* species complex) obtained from combined *EF-1α*, ITS, LSU and *RPB2* sequences. Branch lengths are proportional to distance. Numbers on the nodes are ML bootstrap values greater than 55%; and Bayesian posterior probability values greater than 0.95. Full supported branches and isolates obtained from ornamentals plants are indicated in bold. Ex-type and ex-epitype strains are indicated with ^{T} and ^{ET} respectively.

been described as the causal agent of blight, dieback and wilt of several palms belonging to the genera *Chamaerops*, *Phoenix*, *Ravenea*, *Trachycarpus* and *Washingtonia* (Polizzi and Vitale, 2003; Armengol *et al.*, 2005).

In the present study, two *Neocosmospora* and 31 *Fusarium* isolates were recovered from eight ornamental species in Sicily over a 5-year period. Disease symptoms were observed in five ornamental nurseries, and included damping-off, crown and/or root rot, and wilt. The isolates obtained from symptomatic tissues were identified based on single and multilocus phylogenetic analyses of seven loci (*EF-1α*, IGS, ITS, LSU, *RPB1*, *RPB2*, and *TUB*), as well as morphological characters. Our study revealed considerable diversity in the composition of *Fusarium*-like fungal populations recovered from nurseries.

As confirmed in the pathogenicity tests, all the *F. oxysporum* isolates caused symptoms except on *D. longissimum*. However, this host showed disease symptoms when inoculated with *F. anthophilum*. The inoculated isolate of *F. fujikuroi* produced no symptoms on *T. princeps*, while the remaining *Fusarium* species investigated, *F. agapanthi*, *F. proliferatum* and *N. solani*, were pathogenic to the respective tested hosts from which they were isolated.

Fusarium and *Neocosmospora* species are widespread in nurseries in Italy (Polizzi *et al.*, 2003; 2010a; 2010b; 2011; Bertoldo *et al.*, 2015), where they represent a limiting factor for production of ornamental plants cultivated in Sicily. These pathogenic species have very broad host ranges worldwide (Farr and Rossman, 2018). However, there are no known reports of diseases caused by *F. anthophilum* on *D. longissimum*. Moreover, *F. agapanthi* was originally described as pathogenic on *Agapanthus praecox* in Australia and Italy (Edwards *et al.*, 2016). However, this pathogen was isolated in the present study, causing serious seedling damping-off of *A. africanus*, suggesting that it may also be more prevalent on other species of *Agapanthus*. Previous studies have reported *F. proliferatum* associated with palms belonging to the genera *Chamaerops*, *Phoenix*, *Trachycarpus* and *Washingtonia* (Polizzi and Vitale, 2003; Armengol *et al.*, 2005). Our study presents a new report for *F. proliferatum* as a pathogen of *T. princeps*.

The FOSC includes soil-borne pathogens responsible for vascular wilts, stem cankers, rots, and damping-off of a wide range of agronomical and horticulturally important crops (Baayen *et al.*, 2000; Michielse and Rep, 2009; O'Donnell, 2009). Members of this complex collectively represent the most commonly found and economically important species complex within *Fusarium*. *Fusarium oxysporum* was the predominant species found in all the nurseries sampled, and unlike other species, it was recovered from multiple hosts. Recently, Polizzi *et al.* (2010a; 2010b; 2011) identified *F. oxysporum* associated with wilt diseases of *B. glabra*, *E. laanii* and *P. myoporoides*. However, no reports were previously known of diseases caused by *F. oxysporum* on *C. australis*, as reported here.

The present study is also the first report of *N. solani* causing crown and root rot of *F. carica* cuttings. This plant species is often cultivated for fruit production, and thousands of cuttings cultivated for ornamental purposes were investigated because serious losses were observed from of crown and root rot, leading to plant death. *Neocosmospora* is a species-rich genus containing at least 60 phylogenetically distinct species (O'Donnell, 2000; Zhang *et al.*, 2006; O'Donnell *et al.*, 2008; Nalim *et al.*,

2011). These fungi generally cause crown and/or root rot of infected host plants, while symptoms on aboveground plant portions may manifest as cankers, wilting, stunting and chlorosis, or as lesions on stems and/or leaves (Coleman, 2016; Guarnaccia *et al.*, 2018).

The high disease incidence observed in the investigated ornamental nurseries probably depends on the prevailing climatic conditions, farming practices and environmental conditions such as temperature, humidity, irrigation systems or the use of non-disinfected plant growth substrates. Potted plant production could promote infections, since plants are frequently stressed due to being containerised during the production processes. Moreover, several wounds can occur during transplanting. Thus, prevention is a major strategy to control *Fusarium* diseases, and an accurate diagnosis of *Fusaria* species occurring in a particular area is significant for the selection of effective disease management strategies.

This study provides the first overview of *Fusarium* and *Neocosmospora* diversity associated with diseased ornamental plants in Southern Italy, and includes information on the pathogenicity of these fungi. It also provides the first reports of several new pathogen/host combinations, such as *N. solani* associated with crown and root rot of *F. carica*, and *F. agapanthi*, *F. anthophilum*, *F. oxysporum* and *F. proliferatum* as pathogens, respectively, of *A. africanus*, *D. longissimum*, *C. australis* and *T. princeps*.

ACKNOWLEDGEMENTS

The authors thank Dr Pietro Formica for support with specimen collection and fungal isolations. This research was funded by Research Project 2016-2018 "Emergent Pests and Pathogens and Relative Sustainable Management Strategies", financed by University of Catania.

LITERATURE CITED

Aiello D., Guarnaccia V., Vitale A., Cirvilleri G., Granata G., ... Crous P.W., 2014. *Ilyonectria palmarum* sp. nov. causing dry basal stem rot of Arecaceae. *European Journal of Plant Pathology* 138: 347–359.

Aiello D., Guarnaccia V., Epifani F., Perrone G., Polizzi G., 2015. '*Cylindrocarpon*' and *Ilyonectria* species causing root and crown rot disease of potted laurustinus plants in Italy. *Journal of Phytopathology* 163: 675–680.

Aiello D., Polizzi G., Crous P.W., Lombard L., 2017. *Pleiocarpon* gen. nov. and a new species of *Ilyonectria* causing basal rot of *Strelitzia reginae* in Italy. *IMA Fungus* 8: 65–76.

Aiello D., Hansen Z.R., Smart C.D., Polizzi G., Guarnaccia V., 2018. Characterisation and mefenoxam sensitivity of *Phytophthora* spp. from ornamental plants in Italian nurseries. *Phytopathologia Mediterranea* 57: 245–256.

Aoki T., O'Donnell K., Homma Y., Lattanzi A.R., 2003. Sudden-death syndrome of soybean is caused by two morphologically and phylogenetically distinct species within the *Fusarium solani* species complex - *F. virguliforme* in North America and *F. tucumaniae* in South America. *Mycologia* 95: 660–684.

Armengol J., Moretti A., Perrone G., Vicent A., Bengoechea J.A., García-Jiménez J., 2005. Identification, incidence and characterization of *Fusarium proliferatum* on ornamental palms in Spain. *European Journal of Plant Pathology* 112: 123–131.

Baayen R.P., O'Donnell K., Bonants P.J.M., Cigelnik E., Kroon L.P.N.M., ... Waalwijk C., 2000. Gene genealogies and AFLP analyses in the *Fusarium oxysporum* complex identify monophyletic and nonmonophyletic *formae speciales* causing wilt and rot diseases. *Phytopathology* 90: 891–900.

Bertoldo C., Gilardi G., Spadaro D., Gullino M.L., Garibaldi A., 2015. Genetic diversity and virulence of Italian strains of *Fusarium oxysporum* isolated from *Eustoma grandiflorum*. *European Journal of Plant Pathology* 141: 83–97.

Coleman J.J., 2016. The *Fusarium solani* species complex: ubiquitous pathogens of agricultural importance. *Molecular Plant Pathology* 17: 146–158.

Edwards J., Auer D., de Alwis S.K., Summerell B., Aoki T., ... O'Donnell K., 2016. *Fusarium agapanthi* sp. nov., a novel bikaverin and fusarubin-producing leaf and stem spot pathogen of *Agapanthus praecox* (African lily) from Australia and Italy. *Mycologia* 108: 981–992.

Elliott M.L., Broschat T.K., Uchida J.Y., Simone G.W., 2004. Compendium of ornamental palm diseases and disorders. St. Paul, MN, APS Press.

Engelhard A.W., Woltz S.S., 1971. Fusarium wilt of *chrysanthemum*: symptomatology and cultivar reactions. In: *Proceedings of the Florida State Horticultural Society*, 84: 351–354.

Farr D.F., Rossman A.Y., 2018. Fungal Databases, U.S. National Fungus Collections, ARS, USDA. Retrieved February 05, 2018, from https://nt.ars-grin.gov/fungaldatabases.

Gordon T.R., Martyn R.D., 1997. The evolutionary biology of *Fusarium oxysporum*. *Annual Review of Phytopathology* 35: 111–128.

Guarnaccia V., Sandoval-Denis M., Aiello D., Polizzi G., Crous P.W., 2018. *Neocosmospora perseae* sp. nov., causing trunk cankers on avocado in Italy. *Fungal Systematics and Evolution* 1: 131–140.

Gullino M.L., Garibaldi A., 2006. Evolution of fungal diseases of ornamental plants and main implications for their management. In: *Floriculture, Ornamental and Plant Biotechnology: Advances and Topical Issues* (J.A. Teixeira da Silva, ed.), Global Science Books, London, UK, 464–471.

Gullino M.L., Katan J., Garibaldi A., 2012. The genus *Fusarium* and the species that affect greenhouse vegetables and ornamentals. In: *Fusarium wilts of greenhouse vegetable and ornamental crops* (M.L. Gullino, J. Katan, A. Garibaldi, ed.), APS Press, St. Paul, MN, USA, 5–9.

Gullino M.L., Daughtrey M.L., Garibaldi A., Elmer W.H., 2015. Fusarium wilts of ornamental crops and their management. *Crop Protection* 73: 50–59.

Horst R.K., Nelson P.E., 1997. Compendium of chrysanthemum diseases. St. Paul, MN, APS Press.

Linderman R.G., 1981. Fusarium disease of flowering bulb crops. In: *Fusarium: Disease, Biology, and Taxonomy* (P.E. Nelson, T.A. Toussoun, R.J. Cook, ed.), The Pennsylvania State University Press, University Park, PA, USA, 129–142.

Liu Y.J., Whelen S., Hall B.D., 1999. Phylogenetic relationships among ascomycetes: evidence from an RNA polymerse II subunit. *Molecular Biology and Evolution* 16: 1799–1808.

Lombard L., Van der Merwe N.A., Groenewald J.Z., Crous P.W., 2015. Generic concepts in Nectriaceae. *Studies in Mycology* 80: 189–245.

Lupien S.L., Dugan F.M., Ward K.M., O'Donnell K., 2017. Wilt, crown, and root rot of common rose mallow (*Hibiscus moscheutos*) caused by a novel *Fusarium* sp. *Plant Disease* 101: 354–358.

Michielse C.B., Rep M., 2009. Pathogen profile update: *Fusarium oxysporum*. *Molecular Plant Pathology* 10: 311–324.

Nalim F.A., Samuels G.J., Wijesundera R.L., Geiser D.M., 2011. New species from the *Fusarium solani* species complex derived from perithecia and soil in the Old World tropics. *Mycologia* 103: 1302–1330.

O'Donnell K., 2000. Molecular phylogeny of the *Nectria haematococca-Fusarium solani* species complex. *Mycologia* 92: 919–938.

O'Donnell K., Sutton D.A., Fothergill A., McCarthy D., Rinaldi M.G., ... Geiser D.M., 2008. Molecular phylogenetic diversity, multilocus haplotype nomenclature, and *in vitro* antifungal resistance within the *Fusarium solani* species complex. *Journal of Clinical Microbiology* 46: 2477–2490.

O'Donnell K., Gueidan C., Sink S., Johnston P.R., Crous P.W., ... Sarver B.A., 2009. A two-locus DNA sequence database for typing plant and human pathogens within the *Fusarium oxysporum* species complex. *Fungal Genetics and Biology* 46: 936–948.

O'Donnell K., Sutton D.A., Rinaldi M.G., Sarver B.A., Balajee S.A., ... Geiser D.M., 2010. Internet-accessible DNA sequence database for identifying fusaria from human and animal infections. *Journal of Clinical Microbiology* 48: 3708–3718.

Polizzi G., Vitale A., 2003. First report of Fusarium blight on majesty palm caused by *Fusarium proliferatum* in Italy. *Plant Disease* 87: 1149.

Polizzi G., Vitale A., Aiello D., Dimartino M.A., Parlavecchio G., 2007. First report of damping-off and leaf spot caused by *Cylindrocladium scoparium* on different accessions of bottlebrushes cuttings in Italy. *Plant Disease* 91: 769.

Polizzi G., Aiello D., Guarnaccia V., Vitale A., Perrone G., Epifani F., 2010a. First report of Fusarium wilt on *Eremophila* spp. caused by *Fusarium oxysporum* in Italy. *Plant Disease* 94: 1509.

Polizzi G., Aiello D., Guarnaccia V., Vitale A., Perrone G., Stea G., 2010b. First report of Fusarium wilt of paper flower (*Bougainvillea glabra*) caused by *Fusarium oxysporum* in Italy. *Plant Disease* 94: 483.

Polizzi G., Aiello D., Guarnaccia V., Vitale A., Perrone G., Stea G., 2011. First report of Fusarium wilt on *Philotheca myoporoides* caused by *Fusarium oxysporum* in Italy. *Plant Disease* 95: 877.

Polizzi G., Vitale A., Aiello D., Guarnaccia V., Crous P.W., Lombard L., 2012. First report of *Calonectria ilicicola* causing a new disease on Laurus (*Laurus nobilis*) in Europe. *Journal of Phytopathology* 160: 41–44.

Sandoval-Denis M., Guarnaccia V., Polizzi G., Crous P.W., 2018. Symptomatic citrus trees reveal a new pathogenic lineage in *Fusarium* and two new *Neocosmospora* species. *Persoonia* 40: 1–25.

Sinclair W.A., Lyon H.H., 2005. Diseases of Trees and Shrubs. Cornell University Press, Ithaca, NY, p. 660.

Sung G.H., Sung J.M., Hywel-Jones N.L., Spatafora J.W., 2007. A multi-gene phylogeny of Clavicipitaceae (Ascomycota, Fungi): Identification of localized incongruence using a combinational bootstrap approach. *Molecular Phylogenetics and Evolution* 44: 1204–1223.

Vilgalys R., Hester M., 1990. Rapid genetic identification and mapping of enzymatically amplified ribosomal DNA from several *Cryptococcus* species. *Journal of Bacteriology* 172: 4238–4246.

Vilgalys R., Sun B.L., 1994. Ancient and recent patterns of geographic speciation in the oyster mushroom *Pleu-*

rotus revealed by phylogenetic analysis of ribosomal DNA sequences. In: *Proceedings of the National Academy of Sciences of the United States of America* 91: 4599–4603.

Vitale A., Aiello D., Castello I., Dimartino M.A., Parlavecchio G., Polizzi G., 2009. Severe outbreak of crown rot and root rot caused by *Cylindrocladium pauciramosum* on strawberry tree in Italy. *Plant Disease* 93: 842.

White T.J., Bruns T., Lee S., Taylor J.L., 1990. Amplification and direct sequencing of fungal ribosomal RNA genes for phylogenetics. In: *PCR protocols: a guide to methods and applications* (M.A. Innes, D.H. Gelfand, J.J. Sninsky,T.J. White, ed.), USA: Academic Press, NY, 315–322.

Woudenberg J.H.C., Aveskamp M.M., de Gruyter J., Spiers A.G., Crous P.W., 2009. Multiple *Didymella* teleomorphs are linked to the *Phoma clematidina* morphotype. *Persoonia* 22: 56–62.

Zhang N., O'Donnell K., Sutton D.A., Nalim F.A., Summerbell R.C., … Geiser D.M., 2006. Members of the *Fusarium solani* species complex that cause infections in humans and plants are common in the environment. *Journal of Clinical Microbiology* 44: 2186–2190.

7

Emerging and re-emerging fungus and oomycete soil-borne plant diseases in Italy

SANTA OLGA CACCIOLA[1,*], MARIA LODOVICA GULLINO[2]

[1] *Department of Agriculture, Food and Environment (Di3A), University of Catania, V. S. Sofia 100, 95123 Catania, Italy*
[2] *Centre for Innovation in the Agro-Environmental Sector, AGROINNOVA, University of Torino, Largo Braccini 2, 10095 Grugliasco (TO), Italy*
*Corresponding author: olgacacciola@unict.it

Summary. A disease is recognized as emerging if it is new, it occurs in a new host, there is an unexpected outbreak, its economic importance increases, if it attracts public opinion and the scientific community regardless of economic importance, or if it appears in an area for the first time (referred to as geographic emergence). This review deals with major driving factors of the emergence of plant diseases caused by soil-borne fungi and oomycetes (here indicated as "fungi"), in Italy during recent years. These factors include: accidental introduction of alien pathogens by human activities; effect of climate change; unusually severe weather events; favourable environmental and ecological conditions; pathogen genetic variation; host shifts and expansion of host ranges; introduction or expansion of the geographic range of a susceptible plant species or variety; limited availability of fungicides or development of fungicide-resistance pathogen strains; changes of cropping systems; and/or increased pathogen in soil as a consequence of intensive monoculture of crops. Although in most cases more than a single driving factor contributes to the emergence of an infectious disease, there are examples where a determinant may prevail over others. The case studies reviewed include pathogens belonging to major genera of soil-inhabiting fungi and oomycetes, including *Armillaria, Calonectria, Coniella, Fusarium sensu lato, Ilyonectria, Monosporoascus, Plectosphaerella, Rhizoctonia, Rosellinia, Sclerotinia, Sclerotium, Verticillium, Pythium* and *Phytophthora*. The examples encompass natural and forest ecosystems, economically important agricultural crops including citrus, fruit trees, olive, legumes, vegetables, and ornamentals, as well as exotic or expanding minor crops, such as avocado, goji berry, and pomegranate. Whatever the prevailing driving factor(s) these case studies all show that the large-scale emergence of soil-borne fungal diseases of plants is the consequence of human activities.

Keywords. Exotic pathogens, endemic pathogens, agricultural crops, minor crops, forest ecosystems.

Competing Interests: The Author(s) declare(s) no conflict of interest.

Editor: Josep Armengol.

INTRODUCTION

There are different definitions of the term soil-borne when referring to plant pathogens. According to Koike *et al.* (2003), this includes pathogens that infect the plant through the soil while Katan (2017) broadened the concept and defined as soil-borne those pathogens surviving and acting in the soil, at least during part of their lives. Major genera of fungi and oomycetes (Chromista) recognized as typical soil-borne plant pathogens include *Armillaria*, *Fusarium sensu lato* (including *Neocosmospora* spp., formerly the species complex *Fusarium solani*, and other *Fusarium*-like genera recently segregated from *Fusarium sensu lato*), *Gaeumannomyces*, *Macrophomina*, *Cylindrocarpon*-like asexual morphs, *Monosporascus*, *Phytophthora*, *Phytopythium*, *Plectosphaerella*, *Pyrenochaeta*, *Pythium*, *Rhizoctonia*, *Rosellinia*, *Sclerotinia*, *Sclerotium*, *Thielaviopsis* and *Verticillium*. These pathogens are well-studied because of their widespread occurrence in many important world food and fibre crops. However, other genera of plant pathogens, such as *Colletotrichum* as causal agents of anthracnose of many host plants, include species living in soil and affecting important crops (Gilardi *et al.*, 2014c). Whatever the definition, soil-borne fungal pathogens have the following characteristics: they have adapted to living in terrestrial habitats; they infect plants through belowground organs but are also able to cause infections of aboveground parts of the plants; they are saprobic, hemibiotrophic or parasitic on roots, stems and/or leaves of herbaceous or woody plants; and they cause monocyclic, in some cases polyetic and, more rarely, polycyclic diseases.

Soil-borne pathogens causing polycyclic leaf diseases have been frequently reported in Italy in nurseries of ornamentals, as consequences of conducive environments. They include *Calonectria* (Saracchi *et al.*, 2008; Vitale *et al.*, 2013) and *Phytophthora* species (Pane *et al.*, 2007a; Cacciola *et al.*, 2011b; Ginetti *et al.*, 2014). Microsclerotia of *Calonectria* species are the primary inocula and the survival form of these pathogens in the soil, while aerial dissemination occurs through conidia of their *Cylindrocladium*-like anamorphic stage and ascospores. As far as *Phytophthora* is concerned, few species, like *P. infestans*, have adapted to almost completely aerial lifestyles, while the majority of *Phytophthora* species are typically terricolous. A number of species that produce deciduous sporangia have partially adapted to aerial lifestyles (Jung *et al.*, 2018). *Phytophthora palmivora*, in particular, has developed evolutive transitional dispersal conidium-like propagules (sporocysts). Sporocysts are non-papillate caducous sporangia that can be easily detached and dispersed by rain.

A second aspect to be defined is when an infectious disease can be considered to be emerging. In this review, we recognize diseases as "emerging" when they are new, occurring in an area for the first time (geographic emergence), if they occur in new hosts, if there is an unexpected outbreak, if their economic importance increases, or if for some reason they attract public or scientific attention.

In recent years there has been considerable increase in the number of new *Phytophthora* species discovered. The number of validly described and recognized species has increased from approx. 40 in 1990 101 by 2011 (Kroon *et al.*, 2011) and now exceeds 150 (Ruano-Rosa *et al.*, 2018b, Albuquerque Alves *et al.*, 2019). The increase in the number of newly described *Phytophthora* species probably results from the scientific interest for these Oomycetes, their spread through the nursery plant trade and the rapid progress and evolution of molecular diagnostic techniques and taxonomy (Cooke *et al.*, 2007; Lamour *et al.*, 2007; Blair *et al.*, 2008; Martin *et al.*, 2012; Scibetta *et al.*, 2012; Bilodeau *et al.*, 2014; Jung *et al.*, 2016a; Yang *et al.*, 2017).

Molecular studies have also revealed the complexity of *Fusarium*, which encompasses numerous soil-borne pathogens of economically important crops. The taxonomy of this genus has undergone substantial revision and numerous new cryptic species have been described which differ in subtle morphological details and were previously assigned to the species complexes of *F. fujikuroi*, *F. incarnatum-equiseti*, *F. oxysporum*, *F. sambucinum* and *F.solani*. These have been separated or reallocated in the genus *Neocosmospora* on the basis of multi-locus phylogenetic analyses (O'Donnell *et al.*, 2015; Schroers *et al.*, 2016; Sandoval-Denis and Crous, 2018; Sandoval-Denis *et al.*, 2018; Lombard *et al.*, 2019; Maryani *et al.*, 2019).

Major determinants of the emergence of infectious plant diseases have been reviewed elsewhere (Fisher *et al.*, 2012; Gonthier and Garbelotto, 2011; Santini *et al.*, 2013; Gilardi *et al.*, 2018a,b; Moricca *et al.*, 2018). These determinants include anthropogenic introduction of alien pathogens, climate change, severe weather events, favourable environmental and ecological conditions, pathogen genetic recombination or mutation, host shifts or expansion of host range, introduction or expansion of the geographic ranges of susceptible plant species or varieties, limited availability of fungicides and development of resistance, changes of cropping systems, and increases in soil inoculum consequences of monoculture. These factors may be involved individually or together, instantaneously or continuously, simultaneously or in succession.

There is a body of direct or circumstantial evidence indicating that emergence of infectious plant diseases results from human activities (Andjic *et al.*, 2011; Mammella *et al.*, 2013; Santini *et al.*, 2013; Barnes *et al.*, 2014; Biasi *et al.*, 2016; Engelbrecht *et al.*, 2017), and several driving factors usually concur to cause emergence of a disease (Desprez-Loustau *et al.*, 2010; Stenlid *et al.*, 2011; Garbelotto and Pautasso, 2012). The recent study by Serrano *et al.* (2019), of the genetic structure of *P. cinnamomi* isolates of worldwide origin, using four microsatellite markers, showed that identical genotypes of this pathogen were associated with the same hosts on different continents. This indicated long-distance transport by man, while the presence of identical genotypes in agricultural settings and neighbouring wildlands would suggested that specific commodities may have been the common sources of recent infestations caused by new invasive genotypes.

The present review considers emerging soil-borne plant diseases reported in Italy during recent years. These examples exemplify prominent roles of disease emergence drivers.

INTRODUCTION OF ALIEN PATHOGENS

In the modern era, the most devastating plant disease epidemics have been caused by exotic invasive pathogens. The causal agents of several devastating *Phytophthora* epidemics in Europe, Australia, and North America, including those caused by *P. ×cambivora, P. cinnamomi, P. lateralis, P. plurivora* and *P. ramorum*, have probably originated in South-east Asia (Jung *et al.*, 2016a; Jung *et al.*, 2018). Intensification of international trade of plant material and increased efficiency and speed of transport have favoured the introduction of pathogens into new areas. Most alien emerging pathogens have been inadvertently introduced with exotic plants or imported plant materials. Seeds and timber can be vectors of soilborne plant pathogens, but the nursery trade has been increasingly identified as responsible for the movement of this group of pathogens, despite phytosanitary regulations of international trade of plants and commodities (Migliorini *et al.*, 2015; Jung *et al.*, 2016b; Simamora *et al.*, 2017; Jung *et al.*, 2019). During visual inspections, soil-borne pathogens often go undetected because they induce obvious symptoms only in advanced stages of infection. The most commonly moved soil-borne pathogens globally are species of *Phytophthora* (Brasier, 2008). Genetic analysis of the variability of large numbers of isolates of the cosmopolitan and polyphagous plant pathogen *P. nicotianae*, from a wide range of geographic origins and hosts including ornamentals and agricultural crops, revealed that isolates from *Citrus* spp. were genetically related. This was regardless of their geographic origin, and these pathogens were characterized by genetic uniformity and high inbreeding coefficients (Mammella *et al.*, 2013; Biasi *et al.*, 2016). Greater variability was observed for populations from other hosts and a significant geographical structuring was found only for isolates from *Nicotiana* and *Solanum* spp. These differences were possibly related to the propagation systems for different crops. Isolates obtained from *Citrus* spp. are more probably distributed worldwide with infected nursery plants, whereas *Nicotiana* and *Solanum* spp. are propagated by seeds, which would not contribute to the spread of the pathogen and result in a greater opportunity for geographic isolation of different lineages. For ornamental species in nurseries, the high genetic variation is likely to result from mixtures of diverse pathogen genotypes through the trade of infected plant material from various geographic origins, the presence of several host plants in the same nursery, and genetic recombination through sexual reproduction of this heterothallic pathogen (Mammella *et al.*, 2013; Biasi *et al.*, 2016).

Restoration plantings or afforestation with nursery plants are pathways for the introduction and spread of exotic *Phytophthora* species in natural habitats and forests (Sims *et al.*, 2019a, b). In a recent survey of protected natural areas and water courses crossing these areas in Sicily, 13 of 20 recovered *Phytophthora* species were exotic, while only seven, including *P. tyrrhenica* and *P. vulcanica* associated with *Fagaceae* hosts, could be regarded as endemic to Europe (Jung *et al.*, 2017; Jung *et al.*, 2019). Several species found in this survey, including *P. cactorum, P. citrophthora, P. megasperma, P. multivora, P. plurivora* and *P. ×cambivora*, are well-known invasive pathogens with wide host ranges and aggressiveness towards many cultivated and native European plant species. The presence of these pathogens indicates that they constitute threats to the homeostasis and resilience of these ecosystems, and protected natural areas are reservoirs of inoculum of potential pathogens for economically important crops. Garbelotto and Hayden (2012) highlighted the link between the ornamental plant industry and the introduction of *P. ramorum*, which causes lethal cankers on oak species native to California, into the wildlands in North America. *Phytophthora ramorum* is a quarantine pathogen also in the EPPO region (in the A2 list from 2013), has been repeatedly intercepted in nurseries of ornamentals in Italy, but was promptly eradicated (Ginetti *et al.*, 2014). It has also been detected as an operational taxonomic unit (OTU) by metabarcoding analyses of soil samples sourced from chestnut stands in central Italy and ornamental nurseries in southern Italy (Vannini *et al.*, 2013; Prigigallo *et al.*, 2016). However, in both cases, the presence of the pathogen in samples was not confirmed by the isolation of living cultures. The official status of *P.*

ramorum in Italy, based on information updated to 2014, is "transient, under eradication". *Phytophthora nierderhauserii*, another invasive polyphagous species, has recently emerged in many countries (Abad *et al.*, 2014). The occurrence of this species in natural ecosystems in Australia, in vineyards (*Vitis vinifera*) in South Africa and in almond (*Prunus dulcis*) trees in California, Spain and Turkey, and its capability to infect shrubs and herbaceous ornamentals in several unrelated families indicates *P. nierderhauserii* has wide ecological adaptability and may threaten agricultural and natural ecosystems. There is evidence indicating that after the first detections of *P. niederhauserii* in Italy (Brasier, 2008; Cacciola *et al.*, 2009 a, b), this species has been spreading in nurseries of ornamental plants (Faedda *et al.*, 2013a; Prigigallo *et al.*, 2015; Aiello *et al.*, 2018). *Phytophthora tentaculata* is another emerging but less invasive, exotic *Phytophthora* species that has been reported in northern Italy and on origanum (*Origanum vulgare*) and on loof chicory (*Cichorium intibus*) in central Italy (Martini *et al.*, 2009; Garibaldi *et al.*, 2010). *Phytophthora capsici* remains a serious and economically important pathogen on bell pepper, tomato, eggplant and cucurbits in many countries. This pathogen occurs in protected crops and in open fields, causing severe losses (Hausbeck and Lamour, 2004). The type culture of *P. capsici* is from Italy and was deposited in 1927, suggesting this species was introduced into this country in the 20th century. The long-standing presence of *P. capsici* in Italy is reflected by high levels of genetic variability of the Italian population of this pathogen but its centre of origin remains unknown (Quesada-Ocampo *et al.*, 2011).

Fusarium oxysporum f. sp. *radicis-cucumerinum*, which causes wilting, root and stem rots on cucumber, was first observed in Greece and then in Spain in 2001 and Turkey in 2009, and has been recently reported in Italy on farms that have repeatedly grown cucumber in the same soil for 10 years (Garibaldi *et al.*, 2016). This pathogen can also infect other cucurbit crops, including melon (*Cucumis melo*), watermelon (*Citrullus lanatus*) and sponge gourd (*Luffa aegyptiaca*), while the interspecific hybrids of *Cucurbita maxima* × *C. moschata* and zucchini are not susceptible (Vakalounakis *et al.*, 2005).

Fusarium solani f. sp. *cucurbitae* is responsible for severe losses, causing root and stem rots, particularly on cucumbers in greenhouses. This pathogen is spreading in Spain, where it was reported for the first time in Europe (Gómez *et al.*, 2014). Significant economic damage caused by this pathogen can be expected in Italy in intensive cultivation systems as it can be transmitted by seeds (Vannacci, 1980), and has several cucurbit hosts, including *Cucurbita* hybrids (*C. maxima* × *C. moschata*) used as rootstocks for watermelon (Armengol *et al.*, 2008).

Fusarium wilt of lettuce (*Lactuca sativa*) caused by *F. oxysporum* f. sp. *lactucae* has recently become the most important disease of lettuce in cultivation areas in many countries, and its spread is favoured by seed transmission (Cabral *et al.*, 2018). Until recently, three races of *F. oxysporum* f. sp. *lactucae* were described and reported: race 1 in Europe, USA and South America and races 2 and 3 in Japan. However, in 2017, a new race, race 4, was detected in the Netherlands, identified through biological assays and molecular tools (Gilardi *et al.*, 2017). This new race is apparently spreading rapidly and has been observed in several European countries, including Belgium, the United Kingdom, Ireland (Taylor *et al.*, 2019) and Italy (Gilardi *et al.*, 2019). The presence of this new race is posing serious threats to growers and plant breeders. Before resistant varieties can be developed, preventative management measures, such as the use of healthy seeds and/or seed treatments, are required to reduce the risk of its rapid spread to new areas. Gilardi *et al.* (2017) speculated that race 4 may have evolved due to high selection pressure as a consequence of lettuce monoculture, or could have been introduced from a foreign source through infected seeds or seedlings.

Careful monitoring of the race situation in the field would be useful for the efficient use of host resistance for disease management. Specific molecular markers can also provide successful detection and identification of *formae speciales* and races of *F. oxysporum* from seeds, plants, and soil samples.

Three vegetative compatibility groups (VCGs), VCG-300, VCG-301 and VCG-302, corresponding, respectively, to races 1, 2 and 3, have been reported in *F. oxysporum* f. sp. *lactucae* (Pintore *et al.*, 2017). The Arizona, California and type 1 isolates from Taiwan all belong to the same VCG as race 1 isolates from Japan, and all the race 1 isolates from Arizona, California and Japan have identical mtSSU and EF-1α sequences and almost identical intergenic spacer (IGS) region sequences. This indicates a common origin of the three races. However, the possibilities of the pathogen being introduced from a foreign source through infected seeds or seedlings, or of evolution from non-pathogenic *F. oxysporum*, cannot be excluded. The isolates obtained from lettuce in the Netherlands belong to VCG-303 (Pintore *et al.*, 2017). Further studies are needed to elucidate why and how a new race of this pathogen has developed in Northern Europe.

Melon collapse, commonly considered a synonymous of Monosporascus root rot of melon and vine decline (MRRVD), one of the most important disease of melon and watermelon (Martyn and Miller, 1996;

Cohen *et al.*, 2000), has emerged in the last 15 years in Italy. The exotic Ascomycete *Monosporascus cannonballus* has been regarded as the main cause of the disease in several countries (Stanghellini *et al.*, 2003; Chilosi *et al.*, 2008). However, lines of evidence indicate that other soilborne pathogens, such as *Acremonium cucurbitacearum*, the most common fungal species associated to this disease in Spain (García-Jiménez *et al.*, 2000), *Olpidium* and *Plectosphaerella* species, *Macrophomina phaseolina* and *Rhizoctonia* species, are also involved in melon collapse (Bruton, 2000; García-Jiménez *et al.*, 2000; Stanghellini *et al.*, 2010; Stanghellini and Misaghi 2011; Ben Salem *et al.*, 2013; Felipe *et al.*, 2018). Recently in central Italy, *M. cannonballus*, *O. bornovanus*, and *O. virulentum* were recovered from a melon greenhouse soil with a history of severe infections of *Melon necrotic spot virus* (MNSV), which is vectored by *Olpidium* spp. In pathogenicity tests, all three fungi induced symptoms of root rot and vine decline, confirming a complex aetiology of MRRDV (Aleandri *et al.*, 2017). A binucleate *Rhizoctonia* AG-F was reported to be responsible for watermelon vine decline in Sicily (Aiello *et al.*, 2012). Carlucci *et al.* (2012) found several species of *Plectosphaerella* associated with melon collapse in southern Italy, including an already known species, *Pa. cucumerina* (= *Plectosporum tabacinum*), along with four new species, *Pa. citrullae*, *Pa. pauciseptata*, *Pa. plurivora* and *Pa. ramiseptata*. Carlucci *et al.* (2012) also showed that *A. cucurbitacearum* was a synonym of *Nodulisporium melonis*, and transferred it to the genus *Plectosphaerella* as *Plectosphaerella melonis* comb. nov. This increased to six the number of *Plectosphaerella* species reported on melon and watermelon in Italy. However, the roles of new *Plectosphaerella* species in MRRVD have not yet been clarified (Carlucci *et al.*, 2012). Some were reported to be responsible for root rot of other vegetable crops such as tomato, pepper, parsley and basil (Raimondo and Carlucci, 2018a and b). *Monosporascus cannonballus* has been found only on *Cucurbitaceae*, commonly in arid, hot climates (Stanghellini *et al.*, 1996). This pathogen is widespread in major melon-producing countries in North and Central America, Asia, North Africa, and Europe (Martyn and Miller, 1996; Cohen *et al.*, 2000). After the phase-out of methyl bromide fumigation and application of fungicides through irrigation systems, crop rotation with non-susceptible host plants, breeding for disease resistance and grafting on resistant rootstocks have been regarded as alternatives for the management of melon vine decline (Cohen *et al.*, 2000). Grafting on resistant rootstocks is an effective means for the management of a number of soil-borne diseases of vegetables (Cohen *et al.*, 2007; Davis *et al.*, 2008), and is increasingly used in Italy in commercial *Solanum* and cucurbit crops (Gilardi *et al.*, 2011; Colla *et al.*, 2012; Gilardi *et al.*, 2014a, b, c). However, no commercial rootstocks of melon and watermelon combine multiple resistance to melon collapse and Fusarium wilt with good production performance (Gilardi *et al.*, 2013).

CLIMATE CHANGE AND SEVERE WEATHER EVENTS

The role of climate change in the emergence of infectious plant diseases, its impact on endemic plant pathogens and the interactions between climate change and the introduction of exotic pathogens as a result of globalization have been the subjects of study and extended debate within the international scientific community (La Porta *et al.*, 2008; Pautasso *et al.*, 2012; Ramsfield *et al.*, 2016; Gilardi *et al.*, 2018b). Several studies have concerned effects of climate change on soil microbial communities and, in particular, on soil-borne pathogens (Dukes *et al.*, 2009; Manici *et al.*, 2014; Kubiak *et al.*, 2017; Gilardi *et al.*, 2018a). Climate change can have direct and indirect effects on plant disease epidemics, as climate affects host susceptibility, the survival of pathogen inoculum, the rate of disease progress and epidemic duration. Frequencies of extreme climatic conditions, such as droughts, floods and hurricanes, as well as damage caused by wind, snow and hail, are also likely to increase due to climate change (IPCC, 2019). According to projections, the increase in temperatures will favour expansion of the geographical ranges of mesophilic and thermophilic pathogens at their northern limits in the Northern hemisphere. More generally, warming is expected to cause pole-ward range shifts of plants and their pathogens affecting natural and managed ecosystems (Chakabortry, 2013).

Jung (2009) imputed the decline of European beech (*Fagus sylvatica* L.) in Central Europe to the interaction between *Phytophthora* spp. infections and climatic extremes. On the basis of a survey of natural parks and reserves in Sicily, it has been assumed that the interaction between climate change and root infections by *P.* × *cambivora* and other less frequent or less aggressive *Phytophthora* species (Jung *et al.*, 2017; Jung *et al.*, 2019) is the main driving factor of the decline of European beech in the Nebrodi regional park (Sicily, southern Italy). This is the extreme southern limit of the natural geographical range of European beech in Europe. *Phytophthora* × *cambivora* is also the prevalent species responsible for the resurgence of ink disease in central Italy. Although the effects of climate change on the complex interactions between the diverse components of pathobiomes in agricultural and natural ecosystems are not fully known, there is abundant literature predicting altered geographic distribution of pathogens with changing climate suitability and host distribution (Shaw and Osborne, 2011). For example, a model for predicting the global distribution of *P. cinnamomi*, which is a generalist soil-borne pathogen with a very wide host range, has been developed. This pathogen is considered one of the 100 worst invasive alien species in many countries (Hardham and Blackman, 2018). Since it was first detected on avocado (*Persea americana*) (Cacciola *et al.*, 1998), the number of records of this pathogen in Italian ornamental and forest nurseries, agricultural crops, plantation forests and native woodlands has been increasing (Scanu *et al.*, 2013; Pilotti *et al.*, 2014; Frisullo *et al.*, 2018; Vitale *et al.*, 2019). The model is based on the response of the pathogen to temperature and moisture, and incorporates extensive empirical evidence on the presence of *P. cinnamomi* in the soil (Burgess *et al.*, 2017). Consistently with the model, the comprehensive global map of the *P. cinnamomi* distribution also includes Italian regions with temperate climates, where *P. cinnamomi* has been recently reported as the main stressor threatening the forest stands of evergreen Mediterranean oaks and the Mediterranean maquis vegetation (Scanu *et al.*, 2015; Moricca *et al*, 2016; Frisullo *et al.*, 2018). A study of diversity and distribution of *Phytophthora* species in chestnut (*Castanea sativa*) stands, and their association with ink disease in Europe, showed that *P. cinnamomi* had peculiar ecological requirements compared to the other species occurring in the chestnut rhizosphere (Vettraino *et al.*, 2005). The pathogen was never detected from sites characterized by minimum and maximum temperatures, respectively, below 1.4°C and above 28°C. This confirms that climate change is a major driving factor conditioning the geographical distribution and emergence of this pathogen that is inhibited by low soil temperature. The rise in temperatures as a result of climate change, along with other factors, may have favoured the emergence of Pythium root rot caused by thermophilic *Pythium* species, in leafy vegetables crops in Northern and Southern Italy (Garibaldi *et al.*, 2010; Garibaldi *et al.*, 2015; Gilardi *et al.*, 2018c).

The combined effects of temperature and atmospheric CO_2 concentration on the severity of infections by fungal pathogens in vegetable crops, including the soil-borne pathogen *F. oxysporum*, has been tested in controlled environment conditions simulating global warming (Ferrocino *et al.*, 2013; Chitarra *et al.*, 2015; Gullino *et al.*, 2018). The information provided by these experiments can be useful for developing provisional models to forecast and counteract the effects of climate change on plant diseases caused by soil-borne pathogens.

A consequence of climate change is the increase in the frequency, extent, and intensity of extreme weather events, although these events are not necessarily linked to global warming. Climate change is believed to be responsible for the increasing frequency of medicanes, which are hurricanes of the tropical type occurring in the Mediterranean Sea. Extreme, severe weather events may trigger the emergence of soil-borne plant diseases as they predispose host plants to the infections and favour pathogen spread. The resurgence of mal secco disease (caused by *Plenodomus tracheiphilus*) in lemon (*Citrus* ×*limon*) orchards in the Syracuse province during the last few years may be imputed to severe hailstorms and the high susceptibility of 'Femminello Siracusano 2Kr' (Migheli *et al.*, 2009). This is the prevalent cultivar in new plantings of this typical lemon growing area of Sicily. Medicanes were the drivers of epidemic outbreaks of citrus fruit brown rot and foliage blight in Sicily, caused by *P. citrophthora* (De Patrizio *et al.*, 2012). This is a common soil-borne pathogen which has adapted to an occasional aerial lifestyle.

ENVIRONMENTAL AND ECOLOGICAL CONDITIONS

Environmental and ecological conditions, including warm temperatures, high relative humidity and conducive soil conditions, such as low pH, presence of residues of previous susceptible crops and overwatering, may be major drivers of the emergence of diseases caused by polyphagous soil-borne pathogens. The following case studies illustrate these interactions.

Damping-off, caused by *Pythium ultimum,* has been observed with increasing frequency in Italy on lettuce, wild rocket and lamb's lettuce at temperatures between 15 and 25°C (Gilardi *et al.*, 2018a, b). *Pythium ultimum* has also been reported in northern Italy on coriander (*Coriandrum sativum*) for the first time in the world (Garibaldi *et al.*, 2010a). The presence of new species of *Pythium* in Italy, such as *P. aphanidermatum* on spinach and on swiss chard, *P. irregulare* on lamb's lettuce and *Pythium* Cluster B2a (*P. dissotocum, P. coloratum, P. diclinum, P. dictyosporum, P. lutarium, P.* sp. 'Group F' and *P.* sp. *tumidum*) on lettuce is particularly important at warm temperatures (Garibaldi *et al.*, 2015d; Gilardi *et al.*, 2018b, c). Web blight and damping-off of seedlings, caused by the *Rhizoctonia solani* complex, have recently been found on many hosts of different families, including species of *Campanula* (C. *trachelium*, C. *rapunculodes* and C. *carpatica*), *Rebutia perplexa*, *Nigella dama-*

scena, Lavandula officinalis, Origanum vulgare, L. stoechas, Rosmarinus officinalis, Satureja montana, Dodonea viscosa, Coprosma repens and *C. lucida, Viburnum tinus, Murraya paniculata, Streptosolen jamesonii, Thyptomene saxicola, Chamaerops humilis, Passiflora mollissima* and *Tabebuia impetiginosa* (Garibaldi *et al.*, 2015a, c, e, g; Bertetti *et al.*, 2017; Aiello *et al.*, 2017b; Bertetti *et al.*, 2018a,b). Southern blight caused by *Sclerotium rolfsii* (teleomorph *Athelia rolfsii*) has been reported on numerous host plants grown either in pots or in fields, in greenhouses or in the open air, in northern and southern Italy. These host include *Dichondra repens*, potato, common bean, *Stevia rebaudiana, Hedera helix, Cannabis sativa*, ornamental *Citrus* species, *Convolvulus cneorum* and young seedlings of several other ornamental plants (Pane *et al.*, 2007b, c; Polizzi *et al.*, 2007; Pane *et al.*, 2008; Polizzi *et al.*, 2010; Garibaldi *et al.*, 2013). *Sclerotinia sclerotiorum*, a necrotrophic pathogen known to infect over 400 species of plants from 75 families (Grabowski, 2017), has been observed on many aromatic and ornamental plants, such as thyme, sage, borage, mint, rosemary, aquilegia, petunia, paris daisy, lavender, and gaillardia (Garibaldi *et al.*, 2008; Garibaldi *et al.*, 2015b, f; Garibaldi *et al.*, 2017). Several species of *Fusarium, Cylindrocarpon*-like asexual morphs (*Ilyonectria, Pleiocarpon)* and *Neocosmospora*, have been reported on ornamental plants, including *Agapanthus africanus, Bougainvillea glabra, Cordyline australis, Dasylirion longissimum, Eremophila* spp., *Philoteca myoporoides, Strelitzia reginae, V. tinus* as well as *Trachycarpus princeps* and various other *Arecaceae* species (Aiello *et al.*, 2014; Aiello *et al.*, 2017a; Guarnaccia *et al.*, 2019). Climatic conditions, farming practices, and conducive environmental conditions have been indicated as major drivers of the emergence of these soil-borne pathogens of ornamental plants.

PATHOGEN GENETIC RECOMBINATION OR MUTATION

Successful invasion by an exotic pathogen may depend on its evolutionary potential which allows it to emerge, adapt to new hosts and environments and persist in populations of host plants. It is generally assumed that asexual organisms may exhibit lower invasion success compared to sexually reproducing organisms, due to their inability to generate meiotic progeny which can rapidly adapt to new hosts and environments. Despite, mechanisms generating variation, such as genetic recombination and mutations, are not always associated with fitness benefits. There are several examples showing that clonality does not necessarily reduce invasiveness (Prospero and Cleary, 2017). One of the best documented cases of genetic recombination of an invasive soil-borne plant pathogen is that of *Heterobasidion irregulare* native to North America. This pathogen was accidentally introduced into Italy, and has become invasive in Latium (central Italy), where the Italian stone pine (*Pinus pinea*) is the sole or major pine species. In this invasion area *H. irregulare* prevails on the native sibling species *H. annosum*, and, unlike *H. annosum*, is also able to colonize as a saprophyte pure oak stands (Gonthier *et al.* 2012; Giordano *et al.* 2013; Garbelotto *et al.* 2013). There is evidence in the invasion area, where *H. irregulare* and *H. annosum* are in sympatry, of interspecific pathogen hybridization and introgression of genes, mostly from the native species into the invasive one, suggesting rapid, possibly adaptive, evolution of *H. irregulare* (Gonthier *et al.* 2007; Linzer *et al.*, 2008; Gonthier *et al.* 2015). Epidemiological consequences of this evolution are unpredictable, and no studies have determined how gene introgression may affect fitness and virulence of either the invasive or the native species. However, examples from other organisms, including several plant pathogens, indicate gene introgression is an important evolutionary mechanism, increasing adaptation and pathogenicity of the species involved (Brasier, 2001; Depotter *et al.*, 2016). Interspecific hybrids of *Phytophthora* occur frequently in natural ecosystems and may become invasive (Brasier *et al.*, 2004; Ioos *et al.*, 2006; Burgess, 2015). Hybridization and polyploidy are assumed to be the genetic mechanisms for adaptation to new hosts and speciation in these oomycetes (Bertier *et al.*, 2013). The interspecific hybrid nature, the mating reproduction system (both A1 and A2 mating types occur in Europe) and the intraspecific variability of *P.* ×*cambivora* may explain the ability of this heterothallic oomycete to adapt to different environmental conditions, ranging from rainforests of Southeast Asia, where the species probably originated, to temperate deciduous forests of central Europe and Southern Italy (Jung *et al.* 2017). In nurseries of ornamental plants, the sympatric occurrence of diverse *Phytophthora* species from different geographic origins favours genetic recombination through sexual reproduction between species that have evolved separately and have not developed premating barriers. A natural hybrid between *P. nicotianae* and *P. cactorum*, referred to as *P.* x *pelgrandis*, was reported in Italy as the causal agent of root rot of potted lavender (*L. stoechas*) plants. This interspecific hybrid showed a unique combination of morphological, biological and ecological characteristics inherited from both parental species, some of which may have epidemiological implications (Faedda *et al.* 2103b).

The frequency of natural *Phytophthora* interspecific hybrids in nurseries is possibly greater than expected from the low number of reports. Hybrids, may go unnoticed as they do not always have distinctive morphological traits. Sequencing of ITS-rDNA regions after PCR amplification with universal primers ITS6 and ITS4, often used for molecular identification of *Phytophthora* species, may fail to discriminate these species from their parents. In many cases, application of diverse molecular techniques is required to ascertain the hybrid nature of the isolates (Faedda *et al.*, 2013b). Very recently a rapid High-Resolution Melting (HRM) diagnostic method has been proposed to distinguish *Phytophthora* hybrids from their parental species (Ratti *et al.* 2019).

The high genetic variability of soil-borne vegetable crop pathogens, such as *Fusarium oxysporum* and *P. capsici*, is a challenge and limits application of genetic resistance for the management of the diseases they cause. Evolution or accidental introduction of new pathogen physiological races have been indicated as possible causes of breakdown of rootstock resistance in grafted *Solanum* or cucurbit plants.

Genetic plasticity of *F. oxysporum* and the problem this poses for farmers and plant breeders are exemplified in *F. oxysporum* f. sp. *lactucae*, the cause of Fusarium wilt of lettuce. Until recently, three races (1, 2 and 3) of the pathogen had been identified by their ability to cause disease on differential lettuce cultivars, and using molecular tools. Race 4 was identified in the Netherlands and in Italy (Gilardi *et al.*, 2019). Pathogenicity tests in controlled environmental conditions showed that none of the commercial lettuce cultivars popular in Italy are completely resistant to race 1 of *F. oxysporum* f. sp. *lactucae*, and only 57%of the tested cultivars were resistant to race 2, and 21% were resistant to race 3.

Almost all currently available commercial hybrids of muskmelon (*Cucumis melo*) possess FOM-1 and FOM-2 genes, conferring resistance to the physiological races 0, 1 and 2 of *F. oxysporum* f. sp. *melonis*. However, none of these hybrid hosts are resistant to the race 1-2 of this pathogen, which has occurred in all major production areas in Italy since the 1990s. No commercial hybrids of pepper are completely resistant to *P. capsici* due to the variability of this pathogen, which in Italy is also a major pathogen of tomato and cucumber grown in plastic greenhouses. *Fusarium oxysporum*, because of genetic variability and ability to evolve, encompasses *formae speciales* infecting aromatic flower plants (Gullino *et al.* 2012; Gullino *et al.*, 2015). New *formae speciales* of this pathogen have been frequently reported in nurseries of ornamentals in Italy (Garibaldi *et al.*, 2012; Matic *et al.*, 2018; Ortu *et al.* 2018).

HOST SHIFTS AND EXPANSION OF HOST RANGES

Polyphagous soil-borne plant pathogens, including *S. sclerotiorum*, *S. rolfsii*, *P. cinnamomi*, *P. nicotianae*, *P. palmivora* and *P. niederhauserii*, infecting very many host plant species, usually widen their host ranges when they invade new areas rich in biodiversity. These areas include natural ecosystems, nurseries of ornamentals, and complex agricultural systems, so the pathogens come into contact with potential new host plants. General pathogens are better invaders than specialists, due to their non-selective ability to seek new hosts in a new environment (Navaud *et al.*, 2018; Thines 2019). Although specialized pathogens may infect new hosts less frequently, the host range expansion of *formae speciales* of *F. oxysporum* to include plant species of the same family is quite common. A new *forma specialis* (f. sp.) of *F. oxysporum*, *F. oxysporum* f. sp. *papaveris*, was discovered in the Liguria region (northern Italy) on Iceland poppy (*Papaver nudicaule*) a plant native to Arctic regions of North America and Eurasia (Garibaldi *et al.* 2012; Ortu *et al.* 2015). The pathogen was initially thought to have been introduced through contaminated seeds (Bertetti *et al.* 2015). However, the high susceptibility of artificially inoculated *Chelidonium majus* and *P. rhoeas*, two species of *Papaveraceae* endemic in Italy, to *F. oxysporum* f. sp. *papaveris*, supports the hypothesis that this f. sp. was already present in Liguria, and shifted onto Iceland poppy from wild relatives (Bertetti *et al.*, 2018a).

The report of *F. oxysporum* f. sp. *chrysanthemi* on orange coneflower (*Rudbeckia fulgida*) in Northern Italy, has expanded the list of hosts of this f. sp., including several other ornamental plants of the *Asteraceae* such as chrysanthemum, Paris daisy, African daisy, and gerbera (Matic *et al.* 2018).

INTRODUCTION, OR GEOGRAPHIC RANGE EXPANSION OF SUSCEPTIBLE HOST PLANTS

When an exotic plant is introduced into a new geographical area, there is justified concern about the risk of it being a vehicle for alien pathogens (Eschen *et al.* 2019), while its susceptibility to pathogens already established in the area is often overlooked. It is more likely for new hosts to be susceptible to native pathogens if components of the resident flora are closely related to the introduced host or if polyphagous pathogens are present in the area. One example is avocado (*P. americana*), whose culture is rapidly expanding in southern Italy. This tropical fruit tree is susceptible to white root rot caused by *Rosellinia necatrix*, which is a well-known and widespread pathogen of olive and fruit trees in Italy (Schena *et al.*, 2008; Pasini *et al.*, 2016). White root rot is regarded as a major disease of avocado in Spain, the most important avocado producing country in the Mediterranean Basin. Selection for tolerant rootstocks was carried out, but this was probably not completely successful because integrated management strategies are being sought (Ruano-Rosa *et al.*, 2018a). The disease is a serious threat to the avocado industry in Italy, that currently relies on *P. cinnamomi*–tolerant rootstocks for sustainable production. Molecular diagnostic methods to detect the pathogen in soil and host tissues are available (Schena *et al.*, 2002; Schena *et al.*, 2013), and these could be useful tools to prevent introduction with infected nursery plants or to select non-infested planting sites.

Verticillium wilt caused by *V. dahliae* and Phytophthora crown and root rot caused by *Phytophthora* spp. are common diseases in several traditional crops in Italy, and these pathogens have been recently reported on goji (*Lycium barbarum* L.). This plant, producing edible fruit, is native to China, and was introduced into Italy and grown commercially from only a few years ago. Although goji can is a minor crop, it has become very popular due to the vaunted health benefits of its berries. Verticillium wilt of young plants of goji has been reported in Calabria (Ruano-Rosa *et al.* 2017), while crown and root rot caused by *P. nicotianae* have been reported in Apulia (Cariddi *et al.* 2018).

AVAILABILITY OF FUNGICIDES AND DEVELOPMENT OF FUNGICIDE RESISTANCE

Limited availability of fungicides and the development of fungicide resistance can be relevant as factors fostering emergence of soil-borne diseases in nurseries and agricultural systems. Insensitivity to metalaxyl among isolates of *P. capsici* causing Phytophthora blight of pepper has been reported in southern Italy since the 1990s (Pennisi *et al.*, 1998). In a European-wide sample of 77 *P. ramorum* isolates collected in 2004, 24% were resistant to mefenoxam (Brasier, 2008). Phenylamide fungicides, including compounds such as metalaxyl, metalaxyl-M (mefenoxam) and benalaxyl, are frequently used against damping-off caused by *Pythium*. Resistance to phenylamides appeared shortly after their commercialization, in populations of various plant pathogenic oomycetes, including several *Pythium* spp. such as *P. aphanidermatum, P. dissotocum, P. heterothallicum, P. irregulare, P. cylindrosporum, P. splendens, P. torulosum* and *P. ultimum* (Moorman *et al.*, 2002; Taylor *et al.*, 2002).

A recent study by Matic *et al.* (2019), with 53 isolates of six *Pythium* species (*P. ultimum, P. aphanidermatum, P. irregulare* complex, *P. sylvaticum*, and *Pythium* ClusterB2a sp.) obtained from different vegetable hosts, showed that they were all sensitive to azoxystrobin, with small variations in their species-specific baseline sensitivity. As a consequence, this fungicide may be effectively applied regardless of the pathogen or host species involved. Conversely, precise *Pythium* species identification and sensitivity tests of isolates may be crucial for reliable use of mefenoxam, as baseline sensitivity to this fungicide varies greatly among species, and resistant isolates may occur in field populations of a sensitive species (Matic *et al.*, 2019).

A relevant example of the emergence of diseases due to restrictions in the use of fungicides is the emergence or re-emergence of endemic soil-borne pathogens of vegetable crops after the phasing out of methyl bromide. Without soil fumigation, soil-borne pathogens considered minor, such as *C. coccodes* on tomato and pepper, have emerged, and well-known major pathogens, such as *P. capsici* on pepper and tomato, *S. sclerotiorum*, *R. solani*, *F. oxysporum* f. sp. *lactucae* and *Verticillium dahliae* on lettuce, and *S. sclerotiorum* and *R. solani* on *Solanum* or cucurbit hosts, have re-emerged (Garibaldi *et al.*, 2008; Gilardi *et al.* 2014a,b). The implications of phasing-out of methyl bromide for the management of diseases of vegetable crops in Europe have been addressed by exhaustive reviews (Lazarovits and Subbarao, 2010; Colla *et al.*, 2012).

In the last 10 years, cases of resistance to diverse classes of fungicides have been reported in populations of *Calonectria* and *Phytophthora* species in nurseries of ornamentals in Southern Italy (Vitale *et al.* 2009; Guarnaccia *et al.* 2014; Aiello *et al.* 2018). These cases document the failures to control already established soil-borne polycyclic diseases as results of excessive and improper use of fungicides to control diseases of ornamental plants. The emergence of fungicide resistance in nursery populations of pathogens that have wide host ranges, such as *Phytophthora* and *Calonectria*, is a potential threat for other agricultural systems, because these fungicides are part of integrated pest management (IPM) strategies for many horticultural crops. More ecological approaches, such as the use of pathogen-suppressive soil mixtures or substrates (Hoitink *et al.*,1997; Hoitink and Boehm, 1999; Raviv, 2008; Pugliese *et al.*, 2012; Cesarano *et al.*, 2017a; Pascual *et al.*, 2018; De Corato *et al.*, 2019), and systems-based methods, can be alternatives to the intensive use of fungicides for the management of polycyclic soil-borne diseases in ornamental plant nurseries. Systems-based approaches are disease management strategies that have evolved from IPM concepts, and these incorporate agricultural practices fostering disease suppressive soil microbial communities into the cropping system design (Chellemi *et al.*, 2016).

CHANGES OF CROPPING SYSTEMS

Shifts in farming practices and agricultural techniques, such as propagation methods, soil management types, planting densities and irrigation systems, or the substitution of cultivars or rootstocks, may trigger the emergence or re-emergence of soil-borne plant diseases. The disease consequences of the introduction and the large-scale use of grafted plants in intensive and soilless vegetable crops in Italy, fostered by the phasing-out of methyl-bromide, have changed cropping systems, and this can be a major driving factor of the emergence or resurgence of soil-borne fungal diseases (Gilardi *et al.*, 2013; Gilardi *et al.*, 2014 a, b, c,). Although most rootstocks of *Solanum* or cucurbit crops possess multiple resistances, none tolerate all the potential soil-borne pathogens of a particular crop. Resistance of different rootstocks to a single pathogen also varies greatly, and in some cases resistance may be overcome by high inoculum pressure and environmental conditions conducive for the disease. This has occurred for infections from *V. dahliae* on eggplant grafted on *S. nigrum*, and for *P. nicotianae* on tomato grafted on tomato hybrid rootstocks 'Beaufort' and 'He Man' (*Solanum lycopersicum* x *S. hirsutum*) (Gilardi *et al.*, 2011).

Change and intensification of cultivation systems may favour emergence of soil-borne fungal diseases in traditional tree crops. This is the case with Verticillium wilt (*V. dahliae*) and Phytophthora root rot (*Phytophthora* spp.) in olive orchards, as a consequence of the expansion of intensive and super-intensive planting systems (Cacciola *et al.*, 2011a; Jiménez-Diaz *et al.*, 2012). Mixed infections of both diseases have also been reported in Italy (Lo Giudice *et al.*, 2010). Verticillium wilt is the most serious disease of olive in Spain (López-Escudero and Mercado-Blanco, 2011), the most important world olive-producing country. In Italy, Verticillium wilt is regarded as a major disease of olive in Apulia (Nigro *et al.*, 2005), the most important olive-producing region, and is increasingly becoming a serious problem in new plantings. In Calabria and Sicily this disease occurs sporadically. Verticillium wilt also occurs on old trees, but symptoms are usually less severe and transient, as mature trees may recover spontaneously. Particular attention has been paid to irrigation as a factor favouring the increasing incidence and severity of Verticillium wilt in young, intensive and super-intensive olive plantations (Pérez-Rodríguez *et al.* 2016; Santos-Rufo *et al.*, 2017). However, numerous other factors are involved in the emergence of this disease, including infected planting (rooted-cuttings) and propagation material, establishment of new orchards in infested soils, spread of virulent *V. dahliae* strains of the D (defoliating) pathotype, and susceptibility of many popular cultivars, such as 'Arbequina' and 'Arbosana', which are widely used in super-intensive plantings. These and additional factors contributing to the emergence of Verticillium wilt in olive orchards are discussed in detail in the review on this disease by López-Escudero and Mercado-Blanco (2011). In Italy, evaluation of susceptibility of olive cultivars to the disease, and the search for resistant rootstocks, have potential for management of this disease (Bubici and Cirulli, 2012).

Pomegranate (*Punica granatum*) is another traditional fruit tree, native to the region extending from modern-day Iran to northern India. This plant has been cultivated since ancient times throughout the Middle East, Caucasus and Mediterranean region for its edible fruit. Recent intensive cultivation of this deciduous shrub, and establishment of modern commercial orchards for increasing yields, are posing new disease problems that could be serious constraint to expansion of this fruit crop in southern Italy. Crown and root rots caused by *Coniella* (syn. *Pilidiella*) *granati* are emerging soil-borne diseases in irrigated pomegranate orchards. This pathogen has been recently reported in Apulia, Basilicata, and Calabria regions (Pollastro *et al.*, 2016), and occurs in many other pomegranate-producing countries, including China, Iran, Spain, Greece and Turkey. *Coniella granati* may also cause pomegranate fruit rots. No effective management strategy is available for crown and root rot caused by this pathogen.

MONOCULTURE

It is commonly assumed that continuous cultivation of one crop, or crops of the same family, in the same soil results in increased inoculum of soil-borne pathogens, and increased incidence of root diseases, with consequent detrimental effects on crop yields. There is extensive evidence that reduction of biodiversity due to monoculture leads to development of detrimental soil conditions, that limit the cultivation of the same crop. This phenomenon, known in agriculture since ancient times but not fully explained, has been called "soil sickness". More recently, plant ecologists have preferred the more comprehensive term negative plant-soil feedback (NPSF). This concept stresses the negative feedback between plant and soil, and includes agro-ecosystems and natural plant communities (Cesarano *et al.*, 2017b). Soil sickness is a serious concern for staple, cash, vegetable, forage, flower, ornamental and fruit tree crops (Bonanomi *et al.*, 2007; Bonanomi *et al.*, 2011a,b), including major crops such as wheat, maize, rice, sugarcane, alfalfa, soybean, grape, peach, apple, olive, citrus, tea and coffee. Three fundamental hypotheses have been put forward to explain soil sickness. These are: i) soil nutrient depletion or imbalance; ii) release of autotoxic compounds as root exudates or during decomposition of crop residues; and iii) build up of inoculum of soil-borne pathogens and corresponding modification in soil microbiomes, i.e. shifts in the soil microbial community structure from beneficial, including mychorrizae, to detrimental microorganisms (Cesarano *et al.*, 2017b). The effectiveness of soil sterilization or soil treatments with fungicides in restoring crop productivity has been regarded as the most convincing proof supporting the hypothesis that plant pathogens are major determinants of soil sickness (Cesarano *et al.*, 2017b). In Italy, the build-up of inoculum of several soil-borne fungal pathogens has been implicated in apple replanting problems and in the black root rot complex of strawberry (Manici *et al.*, 2003; 2005). Soil sickness due to *Phytophthora* spp. is causing the decline of lentil (*Lens culinaris*) crops on the island of Ustica (Puglisi *et al.*, 2016), a small island in the Tyrrhenian sea 38 km north of Sicily. The local landrace of lentil, appreciated by consumers and recently recognized as a Slow Food Presidium, is grown as a monocrop.

CONCLUSIONS

Based on published reports, most of the recently emerging diseases in Italy have been caused by pathogens introduced from other countries. Although emerging disease occurrence is partly distorted by the greater attention paid to discoveries of exotic pathogens and the rapidity with which the news of the emergence of new pathogens is delivered, this can mainly be imputed to the failure of the current phytosanitary system to prevent the introduction of alien pathogens, and confirms the need to reinforce this system. The EU regulation 2016/2031, which becomes fully effective in December 2019, demonstrates that there is awareness of the problem, and has addressed biosecurity by introducing major changes in phytosanitary regulations. This regulation aims to prevent introduction, establishment, and spread of harmful organisms for plants, and to coordinate and harmonize efforts made by different countries throughout the European Union. Similarly, an efficient phytosanitary certification system for propagation materials and nursery plants is the only effective means to prevent the spread of introduced and native pathogens within individual countries. These systems may be mandatory or voluntary, and must include forestry and landscape nursery plants as there is considerable evidence that these plants may be "Trojan horses" for the introduction and spread of invasive soil-borne plant pathogens in natural and forestry ecosystems. These ecosystems are particularly vulnerable, as eradication or mitigation of the effects of harmful pathogens in these ecosystems are not feasible or is more problematic than in agricultural systems.

Another way to reduce the risk of introduction and spread of exotic pathogens is to develop modern and self-sufficient national nursery production systems. In this regard Italy, which was able to meet domestic demand for nursery plants in strategic sectors, such as citrus, olive, fruit and nut trees crops, increasingly relies on imports from other countries. Plant material for ornamental, forestry, and staple and horticultural crops such as potato and strawberry, is also imported into Italy.

Application of phytosanitary regulations and the success of phytosanitary certification systems depend largely on the availability of adequate diagnostic methods, which have to be continuously updated to accommodate the rapid evolution of nomenclature and molecular taxonomy of fungi (Rossman and Palm-Hernandez, 2008; Groenewald *et al.*, 2011; Stielow *et al.*, 2015). These systems must be specific, sensitive, rapid, reproducible and practical. In the last 20 years, a large number of molecular methods for the detection of soil-borne fungal plant pathogens have been published in the scientific literature (Schena *et al.*, 2013). However, very few of these methods have been validated or applied routinely on large-scales, while some lack specificity which is a basic requirement (Blomquist *et al.*, 2005; Kunadiya *et al.*, 2017). Development of new and more specific molecular diagnostic methods is facilitated by the availability of whole genomes for a growing number of fungi and oomycetes (Feau *et al.*, 2019). Innovative and promising approaches for diagnoses of soil-borne pathogens include multiplex assays for genus- and species-specific detection of *Phytophthora* in environmental samples (Scibetta *et al.*, 2012; Bilodeau *et al.*, 2014), mRNA-based protocols to circumvent the problem of false positives due to the detection of DNA of non-viable propagules (Chimento *et al.*, 2012; Kunadiya *et al.*, 2019), isothermal amplification assays for *in situ* detection of *Phytophthora* spp. in plant tissues (Miles *et al.*, 2014), and PCR-based methods for the identification of *formae speciales* and races of *F. oxysporum* from seeds, plants and soil samples (Pasquali *et al.*,2008; Mbofung and Pryor, 2010; Srinivasan *et al.*, 2010; Lievens *et al.*, 2012; Gilardi *et al.*, 2017; Thomas *et al.*, 2017). Recently, a loop-mediated-isothermal amplification (LAMP) assay using a panel of target and non-target species was developed for detection of *Fusarium oxysporum* f. sp. *lactucae* in soil, lettuce seeds and plants. This assay is a significant advantage over the traditional methods, which do not allow clear discrimination of the *formae speciales* of *F. oxysporum* (Franco Ortega *et al.*, 2018). The taxonomy of *F. oxysporum* is evolving, so this will affect evaluation and interpretation of results obtained with these new diagnostic methods. Very recently, this taxonomy has been substantially modified and several cryptic species (so far 15) have been resolved within this species complex (Lombard *et al.*, 2019).

Next-generation sequencing approaches, which identify microorganisms in terms of operational taxonomic units, remain too expensive to be used as routine diagnostic methods, and the results are not precise enough to be used for quarantine or certification purposes. Conversely, as shown for aerial plant pathogens (Mosca *et al.*, 2014; Abdelfattah *et al.*, 2016), high-throughput sequencing could be an appropriate tool for studying the complexity and ecological functions of soil microbiota, its role in soil suppressiveness and negative plant-soil feedback, its interactions with soil-inhabitant pathogens, and effects of soil management practices on microbial population dynamics and functionality (Bonanomi *et al.*, 2016; Gómez Expósito *et al.*, 2017; Schlatter *et al.*, 2017; Ampt *et al.*, 2018).

Climate change, whose most evident effect is the rise in temperatures, has often been presumed to be responsible for the emergence of soil-borne diseases, simply because they were caused by thermophilic or mesophilic pathogens. This over simplification causes direct and indirect effects of climate change to be underestimated, and disease emergence drivers, such as severe weather events or conducive environmental conditions, which here have been considered distinct, may themselves be a consequence of climate change. Natural ecosystems are generally more vulnerable to the effects of climate changes, while in agricultural systems these effects can be mitigated by active interventions, such as irrigation during dry periods or the use of genotypes resistant to biotic or abiotic stress factors. Conversely, other disease emergence drivers, such as the genetic mutation of pathogens, prevail in agricultural systems due to the selective pressure exerted by the genetic uniformity of the host on the pathogen. In general, host genetic uniformity and susceptibility have crucial roles in determining invasion success and spread dynamics of plant pathogens after introduction and establishment, so monocultures of genetically similar or identical plants are severely impacted by invasive pathogens. It is generally assumed that biodiversity of natural ecosystems give them resilience against invasive exotic pathogens, but if the invader is a polyphagous pathogen, such as *P. cinnamomi* or *P. ramorum*, the result is a loss of biodiversity. It is, therefore, not easy to predict and evaluate the impacts of emerging diseases, as the emergence driving factors are numerous and complex, often interacting with each other.

ACKNOWLEDGEMENTS

The authors thank A. Garibaldi for the invaluable suggestions in preparing this review, G.E. Agosteo, P. Capretti, M. Garbelotto, P. Gonthier, G. Magnano di San Lio, S. Moricca, G. Polizzi, A. Santini, B. Scanu and L. Schena for provision of updated information, and Mrs. A. Davies for English revision of the manuscript.

LITERATURE CITED

Abad Z.G., Abad J.A., Cacciola S.O., Pane A., Faedda R., ..., Değirmenci K., 2014. *Phytophthora niederhauserii* sp. nov., a polyphagous species associated with ornamentals, fruit trees and native plants in 13 countries. *Mycologia* 106: 431–447.

Abdelfattah A., Wisniewski M., Li Destri Nicosia M.G., Cacciola S.O., Schena L., 2016. Metagenomic analysis of fungal diversity on strawberry plants and the effect of management practices on the fungal community structure of aerial organs. *PLoS ONE* 11(8): e0160470. https: //doi.org/10.1371/journal.pone.0160470

Aiello D., Vitale A., Hyakumachi M., Polizzi G., 2012. Molecular characterization and pathogenicity of binucleate *Rhizoctonia* AG-F associated to the watermelon vine decline in Italy. *European Journal of Plant Pathology* 134: 161–165.

Aiello D., Polizzi G., Crous P.W., Lombard L., 2017a. *Pleiocarpon* gen. nov. and a new species of *Ilyonectria* causing basal rot of *Strelitzia reginae* in Italy. *IMA Fungus* 8: 65–76.

Aiello D., Guarnaccia V., Vitale A., Cirvilleri G., Granata G., ... Crous P.W., 2014. *Ilyonectria palmarum* sp. nov. causing dry basal stem rot of Arecaceae. *European Journal of Plant Pathology* 138: 347–359.

Aiello D., Guarnaccia V. , Formica P. T. , Hyakumachi M., Polizzi G., 2017b. Occurrence and characterisation of *Rhizoctonia* species causing diseases of ornamental plants in Italy. *European Journal of PlantPathology* 148: 967–982.

Aiello D., Hansen Z.R., Smart C.D., Polizzi G., Guarnaccia V., 2018. Characterisation and mefenoxam sensitivity of *Phytophthora* spp. from ornamental plants in Italian nurseries. *Phytopathologia Mediterranea* 57: 245–256

Albuquerque Alves T.C., Tessmann D.J., Ivors K.L., Ristaino J.B., dos Santos À.F. 2019. *Phytophthora acaciae* sp. nov., a new species causing gummosis of black wattle in Brasil. *Mycologia* 111: 445–455.

Aleandri M.P., Martignoni D., Reda R., Alfaro-Fernández A., Font M.I., Armengol J., Chilosi G., 2017. Involvement of *Olpidium bornovanus* and *O. virulentus* in the occurrence of melon root rotand vine decline caused by *Monosporascus cannonballus* in central Italy. *Journal of Plant Pathology* 99: 169- 176.

Ampt E.A., vanRuijven J., Raaijmakers J.M., Termorshuizen A.J., Mommer L., 2018. Linking ecology and plant pathology to unravel the importance of soil-borne fungal pathogens in species-rich grasslands. *European Journal of Plant Pathology* https: //doi.org/10.1007/s10658-018-1573-x

Andjic V., Dell B., Barber P., Hardy G., Wingfield M., Burgess T., 2011. Plants for planting; indirect evidence for the movement of a serious forest pathogen, *Teratosphaeria destructans*, in Asia. *European Journal of Plant Pathology* 131: 49–58.

Armengol J., José C.M., Moya M.J., Sales R., Vicent A., García-Jiménez J. 2008. *Fusarium solani* f.sp. *cucurbitae* race 1, a potential pathogen of grafted watermelon production in Spain. *Bulletin OEPP/EPPO Bulletin* 30: 179–183.

Barnes I., Wingfield M.J., Carbone I., Kirisits T., Wingfield B.D., 2014. Population structure and diversity of an invasive pine needle pathogen reflects anthropogenic activity. *Ecology and Evolution* 4: 3642–3661.

Ben Salem I., Correia K.C., Boughalleb N., Michereff S.J., León M., Abad-Campos P., García-Jiménez J., Armengol J., 2013. *Monosporascus eutypoides*, a cause of root rot and vine decline in Tunisia, and evidence that *M. cannonballus* and *M. eutypoides* are distinct species. *Plant Disease* 97: 737–743.

Bertetti D., Ortu G.,Gullino M.L. Garibaldi A., 2015. Contamination of seeds of Iceland poppy (*Papaver nudicaule* L.) by *Fusarium oxysporum*. *Phytoparasitica* 43: 189–196.

Bertetti D., Martini P., Pensa P., Gullino M.L., Garibaldi A., 2017. Nuovi parassiti riscontrati su colture aromatiche e ornamentali del Nord Italia. *Protezione delle Colture* 10: 7-13. (in Italian).

Bertetti D., Gullino M.L., Garibaldi A. 2018a. Susceptibility of some Papaveraceae plants to *Fusarium oxysporum* f. sp. *papaveris*. *Journal of Plant Diseases and Protection* 125: 103–108.

Bertetti D., Gilardi G., Gullino M.L., Garibaldi A. 2018b. Nuovi parassiti fungini su colture ornamentali comparsi in Italia settentrionale negli ultimi mesi. *Protezione delle colture* 118: 2-6. (in Italian).

Bertier L., Leus L., D'hondt L, de Cock A.W., Höfte M., 2013. Host adaptation and speciation through hybridization and polyploidy in *Phytophthora*. *PLoS ONE*8: e85385. https: //doi.org/10.1371/journal.pone.0085385.

Biasi A., Martin F.N., Cacciola S.O., Magnano di San Lio G., Grünwald N.J., Schena L., 2016. Genetic analysis of *Phytophthora nicotianae* populations from different hosts using microsatellite markers. *Phytopathology* 106: 1006–1014.

Bilodeau G.J., Martin F.N., Coffey M.D., Blomquist C.L., 2014. Development of a multiplex assay for genus- and species-specific detection of *Phytophthora* a multiplex assay for genus- and species-specific detection of *Phytophthora* based on differences in mitochondrial gene order. *Phytopathology* 104: 733–748.

Blair J.E., Coffey M.D., Park S.Y., Geiser D.M., Kang S., 2008. A multi-locus phylogeny for *Phytophthora* utilizing markers derived from complete genome sequences. *Fungal Genetics and Biology* 45: 266–277.

Blomquist C., Irving T., Osterbauer N., Reeser P., 2005. *Phytophthora hibernalis*: a new pathogen on rhododendron and evidence of cross amplification with two PCR detection assays for *Phytophthora ramorum*. *Plant Health Progress* doi: 10.1094/PHP-2005-0728-01-HN.

Bonanomi G., Del Sorbo G., Mazzoleni S., Scala F., 2007. Autotoxicity of decaying tomato residues affects susceptibility of tomato to fusarium wilt. *Journal of Plant Pathology* 89: 219–226.

Bonanomi G., Antignani V., Barile E., Lanzotti V., Scala F., 2011a. Decomposition of *Medicago sativa* residues affects phytotoxicity, fungal growth and soil-borne pathogen diseases. *Journal of Plant Pathology* 93: 57–69.

Bonanomi G., D'Ascoli R., Antignani V., Capodilupo M., Cozzolino L., …Zoina A., 2011b. Assessing soil quality under intensive cultivation and tree orchards in Southern Italy. *Applied Soil Ecology* 47: 184–194.

Bonanomi G., De Filippis F., Cesarano G., La Storia A., Ercolini D., Scala F,. 2016. Organic farming induces changes in soil microbiota that affect agro-ecosystem functions. *Soil Biology and Biochemistry* 103: 327–336.

Brasier C.M., 2001. Rapid evolution of introduced plant pathogens via interspecific hybridization. *Bioscience* 51: 123.

Brasier C., 2008. The biosecurity treath to the UK and global environment from international trade in plants. *Plant Pathology* 57: 792–808.

Brasier C., Kirk S.A., Delcan J., Cooke D.E.L., Jung T., Man In't Veld W.A. 2004. *Phytophthora alni* sp. nov. and its variants: designation of emerging heteroploid hybrid pathogens spreading on Alnus trees. *Mycological Research* 108: 1172–1184.

Bruton B.D., 2000. Assessment of virulence of *Acremonium cucurbitacearum* and *Monosporascus cannonballus* on *Cucumis melo*. *Plant Disease* 84: 907–913.

Bubici G., Cirulli M., 2012. Control of Verticillium wilt of olive by resistant rootstocks. *Plant Soil* 352: 363–376.

Burgess T.I., 2015. Molecular characterization of natural hybrids formed between five related indigenous clade 6 *Phytophthora* species. *PLoS ONE* 10: e0134225. https: //doi.org/10.1371/journal.pone.0134225. 109.

Burgess T.I., Scott J.K., Mcdougall K.L., Stukely M.J.C., Crane C., … Hardy G.E.St.J., 2017. Current and projected global distribution of *Phytophthora cinnamomi*, one of the world's worst plant pathogens. *Global Change Biology* 23: 1661–1674.

Cabral C.S., de N. Fonseca M.E., Brunelli K.R., Rossato M., Costa H., Boiteux L.S., 2018. Relationships among Brazilian and worldwide isolates of *Fusarium oxysporum* f. sp. *lactucae* race 1 inferred from ribosomal intergenic spacer (IGS-rDNA) region and EF-1α gene sequences. *European Journal of Plant Pathology* 152: 81–94.

Cacciola S.O., Pane A., Davino M., Magnano di San Lio G., 1998. First report of root rot caused by *Phytophthora cinnamomi* on avocado in Italy. *Plant Disease* 82: 1281.

Cacciola S.O., Scibetta S., Martini P., Rizza C., Pane A. 2009b. *Phytophthora* taxon *niederhauserii*, a new root and crown rot pathogen of *Banksia* spp. in Italy. *Plant Disease* 93: 1216.

Cacciola S.O., Scibetta S., Pane A., Faedda R., Rizza C., 2009a. *Callistemon citrinus* and *Cistus salvifolius*, two new hosts of *Phytophthora* taxon *niederhauserii* in Italy. *Plant Disease* 93: 1075.

Cacciola S.O., Faedda R., Pane A., Scarito G., 2011a. Root and crown rot of olive caused by *Phytophthora* spp. In: *Olive Diseases and Disorders*, (L.Schena, G.E. Agosteo, S.O. Cacciola, Magnano di San Lio G. ed.), Transworld Research Network, Trivandrum, Kerala, India, 305–327.

Cacciola S.O., Pane A., Faedda R., Rizza C., Badalà F., Magnano di San Lio G., 2011b. Bud and root rot of windmill palm (*Trachycarpus fortunei*) caused by simultaneous infections of *Phytophthora palmivora* and *P. nicotianae* in Sicily. *Plant Disease* 95: 769.

Cariddi C., Mincuzzi A., Schena L., Ippolito A., Sanzani S.M., 2018. First report of collar and root rot caused by *Phytophthora nicotianae* on *Lycium barbarum*. *Journal of Plant Pathology* 100: 361.

Carlucci A., Raimondo M.L., Santos J., Phillips A.J.L., 2012. *Plectosphaerella* species associated with root and collar rots of horticultural crops in southern Italy. *Persoonia* 28: 34–48.

Cesarano G., De Filippis F., La Storia A., Scala F., Bonanomi G., 2017a. Organic amendment type and application frequency affect crop yields, soil fertility and microbiome composition. *Applied Soil Ecology* 120: 254–264.

Cesarano G., Zotti M., Antignani V., Marra R., Scala F., Bonanomi G. 2017b. Soil sickness and negative plant–soil feedback: a reappraisal of hypotheses. *Journal of Plant Pathology* 99: 545–570.

Chakrabortry S., 2013. Migrate or evolve: options for plant pathogens under climate change. *Global Change Biology* 19: 1985–2000.

Chellemi D.O., Gamliel A., Katan J., Subbarau K.V., 2016. Development and deployment of systems-based approaches for the management of soilborne plant pathogens. *Phytopthology* 106: 216–225.

Chilosi G., Reda R., Aleandri M.P., Camelel., Altieri L.,… Frisullo S., 2008. Fungi associated with root rot and collapse of melon in Italy. *OEPP/EPPO Bulletin* 38: 147–154.

Chimento A., Cacciola S.O., Garbelotto M., 2012. Detection of mRNA by reverse-transcription PCR as an indicator of viability in *Phytophthora ramorum*. *Forest Pathology* 42: 14–21.

Chitarra W., Siciliano I., Ferrocino I., Gullino M.L., Garibaldi A., 2015. Effect of elevated atmospheric CO_2 and temperature on the disease severity of rocket plants caused by Fusarium wilt under phytotron conditions. *PLoS ONE* 10: e0140769.

Cohen R., Pivonia S., Burger Y., Edelstein M., Gamliel A., Katan J. 2000. Toward integrated management of *Monosporascus* wilt of melons in Israel. *Plant Disease* 84: 496–505.

Cohen R., Burger Y., Horev C., Koren A., Hedelstein M., 2007. Introducing grafted cucrbits to modern agriculture. The Israeli experience. *Plant Disease* 91: 916–923.

Colla P., Gilardi G., Gullino M.L. 2012. A review and critical analysis of the European situation of soil-borne disease management in the vegetable sector. *Phytoparasitica* 40: 515–523.

Cooke D.E.L., Schena L., Cacciola S.O., 2007. Tools to detect, identify, and monitor *Phytophthora* species in natural ecosystems. *Journal of Plant Pathology* 89: 13–28.

Davis A.R., Perkins-Veazie P., Sakata Y., López-Galarza S., Maroto J.V., ..., Lee J-M. 2008. Cucurbit grafting. *Critical Reviews in Plant Sciences* 27: 50–74.

De Corato U., Patruno L., Avella N., La Colla G., Cucci G. 2019. Composts from green sources show an increased suppressiveness to soilborne plant pathogenic fungi: Relationships between physicochemical properties, disease suppression, and the microbiome. *Crop Protection* 124: 104870.

De Patrizio A., Faedda R., Pane A., Cacciola S.O. 2014. Air-borne infections of fruit brown rot of citrus in Sicily. P. 113 *in* (T. Jung, C. Brasier, M.E. Sánchez, A. Pérez-Sierra eds.) Phytophthoras in Forest and Natural Ecosystems. *Proceedings of the 6th Meeting of International Union of Forest Research Organizations (IUFRO) Working Party S07-02-09*, 9th-14th September, 2012, Cordoba, Spain. 202 pp.

Depotter J.R.L., Seidl M.F., Wood T.A., Thomma B.P.H.J., 2016. Interspecific hybridization impacts host range and pathogenicity of filamentous microbes.*Current Opinion in Microbiology* 32: 7-13.

Desprez-Loustau M.L., Courtecuisse R., Robin C., Husson C., Moreau P.A., ...Sache I., 2010. Species diversity and drivers of spread of alien fungi (sensu lato) in Europe with a particular focus on France. *Biological Invasions* 12: 157–172.

Dukes J.S., Pontius J., Orwig D., Garnas J.R., Rodgers V.L., ..., Ayres M., 2009. Responses of insect pests, pathogens, and invasive plant species to climate change in the forests of northeastern North America: what can we predict? *Canadian Journal of Forest Research* 39: 231–248.

Engelbrecht J., T.A. Duong, Berg N.V.D., 2017. New microsatellite markers for population studies of *Phytophthora cinnamomi*, an important global pathogen. *Scientific Reports* 7: 17631.

Eschen R., O'Hanlon, Santini A., Vannini A., Roques A., ... Kenis M., 2019. Safeguarding global plant health: the rise of sentinels. *Journal of Pest Science* 92: 29–36.

Faedda R., Cacciola S.O., Pane A., Martini P., Odasso M., Magnano di San Lio G., 2013a. First report of *Phytophthora* taxon *niederhauserii* causing root and stem rot of mimosa in Italy. *Plant Disease* 97: 688.

Faedda R., Cacciola S.O., Pane A., Szigethy A., Bakonyi J., ... Magnano di San Lio G., 2013b. *Phytophthora × pelgrandis* causes root and collar rot of *Lavandula stoechas* in Italy. *Plant Disease* 97: 1091–1096.

Feau N., Ojeda Alayon D.I., Beauseigle S., Bilodeau G., Brar A., ... Hamelin R.C., 2019. Improved detection and identification of the sudden oak death pathogen *Phytophthora ramorum* and the Port Orford cedar root pathogen *Phytophthora lateralis*. *Plant Pathology* doi: 10.1111/ppa.13015.

Felipe A.,García López F.A., González-Eguiarte D.R., Rodríguez-Macías R., Zarazúa-Villaseñor P., Huitrón-Ramírez M.V. 2018. Watermelon production with rootstocks in soils infested with the melon necrotic spot virus. *Revista Mexicana de Ciencias Agrícolas* 9: 577–587.

Ferrocino I., Chitarra W., Pugliese M., Gilardi G., Gullino M.L., Garibaldi A., 2013. Effect of elevated atmospheric CO_2 and temperature on disease severity of *Fusarium oxysporum* f. sp. *lactucae* on lettuce plants. *Applied Soil Ecolology* 72: 1-6.

Fisher M.C., Henk D.A., Briggs C.J., Brownstein J.S., Madoff L.C., McCraw S.L., Gurr S.J., 2012. Emerging fungal threats to animal, plant and ecosystem health. *Nature* 484: 186–194.

Franco-Ortega S., Tomlinson J., Gilardi G., Spadaro D., Gullino M.L., Garibaldi A., Boonham N., 2018. Rapid detection of *Fusarium oxysporum* f. sp. *lactucae* on soil, lettuce seeds and plants using loop-mediated isothermal amplification. *Plant Patholology* 67: 1462–1473.

Frisullo S., Lima G., Magnano di San Lio G., Camele I., Melissano L., ... Cacciola S.O., 2018 *Phytophthora cinnamomi* involved in the decline of holm oak (*Quercus ilex*) stands in southern Italy. *Forest Science* 64: 290–298.

Garbelotto M., Gonthier P., 2013. Biology, epidemiology and control of *Heterobasidion* species worldwide. *Annual Review of Phytopathology* 51: 39–59.

Garbelotto M., Hayden K.J. 2012. Sudden oak death: interactions of the exotic oomycete *Phyotphthora ramorum* with native North American hosts. *Eukaryotic Cell* 11: 1313–1323.

Garbelotto M., Pautasso M., 2012. Impacts of exotic forest pathogens on Mediterranean ecosystems: four case studies. *European Journal of Plant Pathology* 133: 101–116.

Garbelotto M., Guglielmo F., Mascheretti S., Croucher P.J.P., Gonthier P., 2013. Population genetic analyses provide insights on the introduction pathway and spread patterns of the North American forest pathogen *Heterobasidion irregulare* in Italy. *Molecular Ecology* 22: 4855–4869.

García-Jiménez J., Armengol J., Sales R., Jordá C., Bruton B.D., 2000. Fungal pathogens associated to melon collapse in Spain. OEPP Bulletin/EPPO Bulletin 30: 169–173.

Garibaldi A., Gilardi G., Gullino M.L. 2010a. First report of collar and root rot caused by *Pythium ultimum* on coriander in Italy. *Plant Disease* 94: 1167.

Garibaldi A., Gilardi G., Gullino M.L. 2010b. First report of collar and root rot caused by *Phytophthora tentaculata* on witloof chicory (*Cichorium intybus*) in Italy. *Plant Disease* 94: 1504.

Garibaldi A., Pensa P., Gullino M.L. 2008. First report of *Sclerotinia sclerotiorum* on *Argyranthemun frutescens* in Italy. *Plant Disease* 92: 1250.

Garibaldi A., Baudino M., Minuto A., Gullino M.L., 2008. Effectiveness of fumigants and grafting against tomato brown root rot caused by *Colletotrichum coccodes*. *Phytoparasitica* 36: 483–488.

Garibaldi A., Bertetti D., Franco Ortega S., Gullino M.L. 2015a. First report of web blight on NettleLeaved Bellflower caused by *Rhizoctonia solani* AG 1-IB in Italy. *Journal of Plant Pathology* 97: 396.

Garibaldi A., Bertetti D., Ortu G., Gullino M.L., 2015b. First report of Sclerotinia blight caused by *Sclerotinia sclerotiorum* on *Gaillardia* x *grandiflora* in northern Italy. *Plant Disease* 99: 729.

Garibaldi A., Bertetti D., Ortu G., Gullino M.L. 2015c. First report of web blight on Rose Campion (*Lychnis coronaria*) caused by *Rhizoctonia solani* AG 1 - IB in Italy. *Plant Disease* 99: 162.

Garibaldi A., Gilardi G., Ortu G., Gullino M.L., 2015d. Root rot of spinach in Southern Italy caused by *Pythium aphanidermatum*. *Plant Disease* 9: 159.

Garibaldi A., Gilardi G., Ortu G., Gullino M.L. 2016. First report of *Fusarium oxysporum* f. sp. *radicis-cucumerinum* causing wilt on cucumber (*Cucumis sativus*) in Italy. *Plant Disease* 100: 1791.

Garibaldi A., Ortu G., Bertetti D., Gullino M.L. 2015e. First report of web blight on Creeping Bellflower (*Campanula rapunculoides*) caused by Rhizoctonia solani AG 1-IB in Italy. *Plant Disease* 99: 1037.

Garibaldi A., Bertetti D., Pensa P., Franco Ortega S., Gullino M.L. 2015f. First report of *Sclerotinia sclerotiorum* on butterly lavender in Italy. *Journal of Plant Pathology* 97: 397.

Garibaldi A., Bertetti D., Pensa P., Matic S.,. Gullino M.L, 2017. First report of white mould caused by *Sclerotinia sclerotiorum* on rosemary in Italy. *Journal of Plant Pathology* 99: 543.

Garibaldi A., Bertetti D., Pensa P., Ortu G., Gullino M.L., 2015g. First report of web blight on Butterfly Lavender (*Lavandula stoechas*) caused by *Rhizoctonia solani* AG 1 - IB in Italy. *Plant Disease* 99: 1039.

Garibaldi A., Gilardi G., Ortu G., Gullino M.L., Testa M., 2013. First report of southern blight caused by *Sclerotium rolfsii* on common bean (*Phaseolus vulgaris*) in Italy. *Plant Disease* 97: 1386.

Garibaldi A., Martini P., Repetto, L., Odasso, M., Bertetti, D., Poli, A., Gullino M.L., 2012. First report of *Fusarium oxysporum* causing wilt on Iceland poppy (*Papaver nudicaule*) in Italy. *Plant Disease* 96: 1823.

Gilardi G., Garibaldi A., Gullino M.L., 2018a. Emerging pathogens as a consequence of globalization and climate change: leafy vegetable as a case study. *Phytopathologia Mediterranea* 57: 146–152.

Gilardi G., Gullino M.L., Garibaldi A., 2011. Reaction of tomato rootstocks to selected soil-borne pathogens under artificial inoculation conditions. *Acta Horticulturae* 914: 345–348.

Gilardi G., Gullino M.L., Garibaldi A., 2013. Critical aspects of grafting as a possible strategy to manage soil-borne pathogens. *Scientia Horticulturae*, 149: 19–21.

Gilardi G., Gullino M.L., Garibaldi A., 2018b. Emerging foliar and soil-borne pathogens of leafy vegetable crops: a possible threat to Europe. *EPPO Bullettin* 48: 116–127.

Gilardi G., Matic S., Gullino M.L., Garibaldi A., 2018c. First report of root rot caused by *Pythium sylvaticum* on lettuce in Italy. *Plant Disease* 102: 454.

Gilardi G., Demarchi S., Martano G., Gullino M.L., Garibaldi A., 2014a. Success and failures of grafting pepper against soil-borne pathogens. *Acta Horticulturae* 1044: 67–71.

Gilardi G., Pugliese M., Colla P., Gullino M.L., Garibaldi A., 2014b. Management of *Phytophthora capsici* on bell pepper and *Colletotrichum coccodes* on tomato by using grafting and organic amendments. *Acta Horticulturae* 1044: 257–262.

Gilardi G., Colla P., Pugliese M., Baudino M., Gullino M.L., Garibaldi A., 2014c. Control of *Colletotrichum coccodes* on tomato by grafting and soil amendments. *Journal of Phytopathology* 162: 116–123.

Gilardi G., Franco Ortega S., van Rijswick P.C.J., Ortu G., Gullino M.L., Garibaldi A., 2017. A new race of *Fusarium oxysporum* f. sp. *lactucae* of lettuce. *Plant Pathology* 66: 677–688.

Gilardi G., Garibaldi A., Matic S., Senatore M.T., Pipponzi S., Prodi A., Gullino M.L., 2019. First report of *Fusarium oxysporum* f. sp. *lactucae* race 4 on lettuce in Italy. *Plant Disease*, https://doi.org/10.1094/PDIS-05-19-0902-PDN.

Ginetti B., Carmignani S., Ragazzi A., Werres S., Moricca S., 2014. Foliar blight and shoot dieback caused by *Phytophthora ramorum* on *Viburnum tinus* in the Pistoia area, Tuscany, central Italy. *Plant Disease* 98: 423.

Giordano L., Gonthier P., Lione L., Capretti P., Garbelotto M., 2013. The saprobic and fruiting abilities of the exotic forest pathogen *Heterobasidion irregulare* may explain its invasiveness. *Biological Invasions* 14: 803–814.

Gómez J., Serrano Y., Pérez A., Porcel E., Gómez R., Aguilar M.I. 2014. *Fusarium solani* f. sp. *cucurbitae*, affecting melon in Almería Province, Spain. *Australasian Plant Disease Notes* 9: 136.

Gómez Expósito R., de Bruijn I., Postma J., Raaijmakers J.M., 2017. Current insights into the role of rhizosphere bacteria in disease suppressive soils. *Frontiers in Microbiology* 8: 2529.doi: 10.3389/fmicb.2017.02529.

Gonthier P., Garbelotto M., 2011. Amplified fragment length polymorphism and sequence analyses reveal massive gene introgression from the European fungal pathogen *Heterobasidion annosum* into its introduced congener *H. irregulare*. *Molecular Ecology* 20: 2756–2770.

Gonthier P., Lione G., Giordano L., Garbelotto M., 2012. The American forest pathogen *Heterobasidion irregulare* colonizes unexpected habitats after its introduction in Italy. *Ecological Applications* 22: 2135–2143.

Gonthier P., Nicolotti G., Linzer R., Guglielmo F., Garbelotto M., 2007. Invasion of European pine stands by a North American forest pathogen and its hybridization with a native interfertile taxon. *Molecular Ecology*16: 1389–1400.

Gonthier P., Sillo F., Lagostina E., Roccotelli A., Cacciola S.O., Stenlid J., Garbelotto M., 2015. Selection processes in simple sequence repeats suggest a correlation with their genomic location: insights from a fungal model system. *BMC Genomics* 16: 1107. doi: 10.1186/s12864-015-2274-x.

Grabowski M., 2017. Exploring the host range of *Sclerotinia sclerotiorum* in herbaceous ornamental plants. University of Minnesota, PhD Thesis, 106 pp. Retrieved from the University of Minnesota Digital Conservancy, http: //hdl.handle.net/11299/201043.

Groenewald J.Z., Groenewald M., Crous P. W., 2011. Impact of DNA data on fungal and yeast taxonomy. *Microbiology Australia* 32: 100–104.

Guarnaccia V., Aiello D., Polizzi G., Crous P.W., Sandoval-Denis M. 2019. Soilborne diseases caused by *Fusarium* and *Neocosmospora* spp. on ornamental plants in Italy. *Phytopathologia Mediterranea* 58: 127–137.

Guarnaccia V., Aiello D., Polizzi G., Perrone G., Stea G., Vitale A., 2014. Emergence of prochloraz-resistant populations of *Calonectria pauciramosa* and *Calonectria polizzii* in ornamental nurseries of Southern Italy. *Plant Disease* 98: 344–350.

Gullino M. L., Katan J., Garibaldi A. 2012. The genus *Fusarium* and the species that affect greenhouse vegetables and ornamentals. In: *Fusarium wilts of greenhouse vegetable and ornamental crops* (M.L. Gullino, J. Katan, A. Garibaldi eds.). APS Press, St. Paul, MN, USA, 5-9.

Gullino M.L., Daughtrey M.L., Garibaldi A., Elmer W.H., 2015. Fusarium wilts of ornamental crops and their management. *Crop Protection*, 75: 50 – 59.

Gullino M.L., Pugliese M., Gilardi G., Garibaldi A., 2018. Effect of increased CO_2 and temperature on plant diseases: a critical appraisal of results obtained in studies carried out under controlled environment facilities. *Journal of Plant Pathology* 100: 371–389.

Hardham A.R. and L.M. Blackman, 2018. *Phytophthora cinnamomi*. *Molecular Plant Pathology* 19: 260–285.

Hausbeck M.K., Lamour K.H., 2004. *Phytophthora capsici* on vegetable crops: research progress and management challenges. *Plant Disease* 88: 1292–1303.

Hoitink H.A.J., Boehm M.J. 1999. Biocontrol within the context of soil microbial communities: a substrate-dependent phenomenon. *Annual Review of Phytopathology* 37: 427–446.

Hoitink H.A.J., Stone A.G., Han D.Y., 1997. Suppression of plant diseases by composts. *HortScience* 32: 184–187.

Ioos R., Axelle A., Marçais B., Frey P., 2006. Genetic characterization of the natural hybrid species *Phytophthora alni* as inferred from nuclear and mitochondrial DNA analyses. *Fungal Genetic and Biology* 43: 511–529.

IPCC Special report, 2019. Climate Change and Land. https: //www.ipcc.ph. /report/srccl/.

Jiménez-Diaz R.M., Cirulli M., Bubici G., Jimenez-Gasco M. del M., Antoniou P.A., Tjamos E.C., 2012. Verticillium Wilt, A Major Threat to Olive Production: current status and future prospects for its management. *Plant Disease* 96: 304–329.

Jung T. 2009. Beech decline in Central Europe driven by the interaction between *Phytophthora* infections and climatic extremes. *Forest Pathology* 39: 77–94.

Jung T., Pérez-Sierra A., Durán A., Horta Jung M., Balci Y., Scanu B., 2018. Canker and decline diseases caused by soil- and airborne *Phytophthora* species in forests and woodlands. *Persoonia* 40: 182–220.

Jung T., Chang T.T., Bakonyi J., Seress D., Pérez-Sierra A.,...Horta Jung M., 2016a. Diversity of *Phytophthora* species in natural ecosystems of Taiwan and association with disease symptoms. *Plant Pathology* 66: 194–211.

Jung T., Horta Jung M., Cacciola S.O., Cech T., Bakonyi J., ... Scanu B., 2017. Multiple new cryptic pathogenic *Phytophthora* species from *Fagaceae* forests in Austria, Italy and Portugal. *IMA Fungus* 8: 219–244.

Jung T., La Spada F., Pane A., Aloi F., Evoli M., … Cacciola S.O., 2019. Diversity and distribution of *Phytophthora* species in protected natural areas in Sicily. *Forests* 10: 259. https: //doi.org/10.3390/f10030259.

Jung T., Orlikowski L., Henricot B., Abad-Campos P., Aday A.G., ….Peréz-Sierra A., 2016b. Widespread *Phytophthora* infestations in European nurseries put forest, semi-natural and horticultural ecosystems at high risk of Phytophthora diseases. *Forest Pathology* 46: 134–163.

Katan J., 2017. Disases caused by soilborne pathogens: biology, management and challenges. *Journal of Plant Patholology* 9: 305–315.

Koike S.T., Subbarao K.V., Davis R.M., Turini T.A., 2003. Vegetable diseases caused by soilborne pathogens. University of California, Division of Agriculture and Natural Resources. Publ. N. 8099. 13 pp.

Kroon L., Brower H., de Cock A.W.A.M., Summerbell R., 2011. The genus *Phytophthora* anno 2012. *Phytopathology* 102: 348–364.

Kubiak K., Zółciak A., Damszel M., Lech P., Sierota Z., 2017. *Armillaria* pathogenesis under climate changes. *Forests* 8. 100.

Kunadiya M., Dunstan W., White D., Hardy G.E.St.J., Grigg A., Burgess T., 2019. An mRNA protocol designed to establish propagule viability in environmental samples. *Plant Disease* https: //doi.org/10.1094/PDIS-09-18-1641-RE.

Kunadiya M., White D., Dunstan W.A., Hardy G.E. St.J., Andjic V., Burgess T.I., 2017. Pathways to false-positive diagnoses using molecular genetic detection methods; *Phytophthora cinnamomi* a case study. *FEMS Microbiology Letters* 364: fnx009 doi: 10.1093/femsle/fnx009.

La Porta N., Capretti P., Thomsen I.M., Kasanen R., Hietala A.M., Von Weissenberg K. 2008. Forest pathogens with higher damage potential due to climate change in Europe. *Canadian Jornal of Plant Pathology* 30: 177–195.

Lamour K.H., Win J., Kamoun S., 2007. Oomycete genomics: new insights and future directions. *FEMS Microbiology Letters* 274: 1-8.

Lazarovits G., Subbarao K., 2010. Challenges in controlling Vericillium wilt by the use of non chemical methods. In: *Recent developments in management of plant diseases*(U. Gisi, I. Chet, M.L. Gullino eds.), Springer Sciences + Business Media, Berlin/Heidelberg, Germany, 247–264.

Lievens B., Hanssen I.M., Rep M., 2012. Recent development in the detection and identification of *formae speciales* and races of *Fusarium oxysporum*: from pathogenicity testing to molecular biology. In: *Fusarium wilts of greenhouse vegetable and ornamental crops* (M.L. Gullino, J. Katan and A. Garibaldi eds.), APS Press, St Paul, MN, USA, 47–55.

Linzer R.E., Otrosina W.J., Gonthier P., Bruhn J., Laflamme G., Bussières G., Garbelotto M., 2008. Inferences on the phylogeography of the fungal pathogen *Heterobasidion annosum*, including evidence of interspecific horizontal genetic transfer and of human-mediated, long-range dispersal. *Molecular Phylogenetics and Evolution* 46: 844–862.

Lo Giudice V., Raudino F., Magnano di San Lio R., Cacciola S.O., Faedda R., Pane A., 2010. First report of a decline and wilt of young olive trees caused by simultaneous infections of *Verticillium dahliae* and *Phytophthora palmivora* in Sicily. *Plant Disease* 94: 1372.

Lombard L., Sandoval-Denis M., Lamprecht S.C., Crous P.W., 2019. Epitypification of *Fusarium oxysporum* – clearing the taxonomic chaos. *Persoonia* 43: 1–47.

López-Escudero F.J., Mercado-Blanco J., 2011. Verticillium wilt of olive: a case study to implement an integrated strategy to control a soil-borne pathogen. *Plant and Soil* 344: 1-50.

Mammella M.A., Martin F.N., Cacciola S.O., Coffey M.D., Faedda R., Schena L., 2013. Analyses of the population structure in a global collection of *Phytophthora nicotianae* isolates inferred from mitochondrial and nuclear DNA sequences. *Phytopathology* 103: 610–622.

Manici L.M., F. Caputo and G. Baruzzi, 2005. Additional experiences to elucidate the microbial component of soil suppressiveness towards strawberry black root rot complex. *Annals of Applied Biology* 146: 421–431.

Manici L.M., Ciavatta C., Kelderer M., Erschbaumer G., 2003. Replant problems in South Tyrol: role of fungal pathogens and microbial population in conventional and organic apple orchards. *Plant and Soil* 256: 315–324.

Manici L.M., Bregaglio S., Fumagalli D., Donatelli M. 2014. Modelling soil borne fungal pathogens of arable crops under climate change. *International Journal of Biometeorology* .58 : 2071–283.

Martini P., Pane A., Raudino F., Chimento A., Scibetta S., Cacciola S.O. 2009. First report of *Phytophthora tentaculata* causing root and stem rot of oregano in Italy. *Plant Disease* 93: 843.

Martyn R.D., Miller M.E., 1996. Monosporascus root rot and vine decline: An emerging disease of melons world-wide. *Plant Disease* 80: 716–725.

Martin F.N., Gloria Abad Z., Balci Y., Ivors K., 2012. Identification and detection of *Phytophthora*: reviewing our progress, identifying our needs. *Plant Disease* 96: 1080–1103.

Martyn R.D., Miller M.E., 1996. Monosporascus root rot and vine deline. *Plant Disease* 80: 716–725.

Maryani N., Lombard L., Poerba Y.S., Subandiyah S., Crous P.W., Kema G.H.J. 2019. Phylogeny and genetic diversity of the banana Fusarium wilt pathogen *Fusarium oxysporum* f. sp. *cubense* in the Indonesian centre of origin. *Studies in Mycology* 92: 155–194.

Matic S., Gilardi G., Gullino M.L., Garibaldi A., 2018. Evidence for an expanded host range of *Fusarium oxysporum* f. sp. *chrysanthemi*. *Journal of Plant Pathology* 100: 97–104.

Matic S., Gilardi G., Gisi U., Gullino M.L., Garibaldi A., 2019. Differentiation of *Pythium* spp. from vegetable crops with molecular markers and sensitivity to azoxystrobin and mefenoxam. *Pest Management Science* 75: 356–365.

Mbofung G.Y. and B.M. Pryor, 2010. A PCR-based assay for detection of *Fusarium oxysporum* f. sp. *lactucae* in lettuce seed. *Plant Disease* 94: 860–866.

Migheli Q., Cacciola S.O., Balmas V., Pane A., Ezra D., Magnano di San Lio G., 2009. Mal secco disease caused by *Phoma tracheiphila*: a potential threat to lemon production worldwide. *Plant Disease* 93: 852–867.

Migliorini D., Ghelardini L., Tondini E., Luchi N., Santini A., 2015 The potential of symptomless potted plants for carrying invasive soilborne plant pathogens. *Diversity and Distributions* 21: 1218–1229.

Miles T.D., Martin F.N., Coffey M.D., 2014. Development of rapid isothermal amplification assays for detection of *Phytophthora* spp. in plant tissue. *Phytopathology* 105: 265–278.

Moorman G.W., Kang S., Geiser D.M., Kim S.H., 2002. Identification and characterisation of *Pythium* species associated with greenhouse floral crops in Pennsylvania. *Plant Disease*86: 1227–1231.

Moricca S., Ginetti B.T.B., Scanu B., Franceschini A., Ragazzi A., 2016. Endemic and emerging pathogens threatening cork oak trees: management options for conserving a unique forest ecosystem. *Plant Disease* 100: 2184–2193.

Moricca S., Bracalini M., Croci F., Corsinovi S., Tiberi R., Ragazzi A., Panzavolta T., 2018. Biotic factors affecting ecosystem services in urban and peri-urban forests in Italy: the role of introduced and impending pathogens and pests. *Forests* 9: 65.

Mosca S., Li Destri Nicosia M.G., Cacciola S.O., Schena L., 2014. Molecular analysis of *Colletotrichum* species in the carposphere and phyllosphere of olive. *PLoS ONE* 9(12): e114031. https: //doi.org/10.1371/journal.pone.0114031.

Nigro F., Gallone P., Romanazzi G., Schena L., Ippolito A., Salerno M., 2005. Spread of Verticillium wilt of olive in Apulia and genetic diversity of *Verticillium dahliae* isolates from infected trees. *Journal of Plant Pathology* 87: 13–23.

Navaud O., Barbacci A., Taylor A.,. Clarkson J.P., Raffaele S. 2018. Shifts in diversification rates and host jump frequencies shaped the diversity of host range among Sclerotiniaceae fungal plant pathogens. *Molecular Ecology* 27: 1309–1323.

O'Donnell K., Ward T. J., Robert V. A. R.G., Crous P.W., Geiser D. M., Kang S., 2015. DNA sequence-based identification of *Fusarium*: Current status and future directions. *Phytoparasitica* 43: 583–595.

Ortu G., Bertetti D., Martini P., Gullino M.L., Garibaldi A., 2015. *Fusarium oxysporum* f. sp. *papaveris*: a new *forma specialis* isolated from iceland poppy (*Papaver nudicaule*). *Phytopathologia Mediterranea* 54: 76–85.

Ortu G., Bertetti D., Gullino M.L., Garibaldi A., 2018. *Fusarium oxysporum* f. sp. *lavandulae*, a novel *forma specialis* causing wilt on *Lavandula* × *allardii*. *Journal of Plant Pathology* 100: 97–104.

Pane A., Allatta C., Sammarco G., Cacciola S.O., 2007a. First report of bud rot of Canary island date palm caused by *Phytophthora palmivora* in Italy. *Plant Disease* 91: 1059.

Pane A., Cosentino L. S., Copani V., Cacciola S.O., 2007b. First report of southern blight caused by *Sclerotium rolfsii* on hemp (*Cannabis sativa*) in Sicily and Southern Italy. *Plant Disease* 91: 636.

Pane A, Magnano di San Lio G., Raudino F., Cacciola S.O., 2008. Blight caused by *Sclerotium rolfsii* on potted ornamental citrus in Sicily. *Plant Disease* 92: 977.

Pane A.,Raudino F., Adornetto S., Proietto Russo G., Cacciola S.O., 2007c. Blight of English ivy (*Hedera helix*) caused by *Sclerotium rolfsii* in Sicily. *Plant Disease* 91: 635.

Pascual J.A., Ceglie F., Tuzel Y., Koller M., Koren A., Hitchings R., Tittarelli F. 2018. Organic substrate for transplant production in organic nurseries. A review. *Agronomy for Sustainable Development* 38: 35.

Pasini L., Prodorutti D., Herrera D., Pertot I., 2016. Genetic diversity and biocontrol of *Rosellinia necatrix* infecting apple in Nothern Italy. *Plant Disease* 100: 444–452.

Pasquali M., Dematheis F., Gullino M.L., Garibaldi A. 2007. Identification of race 1 of *Fusarium oxysporum* f. sp. *lactucae* on lettuce by inter-retrotransposon sequence-characterized amplified region technique. *Phytopathology* 97: 987–996.

Pautasso M., Doring T., Garbelotto M., Pellis M., Jeger L., 2012. Impacts of climate change on plant diseases – Opinions and trends. *European Journal of Plant Pathology* 133: 295–313.

Pennisi A.M., Agosteo G.E., Cacciola S.O., Pane A., Faedda R. 1998. Insensitivity to metalaxyl among isolates of *Phytophthora capsici* causing coot and crown rot of pepper in Southern Italy. *Plant Disease* 82: 1283.

Pérez-Rodríguez M., Serrano N., Arquero O., Orgaz F., Moral J., López-Escudero F.J., 2016. The effect of short irrigation frequencies on the development of Verticillium wilt in the susceptible olive cultivar 'Picual' under field conditions. *Plant Disease* 100: 1880–1888.

Pilotti M., Di Lernia G., Lumia V., Riccioni L., 2014. *Phytophthora cinnamomi* causing stem canker and root rot of nursery grown *Platanus* × *acerifolia*: First report in the Northern emisphere. *Phytopathologia Mediterranea* 53: 75–82.

Pintore I., Gilardi G.., Gullino M.L., Garibaldi A., 2017. Analysis of vegetative compatibility groups of Italian and Dutch isolates of *Fusarium oxysporum* f. sp. *lactucae. Journal of Plant Pathology* 99: 517–521.

Polizzi G., Aiello D., Castello I., Vitale A., Parlavecchio G., 2007. First report of southern blight on Firewheel Tree, Bay Laurel, Bird of Paradise, Mediterranean Fan Palm, and Liverwort caused by *Sclerotium rolfsii* in Italy. *Plant Disease* 91: 1199.

Polizzi G., Aiello D., Guarnaccia V., Parlavecchio G., Vitale A. 2010. First report of southern blight on silverbush (*Convolvulus cneorum*) caused by *Sclerotium rolfsii* in Italy. *Plant Disease* 94: 131.

Pollastro S., Dongiovanni C., Gerin D., Pollastro P., Fumarola G., De Miccolis Angelini R.M., Faretra F., 2016. First report of *Coniella granati* as a causal agent of pomegranate crown rot in southern Italy. *Plant Disease* 100: 1498.

Prigigallo M.I., Mosca S., Cacciola S.O., Cooke D.E.L., Schena L., 2015. Molecular analysis of *Phytophthora* diversity in nursery-grown ornamental and fruit plants. *Plant Pathology* 64: 1308–1319.

Prigigallo M.I., Abdelfattah A., Cacciola S.O., Faedda R., Sanzani S.M., Cooke D.E., Schena L., 2016. Metabarcoding analysis of *Phytophthora* diversity using genus-specific primers and 454 pyrosequencing. *Phytopathology* 106: 305–313.

Prospero S., Cleary M., 2017. Effects of host variability on the spread of invasive forest diseases. *Forests* 8, 80: 99–119.

Pugliese M., Gullino M.L., Garibaldi A., 2012. Compost suppressiveness against *Phytophthora* spp. on *Skimmia japonica* and azalea. *Communictions in Agricultural and Applied Biological Sciences* 77: 237–240.

Puglisi I., Aloi F., La Spada F., Evoli M.,...Cacciola S.O., 2016. A decline of lentil (*Lens culinaris*) crops in the Ustica island caused by *Phytophthora* species. *Journal of Plant Pathology* 98 (supplement) S60.

Quesada-Ocampo L.M., Granke L.L., Mercier, M.R., Olsen J., Hausbeck, M.K. 2011. Investigating the genetic structure of *Phytophthora capsici* populations. *Phytopathology* 101: 1061–1073.

Raimondo M.L., Carlucci A. 2018a. Characterization and pathogenicity assessment of *Plectosphaerella* species associated with stunting disease on tomato and pepper crops in Italy. *Plant Pathology* 67: 626–641. https://doi.org/10.1111/ppa.12766.

Raimondo M.L., Carlucci A. , 2018b. Characterization and pathogenicity of *Plectosphaerella* spp. collected from basil and parsley in Italy. *Phytopathologia Mediterranea* 57: 284–295.

Ramsfield T.D., Bentz B.J., Faccoli M., Jactel H., Brockerhoff E.G., 2016. Forest health in a changing world: effects of globalization and climate change on forest insect and pathogen impacts. *Forestry* 89: 245–252.

Ratti M.F., Farrer R.A., Cano L., Faedda R., Goss E., 2019. Evaluation of high resolution melting for rapid differentiation of *Phytophthora* hybrids and their parental species. *Plant Disease* https://doi.org/10.1094/PDIS-12-18-2291-RE.

Raviv M. 2008. The use of compost in growing media as suppressive agents against soil-borne diseases. *Acta Horticulture* 779: 39–49.

Raviv M. 2008. The use of compost in growing media as suppressive agents against soil-borne diseases. *Acta Horticulture* 779: 39–49.

Rossman A.Y. and M.E. Palm-Hernández, 2008. Systematics of plant pathogenic fungi: Why it matters. *Plant Disease* 92: 1376–1386.

Ruano-Rosa D., Arjona-Girona I., López-Herrera C.J., 2018a. Integrated control of avocado white root rot combining low concentrations of fluazinam and *Trichoderma* spp. *Crop Protection* 112: 363–370.

Ruano-Rosa D., Cichello A.M., Schena L., Magnano di San Lio G., Agosteo G.E., 2017. First report of *Verticillium dahliae* causing wilt of goji (*Lycium barbarum*) in Italy. *New Disease Reports* 35: 25.

Ruano-Rosa D., Schena L., Agosteo G.E., Magnano di San Lio G., Cacciola S.O., 2018b. *Phytophthora oleae* sp. nov, causing fruit rot of olive in southern Italy. *Plant Pathology* 67: 1362–1373.

Sandoval-Denis M., Crous P.W., 2018. Removing chaos from confusion: assigning names to common human and animal pathogens in *Neocosmospora.* Persoonia 41, 2018: 109–129.

Sandoval-Denis M., Guarnaccia V., Polizzi G., Crous P.W., 2018. Symptomatic *Citrus* trees reveal a new pathogenic lineage in *Fusarium* and two new *Neocosmospora* species. *Persoonia* 40, 2018: 1–25.

Santini A., Ghelardini L., De Pace C., Desprez-Loustau M.L., Capretti P.,... Stenlid J., 2013. Biogeographical patterns and determinants of invasion by forest pathogens in Europe. *New Phytologist* 197: 238–250.

Santos-Rufo A., Vega V., Hidalgo J.J., Hidalgo J.C., Rodríguez-Jurado D., 2017. Assessment of the effect of surface drip irrigation on *Verticillium dahliae* propagules differing in persistence in soil and on verticillium wilt of olive. *Plant Pathology* 66: 1117–1127.

Saracchi M., Rocchi F., Pizzati C., Cortesi P., 2008. Box blight, a new disease of *Buxus* in Italy caused by *Cylindrocladiumbuxicola*. *Journal of Plant Pathology* 90: 581–584.

Scanu B., Linaldeddu B.T., Franceschini A., Anselmi N., Vannini A., Vettraino A. M., 2013. Occurrence of *Phytophthora cinnamomi* in cork oak forests in Italy. *Forest Pathology* 43: 340– 343.

Scanu B., Linaldeddu B.T., Deidda A., Jung T., 2015. Diversity of *Phytophthora* species from declining Mediterranean maquis vegetation, including two new species, *Phytophthora crassamura* and *P. ornamentata* sp. nov.*PLoS ONE* 10(12): e0143234. doi: 10.1371/journal.pone.0143234.e.

Schena L., Li Destri Nicosia M.G., Sanzani S.M., Faedda R., Ippolito A.,. Cacciola S.O, 2013. Development of quantitative PCR detection methods for phytopathogenic fungi and oomycetes. *Journal of Plant Pathology* 95: 7-24.

Schena L., F. Nigro, Ippolito A., 2002. Identification and detection of *Rosellinia necatrix* by conventional and real-time Scorpion-PCR. *European Journal of Plant Pathology* 108: 355–366.

Schena L., F. Nigro, Ippolito A., 2008. Integrated management of *Rosellinia necatrix* root rot on fruit tree crops. In: *Integrated management of diseases caused by fungi, phytoplasma and bacteria. Integrated management of plant pests and diseases, vol 3* (A. Ciancio, K. Mukerji, eds.), Springer, Dordrecht, the Netherlands, 137–158.

Schlatter D., Kinkel L., Thomashow L·, Weller D., Paulitz T., 2017. Disease suppressive soils: new insights from the soil microbiome. *Phytopathology* 107: 1284–1297.

Schroers H.-J., Samuels G.J., Zhang N., Short D.P.G., Juba J., Geiser D.M., 2016. Epitypification of *Fusisporium* (*Fusarium*) *solani* and its assignment to a common phylogenetic species in the *Fusarium solani* species complex. *Mycologia* 108: 806–819.

Scibetta S., Schena L., Chimento A., Cacciola S.O., Cooke D.E.L., 2012. A molecular method to assess *Phytophthora* diversity in environmental samples. *Journal of Microbiological Methods* 88: 356–368.

Serrano M.S., Osmundson, T., Almaraz-Sanchez A., Croucher P.J.P., Swiecki T., Alvarado D., Garbelotto M., 2019. A microsatellite analysis identifies global pathways of movement of *Phytophthora cinnamomi* and the likely sources of wildland infestations in California and Mexico. *Phytopathology* https: //doi.org/10.1094/PHYTO-03-19-0102-R.

Shaw M.W., Osborne T.M., 2011. Geographic distribution of plant pathogens in response to climate change. *Plant Pathology* 60: 31–43.

Simamora A.V, Paap T., Howard K., Stukely M.J.C., Hardy G.E.St.J., Burgess T., 2017. *Phytophthora* contamination in a nursery and its potential dispersal into the natural environment. *Plant Disease* 102: 132–139.

Sims L., Tjosvold S., Chambers D., Garbelotto M., 2019a. Control of *Phytophthora* species in plant stock for habitat through best management practices. *Plant Pathology* 68: 196–204.

Sims L., Chee C., Bourret T., Shannon Hunter S., Garbelotto M., 2019b. Genetic and phenotypic variation of *Phytophthora crassamura* isolates from California nurseries and restoration sites. *Fungal Biology* 123: 159–169.

Srinivasan K., Gilardi G., Spadaro D., Garibaldi A., Gullino M.L., 2010. Molecular characterization through IGS sequencing of *formae speciales* of *Fusarium oxysporum* pathogenic on lamb's lettuce. *Phytopathologia Mediterranea* 49: 309–320.

Stanghellini M.E, Misaghi I.J. 2011. *Olpidium bornovanus*-mediated germination of ascospores of *Monosporascus cannonballus*: a host-specific rhizosphere interaction. *Phytopathology*. 101: 794–796.

Stanghellini M.E., Kim D.H., Rasmussen S. L. 1996. Ascospores of *Monosporascus cannonballus*: germination and distribution in cultivated and desert soils in Arizona. *Phytopathology* 86: 509–514.

Stanghellini M.E., Mathews D.M., Misaghi I.J. 2010. Pathogenicity and management of *Olpidium bornovanus*, a root pathogen of melons. *Plant Disease* 94: 163–166.

Stanghellini M.E., Ferrin D.M., Kim D.H., Waugh M.M., Radewald K.C., ... McCaslin M.A. 2003. Application of preplant fumigants via drip irrigation systems for the management of root rot of melons caused by *Monosporascus cannonballus*. *Plant Disease*. 87: 1176–1178.

Stenlid J., Oliva J., Boberg J.B., Hopkins A.J.M., 2011. Emerging diseases in European forest ecosystems and responses in society. *Forests* 2: 486–504.

Stielow J.B., Levesque C.A., Seifert K.A., Meyer W., Iriny L., ... Robert V., 2015. One fungus, which genes? Development and assessment of universal primers for potential secondary fungal DNA barcodes. *Persoonia - Molecular Phylogeny and Evolution of Fungi* 35: 242–263.

Taylor A., Jackson A.C., Clarkson J.P., 2019. First Report of *Fusarium oxysporum* f. sp. *lactucae* Race 4 Causing Lettuce Wilt in England and Ireland. *Plant Disease*, https: //doi.org/10.1094/PDIS-10-18-1858-PDN.

Taylor R.J., Sales B., Secor G.A., Rivera V., Gumestad N.C., 2002. Sensitivity of North American isolates of *Phytophthora erythroseptica* and *Pythium ultimum* to mefenoxam (metalaxyl). *Plant Disease* 86: 797–802.

Thines M., 2019. An evolutionary framework for host shifts - jumping ships for survival. *New Phytologist* doi: 10.1111/nph.16092.

Thomas J.E., Wood T.A., Gullino M.L., Ortu G., 2017. Diagnostic tools for plant biosecurity. In: *Practical Tools for plant and food biosecurity* (M.L. Gullino, J. Stack, J. Fletcher and J. Mumford eds.), Springer, Dordrecht, The Netherlands 209–226.

Vakalounakis D.J., Doulis A.G., Klinoromou E. 2005. Characterization of *Fusarium oxysporum* f. sp. *radicis-cucumerinum* attacking melon under natural conditions in Greece. *Plant Pathology* 54: 339–346.

Vannacci G., Gambogi P. 1980. *Fusarium solani* f.sp. *cucurbitae* razza 1 su semi di *Cucurbita pepo* L.: reperimento del patogeno e influenza di condizioni colturali sull'andamento della malattia. *Phytopathologia Mediterranea* 19: 103–114. (in Italian).

Vannini A., Bruni N., Tomassini A., Franceschini S., Vettraino A.M., 2013. Pyrosequencing of environmental soil samples reveals biodiversity of the *Phytophthora* resident community in chestnut forests. *FEMS Microbiology Ecology* 85: 433–42.

Vettraino A.M., Morel O., Perlerou C., Robin C., Diamandis S., Vannini A., 2005. Occurrence and distribution of *Phytophthora* species in European chestnut stands and their association with Ink Disease and crown decline. *European Journal of Plant Pathology* 111: 169–180.

Vitale A., Aiello D., Castello I., Polizzi G., 2009. First report of benzimidazole-resistant isolates of *Cylindrocladium scoparium* in Europe. *Plant Disease* 93: 110.

Vitale A., Crous P.W., Lombard L., Polizzi G., 2013. Calonectria diseases on ornamental plants in Europe and the Mediterranean basin: an overview. *Journal of Plant Pathology* 95: 463–476.

Vitale S., Scotton M., Vettraino A.M., Vannini A., Haegi A., … Belisario A. 2019. Characterization of *Phytophthora cinnamomi* from common walnut in Southern Europe environment. *Forest Pathology* 49: e1 2477.

Yang X., Tyler B.M., Hong C., 2017. An expanded phylogeny for the genus *Phytophthora*. *IMA Fungus* 8: 355–384.

8

Occurrence and diversity of *Grapevine leafroll-associated virus 1* in Algeria

Arezki LEHAD[1,*], Ilhem SELMI[2], Meriem LOUANCHI[1], Mouni AITOUADA[1], Naima MAHFOUDHI[2]

[1] *Laboratoire de Phytopathologie et Biologie Moléculaire. Ecole Nationale Supérieure d'Agronomie. Rue Hacen Badi, Belfort, El Harrach, 16000 Alger, Algeria*
[2] *Laboratoire de Protection des Végétaux, Institut National de la Recherche Agronomique de Tunisie, Rue Hedi Karray, 2049 Ariana, Tunisia*
*Corresponding author: a.lehad@ensa.dz

Summary. A survey was conducted of central and western Algerian grape cultivars and a germplasm collection to detect the presence of *Grapevine leafroll-associated virus 1* (GLRaV-1), using DAS-ELISA. The virus was found in 26 of 484 sample (5.4%). No infection was found in the germplasm collection. Analysis of the sequence of a coat protein gene region revealed that of the 17 recognised phylogenetic groups, the Algerian isolates belong to Groups I, II and XVI. This is the first study of genetic diversity of GLRaV-1 in Algeria.

Keywords. GLRaV-1, RT-PCR, genetic variability, HSP70h, CP.

Competing Interests: The Author(s) declare(s) no conflict of interest.

Editor: Nihal Buzkan, Kahramanmaraş Sütçü Imam University, Turkey.

INTRODUCTION

Grapevine leafroll disease (GLD) is caused by a complex of virus species from the family *Closteroviridae*, and this disease occurs in all grape-growing regions of the world (Almeida *et al.*, 2013). It is an economically important virus disease (Alabi *et al.*, 2016). The associated viruses are referred as the grapevine leafroll-associated viruses, including GLRaV-1, -2, -3, -4; -7 and -13. GLRaV-1, GLRaV-3, GLRaV-4 and its strains and GLRaV-13 belong to the genus *Ampelovirus*, GLRaV-2 belongs to the genus *Closterovirus*, and GLRaV-7 belongs to a newly proposed genus *Velarivirus* (Martelli *et al.*, 2012, Ito and Nakaune 2016).

The GLRaV-1 genome size is nearly 19.5 kb (Fazeli *et al.*, 2000), from which partial sequence information was characterized (AF195822). The full-length genomic sequence of two isolates of GLRaV-1 has been determined, at ~18,730 nucleotides (nt) (isolate WA-CH) and ~18,945 nt (isolate WA-PN). Isolates WA-CH and WA-PN encode ten open reading frames, and their overall genome organization is similar to previously reported GLRaV-1 isolates (Donda *et al.*, 2017).

Grapevine leafroll-associated viruses are known to spread by grafting and by mealybug transmission. Several species of mealybugs have been shown to transmit GLRaV-3 (Almeida *et al.*, 2013). GLRaV-1 is restricted to the phloem tissues of host plants and is frequently present with other viruses. Transmission of GLRa-1 by mealybugs follows the "semi-persistent and non circulative" mode (Tsai *et al.*, 2008), and several mealybug and soft scale species have been reported to transmit GLRaV-1 (Naidu, 2017).

Little *et al.* (2001) identified ten genes associated with GLRaV-1, several of which are hypervariable. Genetic diversity of GLRaV-1 has been less investigated compared to other leafroll disease associated viruses. Based on the HSP70h gene region, two phylogenetic groups were designed (Kominek *et al.*, 2005). Based on the heat shock 70 protein homolog (HSP70h) gene region, three phylogenetic groups were identified (Alabi *et al.*, 2011). Based on the coat protein (CP) gene, Esteves *et al.*, (2013) described eight groups. More recently, eignt groups were designated for the CP gene region and seven groups for the HSP70h gene region (Fan *et al* ., 2015).

Here we describe a survey of GLRaV-1 in grapevines in Algeria, and the chararterization of the genetic relationship of Algerian virus isolates with other isolates, in a partial fragment of the CP gene.

MATERIALS AND METHODS

In autumn 2012, 484 grapevine samples were collected from the major grapevine growing areas in the western and central regions of Algeria. Dormant canes from individual vines were randomly collected in commercial vineyards, including from eight introduced cultivars (355 samples) and two local cultivars (100 samples), as well as 18 cultivars (29 samples) from the autochthonous grapevine germplasm collection at the Institut Technique de l'Arboriculture Fruitière et de la Vigne (ITAF). All samples were tested by DAS-ELISA (Clark and Adams, 1977) for the presence of GLRaV-1, using a commercial kit (Agritest). Extracts were obtained by macerating phloem tissues of dormant canes in the grinding buffer described by the manufacturer (1 g per 10 mL). Absorbance was recorded at 405 nm using an automatic microplate reader (Labsystems Multiskan Ascent). Samples with absorbance readings equal or exceeding three times those of negative controls were considered as positive.

All positive samples were analyzed by RT-PCR using total nucleic acids extracted from 0.2 g of cortical scrapings, as described by Foissac *et al.* (2001). The negative and positive controls were provided from the virus collection of the INRAT, Tunis. Reverse transcription was performed using MMLV reverse transcriptase (200 units μL^{-1}), and PCR was carried out using the primer pairs HSP70-417F (5'- GAGCGACTTGCGACTTATCGA-3') and HSP70-737R (5'-GGTAAACGGGTGTTCTTCAATTCT-3'), designed by Osman and Rowhani (2006), and the primer GLRaV-1-CP/F (5'-CGCGCTTGCAGAGTTTAAGTGGTT-3') and GLRaV-1 CP/R (5'-TCCGTGCTGCATTGCAACTTTCTC-3') designed by Alabi *et al.* (2011), to amplify fragments of 320 bp for the HSP70h gene region and of 734 bp for the CP gene region. Some virus isolates were randomly chosen for sequencing of the CP and HSP70h gene regions (Supplementary data 1).

Mean genetic distances of the analysed 104 sequences of each group, as well as between the groups, were calculated using MEGA7 with best fit alignment (Kumar *et al.*, 2016). The Dna SP software v.5.10.01 (Librado and Rozas 2009) was used to estimate the mean values of non-synonymous to synonymous substitution ratios (dN/dS) for the different coding regions of GLRaV-1. The ratio of non-synonymous to synonymous substitutions per site (dN/dS) was also determined.The genetic diversity π and the Tajima'D value were calculated using the MEGA7 software package.

Multiple alignments of nucleotide sequences from this study and corresponding sequences available in GenBank (Supplementary data 1) were performed using ClustalW implemented in MEGA7. Using the best fit models, a phylogenetic tree was constructed for the CP gene region using the best fit alignment (585 nt) with the neighbor-joinning method with 1,000 bootstraps (Figure 1)

A search for recombination events was conducted for the CP gene region with the genetic recombination detection method (GARD) (Kosakovsky Pond *et al.*, 2006) implemented in the Datamonkey web server (https://www.datamonkey.org/).

No analyses were performed for the HSP70h gene region in this study due to the small length of the obtained sequences.

RESULTS

Results revealed that GLRaV-1 was present in 26 (5.4%) of the 484 samples tested. For the variety Dattier de Beyrouth, six of 71 samples were infected, seven of 100 for Gros noir and eight of 70 for Muscat (Table 1). No infected samples were found in the autochtonous grapevine germplasm collection at ITAF. According to previous results for the presence of GLRaV-3 in the same samples, mixed infections of GLRaV-1 and GLRaV-3 were detected in 15 samples, and infection of only GLRaV-1 was detected in 11 samples.

In order to determine the diversity of GLRaV-1 isolates from Algeria, a comparison was performed at the

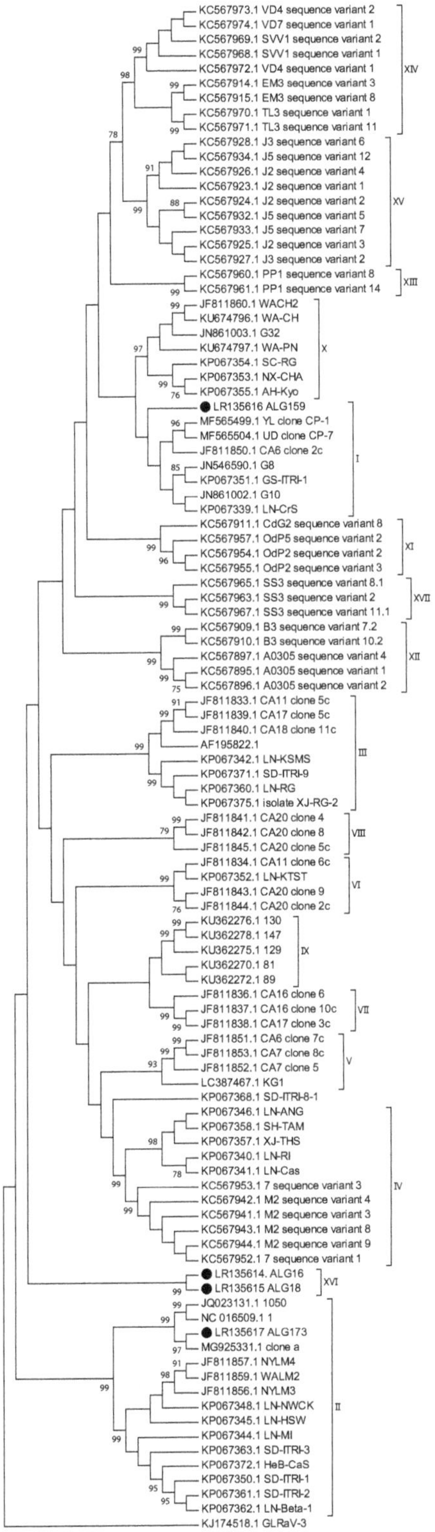

Figure 1. Phylogenetic tree of *Grapevine leafroll associated virus 1*(GLRaV-1) isolates, constructed with sequences of a 585 nt fragment of the viral coat protein gene. Percentages of bootstrap support (≥ 75%) from 1,000 replicates are indicated on relevant tree branches.

Table 1. Rates of virus infections in the surveyed grapevine varieties.

Variety	Number of samples	Number of samples infected
Dattier	71	6
Gros noir	100	7
Cardinal	52	1
Alicante	30	1
King Rubi	9	0
Muscat	70	8
Mersguira	10	0
Çinsaux	40	1
Carignon	31	1
Valensi	42	1
Autochthones	29	0
Total	484	26

nucleotide level of HSP70h and CP gene regions. For HSP70h, proportional nucleotide similarity of Algerian sequences revealed an average of 93%. Isolates ALG5, ALG128, ALG132, ALG270 and ALG420 were closely similar with 95 to 97% similarity. Isolates ALG16 and ALG18 showed 100% similarity between them. A BLASTn search showed that isolates ALG16 and ALG18 had 95% similarity to isolate WA-CH (KU674796.1), while isolates ALG420, ALG132, ALG128, ALG270 and ALG5 had 91-94% similarity between them.

For the sequence of the CP gene region, a comparison at the nucleotide level showed that ALG16 and ALG18 were identical, ALG159 showed less than 87% similarity and ALG173 less than 85% similarity compared to the three Algerian sequences. BLASTn search showed that ALG16 and ALG18 had 92-93% similarity with other sequences available in GenBank. (isolate A0305 variant 4 (KC567897) was very similar).

For the CP gene region, analyses performed with DNA sp 0.6 and MEGA7 showed that the number of non-synonymous sites was 443.91 sites out of 585, and the synonymous site was 141.09 sites out of 585. The mean dN/dS was 0.095 calculated with the Single-Likelihood Ancestor Counting (SLAC) algorithm implemented in the datamonkey server. The genetic diversity analyses gave a value of π = 0.103 and the the D Tajima's test was -0.92. The dN/dS and the Tajima's D test showed that the CP gene was under negative selection.

Tajima's D and Fu and Li's D and F statistics were calculated for groups I- and II-containing Algerian sequences. This gave a negative value for Tajima's D and positive value for Fu and Li's D and F statistics. A value of genetic divertsity π for the two groups was estimated at 0.03282 for Group I and 0.04688 for Group II (Table 2).

Table 2. Representative parameter estimates for the Algerian GLRaV-1 phylogroups.

Phylogroup	π	Tajima's D	Fu & Li's D	Fu & Li's F
I	0.03282	-1.041	1.29777	1.47497
II	0.04688	-0.056	0.26019	0.13310

No recombination event was found in the CP gene region, using the GARD algorithm implementated in the Datamonkey database (https://www.datamonkey.org).

A phylogenetic tree with four Algerian sequences and 104 other sequences downloaded from GenBank was constructed for the CP gene region using the neighbor joining method with MEGA7 software (Kumar *et al.*, 2016). Results confirmed the high diversity of the GLRaV-1 population. The tree revealed that the GLRaV-1 population clustered into 17 groups. Eight clusters were previously reported by Fan *et al.* (2015). The Algerian sequences obtained in this study clustered into three groups. The sequences of ALG16 and ALG18 clustered in a new group (Group XVI), the sequence ALG159 was in G I and ALG173 was in G II.

The within group proportional similarity ranged from 95% to 100%, indicating a close relationship between isolates from the same group, except for the Group VIII which showed 88% similarity. The between group distances ranged from 78% to 96%, showing divergence present between groups especially for some groups which had less than 80% similarity (Groups II and XVI). The tree revealed the absence of correlation between the geographical origins and phylogentic distribution for some groups. Group I was constituted by sequences from France, China and Poland, and Group II with sequences from China, the United States of America and Canada. However, some groups contained only isolates from one country. Group VII only contained sequences from the United States, Group IX contained sequences only from Turkey, and Groups XI, XII XIII, XIV and XV contained sequences only from Portugal (Supplementary data 1).

DISCUSSION

This study describes the occurrence of GLRaV-1 in different regions of Algeria, and the genetic diversity of population of this virus from other countries. There have been few studies of the diversity of this virus, which appears to be very variable.

Our results have demonstrated shown the presence of GLRaV-1 in Algeria. GLRaV-1 is less prevalent in Algeria than GLRaV-2 and GLRaV-3, which were reported in a previous study (Lehad *et al.* 2015). Mixed infections of this virus with GLRaV-3 may explain its parallel transmission by common vectors. Vector transmission of GLRaV-1 has been reported in Algeria (Bisaad *et al.*, 2017). *Parthenolecanium corni* described as vector of this virus was also observed in some vineyard in central Algeria (unpublished data). Other mealybug species, such as *Phenacoccus aceris*, were described as vectors of GLRaV-1 (Alliaume *et al.*, 2015). More research is required on the epidemiology of this virus in Algeria. It is important to determine the efficiency of mealybug transmission of GLRaV-1 compared with other viruses, in order to further understand the epidemiology of grapevine viruses in this country.

Nucleotide comparisons were carried out, and and results for HSP70 showed that the Algerian GLRaV-1 sequences had close genetic relationships ranging from 89% to 100% similarity, when CP gene region sequences were compared. However, some sequences shared less than 85% similarity. This indicates that the the HSP70h gene region was more conserved than the CP gene region. The HSP70h gene region was reported to be highly conserved in the family of *Closteroviridae* (Tian *et al.*, 1996).

The genetic diversity of the Algerian sequences compared to the sequences in NCBI showed an important genetic diversity value for the CP gene region, confirmed by others (Alabi *et al.*, 2011; Esteves *et al.*, 2013). This may be explained by the natural pressure enforced on this protein. The CP gene plays an important role in plant/virus interactions. The Tajima's D test and the mean dN/dS results showed that the CP gene evolves under negative selection. Similar results were obtained for the CP and Rdrp gene of *Grapevine virus A* (Alabi *et al.*, 2014).

The Tajima's D test gave a negative value for the CP gene region. This may indicate that the gene evolved under a low frequency of polymorphism suggesting population size expansion (e.g. after a bottleneck or a selective sweep) and/or purifying selection.

Previously, two groups were reported for HSP70h gene region (Komínek *et al.*, 2005). Alabi *et al.* (2011) reported three clusters, and Fan *et al.* (2015) revealed the presence of eight groups based on the CP gene region and seven based on the HSP70h region.

LITERATURE CITED

Alabi O.J.L., Casassa F., Gutha L.R., Larsen R.C., Henick-Kling T., ... Naidu R. A., 2016. Impacts of Grapevine

Leafroll Disease on Fruit Yield and Grape and Wine Chemistry in a Wine Grape (*Vitis vinifera* L.) Cultivar, *PLOS ONE* DOI:10.1371/journal.pone.0149666.

Alabi O.J., Al Rwahnih M., Mekuria T.A., Naidu R.A., 2014. Genetic diversity of Grapevine virus A in Washington and California vineyards. *Phytopathology* 104: 548–560.

Alabi O.J., Al Rwahnih M., Karthikeyan G., Poojari S., Fuchs M., Rowhani A., Naidu R.A., 2011. *Grapevine leafroll-associated virus 1* occurs as genetically diverse populations. *Phytopathology* 101: 1446–1456.

Alliaume A., Reinbold C., Uzest M., Beuve M., Hommay G., ... Herrbach E., 2015. Transmission studies of the *Grapevine leafroll-associated virus 1* and the *Grapevine virus A* by the mealybug *Phenacoccus aceris*. *Proceedings of the 18th Congress of ICVG*, Ankara, Turkey, 7–11 September 2015.

Almeida R., Daane K., Bell V., Blaisdell G., Cooper M., ... Pietersen G., 2013. Ecology and management of grapevine leafroll disease. *Frontiers in Microbiology* 4: 94.

Bissaad F.Z., Bounaceur F., Doumandji M.B., 2017. Dynamique spatio-temporelle de *Planococcus ficus* (signoret, 1875) dans les vignobles de la mitidja (Algeria), *Lebanese Science Journal* 1: 24–46.

Clark M.F., Adams A.N., 1977. Characteristics of the microplate method of enzyme-linked immunosorbent assay for the detection of plant viruses. *Journal of General Virology* 34: 475–483.

Donda P. B., Jarugula S., Naida RA., 2017. An analysis of the complete genome sequence and subgenomic RNAs reveals unique feature of the Ampelovirus, *Grapevine leafroll associated virus 1*. 9: 1069-1079.

Esteves F., Teixeira Santos M., Eiras-Dias J.E., Fonseca F., 2013. Molecular data mining to improve antibody-based detection of *Grapevine leafroll-associated virus 1* (GLRaV-1). *Journal of Virological Methods* 194: 258–270.

Fan, X., Hong N., Dong Y., Ma Y., Zhang Z.P., ... Wang G., 2015. Genetic diversity and recombination analysis of *Grapevine leafroll-associated virus 1* from China. *Archives of Virology*. 160: 1669–1678.

Fazeli C.F., Rezaian M.A., 2000. Nucleotide sequence and organization of ten open reading frames in the genome of *Grapevine leafroll-associated virus 1* and identification of three subgenomic RNAs. *Journal of General Virology* 81: 605–615.

Foissac X., Svanella-Dumas L., Dulucq M. J., Gentit P., Candresse T., 2001. Polyvalent detection of fruit tree tricho-, capillo- and foveaviruses by nested RT-PCR using degenerated and inosine containing primers (PDO RT-PCR). *Acta Horticulturae* 550: 37–43.

Ito T., Nakaune R., 2016. Molecular characterization of a novel putative ampelovirus tentatively named *Grapevine leafroll-associated virus 13*. *Archives of virology* 161(9): 2555- 2559. doi: 10.1007/s00705-016-2914-8.

Komínek, P., Glasa, M., Bryxiová, M., 2005. Analysis of the molecular variability of *Grapevine leafroll-associated virus 1* reveals the presence of two distinct virus groups and their mixed occurrence in grapevines. *Virus Genes* 31: 247–255.

Kosakovsky Pond S.L., Posada D., Gravenor M.B., Woelk C.H., Frost S.D., 2006. GARD: a genetic algorithm for recombination detection. *Bioinformatics* 22: 3096–3098.

Kumar S., Stecher G., Tamura K., 2016. MEGA7: Molecular Evolutionary Genetics Analysis Version 7.0 for Bigger Datasets. *Molecular Biology and Evolution* 33: 1870–1874.

Lehad A., Selmi I., Louanchi M., Aitouada M., Mahfoudhi N., 2015. Genetic diversity of *Grapevine leafroll-associated virus 3* in Algeria. *Journal of Plant Pathology* 97(1): 203–207.

Librado P., Rozas J., 2009. DnaSP v5: software for comprehensive analysis of DNA polymorphism data. *Bioinformatics* 25: 1451-1452.

Little A., Fazeli C.F., Rezaian M.A., 2001. Hypervariable genes in *Grapevine leafroll-associated virus 1*. *Virus Research* 80: 109–116.

Martelli G.P., Abou Ghanem-Sabanadzovic N., Agranovsky A.A., Al Rwahnih M., Dolja V.V., ... Saldarelli P., 2012. Taxonomic revision of the family *Closteroviridae* with special reference to the grapevine leafroll-associated members of the genus *Ampelovirus* and the putative species unassigned to the family. *Journal of Plant Pathology* 94: 7–19.

Naidu R.A., 2017. *Grapevine leafroll-associated virus 1*. In *Grapevine Viruses: Molecular Biology, Diagnostics and Management* (Meng B., Martelli G.P., Golino D.A., Fuchs M., ed.), Springer, Switzerland, pp. 127–139.

Osman F., Rowhani A., 2006. Application of a spotting sample preparation technique for the detection of pathogens in woody plants by RT-PCR and real-time PCR (TaqMan). *Journal of Virological Methods* 133: 130–136

Tian T., Klaasen V.A., Soong J., Wisler G., Duffus J.E., Falk B.W., 1996. Generation of cDNAs specific to lettuce infectious yellows closterovirus and other whitefly-transmitted viruses by RT-PCR and degenerate oligonucleotide primers corresponding to the closterovirus gene encoding the heat shock protein 70 homolog. *Phytopathology* 86:1167–1173.

Tsai C.W., Chau J., Fernandez L., Bosco D., Daane K.M., Almeida R.P.P., 2008. Transmission of *Grapevine leafroll-associated virus 3* by the vine mealybug (*Planococcus ficus*). *Phytopathology* 98: 1093–1098.

9

Identification and pathogenicity of lignicolous fungi associated with grapevine trunk diseases in Southern Italy

Maria Luisa RAIMONDO[1], Antonia CARLUCCI[1,*], Claudio CICCARONE[1], Abderraouf SADALLAH[1,2,3], Francesco LOPS[1]

[1] *Department of Science of Agriculture, Food and Environment, University of Foggia, Via Napoli, 25, 71122 Foggia, Italy*
[2] *CIHEAM- Mediterranean Agronomic Institute of Bari, Via Ceglie, 9, 70010 Valenzano (Ba), Italy*
3 Department of Soil, Plant and Food Sciences, University of Bari Aldo Moro, Via Amendola 165/A, 70126, Bari, Italy
*Corresponding author: antonia.carlucci@unifg.it

Competing Interests: The Author(s) declare(s) no conflict of interest.

Editor: Lizel Mostert.

Summary. Over the last 10 years, several fungi were isolated from grapevines with grapevine trunk disease (GTD) symptoms, in the Apulia and Molise regions of Italy. Morphological and molecular analyses allowed the identification of species belonging to Botryosphaeriaceae, *Phaeoacremonium* species, *Phaeomoniella chlamydospora*, *Pleurostoma richardsiae* and less-common fungi associated with grapevine trunk diseases, such as *Cadophora*, *Colletotrichum*, *Seimatosporium* and *Truncatella*. These last genera were isolated at significant frequencies, so they were investigated for possible involvement in GTDs. To screen the large numbers of isolates collected, microsatellite-PCR analysis was carried out with the M13 primer, and 29 strains were further studied by amplification of different genes, for multi-locus analyses. Phylogenies and morphological analyses allowed identification, for first time in Italy, of fungi associated with GTDs, including *Cadophora luteo-olivacaea*, *Colletotrichum fioriniae*, *Seimatosporium vitis-vinifera* and *Truncatella angustata*. Pathogenicity assays with these fungi and other fungi known to be pathogens for grapevines (*Lasiodiplodia citricola*, *Phaeoacremonium italicum*, *Pleurostoma richardsiae*) showed that they caused disease symptoms on two Italian grapevine cultivars ('Bombino bianco', 'Nero di Troia'), although with different degrees of severity. Among the fungi isolated for the first time in Italy, *Sei. vitis-vinifera* was the most aggressive, while *C. fioriniae* the least pathogenic. All of these fungi were re-isolated from grapevine, and thus fulfilled Koch's postulates, confirming their pathogenicity on grapevine.

Keywords. *Cadophora luteo-olivacea*, *Colletotrichum fioriniae*, *Seimatosporium vitis-vinifera*, *Truncatella angustata*, phylogenies, artificial inoculation.

INTRODUCTION

Several diseases caused by fungi that have been associated with grapevines over the last 20 years have caused severe yield losses in other grape producing countries (Gramaje *et al.*, 2018; Guerin-Dubrana *et al.*, 2019). Grapevine trunk diseases (GTDs) are considered to be the most destructive and severe diseases of grapevine in Mediterranean countries, including Spain, France, Portugal and Italy, and also in the United States of America, Australia, and Asia (Gubler *et al.*, 2005; Gramaje *et al.*, 2018; Guerin-Dubrana *et al.*, 2019). The main fungi that cause GTDs are species involved in different diseases, which include Esca and Petri disease (Larignon and Dubos, 1997; Mugnai *et al.*, 1999; Gramaje *et al.*, 2011; Navarrete *et al.*, 2011; Bertsch *et al.*, 2013; Carlucci *et al.*, 2015a; Travadon *et al.*, 2015), Botryosphaeria dieback (Urbez-Torres, 2011), Diaporthe and Eutypa diebacks (Larignon and Dubos, 1997; Fourie and Halleen, 2004; Urbez-Torres *et al.*, 2013), and black foot disease (Halleen *et al.*, 2004; Agusti-Brisach and Armengol, 2013; Carlucci *et al.*, 2017).

The main external symptom of GTDs is general decline of affected plants. The specific external symptoms consist of tiger-stripe leaves, stunted shoots and chlorotic leaves which are sometimes cupped and with necrotic margins, flattened areas of the wood without bark, cankered wood and wedge-shaped perennial cankers, black and sunken necrotic lesions on roots, and reddish brown discolouration at the bases of trunks (Gramaje *et al.*, 2018). Internal symptoms include dark-coloured xylem vessels of the grapevine trunks, with exudate from the vessels when the trunks are cut in cross-section, and dark streaks in longitudinal sections (*Phaeomoniella chlamydospora, Phaeoacremonium* spp., *Cadophora* spp.). There can also be black subcortical streaking (*Pleurostoma richardsiae*) and necrosis of the wood tissues. Other symptoms are of rootstock browning in young grapevines (due to black foot fungi). Cordon dieback can also occur, with loss of spurs and internal necrotic wedge-shaped staining in stem cross-sections (*Eutypa lata, Botryosphaeria* spp.), and wood white rot (caused by Basidiomycete fungi) (Gramaje *et al.*, 2018).

Grapevines can be affected by one or more GTDs at the same time, as individual plants can be infected by different pathogens, due to co-occurrence of multiple infections throughout a season, and over years. This produces overlapping of the symptoms described above, which makes their association with the specific responsible fungi particularly difficult to define, and detection of causal pathogens challenging (Gramaje *et al.*, 2018).

High isolation frequency of particular fungal species involved in GTDs from mature, young and nursery grapevines in different countries can be different, due to climatic and geographic conditions, to specific pathogen aggressiveness, and to host cultivar susceptibility (Guerin-Dubrana *et al.*, 2019). Petri and black-foot diseases are mostly detected on planting material and young vines (Rego *et al.*, 2000; Agusti-Brisach and Armengol, 2013; Carlucci *et al.*, 2017). Conversely, apoplexy, Esca and grapevine leaf symptoms, and Phomopsis, Eutypa and Botryosphaeria diebacks are most frequently observed on mature grapevines (Guerin-Dubrana *et al.*, 2019).

To date, up to 138 fungal species belonging to 35 genera have been reported as responsible for GTDs. However, pathogenicity towards grapevine wood has not been tested and/or confirmed for all of these fungi (Gramaje *et al.*, 2018; Berlanas *et al.*, 2020; Brown *et al.*, 2020). For instance, 'Pestalotioides fungi' have been frequently associated with symptomatic and asymptomatic vineyards (Farr and Rossman, 2018; Liu *et al.*, 2019), although no detailed information is available about their involvement in GTD symptoms. In Italy, incidence of Esca, grapevine leaf symptoms and apoplexy is significant and increasing in all grapevine production regions (Guerin-Dubrana *et al.*, 2019). In Apulia, Molise and Sicily, Botryosphaeria dieback has also been reported (Cristinzio, 1978; Burruano *et al.*, 2008; Carlucci *et al.*, 2009; 2015b). Eutypa and Phomopsis diebacks are known to occur, if not frequently, in Italian vineyards (Guerin-Dubrana *et al.*, 2019). *Pleurostoma richardsiae, Dactylonectria torresensis, Ilyonectria liriodendri* and *Thelonectria blackeriella* were reported for the first time in Italy by Carlucci *et al.* (2015a; 2017).

A collection of fungi from a decennial survey carried out in symptomatic vineyards in the Apulia and Molise regions of Italy was subjected to identification and characterisation by morphological and molecular approaches, and pathogenicity testing of representative isolates was carried out to determine their putative involvement in GTDs. The present paper describes results from this research.

MATERIALS AND METHODS

Fungal isolates

Symptomatic grapevine samples were collected and analysed during the years 2009 to 2018. The samples were from many vineyards in the Foggia, Barletta-Trani-Andria and Campobasso provinces in Italy, and were taken from different grapevine cultivars, including 'Sangiovese', 'Montepulciano', 'Nero di Troia', 'Pinot grigio', 'Trebbiano toscano', 'Moscato bianco' and 'Chardonnay' (Table 1).

External symptoms observed on affected grapevine plants included stunting, reduced grapevine vigour, shoot dieback, and leaf discolouration with interveinal chlorosis and necrosis. Internal symptoms included black discolouration of wood under the bark, and necro-

Table 1. Information on vineyards surveyed and sampled in the Apulia and Molise regions (southern Italy).

Survey Year	Location	Vineyard			
		Cultivar	Age (year)	N. samples	GTD Incidence* (%)
2009	Cerignola (FG)	'Sangiovese'	27	8	13.4
	Lucera (FG)	'Nero di Troia'	15	3	11.5
	Lucera (FG)	'Moscato bianco'	10	3	9.8
2011	Canosa di Puglia (BT)	'Montepulciano'	21	4	16.2
	Canosa di Puglia (BT)	'Sangiovese'	12	3	11.3
	Foggia (FG)	'Nero di Troia'	19	5	13.5
	Foggia (FG)	'Pinot grigio'	13	3	9.2
2012	Foggia (FG)	'Moscato bianco'	17	4	6.7
	Campobasso (CB)	'Pinot grigio'	2	3	11.2
	Barletta (BT)	'Sangiovese'	14	3	17.3
	Barletta (BT)	'Trebbiano toscano'	11	3	16.4
2013	San Severo (FG)	'Trebbiano toscano'	29	10	21.4
	San Severo (FG)	'Nero di Troia'	21	6	18.3
	San Severo (FG)	'Pinot grigio'	12	3	16.9
	Termoli (CB)	'Pinot grigio'	5	4	8.5
	Termoli (CB)	'Chardonnay'	5	3	9.2
	Cerignola (FG)	'Sangiovese'	14	3	12.6
	Cerignola (FG)	'Trebbiano toscano'	8	6	9.3
2015	Campomarino (CB)	'Chardonnay'	2	6	11.5
	Campomarino (CB)	'Pinot grigio'	2	6	14.2
	Stornara (FG)	'Sangiovese'	31	8	19.8
	Torremaggiore (FG)	'Trebbiano toscano'	25	6	19.2
	Torremaggiore (FG)	'Nero di Troia'	15	4	14.8
2017	Foggia (FG)	'Sangiovese'	9	3	10.5
	Foggia (FG)	'Trebbiano toscano'	17	4	13.7
	Canosa di Puglia (BT)	'Montepulciano'	21	6	19.4
	Canosa di Puglia (BT)	'Sangiovese'	23	3	18.9
2018	Cerignola (FG)	'Trebbiano toscano'	19	3	13.6
	Barletta (BT)	'Chardonnay'	14	6	11.8
	Barletta (BT)	'Montepulciano'	17	6	10.3

* GTD Incidence was calculated on the basis of vines showing symptoms on 2,500 plants for each surveyed vineyard.

sis of xylem tissues. The samples included grapevine trunks, cordons and woody shoots. These were transported to the laboratory for analyses, where they initially underwent surface sterilization (Fisher *et al.*, 1992). The bark of each sample was removed with a sterile scalpel, and thin wood sections were cut (1 to 3 mm thick). From each section of each sample, five small wood tissue samples were cut and placed onto potato dextrose agar (PDA; 3.9% potato dextrose agar; Oxoid Ltd), and onto malt extract agar (MEA; 2% malt extract, 2% agar; Oxoid Ltd), both of which were supplemented with 500 mg L^{-1} streptomycin sulphate (Oxoid Ltd). After 7 to 10 d of incubation at 22±3°C in the dark, all of the fungal cultures obtained were purified by transferring single germinated conidia or small pieces of hyphae to Petri dishes containing fresh PDA.

Morphological and culture characteristics of isolated fungi were initially used to distinguish genera and species that were isolated from these symptomatic tissues (Crous and Gams, 2000; Mostert *et al.*, 2006; Essakhi *et al.*, 2008; Agustì-Brisach *et al.*, 2013; Phillips *et al.*, 2013; Raimondo *et al.*, 2014; Carlucci *et al.*, 2015a). The isolation frequency (IF; %) for each species was calculated as the number of tissue segments infected by each fungus, divided by the total number of tissue segments incubated.

DNA extraction and microsatellite PCR profiles

Genomic DNA was extracted from the 420 isolates obtained, from 15-d-old cultures grown on PDA (Carlucci *et al.*, 2013). Many of the isolates (339) belonged to Botryosphaeriaceae and *Phaeoacremonium*, *Phaeomoniella*, *Pleurostoma*, *Cadophora*, *Colletotrichum*, *Seimatosporium* and *Truncatella*, so preliminary screening was carried out for each genus based on the M13 minisatellite primers (5'-GAGGGTGGCGGTTCT-3') (Meyer *et al.*, 1993). Microsatellite (MSP)-PCR profiles were generated according to Santos and Phillips (2009). The DNA banding patterns were analysed using the Bionumerics v. 5.1 software (Applied Maths), with calculation of Pearson's correlation coefficients according to the unweighted pair group method with arithmetic means. The reproducibility levels were calculated by comparisons of the banding profiles obtained for the M13 primer. For this purpose, from any cluster, 10% of the strains were chosen at random, and their profiles were repeat.

Molecular characterisation

The MSP dendrogram generated for each genus produced different clades from which representative isolates were chosen for phylogenetic analysis data not shown). Eighty-four representative isolates of Botryosphaeriaceae, *Phaeoacremonium* spp., *Phaeomoniella* spp. and *Pleurostoma* spp. were identified using the keys, descriptions and sequence data from Phillips *et al.* (2013), Mostert *et al.* (2006), Essakhi *et al.* (2008), Raimondo *et al.* (2014), Crous and Gams (2000) and Carlucci *et al.* (2015a). For 41 Botryosphaeriaceae strains, ITS1-5.8S-ITS2 were amplified using the universal primers ITS1 and ITS4 (White *et al.*, 1990), and part of EF1-a was amplified using the primers EF1-688F and EF1-1251R (Alves *et al.*, 2008), according to Carlucci *et al.* (2015b). For 27 *Phaeoacremonium* strains, partial β-tubulin and partial actin genes were amplified using the universal primers T1 (O'Donnell and Cigelnik, 1997), Bt2b (Glass and Donaldson, 1995) and ACT-512F/ACT-783R (Carbone and Kohn, 1999), according to Raimondo *et al.* (2014). For seven *Phaeomoniella* and *Pleurostoma* strains, ITS1-5.8S-ITS2 were amplified using the universal primers ITS1 and ITS4 (White *et al.* 1990), according to Damm *et al.* (2010) and Carlucci *et al.* (2015a).

The other 29 representative strains that belonged to the *Seimatosporium*, *Truncatella*, *Cadophora* or *Colletotrichum* were further studied using molecular and morphological tools. Five loci were amplified for seven *Seimatosporium* and 11 *Truncatella* strains that were representative of the MSP-PCR groups. For large subunit RNA (LSU; ca. 500 bp) were used NL1/NL4 primer pairs (O'Donnell and Gray, 1993); for internal transcribed spacers (ITS) 1 and 2 (including 5.8S of nuclear ribosomal DNA; ca. 500 bp) were used ITS5/ITS4 (White *et al.*, 1990); for the partial β-tubulin gene (*tub*; ca. 680 bp) were used T1 (O'Donnell and Cigelnik, 1997) and Bt2b (Glass and Donaldson, 1995); for the partial translation elongation factor 1-alpha (*tef-1α*; ca. 300 bp) were used EF1-688F and EF1-1251R (Alves *et al.*, 2008); and for the second-largest subunit of DNA-directed RNA polymerase II (*rpb2*; ca. 500 bp) were used RPB2-5f2/RPB2-7cr (Liu *et al.*, 1999; Sung *et al.*, 2007).

The LSU and ITS PCR reactions and conditions were performed according to Carlucci *et al.* (2012), with those for β-tubulin, *tef-1α* and *rpb2* according to Liu *et al.* (2019).

Three loci including ITS (ca. 550 bp), the partial translation elongation factor 1-alpha (*tef-1α*; ca. 420bp) and the partial β-tubulin gene (*tub*; ca. 500 bp) were amplified from six *Cadophora* strains, as representative of the MSP-PCR groups. These amplifications used the following primer pairs: ITS5/ITS4 (White *et al.*, 1990) for internal transcribed spacers (ITS) 1 and 2; EF1-728F and EF1-986R (Carbone and Kohn, 1999) for the partial translation elongation factor 1-alpha; and BTCadF 5' and BTCadR 5' (Travadon *et al.*, 2015) for the partial β-tubulin gene. The ITS PCR reactions and conditions were performed as described above, while those for *tef-1α* and β-tubulin according to Travadon *et al.* (2015).

Six loci were amplified for five *Colletotrichum* strains, as representative of the MSP-PCR groups. These included: the 5.8S nuclear ribosomal gene with the two flanking ITS (ca. 538 bp); β-tubulin (*tub*; ca. 500 bp); partial actin (*act*; ca. 250 bp); the intron of glyceraldehyde-3-phosphate dehydrogenase (*gapdh*; ca. 250 bp), and chitin synthase (*chs-1*; ca. 280 bp). The primer pairs used were ITS5/ITS4 (White *et al.*, 1990) for internal transcribed spacers (ITS) 1 and 2; T1 (O'Donnell and Cigelnik, 1997) and Bt2b (Glass and Donaldson, 1995)) for β-tubulin; ACT-512F/ACT-783R (Carbone and Kohn, 1999) for partial actin; GDF1/GDR1 (Guerber *et al.*, 2003) for the intron of glyceraldehyde-3-phosphate dehydrogenase; and CHS-79F/CHS-345R (Carbone and Kohn, 1999) for chitin synthase. The PCR amplifications and conditions were performed according to Fu *et al.* (2019).

Five microlitres of each amplicon was analysed by electrophoresis, using 1.5% (w/v) agarose gels in 1× TAE buffer (40 mM Tris, 40 mM acetate, 2 mM EDTA, pH 8.0) at 100 V for 30 min. The gels were stained with ethidium bromide and visualised under ultraviolet light (Gel Doc EZ System; BioRad). The PCR products were purified before DNA sequencing (Nucleo Spin Extract II purification kits; Macherey-Nagel), according to the manufacturer instructions. Both strands of the PCR products were sequenced by Eurofins Genomics Service (Milan, Italy).

Phylogenetic analyses

The nucleotide sequences obtained were manually edited using BioEdit version 7.0.9 (http://www.mbio.ncsu.edu/BioEdit). Consensus sequences were compared with those available in the GenBank database, using the Basic Local Alignment Search Tool (BLAST) to confirm the preliminary morphological identification, and to select and download closely related sequences for phylogenetic analyses. GenBank sequences from different species of *Seimatosporium*, *Truncatella*, *Cadophora* and *Colletotrichum* were then selected and added to the sequences dataset obtained (Tables 2, 3).

The sequences were manually concatenated and aligned using the online multiple alignment programme MAFFT v.7 (http://mafft.cbrc.jp/alignment/server/) (Katoh and Standley, 2013). The alignments were visually checked and manually improved where necessary. Multilocus analyses according to maximum parsimony and maximum likelihood were carried out for the LSU, ITS, β-tubulin, *tef-1α* and *rpb2* genes of the *Seimatosporium* and *Truncatella* sequence data.

The maximum parsimony analyses were performed using PAUP, version 4.0b10 (Swofford, 2003), with the heuristic search option with 100 random taxa additions, and tree bisection and reconstruction as the branch swapping algorithm. Branches of zero length were collapsed and all multiple equally parsimonious trees were saved. Bootstrap support values were calculated from 1,000 heuristic search replicates and ten random taxon additions. The tree length (TL), consistency index (CI), retention index (RI), homoplasy index (HI), and rescaled consistency index (RC) were calculated for each, and the resulting trees were visualised with TreeView, version 1.6.6 (Page, 1996). Alignment gaps were treated as missing data for *Seimatosporium* strains, and as fifth characters for *Truncatella* strains.

The maximum likelihood analysis was carried out using RAxML-HPC v.8.2.12 (Stamatakis, 2006; Stamatakis *et al.*, 2008) on the XSEDE Teragrid of the CIPRES Science Gateway (https://www.phylo.org) (Miller *et al.*, 2010), with rapid bootstrap analysis, followed by 1,000 bootstrap replicates. The final trees were selected among the suboptimal trees from each run by comparing the likelihood and bootstrap scores. The outgroups in the *Seimatosporium* multigenic analysis were *Synnemapestaloides juniperi* (CBS 447.77) and *Discosia artocreas* (CBS 124848), and those for *Truncatella* were *Phlogicylindrium eucalypti* (CBS 120080) and *Beltrania pseudorhombica* (CBS 138003).

Multilocus alignment of the *Cadophora* (ITS, *tef-1α*, β-tubulin genes) and *Colletotrichum* (ITS, β-tubulin, *act*, *gapdh*, *chs-1* genes) strains was performed as described above with alignment gaps treated as missing data. *Hyaloscypha finlandica* (CBS 444.86) was used as outgroup in the *Cadophora* analysis, and *Colletotrichum gloeosporioides* (ICMP 17821) for the *Colletotrichum* analysis.

Morphological analyses

For each species identified using molecular tools (as described above), three isolates were used for morphological studies. To enhance sexual sporulation or conidiation, these fungi were grown on MEA in Petri dishes for 10 to 21 d under UV light at 23±2°C. Fungal structures were observed and measured from 100% lactic acid microscope slide mounts by making 30 measurements (at ×400 or ×1,000 magnification), using a measurement module (Leica Application Suite; Leica Microsystems GmbH). Photomicrographs were recorded using a digital camera (DFC320; Leica) on a microscope fitted with Nomarski differential interference contrast optics (DMR; Leica). The morphological features of conidiogenous cells and conidia were also determined in distilled water, by picking mycelium plugs from 30-d-old cultures grown on MEA, with images captured using a microscope (DM5500; Leica) at ×40 magnification.

Pathogenicity tests

Three isolates of each species were used in pathogenicity tests, to determine the infection of grapevine wood tissues by the less-known GTD fungi, and to compare their aggressiveness with the most common and previously determined GTD fungi. The previously determined GTD fungi used were: *Lasiodiplodia citricola* (Carlucci *et al.*, 2015b), *Phaeoacremonium italicum* (Raimondo *et al.*, 2014), and *Pleurostoma richardsiae* (Carlucci *et al.*, 2015a).

Inoculations were carried out in June 2018, on 1-year-old canes (diam. 1.0-2.5 cm) from 10-y-old grapevines of the cultivars 'Nero di Troia' and 'Bombino bianco' in vineyards in an open field. The canes were inoculated at the internodes by wounding, as described by Carlucci *et al.* (2013). The wounds (each 1.0-2.0 cm long) were made on the cane surfaces with a sterile scalpel.

Table 2. Isolate identification numbers, locations, hosts and GenBank accession numbers of the strains of *Seimatosporium* spp., *Truncatella* spp. and *Cadophora* spp. used in the multigenic analyses.

Species	Isolate number[a]	Location	Host	GenBank accession number				
				LSU	ITS	*tub*	*tef-1α*	*rpb2*
Discosia artocreas	CBS 124848 ET[b]	Germany	*Fagus sylvatica*	MH554213	MH553994	MH554662	MH554420	MH554903
Seimatosporium botan	NBRC 104200 HT	Japan	*Paeonia suffruticosa*	AB593731	AB594799	LC047770	–	–
Sei. germanicum	CBS 437.87 HT	Germany	Unknown	MH554259	MH554047	MH554723	MH554482	MH554957
Sei. luteosporum	CBS 142599 HT	USA	*Vitis vinifera*	KY706309	KY706284	KY706259	KY706334	–
Sei. physocarpi	CBS 139968 HT	Russia	*Physocarpus opulifolius*	KT198723	KT198722	MH554676	MH554434	MH554917
	CBS 789.68	The Netherlands	*Physocarpus amurensis*	MH554278	MH554066	MH554742	MH554502	MH554979
Sei. pistaciae	CBS 138865 HT	Iran	*Pistacia vera*	KP004491	KP004463	MH554674	MH554432	MH554915
	CPC 24457	Iran	*Pistacia vera*	MH554331	MH554126	MH554799	MH554561	MH555035
Sei. rosae	CBS 139823 ET	Russia	*Rosa kalmiussica*	KT198727	LT853105	LT853253	LT853203	LT853153
Sei. vitifusiforme	CBS 142600 HT	USA	*Vitis vinifera*	KY706321	KY706296	KY706271	KY706346	–
Sei. vitis-viniferae	CBS 123004 HT	Spain	*Vitis vinifera*	MH554211	MH553992	MH554660	MH554418	MH554901
	CBS 116499	Iran	*Vitis vinifera*	MH554201	MH553984	MH554643	MH554402	MH554884
	CRCC 212[c]	Italy	*Vitis vinifera*	**MN862466**	**MN862459**	**MN862452**	**MN862445**	**MN862473**
	CRCC 213	Italy	*Vitis vinifera*	**MN862467**	**MN862460**	**MN862453**	**MN862446**	**MN862474**
	CRCC 214	Italy	*Vitis vinifera*	**MN862468**	**MN862461**	**MN862454**	**MN862447**	**MN862475**
	CRCC 229	Italy	*Vitis vinifera*	**MN862472**	**MN862465**	**MN862458**	**MN862451**	**MN862479**
	CRCC 247	Italy	*Vitis vinifera*	**MN862469**	**MN862462**	**MN862455**	**MN862448**	**MN862476**
	CRCC 248	Italy	*Vitis vinifera*	**MN862470**	**MN862463**	**MN862456**	**MN862449**	**MN862477**
	CRCC 251	Italy	*Vitis vinifera*	**MN862471**	**MN862464**	**MN862457**	**MN862450**	**MN862478**
Sei. vitis	MFLUCC 14-0051	Italy	*Vitis vinifera*	KR920362	KR920363	–	–	–
	Napa774	Napa County, USA	*Vitis vinifera*	KY706276	KY706301	KY706251	KY706326	–
	Napa772	Napa County, USA	*Vitis vinifera*	KY706275	KY706300	KY706250	KY706325	–
	Napa782	Napa County, USA	*Vitis vinifera*	KY706278	KY706303	KY706253	KY706328	–
	Napa764	Napa County, USA	*Vitis vinifera*	KY706273	KY706298	KY706248	KY706323	–
	Napa759	Napa County, USA	*Vitis vinifera*	KY706282	KY706307	KY706257	KY706332	–
	VMT2_1	Italy	*Vitis vinifera*	–	LS991528	LS997596	LS999502	–
Sporocadus biseptatus	CBS 110324 HT	Unknown	Unknown	MH554179	MH553956	MH554615	MH554374	MH554853
Spo. cornicola	CBS 143889	Germany	*Cornus sanguinea*	MH554326	MH554121	MH554794	MH554555	MH555029
Spo. incanus	CBS 123003 HT	Spain	*Prunus dulcis*	MH554210	MH553991	MH554659	MH554417	MH554900
Spo. lichenicola	CBS 354.90	Germany	*Fagus sylvatica*	MH554252	MH554035	MH554711	MH554470	MH554948
	CPC 24528	Germany	*Juniperus communis*	MH554332	MH554127	MH554800	MH554562	MH555036
	NBRC 32625; IMI 079706 ET	UK	*Rosa canina*	MH883646	MH883643	MH883645	MH883644	MH883647
Spo. mali	CBS 446.70 HT	The Netherlands	*Malus sylvestris*	MH554261	MH554049	MH554725	MH554484	MH554960
Spo. microcyclus	CBS 424.95 HT	Germany	*Sorbus aria*	MH554258	MH554045	MH554721	MH554480	MH554956

(Continued)

Table 2. (Continued).

Species	Isolate number[a]	Location	Host	GenBank accession number				
				LSU	ITS	*tub*	*tef-1α*	*rpb2*
	CBS 887.68	The Netherlands	*Ribes* sp.	MH554280	MH554068	MH554744	MH554504	MH554981
Spo. multiseptatus	CBS 143899 HT	Serbia	*Viburnum* sp.	MH554343	MH554141	MH554814	MH554576	MH555047
Spo. rosarum	CBS 113832	Sweden	*Rosa canina*	MH554189	MH553970	MH554629	MH554388	MH554864
Spo. rosigena	CBS 116498	Iran	*Vitis vinifera*	MH554200	MH553983	MH554642	MH554401	MH554883
	CBS 129166	Latvia	*Rhododendron*	MH554215	MH553996	MH554665	MH554423	MH554905
	CBS 182.50	The Netherlands	*Pyrus communis*	MH554233	MH554013	MH554689	MH554447	MH554926
	CBS 250.49	The Netherlands	*Rubus fruticosus*	MH554245	MH554023	MH554699	MH554457	MH554934
	CBS 466.96	The Netherlands	*Rubus* sp.	MH554265	MH554052	MH554728	MH554487	MH554965
Spo. rotundatus	CBS 616.83 HT	Canada	*Arceuthobium pussilum*	MH554273	MH554060	MH554737	MH554496	MH554974
Spo. sorbi	CBS 160.25	Unknown	Unknown	MH554229	MH554008	MH554684	MH554442	MH554924
Sporocadus sp. 1	CBS 506.71	Italy	*Euphorbia* sp.	MH554268	MH554055	MH554731	MH554490	MH554968
Spo. trimorphus	CBS 114203 HT	Sweden	*Rosa canina*	MH554196	MH553977	MH554636	MH554395	MH554876
Synnemapestaloides juniperi	CBS 477.77 HT	France	*Juniperus phoenicea*	MH554266	MH554053	MH554729	MH554488	MH554966
Bartalinia bella	CBS 464.61 HT	Brazil	Air	MH554264	MH554051	MH554727	MH554486	MH554964
Bar. robillardoides	CBS 122615	South Africa	*Cupressus lusitanica*	MH554207	MH553989	MH554657	MH554415	MH554897
	CBS 122705 ET	Italy	*Leptoglossus occidentalis*	KJ710438	LT853104	LT853252	LT853202	LT853152
Bar. pini	CBS 143891 HT	Uganda	*Pinus patula*	MH554330	MH554125	MH554797	MH554559	MH555033
	CBS 144141	USA	*Acacia koa*	MH554364	MH554170	MH554843	MH554605	MH555067
Beltrania pseudorhombica	CBS 138003	China	*Pinus tabulaeformis*	KJ869215	MH554124	–	MH554558	MH555032
Broomella vitalbae	HPC 1154	Unknown	Unknown	MH554367	MH554173	MH554846	MH554608	MH555069
Diversimediispora humicola	CBS 302.86 HT	USA	Soil	MH554247	MH554028	MH554705	MH554463	MH554941
Heterotruncatella proteicola	CBS 144020 HT	South Africa	*Protea acaulos*	MH554288	MH554077	MH554751	MH554512	MH554989
Het. quercicola	CBS 143895 HT	USA	*Quercus walshii*	MH554337	MH554135	MH554808	MH554570	MH555041
Het. restionacearum	CBS 118150	South Africa	*Restio filiformis*	MH554203	DQ278914	MH554649	MH554407	MH554889
	CBS 119210 HT	South Africa	*Ischyrolepis cf. gaudichaudiana*	DQ278929	DQ278915	MH554653	MH554411	MH554892
Het. spadicea	CBS 118144	South Africa	*Ischyrolepis* sp.	DQ278926	DQ278921	MH554646	MH554404	MH554886
	CBS 118145 ET	South Africa	*Cannomois virgata*	DQ278927	DQ278912	MH554647	MH554405	MH554887
	CBS 118148	South Africa	*Rhodocoma capensis*	DQ278928	DQ278913	MH554648	MH554406	MH554888
	CPC 17911; CMW 22206	South Africa	*Elegia filacea*	MH554308	MH554098	MH554771	MH554532	MH555012
	CPC 28956	Australia	*Sorghum halepense*	MH554353	MH554157	MH554830	MH554592	MH555056
Hymenopleella austroafricana	CBS 143886 HT	South Africa	*Gleditsia triacanthos*	MH554320	MH554115	MH554788	MH554549	MH555023
	CBS 144026	South Africa	*Bridelia mollis*	MH554322	MH554117	MH554790	MH554551	MH555025
	CBS 144027	Zambia	*Combretum hereroense*	MH554324	MH554119	MH554792	MH554553	MH555027

(Continued)

Table 2. (Continued).

Species	Isolate number[a]	Location	Host	GenBank accession number				
				LSU	ITS	*tub*	*tef-1α*	*rpb2*
Hym. polyseptata	CBS 143887 HT	South Africa	*Combretum* sp.	MH554321	MH554116	MH554789	MH554550	MH555024
Hym. hippophaeicola	CBS 113687	Sweden	*Hippophae rhamnoides*	MH554188	MH553969	MH554628	MH554387	MH554863
	CBS 140410 ET	Austria	*Hippophae rhamnoides*	MH554224	KT949901	MH554678	MH554436	MH554919
Hym. subcylindrica	CBS 164.77	India	*Cocos nucifera*	MH554230	MH554009	MH554685	MH554443	MH554925
	CBS 647.74 HT	India	*Gypsophilla* seeds	MH554275	MH554062	MH554739	MH554498	MH554976
Morinia acaciae	CBS 100230	New Zealand	*Prunus salicina* Omega	MH554174	MH553950	MH554609	MH554368	MH554847
	CBS 137994 HT	France	*Acacia melanoxylon*	MH554221	MH554002	MH554673	MH554431	MH554914
Mor. crini	CBS 143888 HT	South Africa	*Crinum bulbispermum*	MH554323	MH554118	MH554791	MH554552	MH555026
Mor. longiappendiculata	CBS 117603 HT	Spain	*Calluna vulgaris*	MH554202	AY929324	MH554644	AY929316	MH554885
Parabartalinia lateralis	CBS 399.71 HT	South Africa	*Acacia karroo*	MH554256	MH554043	MH554719	MH554478	MH554954
Pseudosarcostroma osyridicola	CBS 103.76 HT	France	*Osyris alba*	MH554177	MH553954	MH554613	MH554372	MH554851
Truncatella angustata	CBS 113.11	Germany	*Picea abies*	MH554185	MH553966	MH554625	MH554384	MH554860
	CBS 135.97	Spain	Decaying bark	MH554220	MH554001	MH554671	MH554429	MH554912
	CBS 165.25	Unknown	*Prunus armeniaca*	MH554231	MH554010	MH554686	MH554444	–
	CBS 231.77 = CBS 296.77	Turkey	*Gossypium* sp.	MH554243	MH554021	MH554697	MH554455	MH554932
	CBS 338.32	The Netherlands	*Lupinus* sp.	MH554250	MH554033	MH554709	MH554467	MH554945
	CBS 398.71	Turkey	Soil	MH554255	MH554042	MH554718	MH554477	MH554953
	CBS 144025 NT	France	*Vitis vinifera* Prunelard	MH554318	MH554112	MH554785	MH554546	MH555021
	CBS 449.51	Unknown	*Salix* sp. or *Thuja* sp.	MH554262	MH554050	MH554726	MH554485	MH554961
	CBS 938.70	The Netherlands	*Prunus laurocerasus*	MH554281	MH554070	MH554746	MH554506	MH554982
	CPC 21366	France	*Vitis vinifera* Prunelard	MH554319	MH554113	MH554786	MH554547	MH555022
	CBS 208.80	The Netherlands	Food	MH554239	MH554020	MH554696	MH554454	–
	CBS 443.54	UK	*Picea abies*	MH554260	MH554048	MH554724	MH554483	MH554959
	CPC 21354	France	*Vitis vinifera* Prunelard	MH554317	MH554111	MH554784	MH554545	MH555020
	CBS 642.97	Switzerland	*Heterodera carotae* cyst egg mass, on *Daucus carota*	MH554274	MH554061	MH554738	MH554497	MH554975
	CBS 564.76	Switzerland	*Pyrus malus*	MH554271	MH554057	MH554733	MH554492	MH554970
	CRCC 147	Italy	*Vitis vinifera*	–	–	–	–	–
	CRCC 165	Italy	*Vitis vinifera*	–	–	–	–	–
	CRCC 188	Italy	*Vitis vinifera*	**MN862441**	**MN862439**	**MN862437**	**MN862435**	**MN862443**
	CRCC 189	Italy	*Vitis vinifera*	–	–	–	–	–
	CRCC 195	Italy	*Vitis vinifera*	–	–	–	–	–
	CRCC 199	Italy	*Vitis vinifera*	–	–	–	–	–
	CRCC 201	Italy	*Vitis vinifera*	–	–	–	–	–

(Continued)

Table 2. (Continued).

Species	Isolate number[a]	Location	Host	GenBank accession number				
				LSU	ITS	*tub*	*tef-1α*	*rpb2*
	CRCC 240	Italy	*Vitis vinifera*	–	–	–	–	–
	CRCC 241	Italy	*Vitis vinifera*	–	–	–	–	–
	CRCC 243	Italy	*Vitis vinifera*	–	–	–	–	–
	CRCC 245	Italy	*Vitis vinifera*	**MN862442**	**MN862440**	**MN862438**	**MN862436**	**MN862444**
Phlogicylindrium eucalypti	CBS 120080 HT	*Australia*	*Eucalyptus globulus*	DQ923534	NR_132813	MH704633	MH704607	MH554893
Robillarda africana	CBS 122.75 HT	South Africa	Unknown	KR873281	KR873253	MH554656	MH554414	MH554896
Rob. australiana	CBS 143882 HT	Australia	Unknown	MH554301	MH554091	MH554764	MH554525	MH555005
Rob. terrae	CBS 587.71 HT	India	Soil	KJ710459	KJ710484	MH554734	MH554493	MH554971
Rob. roystoneae	CBS 115445 HT	Hong Kong	*Roystonea regia*	KR873282	KR873254	KR873317	KR873310	MH554880
Strickeria kochii	CBS 140411 ET	Austria	*Robinia pseudoacacia*	KT949918	NR_154423	MH554679	MH554437	MH554920
Cad. gregata	ATCC11073 HT	Unknown	Soybean root	–	U66731	Mf677920	Mf979586	–
Cad. helianthii	CBS 144752 HT	Ukraine	*Helianthus annuus*	–	MF962601	MH733391	MH719029	–
Cad. interclivum	BAP37	Banff, Canada	*Picea glauca*, root	–	MF677930	MF677919	MF979585	–
	BAP33	Banff, Canada	*Picea glauca*, root	–	MF677929	MF677918	MF979584	–
	CBS143323 HT	Banff, Canada	*Carex sprengelii*, root	–	MF677928	MF677917	MF979583	–
Cad. luteo-olivacea	CBS 141.41 HT	Sweden	Unknown	–	AY249066	KM497133	KM497089	–
	A19	California, USA	*Vitis viniferae*	–	KM497038	KM497119	KM497075	–
	A41	California, USA	*Vitis vinifera* 'Chardonnay'	–	KM497039	KM497120	KM497076	–
	A42	California, USA	*Vitis vinifera* 'Chardonnay'	–	KM497040	KM497121	KM497077	–
	U5	California, USA	*Vitis vinifera* 'Sangiovese'	–	KM497041	KM497122	KM497078	–
	U7	California, USA	*Olea europa*	–	KM497044	KM497125	KM497081	–
	U8	California, USA	*Vitis vinifera* 'Semillon'	–	KM497042	KM497123	KM497079	–
	U17	California, USA	*Vitis vinifera* 'Chardonnay'	–	KM497043	KM497124	KM497080	–
	U21	California, USA	*Vitis vinifera*	–	KM497045	KM497126	KM497082	–
	U22	California, USA	*Vitis vinifera* 'Chardonnay'	–	KM497046	KM497127	KM497083	–
	U53	California, USA	*Vitis vinifera* 'Chardonnay'	–	KM497047	KM497128	KM497084	–
	U56	California, USA	*Vitis vinifera* 'Syrah'	–	KM497048	KM497129	KM497085	–
	CRCC 11B	Italy	*Vitis vinifera*	–	–	–	–	–
	CRCC 113A	Italy	*Vitis vinifera*	–	**MN871929**	**MN871925**	**MN871927**	–
	CRCC 122	Italy	*Vitis vinifera*	–	**MN871930**	**MN871926**	**MN871928**	–
	CRCC 131	Italy	*Vitis vinifera*	–	–	–	–	–
Cad. malorum	CBS 165.42 HT	The Netherlands	*Amblystoma mexicanum*	–	AY249059	KM497134	KM497090	–
Cad. melinii	CBS 268.33 HT	Unknown	Unknown	–	AY249072	KM497132	KM497088	–
	U11	California, USA	*Vitis vinifera* 'Sangiovese'	–	KM497032	KM497113	KM497069	–

(Continued)

Table 2. (Continued).

Species	Isolate number[a]	Location	Host	GenBank accession number				
				LSU	ITS	*tub*	*tef-1α*	*rpb2*
	ONC1	Ontario, Canada	*Vitis vinifera* 'Cabernet Franc'	–	KM497033	KM497114	KM497070	–
Cad. meredithiae	CBS143322 HT	Banff, Canada	*Carexsprengelii*, root	–	MF677925	MF677914	MF979580	–
	BAP6	Banff, Canada	*Picea glauca*, root	–	MF677926	MF677915	MF979581	–
	BAP13	Banff, Canada	*Picea glauca*, root	–	MF677927	MF677916	MF979582	–
Cad. novi-eboraci	NYC14 HT	New York, USA	*Vitis labruscana* 'Concord'	–	KM497037	KM497118	KM497074	–
	NYC2	New York, USA	*Vitis labruscana* 'Concord'	–	KM497034	KM497115	KM497071	–
	NYC13	New York, USA	*Vitis vulpina*	–	KM497036	KM497117	KM497073	–
	NYC1	New York, USA	*Vitis vinifera* 'Cabernet' Sauvignon	–	KM497035	KM497116	KM497072	–
Cad. orchidicola	UAMH8152	Alberta, Canada	Northern green orchid, root	–	AF214576	MF677921	MF979587	–
Cad. orientoamericana	NHC1 HT	New Hampshire, USA	*Vitis* hybrid 'Niagara'	–	KM497018	KM497099	KM497055	–
	CTC1	Connecticut, USA	*Vitis vinifera* 'Chardonnay'	–	KM497012	KM497093	KM497049	–
Cad. spadicis	RIC1	Rhode Island, USA	*Vitis vinifera* 'Cabernet' Sauvignon	–	KM497029	KM497110	KM497066	–
	RIC3	Rhode Island, USA	*Vitis hybrid* 'Vidal'	–	KM497030	KM497111	KM497067	–
	QCC1	Quebec, Canada	*Vitis vinifera* 'Gamay'	–	KM497031	KM497112	KM497068	–
	CBS 111743 HT	Italy	*Actinidia chinensis*	–	DQ404351	KM497136	KM497091	–
Cad. viticola	CBS 139517 HT	Spain	*Vitis vinifera* 'Syrah'	–	HQ661096	–	HQ661081	–
	Cme-1	Spain	*Vitis vinifera* 'Syrah'	–	HQ661096	–	HQ661081	–
	Cme-3	Spain	*Vitis vinifera* 'Syrah'	–	HQ661098	–	HQ661083	–
Hyaloscypha finlandica	CBS 444.86 HT	Finland	Unknown	–	AF486119	KM497130	KM497086	–

[a] **ATCC**: American Type Culture Collection, Virginia, USA; BCC: BIOTEC Culture Collection, National Center for Genetic Engineering and Biotechnology (BIOTEC), Khlong Luang, Pathumthani, Thailand. **CBS**: Culture collection of the Westerdijk Fungal Biodiversity Institute, Utrecht, The Netherlands. **CMW**: Culture Collection of the Forestry and Agricultural Biotechnology Institute (FABI), University of Pretoria, Pretoria, South Africa. **CPC**: Culture collection of Pedro Crous, housed at the Westerdijk Institute. **CRCC**: Carlucci and Raimondo Culture Collection, housed at Dept. SAFE of University of Foggia. **HPC**: Herbarium of Pedro Crous, housed at the Westerdijk Institute. **IMI**: International Mycological Institute, CABI-Bioscience, Egham, Bakeham Lane, United Kingdom. **MFLU(CC)**: Mae Fah Luang University Culture Collection. **NBRC**: Biological Resource Center.

[b] Status: status of the strains. ET: ex-epitype. NT: ex-neotype. HT: ex-Holotype.

[c] Strain numbers and newly generated sequences are indicated in bold font.

Table 3. Isolate identification numbers, locations, hosts and GenBank accession numbers of the strains of *Colletotrichum* spp. used in the multigenic analyses.

Species	Isolate number[a]	Location	Host	GenBank accession number				
				ITS	*gapdh*	*chs-1*	*act*	*tub*
C. acerbum	CBS 128530 HT[b]	New Zealand	*Malus domestica*, bitter rot of fruit	JQ948459	JQ948790	JQ949120	JQ949780	JQ950110
C. acutatum	CBS 112996 HT	Australia	*Carica papaya*	JQ005776	JQ948677	JQ005797	JQ005839	JQ005860
C. australe	CBS 116478 HT	South Africa	*Trachycarpus fortunei*	JQ948455	JQ948786	JQ949116	JQ949776	JQ950106
C. brisbanense	CBS 292.67 HT	Australia	*Capsicum annuum*	JQ948291	JQ948621	JQ948952	JQ949612	JQ949942
C. chrysanthemi	IMI 364540, CPC 18930	China	*Chrysanthemum coronarium*, leaf spot	JQ948273	JQ948603	JQ948934	JQ949594	JQ949924
C. cosmi	CBS 853.73 HT	The Netherlands	*Cosmos* sp., seed	JQ948274	JQ948604	JQ948935	JQ949595	JQ949925
C. costaricense	CBS 330.75 HT	Costa Rica	*Coffea arabica*, cv. 'Typica', berry	JQ948180	JQ948510	JQ948841	JQ949501	JQ949831
C. cuscutae	IMI 304802, CPC 18873 HT	Dominica	*Cuscuta* sp.	JQ948195	JQ948525	JQ948856	JQ949516	JQ949846
C. fioriniae	CBS 128517 HT	USA	*Fiorinia externa* (elongate hemlock scale, insect)	JQ948292	JQ948622	JQ948953	JQ949613	JQ949943
	CBS 125396	USA	*Malus domestica*, fruit lesion	JQ948299	JQ948629	JQ948960	JQ949620	JQ949950
	CBS 124958	USA	*Pyrus* sp., fruit rot	JQ948306	JQ948636	JQ948967	JQ949627	JQ949957
	CBS 126526	The Netherlands	*Primula* sp., leaf spots	JQ948323	JQ948653	JQ948984	JQ949644	JQ949974
	IMI 324996, CPC 18880	USA	*Malus pumila*	JQ948301	JQ948631	JQ948962	JQ949622	JQ949952
	CRCC 104[c]	Italy	*Vitis vinifera*	**MN871933**	**MN871939**	**MN871937**	**MN871931**	**MN871935**
	CRCC 140	Italy	*Vitis vinifera*	**MN871934**	**MN871940**	**MN871938**	**MN871932**	**MN871936**
	CRCC 144	Italy	*Vitis vinifera*	–	–	–	–	–
	CRCC 154	Italy	*Vitis vinifera*	–	–	–	–	–
	CRCC 160	Italy	*Vitis vinifera*	–	–	–	–	–
C. godetiae	CBS 133.44 HT	Denmark	*Clarkia hybrida*, cv. 'Kelvon Glory', seed	JQ948407	JQ948738	JQ949068	JQ949728	JQ950058
C. guajavae	IMI 350839, CPC 18893 HT	India	*Psidium guajava*, fruit	JQ948270	JQ948600	JQ948931	JQ949591	JQ949921
C. indonesiense	CBS 127551 HT	Indonesia	*Eucalyptus* sp.	JQ948288	JQ948618	JQ948949	JQ949609	JQ949939
C. johnstonii	CBS 128532 HT	New Zealand	*Solanum lycopersicum*, fruit rot	JQ948444	JQ948775	JQ949105	JQ949765	JQ950095
C. kinghornii	CBS 198.35 HT	UK	*Phormium* sp.	JQ948454	JQ948785	JQ949115	JQ949775	JQ950105
C. laticiphilum	CBS 112989 HT	India	*Hevea brasiliensis*	JQ948289	JQ948619	JQ948950	JQ949610	JQ949940
C. lauri	MFLUCC 17-0205 HT	Italy	*Laurus nobilis*	KY514347	KY514344	KY514341	KY514338	KY514350
C. limetticola	CBS 114.14 HT	USA, Florida	*Citrus aurantifolia*, young twig	JQ948193	JQ948523	JQ948854	JQ949514	JQ949844
C. lupini	CBS 109225 HT	Ukraine	*Lupinus albus*	JQ948155	JQ948485	JQ948816	JQ949476	JQ949806
C. melonis	CBS 159.84 HT	Brazil	*Cucumis melo*, peel of fruit	JQ948194	JQ948524	JQ948855	JQ949515	JQ949845
C. nymphaeae	CBS 515.78 HT	The Netherlands	*Nymphaea alba*, leaf spot	JQ948197	JQ948527	JQ948858	JQ949518	JQ949848
C. orchidophilum	CBS 632.80 HT	USA	*Ascocenda* sp.	JQ948152	JQ948482	JQ948813	JQ949473	JQ949803
C. paxtonii	IMI 165753, CPC 18868 HT	Saint Lucia	*Musa* sp.	JQ948285	JQ948615	JQ948946	JQ949606	JQ949936
C. phormii	CBS 118194 HT	Germany	*Phormium* sp.	JQ948446	JQ948777	JQ949107	JQ949767	JQ950097
C. pyricola	CBS 128531 HT	New Zealand	*Pyrus communis*, fruit rot	JQ948445	JQ948776	JQ949106	JQ949766	JQ950096

(Continued)

Table 3. (Continued).

Species	Isolate number[a]	Location	Host	GenBank accession number				
				ITS	*gapdh*	*chs-1*	*act*	*tub*
C. rhombiforme	CBS 129953 HT	Portugal	*Olea europaea*	JQ948457	JQ948788	JQ949118	JQ949778	JQ950108
C. salicis	CBS 607.94 HT	The Netherlands	*Salix* sp., leaf, spot	JQ948460	JQ948791	JQ949121	JQ949781	JQ950111
C. scovillei	CBS 126529 HT	Indonesia	*Capsicum* sp.	JQ948267	JQ948597	JQ948928	JQ949588	JQ949918
C. simmondsii	CBS 122122 HT	Australia	*Carica papaya*, fruit	JQ948276	JQ948606	JQ948937	JQ949597	JQ949927
C. sloanei	IMI 364297, CPC 18929 HT	Malaysia	*Theobroma cacao*, leaf	JQ948287	JQ948617	JQ948948	JQ949608	JQ949938
C. tamarilloi	CBS 129814 HT	Colombia	*Solanum betaceum*, fruit, anthracnose	JQ948184	JQ948514	JQ948845	JQ949505	JQ949835
C. walleri	CBS 125472 HT	Vietnam	*Coffea* sp., leaf tissue	JQ948275	JQ948605	JQ948936	JQ949596	JQ949926
C. paranaense	CBS 134729 HT	–	–	KC204992	KC205026	KC205043	KC205077	KC205060

[a] **CBS**: Culture collection of the Westerdijk Fungal Biodiversity Institute, Utrecht, The Netherlands. **CPC**: Culture collection of Pedro Crous, housed at the Westerdijk Institute. **CRCC**: Carlucci and Raimondo Culture Collection, housed at Dept. SAFE of University of Foggia. **IMI**: International Mycological Institute, CABI-Bioscience, Egham, Bakeham Lane, United Kingdom. **MFLU(CC)**: Mae Fah Luang University Culture Collection.

[b] Status: status of the strains. ET: ex-epitype. NT: ex-neotype. HT: ex-Holotype.

[c] Strain numbers and newly generated sequences are indicated in bold font.

Agar plugs (diam. 0.5 cm) were taken from 7-d-old fungal cultures grown on water agar at 23±2°C, and the plugs were placed under the cane bark. Wounds were then wrapped with wet sterile cotton wool and sealed with Parafilm. The experimental control canes were inoculated with sterile agar plugs. Each experiment included 18 replicates per treatment.

The canes were examined at 240 d after inoculation, and the lengths of any visible necrotic wood lesions, after removal of the bark, were measured and subjected to mycological analyses. Ten tissue pieces from each inoculated cane were placed on MEA supplemented with streptomycin sulphate at 300 mg L^{-1}, and incubated at 23±2°C in the dark. Resulting fungal colonies were identified to fulfil the Koch's postulates, and the proportions of re-isolation (%) were calculated.

Shapiro-Wilk (W) tests were used to determine whether the data obtained followed normal distributions. Homogeneity of the variances of the dataset was assessed using Levene tests. Statistical analyses were performed using Statistica version 6 (StatSoft). Factorial ANOVA analyses were performed to define the significance of any differences in mean lesion lengths caused by the isolates of each fungal species and the different fungal species, and to detect any interactions between these factors (i.e., isolate × fungal species). One-way ANOVA analyses were performed to evaluate statistically significant differences in the mean brown wood streaking lengths caused by each fungal species inoculated. Fischer's tests were used for the comparisons of the treatment means, at $P < 0.01$.

RESULTS

Fungal isolates

The data related to grapevine trunk disease incidence, recorded during surveys carried out through 10 y in vineyards of different cultivars, in the Apulia and Molise regions, are summarized in Table 1. Isolation frequencies of the fungal taxa isolated from symptomatic grapevine samples affected by GTDs and collected during the 10 y are shown in Figure 1.

The Botryosphaeriaceae (IF = 29.3%) and *Phaeoacremonium* spp. (IF = 19.3%) were the most frequently isolated fungi. *Phaeomoniella chlamydospora* (IF = 5.0%) and *Pleurostoma richardsiae* (IF = 6.4%) were responsible for vascular and subcortical streaking discolouration. The fungal taxa considered as less-known, including *Seimatosporium vitis-vinifera*,

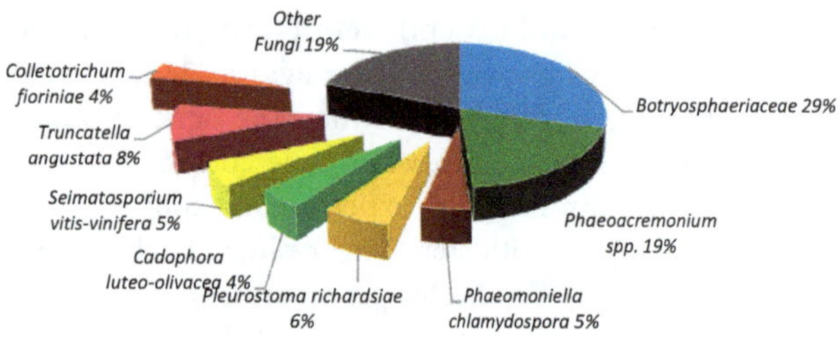

Figure 1. Isolation frequencies of fungal species obtained from symptomatic grapevines during a 10 year survey in the Apulia and Molise regions of Italy.

Truncatella angustata, *Cadophora luteo-olivacea* and *Colletotrichum fioriniae*, were isolated at IFs of 3.6% to 7.9%. The other group denoted here as 'other fungi' had IF of 19.0%, and included several fungal species, including *Alternaria* spp., *Aspergillus* spp., *Epicoccum nigrum*, *Fusarium* spp., *Penicillium* spp. and *Phoma*-like. These were not considered to be the causes of the disease symptoms observed, because these fungi are known common saprophytes.

Molecular identification of representative isolated fungi

Based on the keys, descriptions and sequence of Phillips *et al.* (2013), Mostert *et al.* (2006), Essakhi *et al.* (2008), Raimondo *et al.* (2014), Crous and Gams (2000) and Carlucci *et al.* (2015b), the 84 isolates selected as representative MSP-PCR clades were identified as follows (number of isolates): *Botryosphaeria dothidea* (four); *Diplodia corticola* (one); *D. mutila* (three); *D. seriata* (15); *Lasiodiplodia citricola* (eight); *L. theobromae* (five); *Neofusicoccum parvum* (five); *Phaeoacremonium iranium* (five); *P. italicum* (11); *P. minimum* (six); *P. scolyti* (three); *P. sicilianum* (two); *Phaeomoniella chlamydospora* (seven); and *Pleurostoma richardsiae* (nine) (data not shown).

The data obtained from the phylogenetic studies carried out on the 29 strains that were considered less well-known pathogens, and were representative of MSP-PCR clades related to the *Seimatosporium*, *Truncatella*, *Cadophora* and *Colletotrichum*, are summarized below.

The LSU, ITS, β-tub*ulin*, *tef-1α* and *rpb2* sequences were generated for seven *Seimatosporium* strains selected from the MSP-PCR profiles, and were aligned with 41 sequences retrieved from GenBank (Table 2). The dataset consisted of 48 taxa, which included the outgroup taxa *Synnemapestaloides juniperi* and *Discosia artocreas*. After alignment and exclusion of incomplete portions at either end, the dataset consisted of 3,344 characters (including alignment gaps), of which 2,276 were constant, while 279 were variable and parsimony uninformative. Maximum parsimony analysis of the remaining 789 parsimony-informative characters resulted in the 100 most-parsimonious trees (TL = 2,274; CI = 0.576; RI = 0.826; RC = 0.476; HI = 0.424). The maximum likelihood analysis produced a tree with similar topology (TreeBASE S25531; Figure 2). All of the *Seimatosporium* strains obtained clustered as a single clade with the type sequences of *Sei. vitis-viniferae* (CBS 123004) and *Sei. vitis* (MFLUCC 14-0051) (Figure 2). For the type strain of *Sei. vitis*, only the LSU and ITS sequences were available in GenBank, which were identical to those of *Sei. vitis-viniferae*. However, the isolates analysed here showed β-tubulin, *tef-1α* and *rpb2* sequences identical to those of *Sei. vitis-vinifera*, and therefore the morphological features (conidium dimensions and basal appendages) were used to discriminate between these two species, according to Liu *et al.* (2019).

The LSU, ITS, β-tub*ulin*, *tef-1α* and *rpb2* sequences were generated for 11 *Truncatella* isolates selected from the MSP-PCR profiles, which were aligned with 53 sequences retrieved from GenBank (Table 2). The dataset consisted of 64 taxa, which included two outgroup taxa, *Beltrania pseudorhombica* and *Phlogicylindrium eucalypti*. After alignment and exclusion of incomplete portions at either end, the dataset consisted of 3,983 characters (including alignment gaps), of which 1,124 were constant and 511 were variable and parsimony uninformative. Maximum parsimony analysis of the remaining 2,348 parsimony-informative characters resulted in 35 most-parsimonious trees (TL = 10,415; CI = 0.522; RI = 0.798; RC = 0.417; HI = 0.478). Maximum likelihood analysis produced a tree with similar topology (TreeBASE S25532; Figure 3). All of the *Truncatella* strains obtained in this study clustered with the *ex-neotype* sequences of *T. angustata* (*Stilbospora angustata* CBS 114025) (Figure 3).

The ITS, *tef-1α* and β-tubulin sequences generated for six *Cadophora* strains selected from the MSP-PCR profiles were aligned with 44 sequences retrieved from GenBank (Table 2). The dataset consisted of 60 taxa, which included the outgroup taxon, *Hyaloscypha finlandica*. After alignment and exclusion of incomplete portions at either end, the dataset consisted of 1,613 characters (including alignment gaps), of which 952 were constant, while 167 were variable and parsimony uninformative. Maximum parsimony analysis of the remaining 494 parsimony-informative characters resulted in 100 most-parsimonious trees (TL = 1,255; CI = 0.735; RI = 0.932; RC = 0.686; HI = 0.265). Maximum likelihood analysis produced a tree with similar topology (TreeBASE S25533; Figure 4). All of the *Cadophora* isolates obtained in this study clustered with the type sequences of *Cadophora luteo-olivacea* (CBS 141.41) (Figure 4).

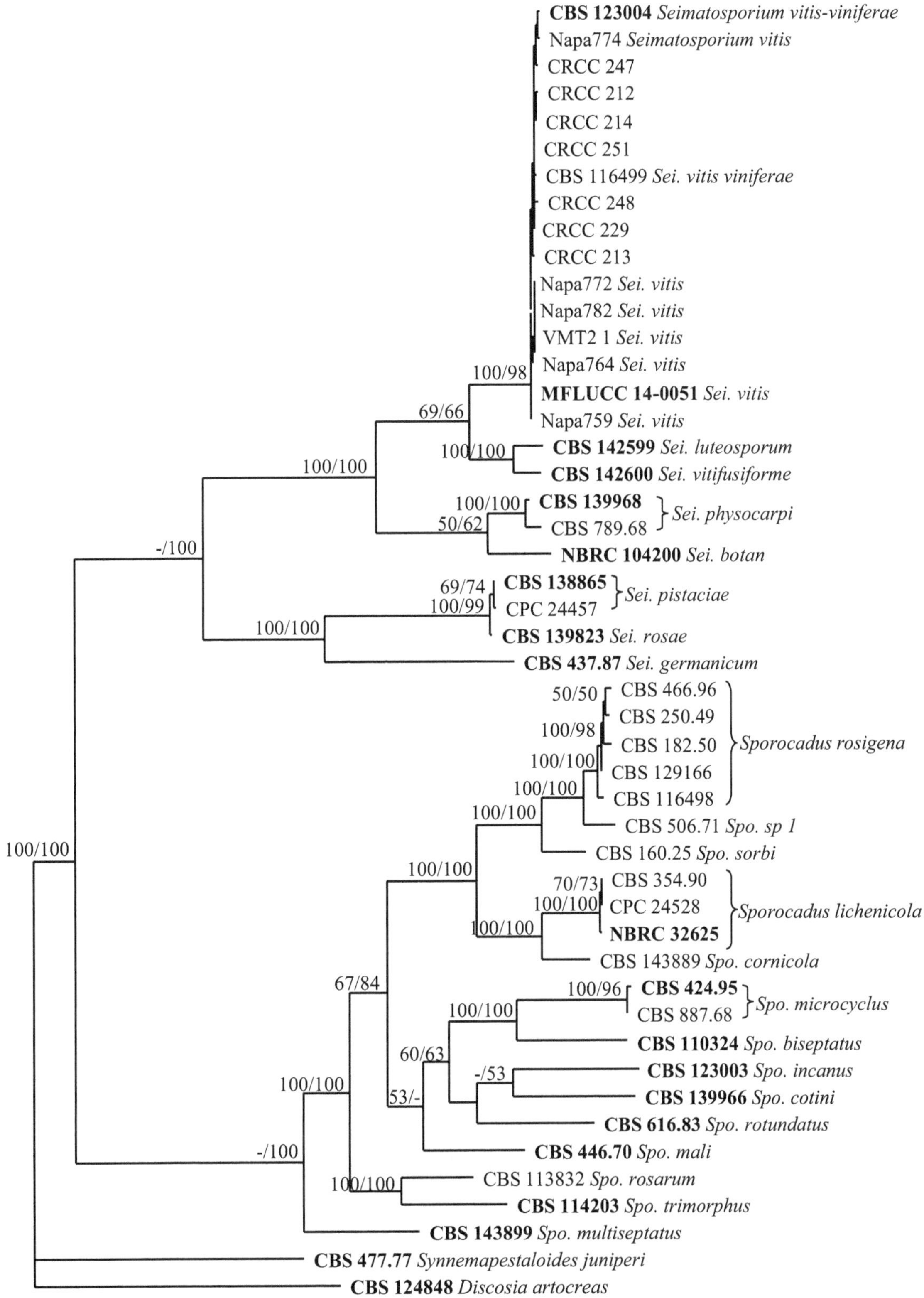

Figure 2. One of the most parsimonious trees obtained from the combined alignment of the LSU, ITS, *tub*, *tef-1α* and *rpb2* sequence datasets of *Seimatosporium* isolates, with bootstrap support values from maximum parsimony/maximum likelihood analyses. Isolates obtained in this study are indicated by blue rectangles. *Ex-type* sequences are given in bold. *Synnemapestaloides juniperi* and *Discosia artocreas* were used as outgroups.

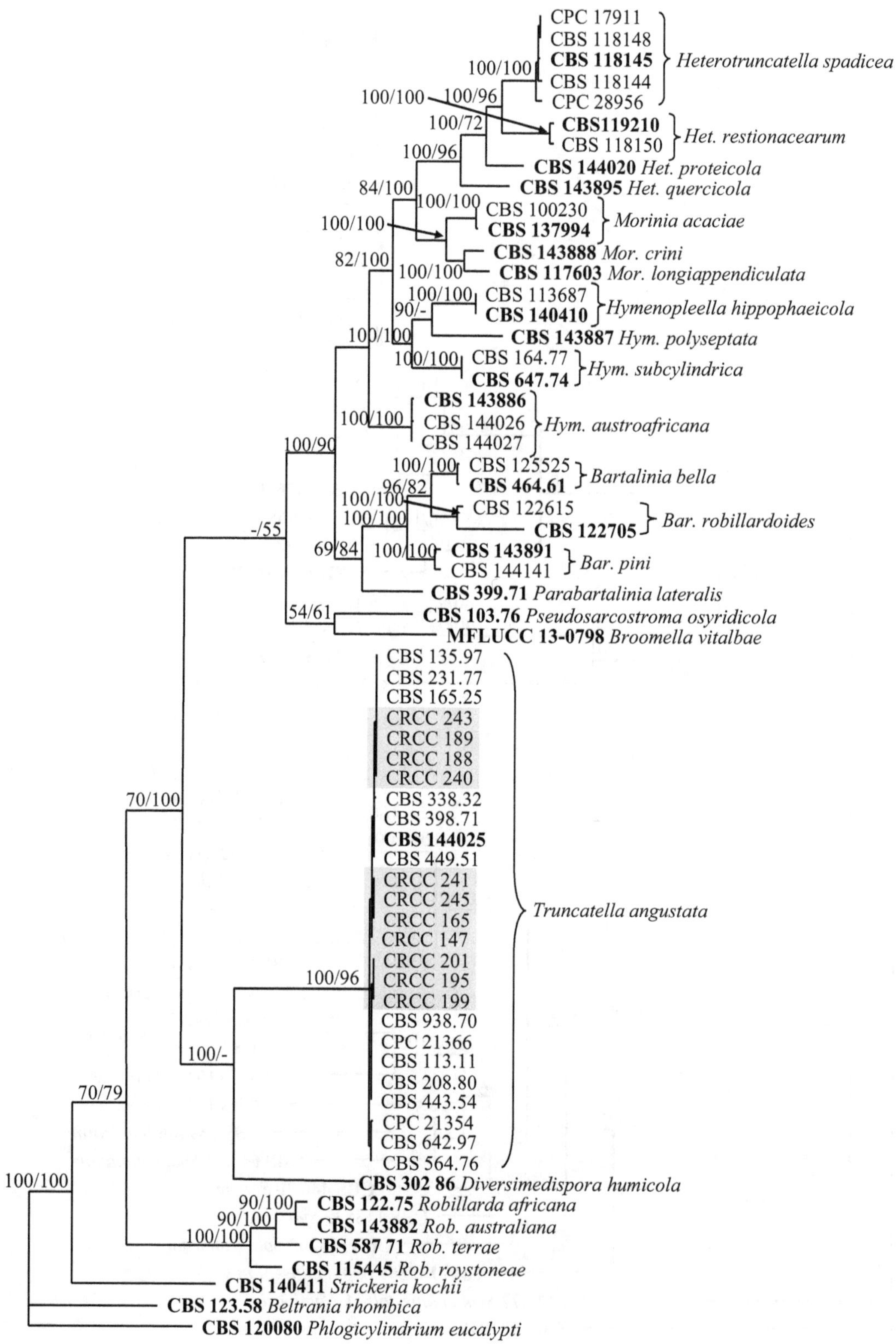

Figure 3. One of the most parsimonious trees obtained from combined alignment of the LSU, ITS, *tub*, *tef-1*α and *rpb2* sequence datasets of *Truncatella* isolates, with bootstrap support values from maximum parsimony/maximum likelihood analyses. Isolates obtained in this study are indicated by pink rectangles. *Ex-type* sequences are indicated in bold. *Beltrania pseudorhombica* and *Phlogicylindrium eucalypti* were used as outgroups.

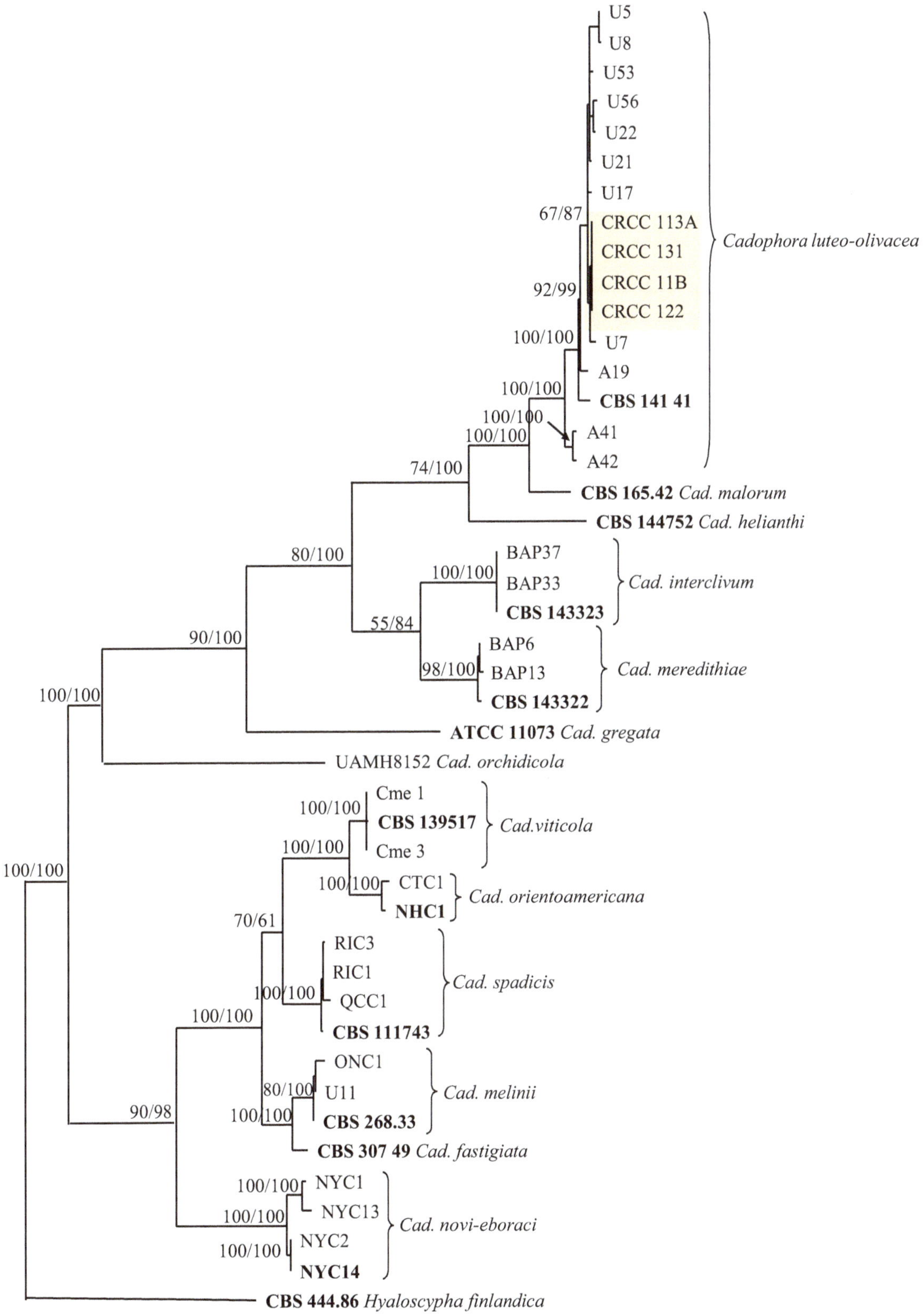

Figure 4. One of the most parsimonious trees obtained from combined ITS, *tef-1α* and *tub* sequence datasets of *Cadophora* isolates, with bootstrap support values from maximum parsimony/maximum likelihood analyses. Isolates obtained in this study are indicated by an orange rectangle. *Ex-type* sequences are indicated in bold. *Hyaloscypha finlandica* was used as outgroup.

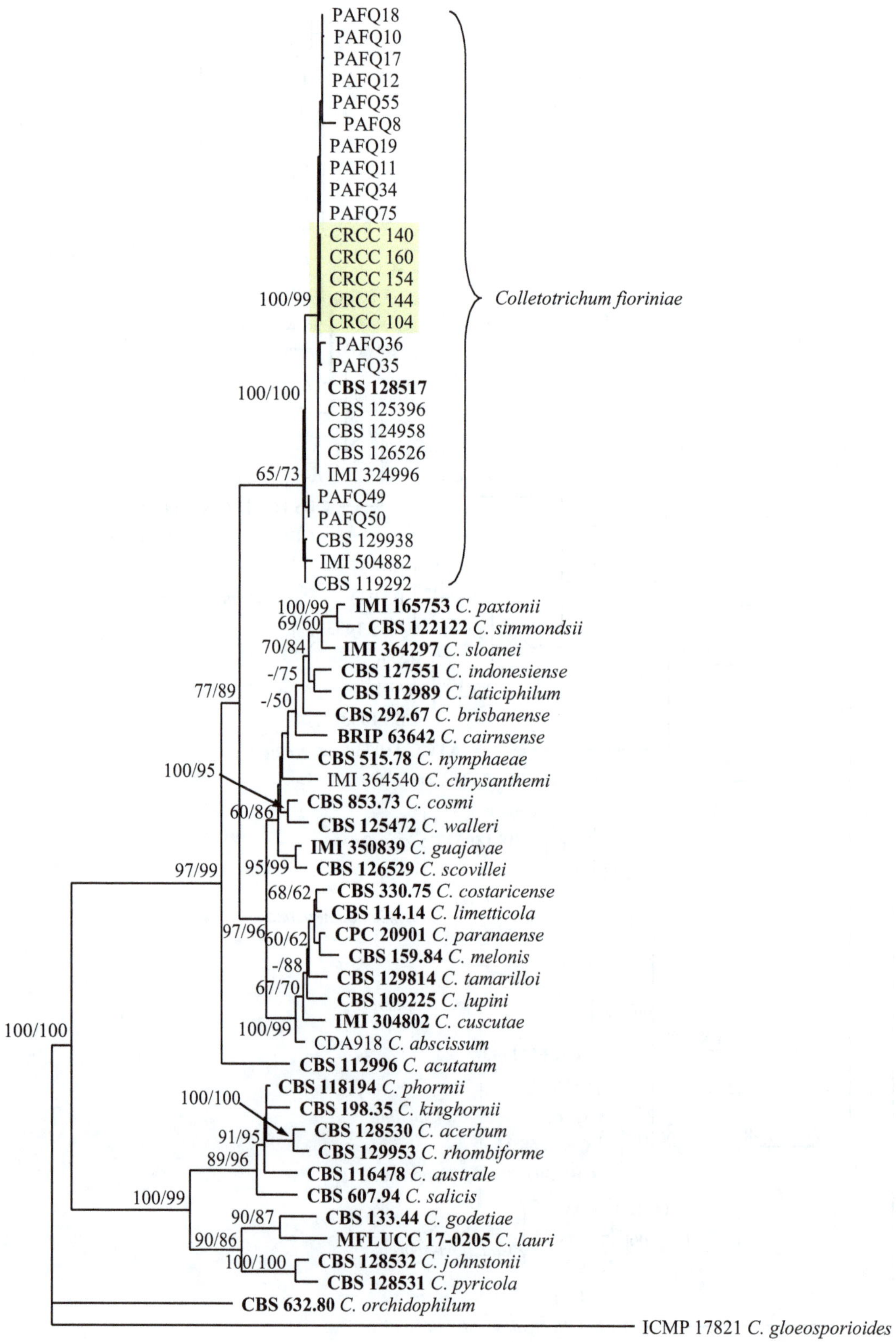

Figure 5. One of the most parsimonious trees obtained from the combined alignment of the ITS, *gapdh*, *chs-1*, *act* and *tub* sequence datasets of *Colletotrichum* isolates, with bootstrap support values from maximum parsimony/maximum likelihood analyses. Isolates obtained in this study are indicated by green rectangles. *Ex-type* sequences are indicated in bold. *Colletotrichum orchidophilum* and *C. gloeosporioides* were used as outgroups.

The ITS, β-tubulin, *act*, *gapdh* and *chs*-1 sequences generated for five *Colletotrichum* strains selected from the MSP-PCR profiles were aligned with 44 sequences retrieved from GenBank (Table 3). The dataset consisted of 61 taxa, which included the two outgroup taxa, *Colletotrichum gloeosporioides* and *C. orchidophilum*). After alignment and exclusion of incomplete portions at either end, the dataset consisted of 1,879 characters (including alignment gaps), of which 1,304 were constant, while 339 were variable and parsimony uninformative. Maximum parsimony analysis of the remaining 236 parsimony-informative characters resulted in 100 most-parsimonious trees (TL = 977; CI = 0.736; RI = 0.872; RC = 0.642; HI = 0.264). Maximum likelihood analysis produced a tree with similar topology (TreeBASE S25534; Figure 5). The *Colletotrichum* isolates obtained in this study clustered in the clade of *Colletotrichum fioriniae* with the *holotype* sequences of *C. fioriniae* (CBS 128517) (Figure 5).

Morphological characterisation of representative isolates

Colonies of the *Seimatosporium* isolates on MEA had entire edges, with brown to purplish grey mycelia, and reached mean diameter of 6.9 cm after 21 d at 23°C. The conidiomata were black and immersed. Conidia were fusoid, 3(-6)-septate, with measurements of 13.8-24.0 × 4.1-5.9 μm. They had truncated basal cells 2.3-3.8 μm long, similar to that of median cells. The median cells (2 -4) were each 3.3-5.1 μm long, and the conidium apical cells were 1.3-4.2 μm long. The majority of conidia each had a single unbranched appendage at both ends (apical appendage, 3.9-11.5 μm long; basal appendage, 3.6-10.3 μm long). On the basis of these culture and morphological features, all of the *Seimatosporium* strains had characteristics similar to those reported by Liu *et al.* (2019) for *Sei. vitis-viniferae.*

Colonies of the *Truncatella* isolates on MEA had entire edges, with white to pale grey mycelia, and reached mean diameter of 7.1 cm after 21 d at 23°C. Conidiomata were black, gregarious, semi-immersed and stromatic. Conidia were fusoid, occasionally slightly curved, mostly 3-septate, and not constricted at the septa (mean, 18.3 ± 1.69 × 6.8 ± 0.50 μm). The basal cells of the conidia had truncate bases, were hyaline to pale brown, 1.3-3.6 μm long, each with two pale to mid-brown doliiform median cells which were pale to mid-brown, each 5.3-7.7 μm long, and the apical cells were conic, hyaline, and 1.9-4.9 μm long. Each conidium had 2 to 4 apical appendages, which were centric, flexuous and branched, 0.6-22 μm long, and were without basal appendages. On the basis of these culture and morphological features, all of the *Truncatella* isolates studied had characteristics similar to those reported by Liu *et al.* (2019) for *Truncatella angustata*, which confirmed the data obtained in the molecular analysis.

Colonies of the *Cadophora* isolates on MEA had entire edges, and the mycelia were white to olivaceous green to grey. Mean colony diameter reached 4.5 cm after 21 d at 23°C. The conidiophores were mostly short, usually unbranched, up to 7-septate and measuring (-11.5) 26-63.90 × 1.78-1.94 (-2.5) μm. The conidiogenous cells were monophialidic, hyaline, terminal or lateral, mostly cylindrical, sometimes elongated ampulliform and attenuated at the base or navicular and tapering towards the apex. These cells measured 7.9-27.3 × 1.4-3.1 μm. The conidia were hyaline, mostly biguttulate, ovoid and aseptate, and measured 3.7-7.3 × 2.1-3.6 μm. On the basis of these culture and morphological features, all of the *Cadophora* strains studied had characteristics similar to those reported by Gramaje *et al.* (2011) and Travadon *et al.* (2015) for *Cadophora luteo-olivacea*, which confirmed the data obtained in the molecular analysis.

Colonies of the *Colletotrichum* isolates on MEA had entire edges, with aerial cottony pink to vinaceous mycelia. Mean colony diameter reached 4.5 cm after 21 d at 23°C. The conidiomata were sparse, with masses of orange conidia. Conidiophores were hyaline to pale brown, septate, branched, and up to 33 μm long. Conidiogenous cells were hyaline to pale brown, cylindrical to elongate ampulliform, monophialic and measured 3.8-11.9 × 2.2-3.9 μm. Conidia were elliptical, hyaline, with both ends acute, and measured 8.0-15.3 × 3.2-4.6 μm. On the basis of these culture and morphological features, all of the *Colletotrichum* isolates studied had characteristics similar to those reported by Damm *et al.* (2012) for *C. fioriniae*, which confirmed the data obtained in the molecular analysis.

Pathogenicity tests

According to Shapiro-Wilk tests, the data from the pathogenicity tests carried out on the grapevine cultivars 'Nero di Troia' and 'Bombino bianco' 240 d after, inoculations followed a normal distribution, with W values, respectively for the cultivars, of 0.96 ($P < 0.01$) and 0.97 ($P < 0.01$). The Levene tests determined for the two cultivars showed that the homogeneity of the variance was significant for 'Nero di Troia' (F = 7.04; $P < 0.01$) and 'Bombino bianco' (F = 4.93; $P < 0.01$). Factorial ANOVA demonstrated that significant differences in pathogenicity were detected among the fungal species inoculated on both 'Nero di Troia' (F = 44.5; $P < 0.01$) and 'Bombino bianco' (F = 83.40; $P < 0.01$). There were no significant

Table 4. Mean lesion lengths from the pathogenicity assays carried out for isolates of seven fungal species on two grapevine cultivars (one-way ANOVA).

Cultivar	Fungal species	Length of brown wood discolouration (cm)			Re-isolation (%)
		Mean	SD	Min–Max[a]	
'Nero di Troia'	Control	0.63 A	0.24	0.30–1.10	0.00
	Colletotrichum fiorinae	8.83 B	2.57	5.00–16.00	73.33
	Cadophora luteo-olivacaea	12.97 C	5.01	11.00–34.00	88.33
	Seimatosporium vitis-vinifera	16.98 D	3.13	13.00–24.70	80.00
	Truncatella angustata	18.01 D	7.31	3.40–29.50	91.67
	Pleurostoma richardsiae	18.53 DE	3.87	6.00–19.90	76.67
	Phaeoacremonium italicum	19.25 DE	3.41	13.60–27.00	93.33
	Lasiodiplodia citricola	21.77 E	4.59	12.00–27.00	86.67
'Bombino bianco'	Control	0.62 A	0.29	0.30–1.10	0.00
	Colletotrichum fiorinae	8.43 B	2.34	4.40–12.40	78.33
	Cadophora luteo-olivacaea	14.13 C	7.08	3.30–38.20	91.67
	Truncatella angustata	15.71 CD	3.20	9.50–19.50	80.00
	Phaeoacremonium italicum	18.41 DE	4.47	14.00–29.60	86.67
	Pleurostoma richardsiae	20.11 E	5.47	9.20–30.50	78.33
	Seimatosporium vitis-vinifera	23.63 F	3.23	13.00–23.00	91.67
	Lasiodiplodia citricola	29.20 G	2.00	24.40–32.00	95.00

[a] Minimum and maximum values detected (18 observations).
Data within each cultivar followed by different capital letters within the column are significantly different ($P < 0.01$; Fischer's tests).

differences in aggressiveness among the isolates of each fungal species used in the artificial inoculations of 'Nero di Troia' ($F = 0.12$; $P = 0.89$) or 'Bombino bianco' ($F = 0.99$; $P = 0.37$).

The mean lengths of vascular discolouration caused by the inoculated, fungal species used in the pathogenicity tests, and examined for one-way analysis of variance, are reported in the Table 4. All of the fungi produced brown wood discolourations on canes of both grapevine cultivars. The most aggressive species was *Lasiodiplodia citricola* towards 'Nero di Troia' and 'Bombino bianco', which produced the longest brown wood discolourations (respective mean lengths = 21.77 and 29.20 cm). *Phaeoacremonium italicum* and *Pleurostoma richardsiae* were pathogenic for both grapevine cultivars, which confirmed their aggressiveness reported by Carlucci *et al.* (2015a) and Raimondo *et al.* (2014). These fungi produced discolourations with mean lengths from 18.41 to 20.11 cm. Among the reference grapevine pathogens used, *Cadophora luteo-olivacea* was less pathogenic than *P. italicum* and *L. citricola*, as it produced mean discolouration lengths of 14.13 and 12.97 cm, respectively, on 'Nero di Troia' and 'Bombino bianco'. *Seimatosporium vitis-vinifera* and *Truncatella angustata* produced variable significant discolouration lengths on both grapevine cultivars, similar to those produced by *Cad. luteo-olivacea*. *Sei. vitis-vinifera* was less aggressive on 'Nero di Troia' (mean discolouration length = 16.98 cm) than on 'Bombino bianco' (mean length = 23.63 cm). *Truncatella angustata* produced different and variable discolouration lengths on 'Nero di Troia' and 'Bombino bianco' of 18.01 and 15.71 cm, respectively. *Colletotrichum fioriniae* was less aggressive, as it produced the least mean discolouration lengths on 'Nero di Troia' and 'Bombino bianco', which were, respectively, 8.83 and 8.43 cm. All of these fungi were re-isolated from the inoculated grapevines, which fulfilled Koch postulates (Table 4).

DISCUSSION

The data obtained in the present study show that vineyards in southern Italy were affected by different fungal species, some of which are known to be responsible for GTDs, such as Esca and Petri disease, and Botryosphaeria dieback. During the survey carried out on symptomatic vineyards over a 10 year period, different fungal species were among the samples collected, including *Botryosphaeria* spp., *Phaeoacremonium* spp. *Phaeomoniella chlamydospora* and *Pleurostoma richardsiae* as the most frequently isolated, and less frequently isolated taxa included *Seimatosporium*, *Truncatella*, *Cadophora* and *Colletotrichum*. The fungi of the first group are spread in most world grape-growing regions, and their pathogenicities and involvement in diseases associated with grapevines are known (Raimondo *et al.*, 2014; Carlucci *et al.*, 2015a; 2015b).

The molecular analysis used in the present study allowed identification of the second group of fungi as *Seimatosporium vitis-viniferae*, *Truncatella angustata*, *Cadophora luteo-olivacea* and *Colletotrichum fioriniae*. The morphological characterisation confirmed the molecular data, and helped in the identification of isolates of *Sei. vitis-viniferae*, for which molecular identification was not discriminatory.

To date, many studies have reported the isolation of "pestalotioides fungi", such as *Seimatosporium* species, from symptomatic grapevines or from dead stems in different countries, initially including Australia (Shivas, 1989), England and France (Sutton, 1980), England and Germany (Nag Raj, 1993) and Pakistan (Ahmad, 1969; Ahmad *et al.*, 1997). More recent reports also include Chile, Hungary, Iran, Italy, Spain and the USA (Castillo-Pando *et al.*, 2001; Sergeeva *et al.*, 2005; Dìaz *et al.*, 2012; Senanayake *et al.*, 2015; Mehrabi *et al.*, 2017; Vàczy, 2017; Lawrence *et al.*, 2018, Camele and Mang, 2019, Liu *et al.*, 2019). However, little information has been provided about their involvement in specific grapevine diseases.

Nine *Seimatosporium* species have been associated with grapevines, including *Sei. botan*, *Sei. hysterioides*, *Sei. lonicerae*, *Sei. luteosporum*, *Sei. macrospermum*, *Sei. parasiticum*, *Sei. vitifusiforme*, *Sei. vitis* and *Sei. vitis-viniferae* (Farr and Rossman, 2018; Liu *et al.*, 2019). However, only four of these have been assessed in standard pathogenicity trials on trunks and canes of vineyard grapevines, to confirm their pathogenicity roles and involvement in GTDs. *Seimatosporium botan* was isolated from symptomatic grapevines in Chile and was reported to be pathogenic on woody canes and trunks of potted grapevines (Dìaz *et al.*, 2012). *Seimatosporium vitis* strains were isolated from symptomatic grapevines in Hungary (Vàczy, 2017), North Carolina, USA (Lawrence *et al.*, 2018) and Italy (Camele and Mang, 2019), and were demonstrated to be pathogenic on green shoots and woody stems of potted grapevines. *Seimatosporium luteosporum* and *Sei. vitifusiforme* were reported as pathogens on woody stems of grapevines in North Carolina, USA (Lawrence *et al.* 2018).

In the present study, the pathogenicity of *Sei. vitis-viniferae* was tested for the first time, which increased the number of *Seimatosporium* species that have been confirmed to be associated with GTDs to five. Based on molecular and morphological studies on the pestalotioides fungi reported by Liu *et al.* (2019), the identification of *Sei. vitis* in some studies appears to have been incorrect. The multilocus analyses performed with LSU, ITS, *tef-1α*, β-tubulin and *rpb2* sequences in the present study demonstrated that the strains of *Sei. vitis* reported by Lawrence *et al.* (2018) and Camele and Mang (2019) all clustered in the clade of *Sei. vitis-viniferae*. The morphological description provided by Lawrence *et al.* (2018) for *Sei. vitis* strains, including conidium dimensions and the presence of appendages at both ends of conidia does not agree with the description of *Sei. vitis* by Senanayake *et al.* (2015), although it does agree with that of Liu *et al.* (2019) for *Sei. vitis-viniferae*. Although the *tef-1α* and β-tubulin sequences of *Sei. vitis* reported by Camele and Mang (2019) were identical to those of *ex-type Sei. vitis-viniferae* described by Liu *et al.* (2019), no detailed morphological information was reported. Therefore, to the best of our knowledge, the present study provides the first report of *Sei. vitis-viniferae* associated with GTD symptoms in Italy.

The genus *Truncatella* is closely related to *Seimatosporium*, which belongs to the pestalotioides fungi, and it has wide distribution and occurs in many hosts, including grapevines (Sutton, 1980). Few reports are available about the association of *Truncatella* with grapevine, and its involvement in GTDs. Nag Raj (1993) reported *T. angustata* and *T. pitospora* (now *Pestalotia pitospora*) on grapevine, but did not include any information on their pathogenicity. Some years later, Casieri *et al.* (2009), in Switzerland, and Gonzalez and Tello (2011), in Spain, reported *T. angustata* as endophytes that were collected from different grapevine cultivars. Urbez-Torrez *et al.* (2009) also isolated *T. angustata* from cankers on grapevines in Texas, and performed pathogenicity tests to demonstrate that this fungus can be a weak and/or opportunistic pathogen on lignified grapevine canes. The pathogenicity of *T. angustata* and its involvement in GTD symptoms were also confirmed by Arzanlou *et al.* (2013) in Iran. Maharachchikumbura *et al.* (2016) and Pintos *et al.* (2018) reported *T. angustata* associated with GTD symptoms on grapevines in France, but no pathogenicity trials were performed. Based on a recent taxonomic revision of the genus *Truncatella* by Liu *et al.* (2019), there is now just one accepted species, as *T. angustata*, while other *Truncatella* species were transferred to different genera, including *Bartalinia*, *Heterotruncatella* and *Morinia*, due to the polyphyletic nature of this genus or to synonymy with *T. angustata*. The pathogenicity tests performed in the present study confirmed the pathogenic behaviour of *T. angustata* and its involvement in GTDs (Arzanlou *et al.* 2013). This is the first report of *T. angustata* associated with GTD symptoms on grapevines in Italy.

To date, seven *Cadophora* species have been reported from grapevines, including *Cad. fastigiata*, *Cad. luteo-

olivacea, *Cad. melinii*, *Cad. novi-eboraci*, *Cad. orientoamericana*, *Cad. spadicis* and *Cad. viticola* (Overton *et al.*, 2005; Halleen *et al.* 2007; Crous *et al.*, 2015; Travadon *et al.*, 2015). Halleen *et al.* (2007) reported *Cad. luteo-olivacea* from grapevines showing decline symptoms, and from apparently healthy plants in commercial nurseries in South Africa. Pathogenicity tests demonstrated that *Cad. luteo-olivacea* caused significant lesions on the trunks and pruned wood of 15-year-old grapevines.

Casieri *et al.* (2009), in Switzerland, and Fischer *et al.* (2016), in Germany, reported *Cad. fastigiata* and *Cad. luteo-olivacea* as fungal species that can cause grapevine diseases. Gramaje *et al.* (2011) reported *Cad. luteo-olivacea* and *Cad. melinii* from nursery grapevines, although pathogenicity tests demonstrated that only *Cad. luteo-olivacea* caused grapevine disease on 1-year-old grapevine cutting rootstock. Travadon *et al.* (2015) confirmed the involvement of *Cad. luteo-olivacea* in GTDs, and associated four other *Cadophora* species with wood decay of grapevines in North America (*Cad. melinii*, and three new species, *Cad. orientoamericana*, *Cad. novi-eboraci* and *Cad. spadicis*). In 2015, Crous *et al.* (2015) described a new species of *Cad. viticola* (previously identified as *Cad. melinii* by Gramaje *et al.*, 2011), which was isolated from grapevine shoots that showed black streaks. *Cad. luteo-olivacea* is the most frequently isolated *Cadophora* species associated with GTD symptoms in different countries, including the USA, France, Germany, New Zealand, South Africa, Spain, Switzerland and Uruguay (Casieri *et al.*, 2009; Manning and Munday, 2009, Gramaje *et al.*, 2011; Travadon *et al.*, 2015; Fischer *et al.*, 2016; Pintos *et al.*, 2018). Isolation of *Cad. luteo-olivacea* in the present study confirms the wide distribution of this species, while the pathogenicity tests performed here confirm the pathogenic behaviour of *Cad. luteo olivacea* and its involvement in GTDs. This is the first report of *Cadophora luteo-olivacea* associated with GTD symptoms in Italy.

Colletotrichum fioriniae was also less frequently isolated that other fungi, and this species is in the *C. acutatum* species complex. The role of *Colletrotrichum* on grapevines is not clear; there have been few reports of species of the *C. acutatum* complex that have described their behaviour on grapevines. *Colletotrichum fioriniae* (Kepner and Swett 2018) and *C. godetiae* (Zapparata *et al.*, 2017) have been associated with grape berry rot, respectively, in the USA and Italy. *Colletotrichum godetiae* has also been reported as a leaf anthracnose agent in the United Kingdom (Baroncelli *et al.*, 2014), and as a saprophyte in China, Italy, Russia and Thailand (Jayawardena *et al.*, 2018). In 2016, Liu *et al.* (2016) reported the first association of species in the *C. acutatum* complex with the wood of grapevines when they described *C. nymphaeae* from twig anthracnose in China.

To date, there has only been one report of *Colletotrichum* spp. associated with GTDs, from a grapevine nursery in France (Pintos *et al.*, 2018), although no specific identification was carried out. In the present study, as *C. fioriniae* produced wood discolouration on both of the grapevine cultivars included, and although this was less severe (shorter discolouration) than for the other fungi inoculated. This fungus can now be considered as a weak pathogen on grapevine wood. This is, therefore, the first report of *C. fioriniae* associated with GTD symptoms.

The study reported in the present paper has demonstrated the presence of *Cadophora luteo-olivacea*, and *Truncatella angustata*, as well as their virulence, also on grapevine in Italy. *Seimatosporium vitis-vinifera*, isolated from grapevine for first time in Italy, when artificially inoculated, was the most aggressive fungus among the less-common fungi assayed here, indicating its involvement in GTDs. *Colletotrichum fioriniae*, although less aggressive among the fungi assayed, was also shown to be another fungus involved in GTDs. These results add to knowledge on the expanding group of fungi involved in the GTD complex.

LITERATURE CITED

Agustí-Brisach C., Armengol J., 2013. Black-foot disease of grapevine: an update on taxonomy, epidemiology and management strategies. *Phytopathologia Mediterranea* 52: 245–261.

Ahmad S., 1969. *Fungi of West Pakistan*. Biological Society of Pakistan Monograph, 5, 1–110.

Ahmad S., Iqbal S.H., Khalid A.N., 1997. Fungi of Pakistan. *Sultan Ahmad Mycological Society of Pakistan*, 248.

Alves A., Crous P.W., Correia A., Phillips A.J.L., 2008. Morphological and molecular data reveal cryptic species in *Lasiodiplodia theobromae*. *Fungal Diversity* 28: 1–13.

Arzanlou M., Narmani A., Moshari S., Khodaei S., Babai-Ahari A., 2013. *Truncatella angustata* associated with grapevine trunk disease in northern Iran. *Archives of Phytopathology and Plant Protection* 46: 1168–1181.

Baroncelli R., Sreenivasaprasad S., Lane C.R., Thon M.R., Sukno S.A., 2014. First report of *Colletotrichum acutatum sensu lato* (*Colletotrichum godetiae*) causing anthracnose on grapevine (*Vitis vinifera*) in the United Kingdom. *New Disease Reports* 29: 26.

Berlanas C., Ojeda S., Lòpez-Manzanares B., Andrés-Sodupe M., Bujanda R., … Gramaje D., 2020. Occurrence and diversity of black-foot disease fungi in symptomless grapevine nursery stock in Spain. *Plant Disease,* 104: 94–104.

Bertsch C., Ramirez-Suero M., Magnin-Robert M., Larignon P., Chong J., … Fontaine F., 2013. Grapevine trunk diseases: complex and still poorly understood. *Plant Pathology* 62: 243–265.

Brown A.A., Lawrence D.P., Baumgartner K., 2020. Role of basidiomycete fungi in the grapevine trunk disease esca. *Plant Pathology,* 69: 205–220.

Burruano S., Mondello V., Conigliaro G., Alfonzo A., Spagnolo A., Mugnai L., 2008. Grapevine decline in Italy caused by *Lasiodiplodia theobromae. Phytopathologia Mediterranea* 47: 132–136.

Camele I., Mang S.M., 2019. First report of *Seimatosporium vitis* associated with grapevine trunk diseases on *Vitis vinifera* in Italy. *Plant Disease* 104: 771.

Carbone I., Kohn L.M., 1999. A method for designing primer sets for speciation studies in filamentous ascomycetes. *Mycologia* 91: 553–556.

Carlucci A., Cibelli F., Lops F., Phillips A.J.L., Ciccarone C., Raimondo M.L., 2015a. *Pleurostomophora richardsiae* associated with trunk diseases of grapevines in southern Italy. *Phytopathologia Mediterranea* 54: 109–123.

Carlucci A., Cibelli F., Lops F., Raimondo M.L., 2015b. Characterization of *Botryosphaeriaceae* species as causal agents of trunk diseases on grapevines. *Plant Disease* 99: 1678–1688.

Carlucci A., Lops F., Mostert L., Halleen F., Raimondo M.L., 2017. Occurrence fungi causing black foot on young grapevines and nursery rootstock plants in Italy. *Phytopathologia Mediterranea* 56: 10–39.

Carlucci A., Lops F., Raimondo M.L., Gentile V., Mucci M., Frisullo S., 2009. The *Botryosphaeria* species from vineyards of Apulia. *Phytopathologia Mediterranea* 48: 180.

Carlucci A., Raimondo M.L., Cibelli F., Phillips A.J.L, Lops F., 2013. *Pleurostomophora richardsiae, Neofusicoccum parvum* and *Phaeoacremonium aleophilum* associated with a decline of olives in southern Italy. *Phytopathologia Mediterranea* 52: 517–527.

Carlucci A., Raimondo M.L., Santos J., Phillips A.J.L., 2012. *Plectosphaerella* species associated with root and collar rots of horticultural crops in southern Italy. *Persoonia* 28: 34–48.

Casieri L., Hofstetter V., Viret O., Gindro K., 2009. Fungal communities living in the wood of different cultivars of young *Vitis vinifera* plants. *Phytopathologia Mediterranea* 48: 73–83.

Castillo-Pando M., Somers A., Green C.D., Priest M., Sriskanthadas M., 2001. Fungi associated with dieback of Semillon grapevines in the Hunter Valley of New South Wales. *Australasian Plant Pathology* 30: 59–63.

Cristinzio G., 1978. Gravi attacchi di *Botryosphaeria obtusa* su vite in provincia di Isernia. *Informatore Fitopatologico* 6: 21–23.

Crous P.W., Gams W., 2000. *Phaeomoniella chlamydospora* gen. et comb. nov., a causal organism of Petri grapevine decline and esca. *Phytopathologia Mediterranea* 39: 112–118.

Crous P.W., Wingfield M.J., Guarro J., Hernandez-Restrepo M., Sutton D.A., … Ercole E., 2015. Fungal Planet description sheets: 320-370. *Persoonia* 34: 167–266.

Damm U., Fourie P.H., Crous P.W., 2010. *Coniochaeta (Lecythophora), Collophora* gen. nov. and *Phaeomoniella* species associated with wood necroses of Prunus trees. *Persoonia* 24: 60–80.

Damm U., Cannon P.F., Woudenberg J.H.C., Crous P.W., 2012. The *Colletotrichum acutatum* species complex. *Studies in Mycology* 73: 37–113.

Diaz G.A., Elfar K., Latorre B.A., 2012. First report of *Seimatosporium botan* associated with trunk disease of grapevine (*Vitis vinifera*) in Chile. *Plant Disease* 96: 1696.

Essakhi S., Mugnai L., Crous P.W., Groenewald J.Z., Surico G., 2008. Molecular and phenotypic characterization of novel *Phaeoacremonium* species associated with Petri disease and esca of grapevine. *Persoonia* 21: 119–134.

Farr D.F., Rossman A.Y., 2018. Fungal Databases, U.S. National Fungus Collections, ARS, USDA. Available at: https://nt.ars-grin.gov/fungaldatabases/.

Fischer M., Schneider P., Kraus C., Molnar M., Dubois C., … Haag N., 2016. Grapevine trunk disease in German viticulture: occurrence of lesser known fungi and first report of *Phaeoacremonium viticola* and *P. fraxinopennsylvanicum. Vitis* 55: 145–156.

Fisher P.J., Petrini O., Petrini L.E., Descals E., 1992. A preliminary study of fungi inhabiting xylem and whole stems of *Olea europaea. Sydowia* 44: 117–121.

Fourie P.H., Halleen F., 2004. Occurrence of grapevine trunk disease pathogens in rootstocks mother plants in South Africa. *Australasian Plant Pathology* 33: 313–315.

Fu M., Crous P.W., Bai Q., Zhang P.F., Xiang J., … Wang G.P., 2019. *Colletotrichum* species associated with anthracnose of *Pyrus* spp. in China. *Persoonia* 42: 1–35.

Glass N.L., Donaldson G.C., 1995. Development of primer sets designed for use with the PCR to amplify conserved genes from filamentous infection due to *Phaeoacremonium* spp. *Journal of Clinical Microbiology* 41: 1332–1336.

Gonzalez V., Tello M.L., 2011. The endophytic mycota associated with *Vitis vinifera* in central Spain. *Fungal Diversity* 47: 29–42.

Gramaje D., Urbez-Torres J.R., Sosnowski M.R., 2018. Managing grapevine trunk diseases with respect to etiology and epidemiology: current strategies and future prospects. *Plant Disease* 102: 12–39.

Gramaje D., Mostert L., Armengol J., 2011. Characterization of *Cadophora luteo-olivacea* and *C. melinii* isolates obtained from grapevines and environmental samples from grapevine nurseries in Spain. *Phytopathologia Mediterranea* 50: 112–126.

Gubler W.D., Rolshausen P.E., Trouillase F.P., Urbez J.R., Voegel T., 2005. Grapevine trunk diseases in California: Research update. *Practical Winery and Vineyard*: 1–9.

Guerber J.C., Liu B., Correll J.C., Johnston P.R., 2003. Characterization of diversity in *Colletotrichum acutatum sensu lato* by sequence analysis of two gene introns, mtDNA and intron RFLPs, and mating compatibility. *Mycologia* 95: 872–895.

Guerin-Dubrana L., Fontaine F., Mugnai L., 2019. Grapevine trunk disease in European and Mediterranean vineyards: occurrence, distribution and associated disease-affecting cultural factors. *Phytopathologia Mediterranea* 58: 49–71.

Halleen F., Schroers H.J., Groenewald J.Z., Crous P.W., 2004. Novel species of *Cylindrocarpon* (*Neonectria*) and *Campylocarpon* gen. nov. associated with black-foot disease of grapevines (*Vitis* spp.). *Studies in Mycology* 50: 431–455.

Halleen F., Mostert L., Crous P.W., 2007. Pathogenicity testing of lesser-known vascular fungi of grapevines. *Australasian Plant Pathology* 36: 277–285.

Jayawardena R.S., Hyde K.D., Chethana K.W.T., Daranagama D.A., Dissanayake A.J., ... Yan J.Y., 2018. Mycosphere notes 102-168: Saprotrophic fungi on *Vitis* in China, Italy, Russia and Thailand. *Mycosphere* 9: 1–114.

Katoh K., Standley D.M., 2013. Multiple sequence alignment software version 7: improvements in performance and usability. *Molecular Biology and Evolution* 30: 772–780.

Kepner C., Swett C.L., 2018. Previously unrecognized diversity within fungal fruit rot pathosystems on *Vitis vinifera* and hybrid white wine grapes in Mid-Atlantic vineyards. *Australasian Plant Pathology* 47: 181–188.

Larignon P., Dubos B., 1997. Fungi associated with esca disease in grapevine. *European Journal of Plant Pathology* 103: 147–157.

Lawrence D.P., Travadon R., Baumgartner K., 2018. Novel *Seimatosporium* species from grapevine in northern California and their interactions with fungal pathogens involved in the trunk-disease complex. *Plant Disease* 102: 1081–1092.

Liu F., Bonthond G., Groenewald J.Z., Cai L., Crous P.W., 2019. Sporocadaceae, a family of coelomycetous fungi with appendage bearing conidia. *Studies in Mycology* 92: 287–415.

Liu Y.J., Whelen S., Hall B.D., 1999. Phylogenetic relationships among ascomycetes: evidence from an RNA polymerase II subunit. *Molecular Biology and Evolution* 16: 1799–1808.

Liu M., Zhang W., Zhou Y., Liu Y., Yan J.Y., ... Hyde K.D., 2016. First report of twig anthracnose on grapevine caused by *Colletotrichum nymphaeae* in China. *Plant Disease* 100: 2530.

Maharachchikumbura S.S.N., Larignon P., Hyde K.D., Al-Sadi A.M., Liu Z.Y., 2016. Characterization of *Neopestalotiopsis*, *Pestalotiopsis* and *Truncatella* species associated with grapevine trunk diseases in France. *Phytopathologia Mediterranea* 55: 380–390.

Manning M.A., Munday D.C., 2009. Fungi associated with grapevine trunk disease in established vineyards in New Zealand. *Phytopathologia Mediterranea* 48: 160–161.

Mehrabi M., Hemmati R., Abdollahzadeh J., 2017. Description of the sexual morph of *Seimatosporium vitis*. *Cryptogamie Mycologie* 38: 3–11.

Meyer W., Mitchell T.G., Freedman E.Z., Vilgalys R., 1993. Hybridization probes for conventional DNA fingerprinting used as single primers in the polymerase chain reaction to distinguish strains of *Cryptococcus neoformans*. *Journal of Clinical Microbiology* 31: 2274–2280.

Miller M.A., Pfeiffer W., Schwartz T., 2010. *Creating the CIPRES Science Gateway for inference of large phylogenetic trees*. In: Gateway Computing Environments Workshop (GCE), New Orleans, LA, November 14, 1–8.

Mostert L., Groenewald J.Z., Summerbell R.C., Gams W., Crous P.W., 2006. Taxonomy and pathology of *Togninia* (*Diaporthales*) and its *Phaeoacremonium* anamorphs. *Studies in Mycology* 54: 1–115.

Mugnai L., Graniti A., Surico G., 1999. Esca (black measles) and brown wood streaking: two old and elusive diseases of grapevines. *Plant Disease* 83: 404–416.

Nag Raj T.R., 1993. *Coelomycetous anamorphs with appendage-bearing conidia*. Mycologue Publications, Waterloo, Ontario, 1101 pp.

Navarrete F., Abreo E., Martìnez S., Bettucci L., Lupo S., 2011. Pathogenicity and molecular detection of Uruguayan isolates of *Greeneria uvicola* and *Cadophora luteo-olivacea* associated with grapevine trunk diseases. *Phytopathologia Mediterranea* 50: S166–S175.

O'Donnell K., Cigelnik E., 1997. Two divergent intragenomic rDNA ITS2 types within a monophyletic lineage of the fungus *Fusarium* are nonorthologous. *Molecular Phylogenetic and Evolution* 7: 103–116.

O'Donnell K., Gray L.E., 1993. Phylogenetic relationships of the soybean sudden death syndrome pathogen *Fusalium solani* f. sp. *phaseoli* inferred from rDNA sequence data and PCR primers for its identification. *Molecular Plant Microbe Interactions* 8: 709–716.

Overton E.B., Steward E.L., Wenner N.G., 2005. Molecular phylogenetics of grapevine decline fungi from Pennsylvania and New York. *Phytopathologia Mediterranea* 44: 90–91.

Page R.D., 1996. TreeView: An application to display phylogenetic trees on personal computers. *Computer Applications in the Biosciences* 12: 357–358.

Phillips A.J.L., Alves A., Abdollahzadeh J., Slippers B., Wingfield M.J., ... Crous P.W., 2013. The Botryosphaeriaceae: Genera and species known from culture. *Studies in Mycology* 76: 51–167.

Pintos C., Redondo V., Costas D., Aguin O., Mansilla P., 2018. Fungi associated with grapevine trunk diseases in nursery-produced *Vitis vinifera* plants. *Phytopathologia Mediterranea* 57: 407–424.

Raimondo M.L., Lops F., Carlucci A., 2014. *Phaeoacremonium italicum* sp. nov., a new species associated with Esca of grapevine in southern Italy. *Mycologia* 106: 1119–1126.

Rego C., Oliveira H., Carvalho A., Phillips A.J.L., 2000. Involvement of *Phaeoacremonium* spp. and *Cylindrocarpon destructans* with grapevine decline in Portugal. *Phytopathologia Mediterranea* 39: 76–79.

Santos J.M., Phillips A.J.L., 2009. Resolving the complex of *Diaporthe* (*Phomopsis*) species occurring on *Foeniculum vulgare* in Portugal. *Fungal Diversity* 34: 111–125.

Senanayake I.C., Maharachchikumbura S.S.N., Hyde K.D., Bhat J.D., Jones E.B.G., ... Camporesi E., 2015. Towards unraveling relationships in Xylariomycetidae (Sordariomycetes). *Fungal Diversity* 73: 73–144.

Sergeeva V., Priest M., Nair N.G., 2005. Preliminary studies on *Pestalotiopsis* species from Southern Hemisphere conifers in Australasia and South Africa. *Australasian Plant Pathology* 34: 255–258

Shivas R.G., 1989. Fungal and bacterial diseases of plants in Western Australia. *Journal of the Royal Society of Western Australia* 72: 1–62.

Stamatakis A., 2006. RAxML-VI-HPC: maximum likelihood-based phylogenetic analyses with thousands of taxa and mixed models. *Bioinformatics* 22: 2688–2690.

Stamatakis A., Hoover P., Rougemont J., 2008. A rapid bootstrap algorithm for the RAxML web servers. *Systematic Biology* 57: 758–771.

Sung G.H., Sung J.M., Hywel-Jones N.L., Spatafora J.W., 2007. A multigene phylogeny of Clavicipitaceae (Ascomycota, Fungi): identification of localized incongruence using a combinational bootstrap approach. *Molecular Phylogenetics and Evolution* 44: 1204–1223.

Sutton B.C., 1980. *The Coelomycetes. Fungi imperfecti with pycnidia, acervuli and stromata.* Commonwealth Mycological Institute, Kew, Surrey, England, 696 pp.

Swofford D.L., 2003. *PAUP**. Phylogenetic Analysis Using Parsimony, Version 4. Sinauer Associates, Sunderland, UK.

Travadon R., Lawrence D.P., Rooney-Latham S., Gubler W.D., Wilcox W.F., ... Baumgartner K., 2015. *Cadophora* species associated with wood-decay of grapevine in North America. *Fungal Biology* 119: 53–66.

Ùrbez-Torres J.R., 2011. The status of Botryosphaeriaceae species infecting grapevines. *Phytopathologia Mediterranea* 50: S5–S45.

Ùrbez-Torres J.R., Peduto F., Smith R.J., Gubler W.D., 2013. Phomopsis dieback: a grapevine trunk disease caused by *Phomopsis viticola* in California. *Plant Disease* 97: 1571–1579.

Úrbez-Torres J.R., Adams Kamas J., Gubler W.D., 2009. Identification, incidence, and pathogenicity of fungal species associated with grapevine dieback in Texas. *American Journal of Enology and Viticulture* 60: 497–507.

Vaczy K.Z., 2017. First report of *Seimatosporium vitis* associated with grapevine trunk disease symptoms in Hungary. *Plant Disease* 101: 253–254.

White T.J., Bruns T., Lee S., Taylor J., 1990. Amplification and direct sequencing of fungal ribosomal RNA genes for phylogenetics. In: *PCR Protocols: A Guide to Methods and Applications.* (Innis MA, Gelfand DH, Sninsky JJ, White TJ, eds). Academic Press Inc., New York, NY, USA, 315–322.

Zapparata A., Da Lio D., Sarrocco S., Vannacci G., Baroncelli R., 2017. First report of *Colletotrichum godetiae* causing grape (*Vitis vinifera*) berry rot in Italy. *Plant Disease* 101: 1051–1052.

10

Factors affecting *Venturia oleaginea* infections on olive and effects of the disease on floral biology

Tahreer ISSA[1], Leen ALMADI[2], Samer JARRAR[3], Michele TUCCI[1], Roberto BUONAURIO[2], Franco FAMIANI[2,*]

[1] *Agronomic Mediterranean Institute of Bari (IAMB), Valenzano (BA), Italy*
[2] *Dipartimento di Scienze Agrarie Alimentari e Ambientali, Università degli Studi di Perugia, Perugia, Italy*
[3] *Canaan Center for Organic Research and Extension (CORE), Canaan, Palestine*
*Corresponding author: franco.famiani@unipg.it

Summary. The disease olive peacock eye was evaluated in organic olive orchards in Palestine (Asira, Burquin, Qabatia and Sir). In each orchard, six trees were randomly chosen (three pruned and three unpruned). Severity was estimated at 20 d intervals by determining the percentage of infected leaves in the upper, mid and lower parts of the canopy of each tree. Proportions were estimated of symptomatic leaves, and those with asymptomatic infections using the NaOH test. In spring, inflorescence length was determined just before flowering (white stage). The proportions of fruit set were also calculated on small, labeled branches. Among the different sites/orchards tested, there were significant differences in the infection level, which were related to climatic conditions. Pruning reduced the susceptibility of the trees to the disease. The infections were much greater in the lower parts of the canopy. Peacock eye greatly reduced inflorescence length and fruit set. *Venturia oleaginea* severity was negatively correlated with inflorescence development and fruit set, due to defoliation caused by the pathogen. Based on effects on fruit set, the disease could severely reduce olive yields. This is the first report showing quantitative relationship between olive defoliation caused by peacock eye and reproductive activity.

Key words. *Olea europaea* L., organic olive growing, pruning, peacock eye, *Spilocaea oleaginea*, Palestine.

Competing Interests: The Author(s) declare(s) no conflict of interest.

Editor: Juan A. Navas-Cortes, Spanish National Research Council (CSIC), Cordoba, Spain.

INTRODUCTION

Peacock eye (olive leaf spot), caused by *Venturia oleaginea* (Castagne) Rossman & Crous (= *Spilocaea oleaginea*), provokes severe damage to olive in all olive-growing countries. In recent years, the incidence and severity of this disease have increased in Palestine, likely resulting from climate change (Salman *et al.*, 2011). Peacock eye is considered the most important disease attacking olive trees in Palestine, where olive production is very important, contributing approx. 13% to the national income (Salman *et al.*, 2011; Abuamsha *et al.*, 2013; UNCTAD, 2015).

Venturia oleaginea mainly attacks olive leaves, sometimes fruit peduncles and rarely tender shoots and fruits (Agosteo and Schena, 2011). The first symptoms of the disease are small, roundish, black spots, which may be arranged in concentric circles with an olive-green to dark olivaceous centre and peripheral rings with reddish-brown to yellow or green halos (Agosteo and Schena, 2011). The infected leaves fall prematurely and defoliation negatively affects the vegetative and reproductive activities of the trees (Graniti, 1993). Tree susceptibility to the disease largely depends on environmental conditions (it is favored by high humidity of the air), plantation density, cultivar and leaf age (old leaves are less susceptible) (Graniti, 1993; Obanor *et al.*, 2008, 2011). Favourable conditions for disease development are temperatures between 15 and 20 °C and persistent rain. At these temperatures, duration of leaf wetness required for infections is the least. Progress of the disease is reduced by hot, dry conditions (Obanor *et al.*, 2008, 2011).

Although pruning is known to reduce humidity and shading within olive tree canopies, and consequently disease incidence and severity, no data are available on these effects. Therefore, the positive role of pruning for reducing peacock eye is hypothetical, but has never been experimentally demonstrated. Little information is available on the quantitative effects of factors that favour the disease, and on effects of the disease on the reproductive activity of the trees. Furthermore, no clear relationships are known regarding the effects of the disease on flowering, fruit set and final olive yields. It is known that the disease mainly attacks the lower parts of tree canopies, but no quantitative data are available to evaluate its distribution in the different parts of the canopies.

The aims of the present study were to: 1) acquire more knowledge on peacock eye in Palestine, and on environmental conditions which affect development and presence of the disease in different areas of the country; 2) establish and quantify the effects of pruning on the disease; 3) determine and quantify the distribution of the disease in the different parts of olive tree canopies; and 4) measure and quantify the effects of the disease on floral biology (inflorescence growth and fruit set). The acquired knowledge of factors affecting the development and severity of peacock eye will assist the definition of disease management strategies, particularly in relation to the different conditions occurring in olive producing countries of the Mediterranean area.

MATERIALS AND METHODS

This study was carried out in the West Bank, which is a landlocked territory forming the bulk of the Palestinian territories and the State of Palestine. The West Bank climate is characterized by long, hot and dry summers with no or little rainfall from the end of April-beginning of May to October. However, rainfall and temperature vary according to altitude and location (Lodolini *et al.*, 2016, 2017). Four representative organic olive groves were chosen. They were located in Jenin and Nablus, which are the two provinces with the greatest concentration of cultivated olive. The groves were chosen to represent the diversity in topography, altitude and climatic conditions (temperature, relative humidity, rainfall), which characterize the study area. Two orchards were located in sub-areas (Sir and Qabatia) which are known to develop low levels of infection, and the other two were located in sub-areas with high levels of infection (Northern Asira and Burqin). The altitudes of these four areas are: Northern Asira 120 m above sea level, Sir 178 m, Burqin 233 m and Qabatia 268 m (PCBS, 2014). The olive cultivar in all four orchards was Nabali, which is the most important in Palestine. The trees were adult, spaced at 7 × 7 m and trained to the globe system. The orchards were managed with two to three tillages each year and organic fertilization. There were no treatments applied to control peacock eye.

Some climatic data (temperature, relative humidity of the air and rainfall) were recorded at meteorological stations near the olive orchards. In each orchard, six trees were randomly chosen. Three of the trees in each orchard were subjected to pruning before starting the investigation, while the other three were not pruned. This was to allow evaluation of the effects of pruning on peacock eye disease development. On each tree, nine branches were selected as follows: three in the upper part of the canopy (the upper third), three in the middle part (the middle third) and three in the lower part (the lower third). For each branch selected, all of the leaves were counted, and peacock eye symptomatic leaves were counted. These observations were repeated at approx. 20 d intervals. At the same time, the NaOH test was used to determine asymptomatic disease in leaves collected from branches close to the labelled branches. From each designated part of the canopy, three samples (one per tree) of 20 leaves were used. These leaves were immersed in a solution of 5% NaOH at 60–70°C for 2–4 min. The leaves were then examined and the numbers of infected leaves (recognizable by the appearance of dark circular spots) was recorded. The data collected from the labelled branches and the determination of the amounts of asymptomatic infection on leaves were used to calculate the proportions of fallen leaves (defoliation), symptomatic leaves and infected asymptomatic leaves for each of the three parts of the canopy. The observations on leaves were carried out from January to the first days of May 2016, and only leaves grown in 2014 and 2015 were considered. In particular, the defoliation caused by the disease on the labeled branches was determined by counting the number of nodes on branches grown in 2014 and 2015. Because each node produces two leaves, it was possible to calculate the number of leaves present on each branch in the absence of leaf fall. Leaves survive for approx. 2.5 years, so it was hypothesized that the leaves that were no longer on the labeled branches had fallen mainly because of peacock eye. Therefore, fallen leaves caused by peacock eye were taken as the difference between the number of leaves derived from the multiplication of the number of nodes × 2 and the leaves that were present at the last sampling (effective defoliation). Because all the *V. oleaginea*-infected leaves fall, the total potential percent defoliation was calculated as the effective percent defoliation plus the percent symptomatic leaves plus the percent infected asymptomatic leaves.

The new leaves that developed from March of 2016 were not considered because of the short period from bud sprouting to the last sampling at the beginning of May.

In spring, the number of inflorescences on the labeled branches was determined, and approx. 1 month after flowering the number of fruits was also determined. This was to allow calculation of the ratio between the number of fruits and inflorescences, to give the amount of fruit set.

Immediately before flowering (when the inflorescences were white), three samples of ten inflorescences per tree were collected from each of the three parts of the canopy, and inflorescence lengths were measured.

Data were statistically analyzed by ANOVA according to a factorial experimental design (site/time of observation/pruning/part of the canopy), and the averages were compared by the Student-Newman-Keuls Test. Some data were also presented as simple linear regressions of some variables.

RESULTS

Climatic parameters of the areas studied

Some climatic parameters of the experimental sites are summarized in Table 1. Temperature (T) and relative humidity (RH) of the air were available only for the Northern Asira and Sir sites, whereas rainfall was available for all four sites. Northern Asira had lower T and higher RH than Sir. Asira and Burquin had the highest rainfall amounts.

Table 1. Climatic data of the experimental sites in Palestine examined in this study. Temperature and relative humidity values are means ± standard errors.

Area	Average air temperature (°C, Dec 2015 – May 2016)	Relative humidity of the air (%, Dec 2015 – May 2016)	Cumulated rainfall (mm, Sept 2015 – May 2016)
Northern Asira	15.1 ± 1.9	71.9 ± 4.7	560.7
Burqin			493.5
Qabatia			409.8
Sir	18.1 ± 2.1	61.7 ± 4.3	409.2

Amounts of leaf infection

Data on the effects of the factors (site/environment, time of observation, pruning, part of the canopy) on the amounts of leaf infection were statistically analyzed using a factorial design. There were statistically significant effects for all the factors, but significant interactions between factors were very few, so these are not presented.

Effects of the different sites

All the parameters used to evaluate the amounts of infection were significantly affected by the site/environment (Figure 1). The percentages of symptomatic leaves and infected asymptomatic leaves were greatest in Northern Asira, followed by much lower values in Burqin and then lower still for Qabatia and Sir. A similar trend was observed for effective and total potential defoliation. The amounts of effective and total potential defoliation were very high in Northern Asira and high in Burqin.

Effects of time

The percentage of symptomatic leaves decreased from the beginning of the experiment (January) to February, and then increased and reached a maximum in March before decreasing thereafter (Figure 2). A similar pattern was observed for the percentage of infected asymptomatic leaves.

Effects of pruning

Pruned trees had lower proportions of symptomatic and infected asymptomatic leaves than unpruned trees (Figure 3). A similar result was observed for both the

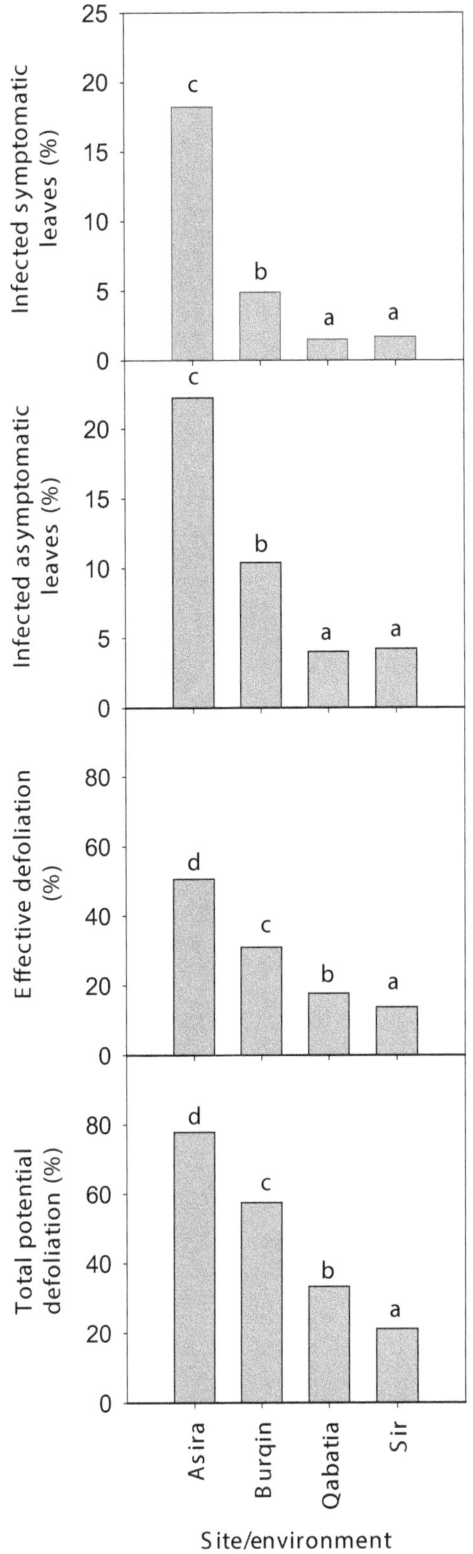

Figure 1. Means of olive tree defoliation and *Venturia oleaginea* infections as influenced by different sites/environments in Palestine. Means accompanied by the same letter are not significantly different ($P \leq 0.05$).

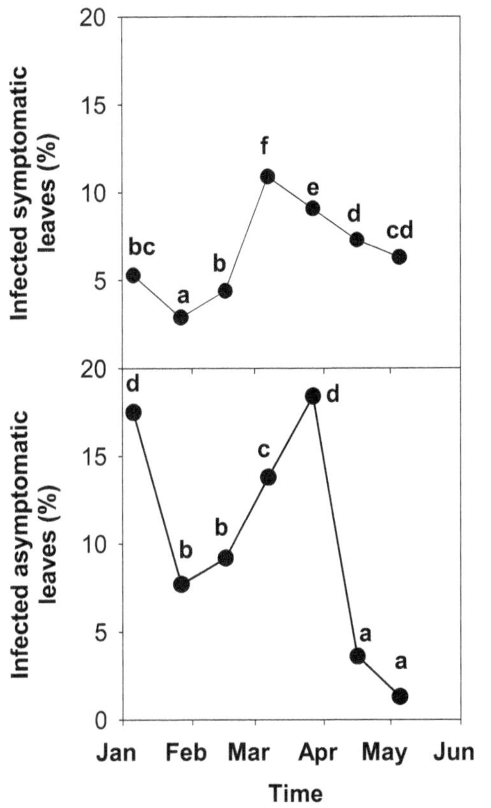

Figure 2. Mean proportions of olive leaf infections by *Venturia oleaginea* as influenced by the time of observation. Means accompanied by the same letter are not significantly different ($P \leq 0.05$).

effective and total potential defoliation proportions.

Effect of leaf position in the canopy

The proportions of symptomatic leaves and infected asymptomatic leaves were both affected by the position of the leaves in the tree canopies (Figure 4). The greatest infection percentages were always in the lower parts of the canopies and the least infection was in the upper parts. The greatest percentages of effective and total potential defoliation were in the lower parts of the canopies.

Floral biology and fruiting

All the factors considered in this study affected the mean length of the olive inflorescences (Figure 5). The longest inflorescences were found in Qabatia, whereas the shortest were in Northern Asira. Intermediate inflorescence length occurred in Burqin and Sir. The inflorescences were longer on pruned trees and in the upper parts of the canopies. Fruit set was significantly affected

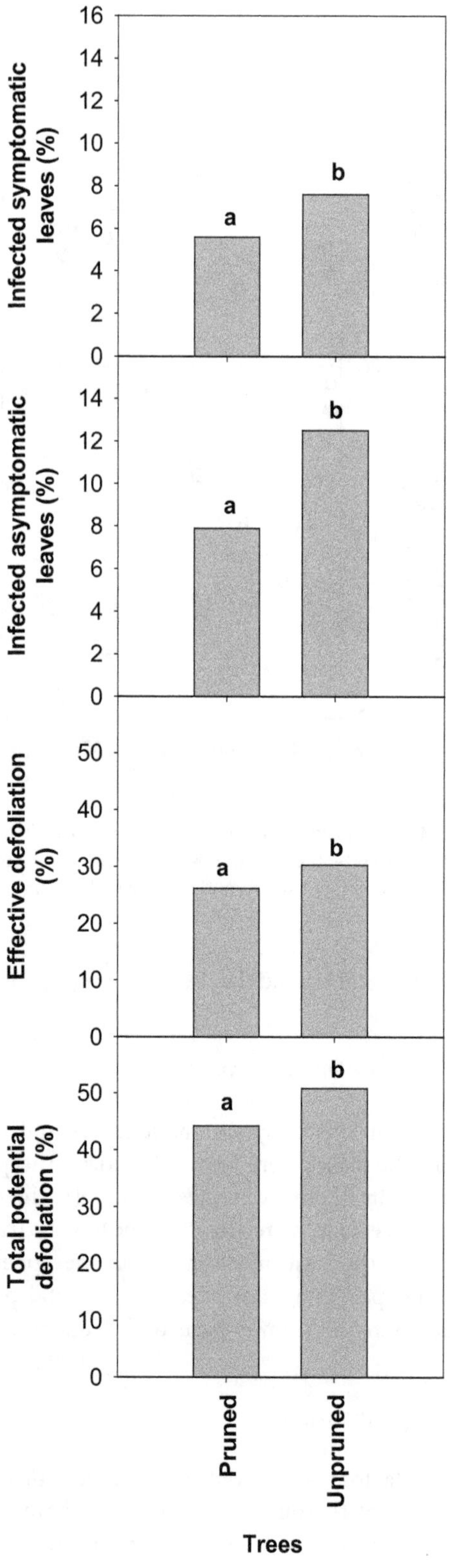

Figure 3. Means of olive tree defoliation and *Venturia oleaginea* infections as influenced by pruning. Means accompanied by the same letter are not significantly different ($P \leq 0.05$).

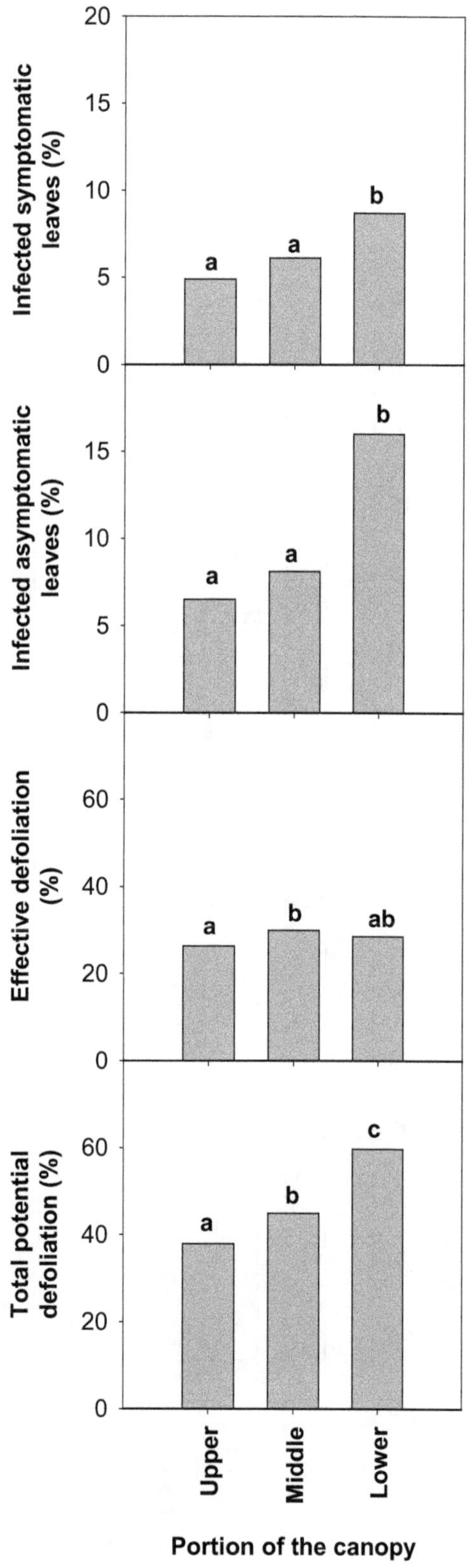

Figure 4. Means of olive tree defoliation and *Venturia oleaginea* infections as influenced by three different tree canopy positions (upper, middle and lower). Means accompanied by the same letter are not significantly different ($P \leq 0.05$).

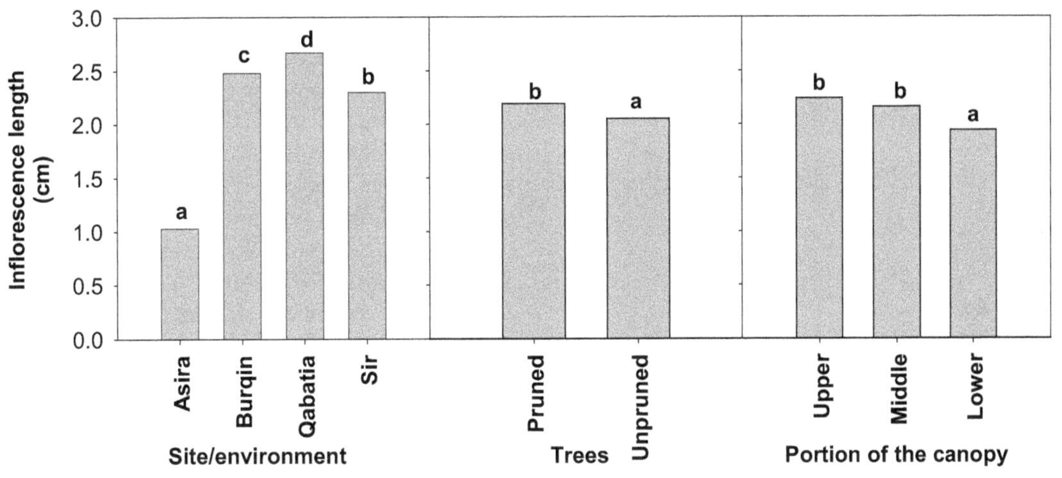

Figure 5. Mean olive inflorescence lengths as influenced by the site/environment, pruning and portion of tree canopy. Means of each factor accompanied by the same letter are not significantly different ($P \leq 0.05$).

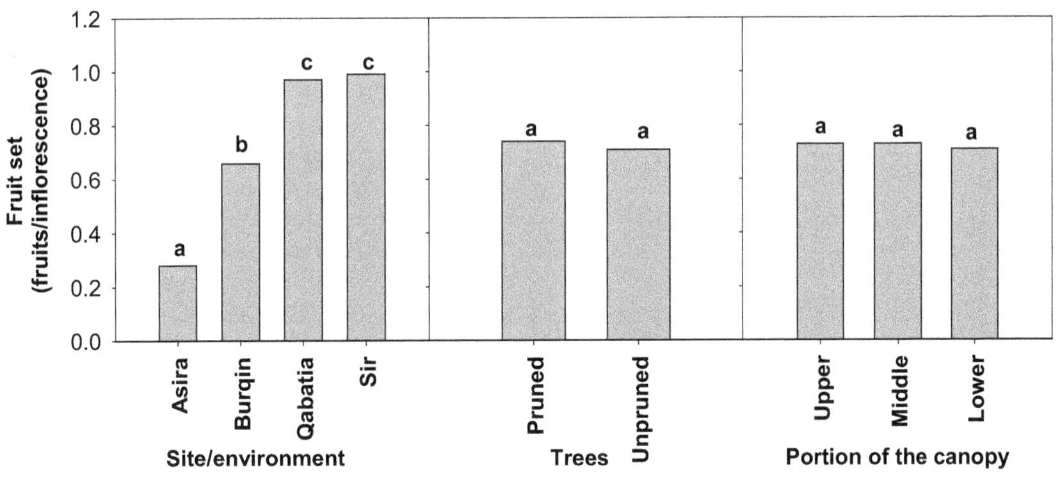

Figure 6. Mean fruit set values for olive trees as influenced by the site/environment, pruning and portion of tree canopy. Means of each factor accompanied by the same letter are not significantly different ($P \leq 0.05$).

by the site/environment, but was not affected by pruning or position in the tree canopies (Figure 6).

DISCUSSION

This investigation has shown a significant influence of all the considered factors on the severity of peacock eye occurring in olive orchards.

The effects of site/environment were most likely due to the different climatic conditions at the different experimental sites. In particular, Northern Asira, where the greatest amounts of infected leaves and defoliation were found, had the greatest rainfall and high values of RH, along with relatively cool temperatures. These conditions are particularly favourable for peacock eye development, as conidium production by *V. oleaginea* is optimal at 15°C, especially when humidity is high (Azeri, 1993; Guechi and Girre, 1994; Obanor *et al.*, 2008). In contrast, the lowest amounts of infected leaves and defoliation were observed at Sir, where the lowest rainfall and RHs, and the highest temperatures, were recorded. Using the rainfall data, which was available for all four sites, the important and positive influence of high humidity resulting from high rainfall on disease development was demonstrated. There was a significant positive linear correlation between the cumulated rainfall at the four sites and the levels of infection and defoliation that were recorded (Figure 7). For rainfall, the period

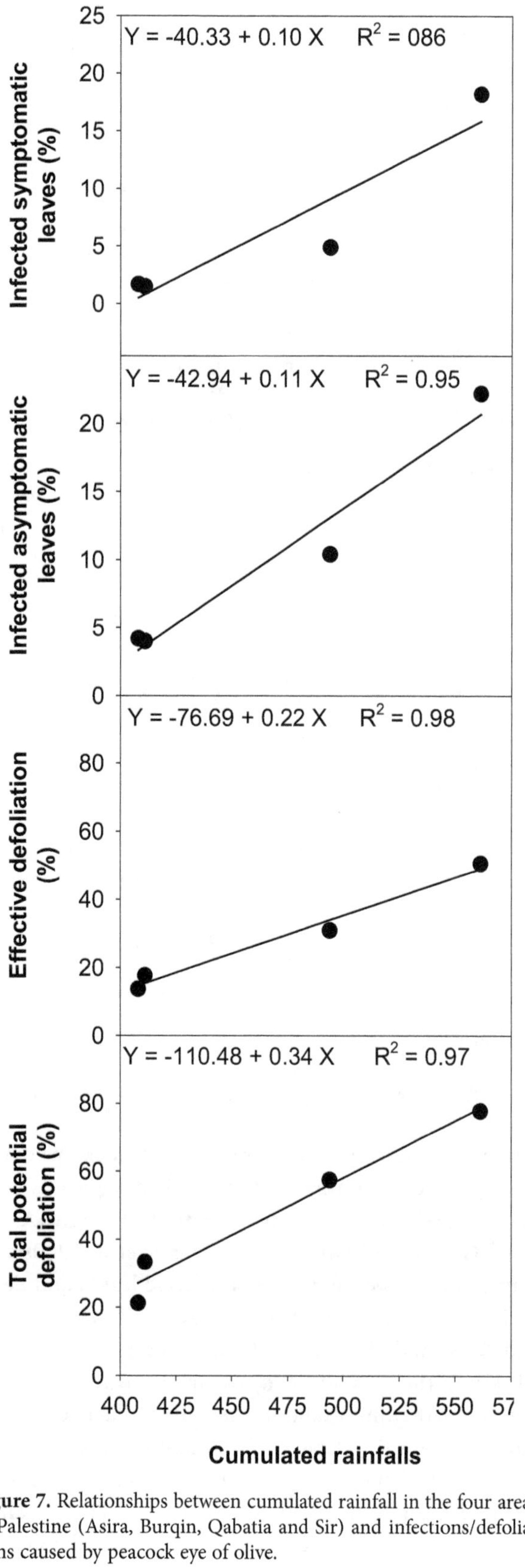

Figure 7. Relationships between cumulated rainfall in the four areas of Palestine (Asira, Burqin, Qabatia and Sir) and infections/defoliations caused by peacock eye of olive.

from September to May was considered as it is when most of the rain occurs and so it is representative of the climatic conditions of the area. Therefore, environmental factors, in particular RH and T, appear to be the major elements in determining differences in the infection and spread of peacock eye disease among the considered sites. Indeed, high humidity and relatively cooler conditions are the most suitable for the disease (Salerno, 1966; Saad and Masri, 1978; Graniti, 1993; Obanor *et al.*, 2008; 2011). These results are similar to the those of previous observations in Palestine, which showed high levels of *V. oleaginea* infection in Nablus where climatic conditions were similar to Northern Asira, and low levels of infection in Tubas where climatic conditions are similar to Sir (Salman *et al.*, 2011).

The trends of infected leaves, both symptomatic and asymptomatic, showed maximum values from the end of winter to beginning of spring. This is when very favourable temperature and humidity conditions support rapid development of peacock eye (Guechi and Girre, 1994). The subsequent decreases in the percentages of diseased leaves were probably due to losses of the infected leaves from the trees.

Lowest levels of infection were observed in pruned trees compared with unpruned ones. This can be explained by considering that thinning of the vegetation creates better conditions for lightening and aeration of the internal parts of the canopy, with a consequent decrease in RH, which is a very important factor in favoring leaf attack by *V. oleaginea* (Obanor *et al.*, 2008; 2011; Al-Jabi, 2013). However, even if the effect of pruning was supposed, the present study has given the first experimental evidence that pruning can reduce the severity of the peacock eye in olive.

Our results clearly showed gradients of peacock eye severity from the lower to upper parts of olive tree canopies. Amounts of infected leaves and defoliation decreased from the lowest to the upper parts of the canopies. This is likely to be because the lower canopy parts retain the greatest air humidities, while upper canopy parts have more light and are well-ventilated.

All the factors investigated in this study influenced the length of olive inflorescences. To determine if these differences were associated with pathogen infections, the relationship between the percentage of effective defoliation and inflorescence length was evaluated by using the average values of the combinations of the following factors: site/pruning/part of the canopy. A significant linear correlation (R^2 = 0.66) was found between these two parameters (Figure 8). This indicates that olive inflorescence length was related to the number of leaves and that the shorter length of the inflorescences in Northern

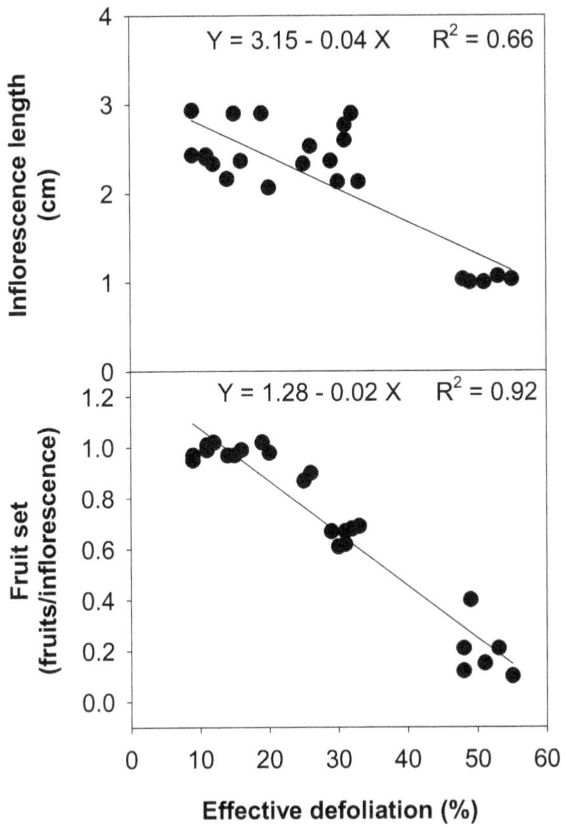

Figure 8. Relationships between effective olive tree defoliation caused by peacock eye with inflorescence lengths and fruit set. Each value represents the average of all the combinations of three factors (site, pruning, portion of canopy).

Asira, in unpruned trees and in the lower parts of the tree canopies were associated with the higher levels of defoliation caused by peacock eye.

The differences in fruit set can also be analyzed by evaluating the relationships between the proportions of effective defoliation and fruit set using the average values of the combinations of the following factors: site/pruning/part of the canopy. Again, a significant, negative, linear correlation (R^2 = 0.92) was found between these two parameters. This means that the differences in fruit set were also significantly associated with defoliation caused by the peacock eye disease. Examination of this relationship also indicates the effect of defoliation on the potential yield of the trees and thus the economic damage caused by the disease. Indeed, it is possible to see that depending on the level of defoliation, the number of fruits can be reduced up to one fifth (20%), when defoliation reaches values around 50%, compared to those obtainable in conditions of very low defoliation. This means that the impact of the disease on the potential production of the trees could be very high. Further investigations are required to determine the effects of peacock eye on olive fruit growth and possible fruit drop.

CONCLUSIONS

The results of this study indicate that variability of the environmental conditions in the different areas of Palestine is enough to determine differences in the susceptibility of olive to peacock eye attacks. Therefore, even though Palestinian Territories have a limited extension, there are significant environmental differences among the different areas, which determine significant changes in the susceptibility to peacock eye, which must be considered in applying site-dependent strategies for disease control.

Pruning can reduce the susceptibility of olive trees to peacock eye. Infection was generally greater in unpruned than pruned trees. To the best of our knowledge, this is the first experimental evidence showing that pruning has a positive role in reducing this disease. In Palestine the importance of olive pruning is often underestimated. Hence efforts have to be made to train farmers to apply this practice in an optimal/regular way, also taking into consideration its importance in controlling important biotic adversities, such as peacock eye disease.

There is a large gradient within olive tree canopies for susceptibility of leaves to *V. oleaginea*. Disease severity is much greater in lower than upper canopy parts. This information is important for optimization of disease control methods. Treatments must ensure uniform and complete coverage especially of the middle and lower parts of the tree canopies.

Peacock eye can reduce the lengths of olive inflorescences and fruit set. This is the first report showing direct and linear influence of defoliation caused by peacock eye on inflorescence length and fruit set, and these negative effects are associated with defoliation caused by the disease. Based on the effects of the disease on fruit set, it can be concluded that peacock eye can cause severe reductions in olive fruit yields.

The results of the present study increase knowledge of factors that affect the susceptibility of olive to infection by *V. oleaginea,* and of effects of defoliation caused by the pathogen on olive tree production. This information, besides being important for a better knowledge of the biology of peacock eye in Palestine and other Mediterranean countries, can be useful for the development of effective strategies for its control.

ACKNOWLEDGMENT

This paper was developed from the Master of Science thesis of Tahreer Issa.

LITERATURE CITED

Abuamsha R., Abueid M., Hajjeh H., Salman M., 2013. Evaluation of the incidence and severity of olive leaf spot caused by *Spilocaea oleagina* in different olive cultivars in Palestine. *Journal of Agriculture and Environment for International Development* 107: 201–212.

Agosteo G.E., Schena L., 2011. Olive leaf spot. In: *Olive Diseases and Disorders* (L. Schena, G.E. Agosteo, S.O. Cacciola, ed.). India Transworld Research Network. 143–176.

Azeri T., 1993. Research on olive leaf spot, olive knot and *verticillium* wilt of olive in Turkey. *European and Mediterranean Plant Protection Organization Bulletin* 23: 437–440.

Graniti A., 1993. Olive scab: a review. *European and Mediterranean Plant Protection Organization Bulletin* 23: 377–384.

Guechi A.,Girre L., 1994. Sources of *Cycloconium oleaginum* (Cast.) conidia for infection of olive leaves and conditions determining leaf spot disease development in the region of Sétif, Algeria. *Mycopathologia* 125: 163–171.

Lodolini E.M., Polverigiani S., Ali S., Mutawea M., Qutub M., Pierini F., Neri D., 2016. Effect of complementary irrigation on yield components and alternate bearing of a traditional olive orchard in semi-arid conditions. *Spanish Journal of Agricultural Research* 14: e1203.

Lodolini E.M., Polverigiani S., Ali S., Mutawea M., Qutub M., Arabasi T., Pierini F., Abed M., Neri D., 2017. Oil characteristics of four Palestinian olive varieties. *Journal of Oleo Science* 66: 435-441.

Obanor F.O., Walter M., Jones E.E., Jaspers M.V., 2008. Effect of temperature, relative humidity, leaf wetness and leaf age on *Spilocaea oleagina* conidium germination on olive leaves. *European Journal of Plant Pathology* 120: 211–222.

Obanor F.O., Walter M., Jones E.E., Jaspers M.V., 2011. Effects of temperature, inoculum concentration, leaf age, and continuous and interrupted wetness on infection of olive plants by *Spilocaea oleagina*. *Plant Pathology* 60: 190–199.

PCBS, Palestinian Central Bureau of Statistics., 2014. Ramallah, Palestine, www.pcbs.ps.

Saad A.T., Masri S., 1978. Epidemiological studies on olive leaf spot incited by *Spilocaea oleaginea* (Cast.) Hughes. *Phytopathologia Mediterranea* 17: 170–173.

Salerno M., 1966. The olive leaf spot *Spilocaea oleagina*(Cast.) Hugh. Considerations on biology and control. *Review of Applied Mycology* 45: 184.

Salman M., Hawamda A., Amarni A., Rahil M., Hajjeh H., ... Abuamsha R., 2011. Evaluation of the incidence and severity of olive leaf spot caused by *Spilocaea oleagina* on olive trees in Palestine. *American Journal of Plant Sciences* 2: 457–460.

UNCTAD, United Nations Conference on Trade And Development., 2015. The Besieged Palestinian Agricultural Sector. New York and Geneva.

11

Xylella fastidiosa subsp. *pauca* on olive in Salento (Southern Italy): infected trees have low *in planta* micronutrient content

MARCO SCORTICHINI[1,*], DANILO MIGONI[2], FEDERICA ANGILÈ[2], LAURA DEL COCO[2], CHIARA ROBERTA GIRELLI[2], LUIGI ZAMPELLA[1], FRANCESCO MASTROBUONI[1], FRANCESCO PAOLO FANIZZI[2]

[1] *Council for Agricultural Research and the Analysis of Agricultural Economics (CREA), Research Centre for Olive, Fruit Trees and Citrus, Via Torrino 3, I-81110 Caserta, Italy*
[2] *Department of Biological and Environmental Sciences and Technology (Di.S.Te.B.A.), University of Salento, Campus Ecotekne, Via Prov.le Lecce-Monteroni, I-73100 Lecce, Italy*
*Corresponding author: marco.scortichini@crea.gov.it

Competing Interests: The Author(s) declare(s) no conflict of interest.

Editor: Jesus Murillo, Universidad Publica de Navarra, Spain.

Abstract. *Xylella fastidiosa* subsp. *pauca* is associated with the olive quick decline syndrome in Salento (Apulia region, Southern Italy). The first outbreak of the disease was noticed in the Gallipoli district, from where it subsequently reached nearby areas. To date, no specific study has verified if abnormal soil or leaf micronutrients is associated with the disease. Soil and leaf samples were taken from 23 olive farms showing symptoms of the syndrome located in Gallipoli and nearby areas. Each sample was analyzed for magnesium and micronutrients content using inductively coupled plasma atomic emission spectroscopy. Real-time PCR indicated that *X. f.* subsp. *pauca* was present in each sampled tree. There is a general lack of molybdenum in soil and low bioavailability of copper and molybdenum in tree leaves. Low content of manganese in soil was also found in some farms located in Gallipoli, Galatone and Trepuzzi. Olive trees grown in Gallipoli area also had low content of boron. Principal component analysis showed that soil and leaf samples from this area had lower micronutrient contents compared with other areas. General copper depletion in leaves was uncommon and has not been previously recorded in Italy in a large area. This could indicate that *X. f.* subsp. *pauca* infection causes a depletion of copper within olive leaves. Reduced copper content has been previously recorded in leaves infected with *X. f.* subsp. *multiplex* and for other bacterial pathogens. The role of copper in relation to *X. fastidiosa* infection is discussed.

Keywords. Olive Quick Decline Syndrome, quarantine bacteria, micronutrients, Inductively Coupled Plasma Atomic Emission Spectroscopy Analysis.

INTRODUCTION

First evidence of a severe and spreading decline of olive trees (shoot, twig and branch dieback, and tree death) growing in Gallipoli area of the Salento peninsula, Apulia, Southern Italy) was noticed by farmers during 2008 (Martelli *et al.*, 2016). Some fungi, such as *Phoeoacremonium parasiticum*, *P. rubrigenum*, *P. aleophilum* and *P. alvesi* as well as *Phaemoniella* spp. were reported to be associated with the diseased olive trees (Nigro *et al.*, 2013). DNA sequences belonging to *Xylella fastidiosa* were detected at the same time from olive trees showing symptoms of decline collected in the same area (Saponari *et al.*, 2013). The area affected by the disease was estimated to be from 8000 to 10000 ha (Nigro *et al.*, 2013; Martelli *et al.*, 2016). Subsequently, *X. f.* subsp. *pauca* was isolated from diseased olive specimens, and the disease was named as the "olive quick decline syndrome" (OQDS) (Cariddi *et al.*, 2014; Loconsole *et al.*, 2014). Once colonizing host tissues, *X. fastidiosa* can be ransmitted and spread by insect vectors (Almeida *et al.*, 2005), and many wild plants are reservoirs of the pathogen (EFSA PLH Panel, 2018). After possible introduction from abroad (Marcelletti and Scortichini, 2016; Giampetruzzi *et al.*, 2017) and the colonization of olive trees, vectors and wild plant species have played fundamental roles in the subsequent spread of *X. f.* subsp. *pauca* from the Gallipoli district to neighboring areas of Salento (Strona *et al.*, 2017; White *et al.*, 2017).

In recent years, several studies aimed at elucidate the genomic structure of *X. fastidiosa*, its origin and detection, the host range where the bacterium can survive, the role of the insect vectors and the potential distribution of *X. fastidiosa* in the Mediterranean basin (EFSA PLH Panel, 2018) were performed. However, no specific study has verified if abnormal soil or leaf availability of micronutrients is associated with the outbreak of *X. f.* subsp. *pauca* on olive in the Salento peninsula.

The area where the OQDS was first observed (i.e., Gallipoli and nearby municipalities) is characterized by distinctive lithological types. According to the geological map of Italy (i.e., leaf 214), in that area there are four types, namely QP, Q^1S-Q^1C and C^{11-7}. QP indicates the "Salento calcarenite" (i.e., chalky and sandy-chalky substrates), Q^1S and Q^1C indicates the so-called "Gallipoli formation" (i.e., Q^1S, sandy-clay and marly substrates; Q^1C, sandy-clay soils mixed with lithic calcarenite and arenite), whereas C^{11-7} is the calcareous substrates of the "Galatina dolomite" (Largaiolli *et al.*, 1969). The soils of this area are also distinct from those of other areas of Apulia (Costantini *et al.*, 2012). However, all the Salento and most of Apulia soils are characterized by their calcareous substrates (Costantini *et al.*, 2012). The soils of Gallipoli area have good texture (no of silty or very sandy soils), and are from moderate to good depth, with presence of rock fragments, high drainage capacity and low salinity (Ancona *et al.*, 2010).

Recent studies have stressed the importance of the element composition in plant tissues and the host ion changes in relation to *X. fastidiosa* virulence and the development of disease symptoms (Cobine *et al.*, 2013; De La Fuente *et al.*, 2013; Oliver *et al.*, 2014; 2015; Navarrete and De La Fuente, 2014; 2015). Consequently, the element content and availability in the soil and in the host leaves could be linked with olive tree infection by *X. fastidiosa*. In the present study, soil and leaf content was assessed for nutrients that have basic physiological roles in plant metabolism. The putative relationships and role of nutrients with *X. f.* subsp. *pauca* infection in Salento area is also discussed.

MATERIALS AND METHODS

Sampling, and soil and leaf analyses

To verify the micronutrient content of soils and nutrient availability to olive trees, soil and leaf samples (from cv. Ogliarola salentina) were taken during summer of 2018, and analyzed. The samples were from the Gallipoli area and from other districts where *X. f.* subsp. *pauca* was found during last 5 years, and where OQDS symptoms were observed since at least 8-10 years. In these areas, the incidence of *X. f.* subsp. *pauca* is very high (Boscia *et al.*, 2017). Samples were taken from 23 farms located in the following municipalities: Alezio, Gagliano del Capo, Galatone, Gallipoli, Leverano, Presicce, Sannicola, Trepuzzi (Salento, Lecce province). All farms had trees with visible symptoms putatively attributable to OQDS (leaf scorching, twig and branch dieback). The following elements were analyzed in soil and leaf tissue: magnesium (Mg), iron (Fe), zinc (Zn), copper (Cu), boron (B), manganese (Mn) and molybdenum (Mo). Leaf calcium content and soil pH were also assessed. Soil pH was measured in bi-distilled water using a suspension of 1:5 solid to liquid phase. For each farm, leaf samples (four subsamples from each tree) were taken from three olive trees, following a distribution of a triangle on the farm (Papadia *et al.*, 2011) and according to the methods suggested by Sanzani *et al.* (2012). Samples were taken from part of the crown of each tree not showing any visible symptom of OQDS. In addition, for beneath each tree from which leaves were collected, four soil slice samples were taken from within the external limit of the foliage projection (Papadia *et al.*, 2011). Samples were taken at 30–35 cm of depth where there was abundance of tree roots (Fernandez and Moreno, 1999), avoiding collection of rocks. This procedure was adopted to ascertain the soil and leaf composition related to the trees assessed.

All samples were placed in plastic bags and transported to the laboratory for preparation. Leaves were washed with distilled water to remove all the soil particles and then dried. Each soil (1 g) and leaf (0.5 g) subsample was analyzed separately at the University of Salento by using the Inductively Coupled Plasma Atomic Emission Spectroscopy (ICP-AES), by following the standard procedures. Briefly, samples of known dry weight were mixed with 4 mL H_2O_2 and 6 mL HNO_3 at 180 °C for 10 min, using a microwave digestion system (Milestone START D). They were then cooled, diluted with ultrapure water to a final volume of 20 mL, filtered through Whatman No. 42 filter papers, and measured for elemental content using an inductively coupled plasma atomic emission spectrometer (ThermoScientific iCap 6000 Series). Results were expressed as mg·kg^{-1} dry weight (ppm).

Statistical and principal component analysis

The Shapiro-Wilk test was applied for checking the normality of the data obtained from the ICP-AES analyses, using the R statistical environment, version 3.5.1 (R Development Core Team, 2013), and correlation matrix based on Pearson's coefficient was calculated for all the measured elements using MetaboAnalyst 4.0, a web-based tool for visualization of metabolomics (Xia *et al.*, 2009; Xia and Wishart, 2016). This approach was used to assess possible linear associations between the variables of the two datasets considered. Multivariate statistical analyses and graphics were obtained using SIMCA-P software SIMCA 14 software, (Sartorius Stedim Biotech, Umeå, Sweden) (Bro and Smilde, 2014). Exploratory data analyses were carried out using Principal Component Analysis (PCA), applying well-established procedures (Papadia *et al.*, 2011), and used to obtain a general overview of the natural data grouping. The original dataset was rearranged in a new multivariate coordinate space where the dimensions were ordered by decreasing variance in the data. The principal components were displayed as a set of scores, which highlights clustering or outliers, and a set of loadings (p), which explain the influence of input variables on principal components. Autoscaling, also known as unit or unit variance scaling, was applied to the data (Van den Berg *et al.*, 2006). The PCA models were validated using the internal cross-validation default method (7-fold), and were further evaluated with the permutation test (400 permutations) (Trygg and Wold, 2002; Triba *et al.*, 2015). The quality of the models was described by R^2 and Q^2. The R^2 value is a cross validation parameter defined as the proportion of variance in the data explained by the models, while the Q^2 parameter is an internal cross validation parameter, which indicates the predictability of the model. Loadings plots were evaluated to investigate the role of the measured variables in the models (Van den Berg *et al.*, 2006; Wheelock and Wheelock, 2013).

Occurrence of Xylella fastidiosa *in olive farms*

The occurrence of *X. fastidiosa* in the sampled olive tree leaves was assessed using real-time PCR (Harper *et al.*, 2010) following the procedures described by Modesti *et al.* (2017) and Scortichini *et al.* (2018). For these analyses, sampled leaves were taken from branches not showing disease symptoms (i.e., leaf wilting or twig dieback).

RESULTS

Soil and leaf analyses

In the soil of the farms located in the area where the first sign of OQDS was noticed (i.e., the Gallipoli area) there were low contents of some micronutrients. In particular, according to the indicated normal mean value content in soil of each element assessed (Hodgson, 1963; Alloway, 1995; McLennan and Taylor, 1999; Kaiser *et al.*, 2005; Pendias, 2010; Reimann *et al.*, 2014; Noulas *et al.*, 2018), manganese and molybdenum contents were low in all the farms assessed (Table 1). The contents of all the other micronutrients were within the range of the mean values. In the leaves sampled from farms of this area, copper, boron and molybdenum contents were below the minimum levels required for a normal olive tree growth (Mahler, 2000; Tittarelli *et al.*, 2002; Sanzani *et al.*, 2012) (Table 2). The analyses for farms located in Galatone district, an area currently showing extensive tree diebacks, also showed general low soil contents of manganese and molybdenum. In addition, there were very low amounts of copper, boron and molybdenum, and, to a lesser extent, of manganese was found into the leaves (Tables 1 and 2). All the other farms located in areas close to Gallipoli, namely Alezio and Sannicola or in districts with olive trees showing the OQDS, namely Gagliano del Capo, Leverano, Presicce, and Trepuzzi, had low amounts of molybdenum in soil and of copper in the leaves. In most of the farms, boron and molybdenum were also in low amount in soil and leaves (Tables 1 and 2).

Soil pH was close to 7.0, or greater than 7.0 or 8.0, for all farms, except for two in Galatone area that had soil pH of approx. 6.0 (Table 1). Real-time PCR showed that *X. f.* subsp. *pauca* was present in all the sampled trees.

Table 1. Magnesium and micronutrient contents, and pH values for soils sampled in Gallipoli, Galatone and other districts of Lecce province, from farms with olive quick decline syndrome and infections by *Xylella fastidiosa* subsp. *pauca*. For each farm, means (± standard deviation) are presented for 12 samples collected in three different sites.

District	Mg (%)	Fe (%)	Mn (ppm)	Cu (ppm)	Zn (ppm)	B (ppm)	Mo (ppm)	pH (H_2O)
Gallipoli 1	0.41±0.03	3.58±0.30	400.89±4.99	48.61±1.44	49.58±0.29	25.00±0.19	0.064±0.004	7.42±0.21
Gallipoli 2	0.32±0.01	1.78±0.06	262.22±3.30	19.22±0.13	34.33±0.23	16.92±0.14	0.123±0003	8.26±0.11
Gallipoli 3	0.25±0.02	1.77±0.14	269.22±3.72	41.24±0.38	26.65±0.16	16.55±0.18	0.065±0.006	7.72±0.21
Gallipoli 4	0.35±0.02	2.39±0.19	294.32±4.51	37.84±0.30	31.24±0.21	20.09±0.11	0.046±0.007	8.13±0.13
Gallipoli 5	0.32±0.02	2.14±0.12	278.12±1.86	88.16±0.75	32.23±0.15	16.36±0.08	0.065±0.006	6.81±0.22
Gallipoli 6	0.30±0.02	2.74±0.19	395.33±5.47	21.48±0.16	34.05±0.21	23.49±0.21	0.033±0.003	7.92±0.16
Gallipoli 7	0.29±0.03	2.67±0.27	275.82±4.30	30.53±0.41	36.08±0.18	19.77±0.13	0.122±0.006	7.93±0.14
Gallipoli 8	0.28±0.02	1.98±0.19	239.39±3.41	32.67±0.25	24.81±0.12	17.20±0.14	0.113±0.006	8.25±0.20
Galatone 1	0.14±0,01	1.79±0.14	141.42±1.18	54.13±0.41	20.38±0.18	12.08±0.10	0.472±0.008	6.12±0.34
Galatone 2	0.45±0.02	4.06±0.20	372.40±5.05	30.62±0.25	47.99±0.29	26.89±0.21	0.476±0.009	6.23±0.24
Galatone 3	0.42±0.03	2.48±0.16	175.43±2.21	46.00±0.51	28.51±0.15	23.94±0.11	0.569±0.002	8.63±0.19
Galatone 4	0.36±0.02	2.64±0.14	335.86±3.13	63.14±0.36	38.14±0.28	21.25±0.18	0.688±0.004	8.04±0.21
Galatone 5	0.35±0.02	2.29±0.13	221.03±2.59	21.11±0.20	37.48±0.14	16.91±0.14	0.218±0.004	8.25±0.34
Galatone 6	0.40±0.01	2.66±0.05	312.99±5.12	30.03±0.25	41.70±0.15	18.33±0.10	0.317±0.009	8.48±0.28
Galatone 7	0.35±0.02	2.07±0.17	256.48±2.15	37.60±0.20	28.21±0.13	16.67±0.13	0.511±0.004	8.41±0.33
Trepuzzi 1	0.36±0.03	1.72±0.19	601.43±66.38	86.43±0.58	32.59±0.11	22.00±0.10	0.027±0.005	8.61±0.29
Trepuzzi 2	0.33±0.02	1.49±0.11	530.67±34.91	62.70±0.27	32.96±0.07	20.01±0.18	0.031±0.006	8.72±0.19
Leverano 1	0.42±0.03	3.14±0.24	1322.25±117.87	31.49±0.13	44.73±0.19	42.29±0.35	0.178±0.009	8.28±0.09
Leverano 2	0.33±0.03	3.52±0.40	1141.06±132.51	18.58±0.06	50.02±0.15	52.65±0.39	0.024±0.011	7.95±0.10
Alezio	0.34±0.04	2.98±0.35	899.04±104.26	24.15±0.13	40.99±0.16	43.70±0.44	0.037±0.001	7.64±0.13
Sannicola	1.06±0.08	3.62±0.31	813.66±74.71	27.28±0.11	59.44±0.18	49.90±0.46	0.302±0.012	8.14±0.16
Presicce	0.29±0.02	3.38±0.32	872.89±76.77	35.61±0.22	55.87±0.23	48.74±0.24	0.081±0.009	8.10±0.21
Gagliano	0.30±0.02	3.60±0.36	1660.86±164.49	43.47±0.37	65.50±0.27	49.38±0.54	0.173±0.008	6.78±0.20

Statistical and principal component analyses

For elements measured in the soil samples, multivariate statistical analysis (PCA) and the related Pearson correlation matrix were carried out. The first two principal components used for the PCA model and explained about 60% of the total variance ($R^2 = 0.39$ for t[1] and 0.21 for t[2]). A specific trend of clustering among the eight groups was observed. In the t[1]/t[2] PCA score plot (Figure 1a), samples from Trepuzzi and Gallipoli districts were almost clustered at positive values of t[1], while samples collected in Leverano, Alezio, Gagliano del Capo and Presicce districts were grouped at negative values of the same t[1] component. Despite the presence of some outliers, the samples collected in the Galatone district were an homogeneous cluster, located in the upper right of the PCA score plot. In this case, the study of the variables (loading plot of Figure 1b) responsible for the class distribution along the first and the second principal components explained the arrangement of the samples in the PCA score plot of Figure 1a. In particular, Fe, Zn and Mn characterized soil samples from Leverano, Alezio, Gagliano del Capo and Presicce districts, while high contents of Mg, Ca, Mo and Cu were observed in the samples collected in the Galatone district. The cluster of samples from the Gallipoli area showed lower contents of all the measured elements. Pearson correlation matrix for soil samples (Table 3) revealed some positive and negative correlations with a high level of significance ($P < 0.001$) for some nutrients (in particular B with Mn, Zn, Fe and Ca; Mn with Zn, Fe, Ca; Zn with Fe and Ca; and Fe with Ca).

For the leaf samples, PCA gave about 55% of total explained variance ($R^2 = 0.37$ for t[1] and 0.18 for t[2]). This model was weakly predictive but sufficiently descriptive, showing a degree of separation among groups. Despite the different number of samples for each group, the t[1]/t[2] PCA score plot (Figure 2a) highlighted some differences among the studied samples, coming from eight different sites, especially for the first principal component t[1]. In particular, leaves samples from Leverano and Gallipoli districts were clear and homogeneous clusters, grouped, respectively, at positive and negative values of t[1]. In contrast, samples from the Galatone district were a scattered group in the t[1]/t[2] score plot. The study of the variables (loading plot of

Table 2. Calcium, magnesium and micronutrient contents in olive leaves sampled in Gallipoli, Galatone and other districts of Lecce province, from farms showing the olive quick decline syndrome and infections by *Xylella fastidiosa* subsp. *pauca*. For each farm, data show the means (± standard deviation) of 12 samples collected from three different trees. Data are referred to leaf dry weight.

District	Ca (%)	Mg (%)	Fe (ppm)	Mn (ppm)	Cu (ppm)	Zn (ppm)	B (ppm)	Mo (ppm)
Gallipoli 1	1.07±0.13	0.108±0.001	63.40±0.49	24.69±0.20	8.71±0.05	19.33±0.10	8.65±0.05	0.182±0.003
Gallipoli 2	0.60±0.01	0.062±0.001	49.38±0.40	17.34±0.19	5.31±0.04	19.13±0.13	9.68±0.06	0.103±0.003
Gallipoli 3	0.80±0.01	0.079±0.001	61.71±0.34	17.21±0.14	5.38±0.03	12.28±0.07	10.91±0.09	0.106±0.003
Gallipoli 4	0.96±0.06	0.106±0.008	65.46±0.38	24.02±0.22	6.40±0.05	16.04±0.11	11.02±0.09	0.113±0.004
Gallipoli 5	1.12±0.07	0.109±0.006	70.71±0.73	32.71±0.37	4.99±0.07	14.28±0.09	10.32±0.15	0.143±0.006
Gallipoli 6	1.23±0.06	0.040±0.006	109.14±1.17	35.20±0.51	4.91±0.01	15.05±0.09	9.58±0.14	0.063±0.002
Gallipoli 7	0.99±0.04	0.101±0.006	73.45±0.96	24.18±0.40	5.65±0.03	15.40±0.09	8.72±0.12	0.098±0.003
Gallipoli 8	1.00±0.05	0.096±0.005	62.01±0.68	18.49±0.28	5.22±0.02	18.30±0.14	8.40±0.11	0.091±0.003
Galatone 1	0.93±0.04	0.121±0.007	110.10±0.76	48.09±0.38	5.11±0.03	31.93±0.12	10.51±0.05	0.092±0.003
Galatone 2	0.93±0.04	0.130±0.005	72.25±0.59	17.88±0.19	6.88±0.03	32.52±0.18	11.34±0.13	0.174±0.004
Galatone 3	0.94±0.05	0.107±0.006	86.28±0.66	24.03±0.18	5.27±0.04	21.36±0.06	10.40±0.09	0.121±0.002
Galatone 4	1.02±0.07	0.104±0.007	79.24±0.60	23.17±0.15	6.29±0.07	17.75±0.07	11.67±0.11	0.267±0.005
Galatone 5	1.93±0.09	0.171±0.007	106.86±0.54	26.98±0.13	19.96±0.10	32.39±0.18	9.87±0.06	0.128±0.004
Galatone 6	1.07±0.05	0.102±0.004	65.11±0.62	23.23±0.21	13.10±0.11	26.87±0.15	22.62±0.16	0.081±0.003
Galatone 7	1.12±0.06	0.108±0.004	68.96±0.69	17.92±0.18	5.20±0.04	17.30±0.11	10.33±0.11	0.147±0.003
Trepuzzi 1	1.37±0.12	0.134±0.001	74.99±0.49	41.13±0.36	6.37±0.02	24.93±0.14	18.50±0.08	0.219±0.005
Trepuzzi 2	0.11±0.02	0.116±0.001	66.86±0.43	53.99±0.57	7.81±0.04	22.97±0.10	18.11±0.11	0.230±0.003
Leverano 1	1.55±0.09	0.172±0.013	80.35±0.49	60.45±0.45	11.76±0.04	21.30±0.07	11.87±0.05	0.223±0.005
Leverano 2	1.08±0.09	0.157±0.011	84.34±0.43	80.85±0.95	18.22±0.07	13.60±0.04	11.74±0.04	0.176±0.003
Alezio	1.15±0.05	0.123±0.005	71.21±0.68	40.20±0.40	5.80±0.04	18.33±0.11	13.34±0.17	0.406±0.072
Sannicola	1.37±0.13	0.165±0.021	72.96±0.40	39.25±0.25	4.65±0.02	20.30±0.07	15.77±0.12	0.405±0.007
Presicce	1.57±0.11	0.149±0.012	97.72±0.64	45.57±0.26	6.53±0.04	12.80±0.04	11.18±0.07	0.073±0.001
Gagliano	1.32±0.03	0.130±0.004	127.56±0.86	44.76±0.37	5.75±0.05	27.25±0.12	10.06±0.05	0.120±0.003

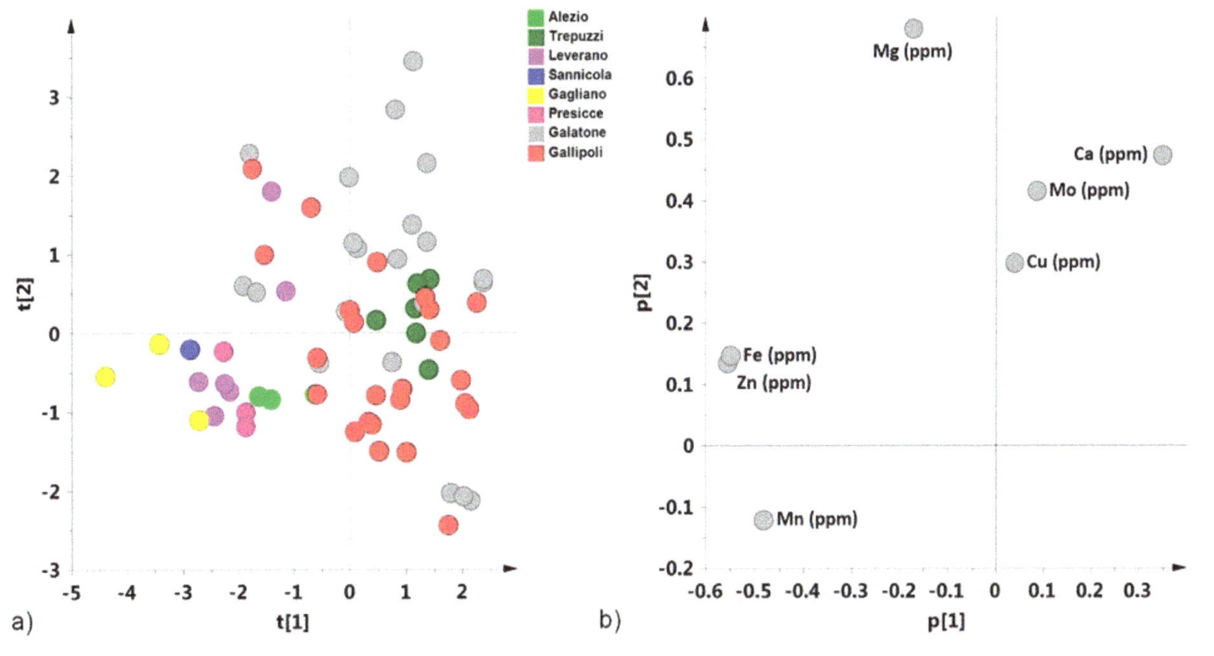

Figure 1. Principal component analysis for soil samples collected from different sites in the Salento area (Southern Apulia). t[1]/t[2] score-plot **(a)** and corresponding p[1]/p[2] loadings plot **(b).** Two principal components (t[1]/t[2]) explained about 60% of total variance (R^2= 0.37 for t[1] and 0.18 for t[2]), with a predictability value of Q^2 = 0.097.

Table 3. Pearson correlation matrix among the variables for soil (A) and olive leaf (B) samples. *,**,*** indicate significance at $P < 0.05$, $P < 0.01$ and $P < 0.001$.

	B ppm	Mo ppm	Zn ppm	Fe ppm	Ca ppm	Mg ppm	Cu ppm
A)							
Mn ppm	0.84***						
Zn ppm	0.81***	0.70***					
Fe ppm	0.74***	0.53***	0.81***				
Ca ppm	-0.43***	-0.33**	-0.34**	-0.51***			
Mg ppm	0.15	0.11	0.24*	0.17	0.17		
Cu ppm	-0.21	-0.14	0.02	-0.05	0.07	-0.05	
Mo ppm	-0.15	-0.19	-0.05	0.05	0.15	0.06	0.05
B)							
Mo ppm	0.28*						
Zn ppm	0.18	-0.01					
Fe ppm	-0.13	-0.05	0.20				
Ca ppm	0.15	0.22	0.22	0.50***			
Mg ppm	0.19	0.27*	0.27*	0.45***	0.80***		
Cu ppm	0.17	-0.09	0.29*	0.15	0.40***	0.42***	
Mn ppm	0.19	0.01	-0.03	0.31**	0.32**	0.55	0.34**

Figure 2b) responsible for the class distribution along the first and the second principal components explained the arrangement of the samples in the PCA score plot of Figure 2a. From the loadings plot, samples from the Gallipoli area had low amounts levels of all the measured elements, while samples from Alezio and Trepuzzi districts clustered together, with a greater contents of Mo and B. These two elements (Mo and B) also characterized samples from Sannicola, which, were mostly dispersed in the upper part of the graph. Samples collected in Leverano district were clustered in the right part of the PCA score-plot, with high contents of elements, in particular for Ca, Mg, Mn and Cu content. Pearson correlation matrix among the variables was used to evaluate the inter-element relationships in leaf samples (Table 3). Most of the elements in leaf samples were positively correlated with each other. In particular, Fe showed high positive correlation with Ca, Mg and Mn; Ca was characterized by highly significant correlation ($P < 0.001$) with Mg, Cu and Mn; Mg with Cu and Cu with Mn ($P < 0.01$), and B and Mo also showed correlation ($P < 0.05$).

DISCUSSION

These results indicate that in the Gallipoli area, where the OQDS was first noticed, as well as in districts similarly characterized by extensive damage to olive trees and the relevant occurrence of *X. f.* subsp. *pauca* (Boscia *et al.*, 2017), there is a general lack of some

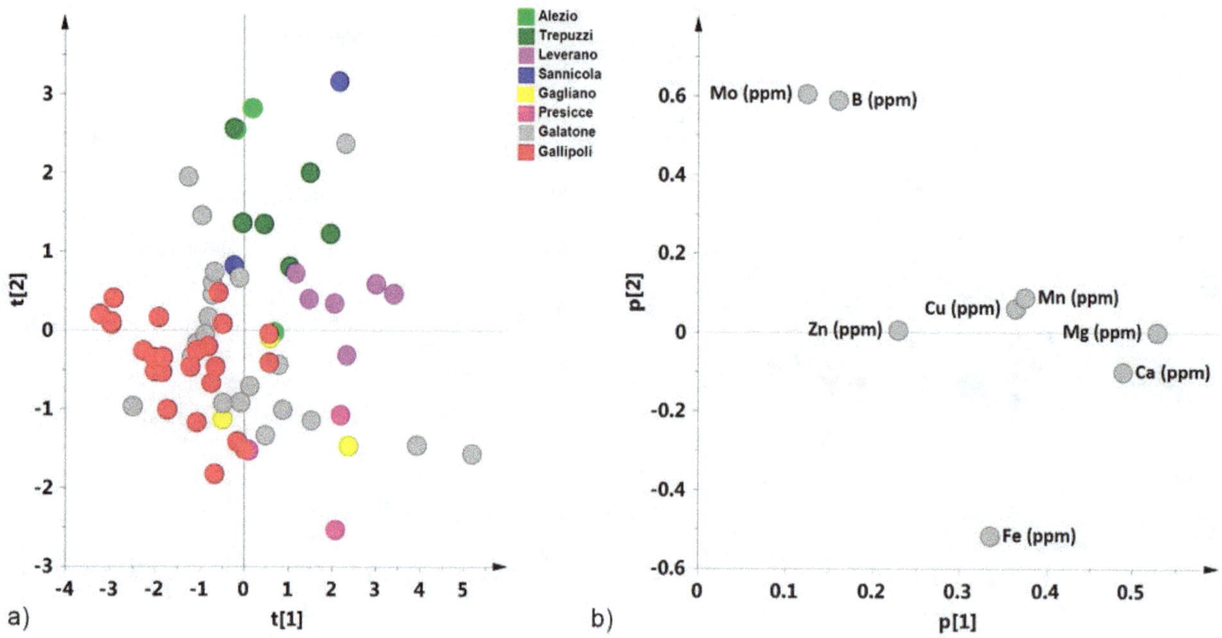

Figure 2. Principal component analysis for leaf samples collected from different sites in the Salento area (Southern Apulia). t[1]/t[2] score-plot **(a)** and corresponding p[1]/p[2] loadings plot **(b).** Two principal components (t[1]/t[2]) explained about 55% of total variance (R^2= 0.39 for t[1] and 0.21 for t[2]), with a predictability value of Q^2 = 0.105.

micronutrients (i.e., manganese and molybdenum) in the soils, and low contents of copper, molybdenum and boron in the leaves of olive trees. In the farms of Gallipoli, Galatone and Trepuzzi, manganese contents in the soils were low, although manganese in the leaves was within the range of normal content. This has been previously found in other studies, and emphasizes that low availability of manganese in soils characterized by a high calcium contents did not reduce amounts of this element in olive leaves (Chatzistathis and Therios, 2009).

No previous data can explain the low molybdenum content in the soil, its low content in many olive trees from the areas studied here where OQDS occurs. However, molybdenum availability should be favoured by pH greater than 5.5 (Kaiser *et al.*, 2005; Bittner, 2014). This micronutrient is an essential part of nitrogenase, the enzyme catalizying nitrogen fixation of atmospheric nitrogen into ammonia (Hoffman *et al.*, 2014). Low boron content within olive leaves is common for olive trees (Rodrigues *et al.*, 2012), and its low bioavailability could be related to the calcareous soil matrix (Goldberg, 1997).

PCA confirmed that soil and leaf samples collected from the Gallipoli district had a lower amounts of all the measured elements compared with the other areas samples. In addition, most of the micronutrients in leaf samples showed positive correlations with each other. The significant relationships between these variables indicates similar sources of input (Islam *et al.*, 2015; Moreira *et al.*, 2016) for these micronutrients. However, detailed investigations in other areas where the OQDS is recorded are necessary to establish robust correlation between individual nutrients. For two farms of Galatone we found soil pHs of approx 6.0, values less than expected for calcareous soils. However, this area is characterized by a high soil salinity (Ancona *et al.*, 2010), and that high concentrations of salts in calcareous soils can considerably reduce the pH (Lai and Stewart, 1990; Al-Busaidi and Cookson, 2003).

In agreement with a thorough survey performed on soils throughout Europe, including the Apulia region, and using the same technique herein applied (Ballabio *et al.*, 2018), the present study also found adequate amounts of copper in the soils of the districts characterized by the occurrence of *X. f.* subsp. *pauca*. In contrast, the copper contents were low in leaves of olive trees infected by the bacterium. Low availability of copper within the olive leaves is unusual and could be related to the infection. The threshold that indicate low copper content in olive leaves varies according to different authors: Tittarelli *et al.* (2002), for Italian olive orchards, indicated 20 to 36 ppm, whereas De Andrés Cantero (2001) reported values ranging from 6 to 10 ppm. However, also considering the lower limit, the values observed in Salento in olive leaves infected by *X. fastidiosa*, especially in the Gallipoli district, were very often below these thresholds. In addition, leaf copper depletion is rarely found in olive and, apparently, it has never been reported in Italy (Tittarelli *et al.*, 2002; Sanzani *et al.*, 2012). A study carried out in an olive orchard free from infection of *X. fastidiosa* in Rende, Calabria, Southern Italy, found normal content of copper in soil and in the leaves, as well as normal concentration of molybdenum in soil and in leaves (Buttafuoco *et al.*, 2016). The regular spraying of copper-based compounds to olive crowns to control pathogenic bacteria and fungi could explain the normal values of this micronutrient usually observed in Apulia soils (Provenzano *et al.*, 2009), thus explaining the unusual observation for leaves in the present study.

Reduced copper content within highbush blueberry (*Vaccinium corymbosum* hybrids) leaves artificially infected by *X. f.* subsp. *multiplex* was found by Oliver *et al.* (2015). Similarly, *Xanthomonas oryzae* pv. *oryzae*, the causal agent of bacterial blight of rice, removes copper from host xylem, and the decreased concentration of this element leads to increased pathogen survival and virulence (Yuan *et al.*, 2010). *Erwinia amylovora*, which causes fire blight, binds copper through exopolysaccharides, so decreasing *in planta* copper toxicity (Ordax *et al.*, 2010).

Copper concentrations in the leaf may play an important role in *X. fastidiosa* infection. This pathogen accumulates copper when forming biofilms, a critical physiological state for its *in planta* survival (Cobine *et al.*, 2013). When it is in the biofilm phase, *X. fastidiosa* has increased resistant to copper (Rodrigues *et al.*, 2008). However, when the copper concentrations are greater than 200 μM, the pathogen loses capability to form biofilms for the relevant toxicity exerted by the element (Cobine *et al.*, 2013). Our data indicate that *X. f.* subsp. *pauca* could decrease the copper contents within olive leaf xylem, thus explaining the general low content of this micronutrient found in olive trees infected by *X. f.* subsp. *pauca* in Salento. It is also known that subinhibitory doses of copper can promote the occurrence of *X. fastidiosa* persister cells (i.e., cells that neither grow nor die in the presence of bactericidal agents). These cells can repopulate the xylem after reduction of the stress condition (Muranaka *et al.*, 2012). Also zinc leaf concentrations play an important role for *X. fastidiosa* aggressiveness. Zinc detoxification is required for commencement of host tissue colonization (Navarrete and De La Fuente, 2014). Similar to copper, zinc doses greater than

0.25 mM inhibit *X. fastidiosa* biofilm formation (Cobine *et al.*, 2013).

The present investigation and previous experimental data corroborate the observed decrease in *X. f.* subsp. *pauca* cell densities and field OQDS symptoms in olive trees repeatedly sprayed with a biocomplex containing zinc (4%), copper (2%) and citric acid (Scortichini *et al.*, 2018). Treatments to olive canopies this compound, characterized by a very effective capability to reach the host xylem networks, the leaf content of zinc and copper remained at greater amount than those that *X. f.* subsp. *pauca* can detoxify to increase its aggressiveness. Such a biocomplex also shows an *in vitro* antibacterial activity againsts *X. f.* subsp. *pauca* isolated from olive trees with OQDS in Salento (S. Loreti, N. Pucci, personal communication).

Restoration of the copper and zinc contents in olive trees foliage coupled with improved soil fertility, if performed during some years, could allow co-existence with *X. f.* subsp. *pauca* in orchards where the infections has not yet greatly reduced the crown of the tree. Restoration and possible maintenance of soil fertility contributes to the uptake of micronutrients for olive trees (Chatzisthatis *et al.*, 2017). For an integrate disease management strategy aiming at reducing the spread of the bacterium, accurate removal of weeds from olive orchard during periods from the end of winter to spring to decrease the juvenile stages of the insect vector populations and the regular, light pruning of the tree crowns are very important. These strategies are about to be carried out on many olive farms of Salento.

ACKNOWLEDGEMENTS

The study was financed by Regione Apulia through the Agreement: "Strategie di controllo integrato per il contenimento di *Xylella fastidiosa* in oliveti pugliesi ed analisi epidemiologiche del complesso del disseccamento rapido dell'olivo (CoDiRO)".

LITERATURE CITED

Al-Busaidi A.S., Cookson P., 2003. Salinity-pH relationships in calcareous soils. *Agricultural and Marine Sciences* 8: 41–46.

Alloway B.J., 1995. *Heavy metals in soil.* 2nd ed. Chapman and Hall, London, UK, 368 pp.

Almeida R.P.P., Blua M.J., Lopes J.R.S, Purcell A.H., 2005. Vector transmission of *Xylella fastidiosa*: applying fundamental knowledge to generate disease management strategies. *Annals of the Entomological Society of America* 98: 775–786.

Ancona V., Bruno D.E., Lopez N., Pappagallo G., Uricchio V.F., 2010. A modified soil quality index to assess the influence of soil degradation processes on desertification risk: the Apulia case. *Italian Journal of Agronomy* 3 Suppl.: 45–55.

Ballabio C., Panagos P., Lugato E., Huang J.-H., Orgiazzi A., ... Montanarella L., 2018. Copper distribution in European topsoils: an assessment based on LUCAS soil survey. *Science of the Total Environment* 636: 282–298.

Bittner F., 2014. Molybdenum metabolism in plants and crosstalk to iron. *Frontiers in Plant Science* 5: 28.

Boscia D., Altamura G., Saponari M., Tavano D., Zicca S., ... Tanielli M., 2017. Incidenza di *Xylella* in oliveti con disseccamento rapido. *L'Informatore Agrario* 27: 47–50.

Bro R., Smilde A.K., 2014. Principal component analysis. *Analytical Methods* 6: 2812–2831.

Buttafuoco G., Guagliardi I., Bastone L., Cipriani M.G., Civitelli D., ... Ricca N., 2016. Mapping soil and leaf micronutrients distribution in an olive (*Olea europea* L.) orchard. *Annals of Agriculture & Crop Sciences* 1: 1005.

Cariddi C., Saponari M., Boscia D., De Stradis A., Loconsole G., ... Martelli G.P., 2014. Isolation of *Xylella fastidiosa* strain infecting olive and oleander in Apulia, Italy. *Journal of Plant Pathology* 96: 425–429.

Chatzisthatis T., Therios I., 2009. Differential distribution within tissues, and use efficiency of manganese, iron, and zinc by olive cultivars Kothreiki and Koroneiki. *HortScience* 44: 1994–1999.

Chatzisthatis T., Papaioannou A., Gasparatos D., Molassiotis A., 2017. From which soil metal fractions Fe, Mn, Zn and Cu are taken up by olive trees (*Olea europea* L. cv. 'Chondrolia Chalkidikis') in organic groves ? *Journal of Environmental Management* 203: 489–499.

Cobine P.A., Cruz L.F., Navarrete F., Duncan D., Tygart M., De La Fuente L., 2013. *Xylella fastidiosa* differentially accumulates mineral elements in biofilm and planktonic cells. *PLoS ONE* 8: e54936.

Costantini E.A.C., L'Abate G., Barbetti R., Fantappiè M., Lorenzetti R., Magini S., 2012. *Carta dei suoli d'Italia, scala 1:1.000.000.* Ministero delle Politiche Agricole, Alimentari e Forestali-Consiglio per la ricerca in agricoltura e l'analisi dell'economia agraria (CREA).

De Andrés Cantero F., 2001. Enfermedades y plagas de l'olivo. 4th ed.. Riquelme y Vargas Ediciones, Jaen, Spain, 646 pp.

De La Fuente L., Parker J.K., Oliver J.E., Granger S., Brannan P.M., ... Cobine P.A., 2013. The bacterial pathogen *Xylella fastidiosa* affects the leaf ionome of plant hosts during infection. *PLoS ONE* 8: e62945.

EFSA (European Food Safety Authority) PLH Panel, Jeger M., Caffier D., Candresse T., Chatzivassiliou E., Dehnen-Schmutz K., ... Bragard C., 2018. Scientific opinion on updated pest categorization of *Xylella fastidiosa*. *EFSA Journal* 16: 5357, 61 pp.

Fernandez J.E., Moreno F., 1999. Water use by the olive tree. *Journal of Crop Protection* 2: 101–162.

Giampetruzzi A., Saponari M., Loconsole G., Boscia D., Savino V.N., ... Saldarelli P., 2017. Genome-wide analysis provides evidence on the genetic relatedness of the emergent *Xylella fastidiosa* genotype in Italy to isolates from Central America. *Phytopathology* 107: 816–827.

Goldberg S., 1997. Reactions of boron with soils. *Plant and Soil* 193: 35–48.

Harper S.J., Wardand L., Clover G.R.G., 2010. Development of LAMP and qPCR methods for the rapid detection of *Xylella fastidiosa* for quarantine and field application. *Phytopathology* 100: 1282–1288.

Hodgson J.F., 1963. Chemistry of the micronutrient elements in soil. *Advances in Agronomy* 15: 119–159.

Hoffman B.M., Lukoyanov D., Yang Z.Y., Dean D.R., Seefeldt L.C., 2014. Mechanism of nitrogen fixation by nitrogenase: the next stage. *Chemical Reviews* 114: 4041–4062.

Kaiser B.N., Gridley K.L., Brady J.N., Phillips T., Tyerman S.D., 2005. The role of molybdenum in agricultural plant production. *Annals of Botany* 96: 745–754.

Islam M.S., Ahmed M.K., Habibullah-Al-Mamun M., Hoque M.F., 2015. Preliminary assessment of heavy metal contamination in surface sediments from a river in Bangladesh. *Environmental Earth Sciences*, 73: 1837–1848.

Lai R., Stewart B.A., 1990. Salt-affected soils. In: *Soil degradation* (Lai R., B.A. Stewart, ed.), Springer-Verlag, New York, U.S.A., 224–247.

Largaiolli T., Martinis B., Mozzi G., Nardin M., Rossi D., Ungaro S., 1969. Note illustrative della carta geologica d'Italia, Foglio 214. Ministero dell'Industria, del Commercio e dell'Artigianato. Libreria dello Stato, Roma, Italy, 64 pp.

Loconsole G., Potere O., Boscia D., Altamura G., Djelouah K., ... Martelli G.P., 2014. Detection of *Xylella fastidiosa* in olive trees by molecular and serological methods. *Journal of Plant Pathology* 96: 7-14.

Mahler R.L., 2000. Molybdenum in Idaho. University of Idaho, College of Agriculture, Cooperative Extension Service, Agricultural Experimental Station, Moscow, CIS 1087.

Marcelletti S., Scortichini M., 2016. *Xylella fastidiosa* CoDiRO strain associated with the olive quick decline syndrome in southern Italy belongs to a clonal complex of the subspecies *pauca* that evolved in Central America. *Microbiology* 162: 2087–2098.

Martelli G.P., Boscia D., Porcelli F., Saponari M., 2016. The olive quick decline syndrome in south-east Italy: a threatening phytosanitary emergence. *European Journal of Plant Pathology* 144: 235–243.

McLennan S.M., Taylor S.R., 1999. Earth's continental crust. In: *Encyclopedia of Biochemistry* (Marshall C.P., R.W. Fairbridge, ed.), Kluwer Academic Publishers, Dordrecht, The Netherlands, 282–292.

Modesti V., Pucci N., Lucchesi S., Campus L., Loreti S., 2017. Experience of the Latium region (central Italy) as a pest-free area for monitoring of *Xylella fastidiosa*: distinctive features of molecular diagnostic methods. *European Journal of Plant Pathology* 148: 557–566.

Moreira-Ascarrunz S.D., Larsson H., Prieto-Linde M.L., Johansson E., 2016. Mineral nutritional yield and nutrient density of locally adapted wheat genotypes under organic production. *Foods* 5: 89.

Muranaka L.S., Takita M.A., Olivato J.C., Kishi L.T., De Souza A.A., 2012. Global expression profile of biofilm resistance to antimicrobial compounds in the plant-pathogenic bacterium *Xylella fastidiosa* reveals evidence of persister cells. *Journal of Bacteriology* 194: 4561–4569.

Navarrete F., De La Fuente L., 2014. Response of *Xylella fastidiosa* to zinc: decreased culturability, increased exopolysaccharide production, and formation of resilient biofilm under flow conditions. *Applied and Environmental Microbiology* 80: 1097–1107.

Navarrete F., De La Fuente L., 2015. Zinc detoxification is required for full virulence and modification of the host leaf ionome by *Xylella fastidiosa*. *Molecular Plant-Microbe Interactions* 28: 497–507.

Nigro F., Boscia D., Antelmi I., Ippolito A., 2013. Fungal species associated with a severe decline of olive in southern Italy. *Journal of Plant Pathology* 95: 668.

Noulas C., Tziouvalekas M., Karyotis T., 2018. Zinc in soils, water and food crops. *Journal of Trace Elements in Medicine and Biology* 49: 252–260.

Oliver J.E., Sefick S.A., Parker J.K., Arnold T., Cobine P.A., De La Fuente L., 2014. Ionome changes in *Xylella fastidiosa*-infected *Nicotiana tabacum* correlate with virulence and discriminate between subspecies of bacterial isolates. *Molecular Plant-Microbe Interactions* 27: 1048–1058.

Oliver J.E., Cobine P.A., De La Fuente L., 2015. *Xylella fastidiosa* isolates from both subsp. *multiplex* and *fastidiosa* cause diseases on southern highbush blueberry (*Vaccinium* sp.) under greenhouse conditions. *Phytopathology* 105: 855–862.

Ordax M., Marco-Noales E., Lopez M.M., Biosca E.G., 2010. Exopolysaccharides favor the survival of *Erwinia amylovora* under copper stress through different strategies. *Research in Microbiology* 161: 549–555.

Papadia P., Del Coco L., Muzzalupo I., Rizzi M., E. Perri E., … Fanizzi F.P., 2011. Multivariate analysis of ^{1}H-NMR spectra of genetically characterized extra virgin olive oils and growth soil correlations. *Journal of American Oil Chemists' Society* 88: 1463–1475.

Pendias H., 2010. *Trace elements in soil and plants*. 4th ed. CRC Press, Boca Raton, FL, USA, 548 pp.

Provenzano M.R., El Bilali H., Simeone V., D. Mondelli D., Baser N., 2009. Monitoring of soil copper concentration in different organic farms over a three-year period in Apulia, south-eastern of Italy. *Italian Journal of Agronomy* 1: 41–51.

Reimann C., Demetriades A., Birke M., Filzmoser P., O'Connor P., … Ladenberger A., 2014. The GEMAS project team, distribution of elements/parameters in agricultural and grazing land soils of Europe. In: *Chemistry of Europe's agricultural soils-Part A: methodology and interpretation of the GEMAS data set*. (Reimann C., M. Birke, A. Demetriades, P. Filzmoser, P. O'Connor ed.). Geologisches Jahrbuch (Reihe B 102), Schweizerbarth: 101–472.

R Development Core Team R, 2013. A language and environment for statistical computing. R Foundation for Statistical Computing, Wien, Austria.

Rodrigues C.M., Takita M.A., Coletta-Filho H.D., Olivato J.C., R. Caserta R., … De Souza A.A., 2008. Copper resistance of biofilm cells of the plant pathogen *Xylella fastidiosa*. *Applied Microbiology and Biotechnology*77: 1145–1157.

Rodrigues M.A., Ferreira I.Q., Claro A.M., Arrobas M., 2012). Fertilizer recommendations for olive based upon nutrients removed in crop and pruning. *Scientia Horticulturae* 142: 205–211.

Sanzani S.M., Schena L., Nigro F., Sergeeva V., Ippolito A., Salerno M.G., 2012. Abiotic diseases of olive. *Journal of Plant Pathology* 94: 469–491.

Saponari M., Boscia D., Nigro F., Martelli G.P., 2013. Identification of DNA sequences related to *Xylella fastidiosa* in oleander, almond and olive trees exhibiting leaf scorch symptoms in Apulia (southern Italy). *Journal of Plant Pathology* 95: 668.

Scortichini M., Chen J.,De Caroli M., Dalessandro G., Pucci N., … Loreti S., 2018. A zinc-copper-citric acid biocomplex shows promise for control of *Xylella fastidiosa* subsp. *pauca* in olive trees in Apulia region (southern Italy). *Phytopathologia Mediterranea* 57 48–72.

Strona G., Carstens C.J., Beck P., 2017. Network analysis reveal why *Xylella fastidiosa* will persist in Europe. *Scientific Reports* 7: 71.

Tittarelli F., Neri U., Poletti P., Lacertosa G., Raus R., 2002. Monitoraggio dello stato nutrizionale dell'olivo. *L'Informatore Agrario* 44: 39–51.

Triba M.N., Le Moyec L. Amathieu R., Goossens C., Bouchemal N., … Savarin P., 2015. PLS/OPLS models in metabolomics: the impact of permutation of dataset rows on the K-fold cross-validation quality parameters. *Molecular BioSystems* 11: 13–19.

Trygg J., Wold S., 2002. Orthogonal projections to latent structures (O-PLS). *Journal of Chemometrics* 16: 119–128.

Van den Berg R. A., Hoefsloot H.C., Westerhuis J.A., Smilde A.K., Van der Werf M.J., 2006. Centering, scaling, and transformations: improving the biological information content of metabolomics data. *BMC Genomics* 7: 142.

Wheelock Å, Wheelock C.E., 2013. Trials and tribulations of 'omics data analysis: assessing quality of SIMCA-based multivariate models using examples from pulmonary medicine. *Molecular Biosystems* 9: 2589–2596.

White S.M., Bullock J.M., Hooftman D.A.P., Chapman D.S., 2017. Modelling the spread and control of *Xylella fastidiosa* in the early stages of invasion in Apulia, Italy. *Biological Invasions* 19: 1825–1837.

Xia J., Psychogios N., Young N., Wishart D.S., 2009. MetaboAnalyst: a web server for metabolomic data analysis and interpretation. *Nucleic Acids Research* 37: W652-W660.

Xia J. and D.S. Wishart, 2016. Using MetaboAnalyst 3.0 for comprehensive metabolomics data analysis. *Current Protocols in Bioinformatics* 55: 14.10. 1-14.10. 91.

Yuan M., Chu Z., Li X., Xu C., Wang S., 2010. The bacterial pathogen *Xanthomonas oryzae* overcomes rice defenses by regulating host copper redistribution. *The Plant Cell* 22: 3164–3176.

12

Fungicide suspensions combined with hot-water treatments affect endogenous *Neofusicoccum parvum* infections and endophytic fungi in dormant grapevine canes

Vedat Görür, Davut Soner Akgül*

Department of Plant Protection, Agriculture Faculty, University of Cukurova, 01330, Balcali, Adana / Turkey

*Corresponding author: sakgul@cu.edu.tr

Competing Interests: The Author(s) declare(s) no conflict of interest.

Editor: Lizel Mostert, Faculty of AgriSciences, Stellenbosch, South Africa.

Summary. *Neofusicoccum parvum* is an important opportunistic fungus causing Botryosphaeria dieback on grapevines. Because of its opportunistic nature, this pathogen spreads to many grape growing areas in a latent phase, and causes serious economic losses. The aims of the study were to examine hot-water treatments combined with fungicides in order to cure latent infections of artificially inoculated *N. parvum* in dormant grapevine canes and to assess the effects of these treatments on endophytic fungi. Artificially *N. parvum*-inoculated canes were dipped into cyprodinil + fludioxonil, tebuconazole or thiophanate-methyl suspensions under the following temperature-duration combinations; 30°C for 12 h, 35°C for 6 h, 40°C for 2 h or 50°C for 30 min (temperature at 50°C was not combined with fungicides). Treated canes were cooled in tap water (18°C) for 1 h, and pathogen re-isolations were immediately attempted from inner wood tissues. Extent to which these applications affected the presence of endophytic fungi were also determined by calculating pre- and post-treatment isolation rates. Hot-water treatments (without fungicides) below 52°C did not reduce *N. parvum* re-isolation rates, and were ineffective. However, these treatments combined with fungicides decreased pathogen incidence at 50°C and below. Maximum reduction (34%) was obtained with heated tebuconazole suspensions at 40°C for 2 h and 50°C for 30 min, and eradicative ability was superior to that of cyprodinil + fludioxonil or thiophanate-methyl. The hot-water treatments reduced incidence of endophytic fungi but greater reduction was observed with the hot-water and fungicide combinations. Fungicide penetration into wood tissues of propagation material could be enhanced by increasing water temperatures in hydration or hot-water treatment tanks, and this approach could be useful method for production of healthy grapevine plants in nurseries.

Keywords. *Vitis vinifera*, tebuconazole, cyprodinil, fludioxonil, thiophanate-methyl.

INTRODUCTION

Neofusicoccum parvum (Pennycook & Samuels) Crous, Slippers & Phillips is one of the most important pathogens causing Botryosphaeria dieback of grapevines. This fungus causes a variety of symptoms in grapevines, including brown wedge-shaped wood cankers, xylem necroses, failure of spring growth and black dead arm (Úrbez-Torres, 2011). Although *N. parvum* pycniospores are the main inoculum sources for short distance dissemination, the pathogen is more likely to spread to different geographical areas through infected propagation materials. When the pathogen infects vines through wounds, natural openings or direct penetration of bark, it can progress systemically in cane vascular and wood tissues, and can cause decline of vines (Amponsah *et al.*, 2012). Isolation of botryosphaeriaceous fungi from asymptomatic canes suggest the importance of latent infections on pathogen dissemination in vineyard areas (Slippers and Wingfield, 2007).

Hot-water treatment (HWT) is a promising and plausible method to reduce endogenous infections of all grapevine trunk disease (GTD) and other pathogens, for production of healthy propagation material (Fourie and Halleen 2004; Halleen *et al.*, 2007). Though some studies have reported that standard HWT (at 50°C for 30 min) can eliminate grapevine pests, some bacterial pathogens (such as *Xylella fastidiosa* and *Rhizobium vitis*) and phytoplasmas (Flavescense dorée) are not completely eradicated (Ophel *et al.*, 1990; Caudwell *et al.*, 1997; Crous *et al.*, 2001). There is also evidence that GTD fungi are not sufficiently eliminated from dormant grapevine canes by these treatments (Rooney and Gubler, 2001; Waite and May, 2005; Gramaje *et al.*, 2010). *Neofusicoccum parvum* is among the GTD fungi which were subjected to HWT studies both in laboratory and controlled conditions. *Diplodia seriata*, *Neofusicoccum luteum*, *N. parvum* and *Spencermartinsia viticola* were found to be the most susceptible species, while *Lasiodiplodia theobromae* and *N. vitifusiforme* were very tolerant to HWT in *in vitro* conditions. Mycelium survival of *N. parvum* was about 15% from HWT at 50°C for 30 min in test tubes containing sterile water (Elena *et al.*, 2015). In this study, when the curative ability of HWT was tested with *N. parvum* inoculated grapevine canes, HWT (at 51°C for 30 min) eradicated the pathogen from dormant canes of 110-Richter rootstock. However, Billones-Baaijens *et al.* (2015) found controversial findings from their study, showing that HWT (at 50°C for 30 min) was less effective or ineffective, respectively, for eradication of *N. luteum* and *N. parvum* from infested cuttings of 101-14 Mgt rootstock. While this treatment reduced the isolation frequency of *N. luteum* by 50%, it did not reduce isolation of *N. parvum*. The expected performance from HWT was obtained just with 53°C for 30 min, but this combination was harmful for bud vitality of the treated rootstock cuttings. Billones-Baaijens *et al.* (2015) suggested that research should be carried out to improve hot-water treatments practicable for grapevine nursery industries. In Turkey, most nurseries avoid HWT because of detrimental effects on plants, such as delayed callusing, bud death, and delayed development and rooting.

Dipping of grapevine cuttings in fungicide suspensions is another control measure to produce healthy grapevine plants in nurseries. Rego *et al.* (2009) determined the efficacy of applications of cyprodinil + fludioxonil, metiram + pyraclostrobin, cyprodinil or fludioxonil suspensions, by hydration (tap water) of the rootstock cuttings prior to grafting. They found that the cyprodinil + fludioxonil reduced the incidence of black foot (by 47%) and botryosphaeriaceous fungi (by 60%) in plants grown in a field that was naturally infested with these pathogens. Halleen and Fourie (2016) developed a combined control strategy to reduce GTD pathogen infections South Africa, showing that benomyl and Sporekill® were useful chemicals for increasing healthy plant rates (up to 100%) in the grapevine nurseries. These studies were conducted in naturally infested fields or mother plants that were known to be infected with some GTD pathogens. Billones-Baaijens *et al.* (2015) showed that tebuconazole and carbendazim soakings (at tap water temperature) were effective for curing superficial bark infections by *N. parvum* or inoculated canes.

Curative performance of fungicide treatments against latent inner tissue infections of canes is important, because outer bark is a natural barrier preventing entry of pesticides into internal host tissues. Fungicide penetration via diffusion is known to be very inefficient (Waite and May, 2005). Therefore, there is a need to increase fungicide penetration into grapevine wood tissues during the soaking stage before grafting or cold storage for production of healthy propagation material.

The combination of HWTs with fungicides may be an effective approach for curing latent infections by GTD fungi in grapevine nurseries. This has previously been studied by many researchers who work on postharvest diseases of some fruits (Wells and Harvey, 1970; Barkai-Golan and Apelbaum, 1991). These studies showed that heat treatment in combinations with fungicides could increase penetration of the pesticides into fruit tissues. Schirra *et al.* (1996) treated lemon fruits with imazalil suspensions (1500 $mg{\cdot}L^{-1}$ at 20°C or 250 $mg{\cdot}L^{-1}$ at 50°C for 3 min) to cure *Penicillium* infections. The second of these treatments was as effective as the first for disease control, and four or five-fold increases in fungicide residues were recorded in albedo tissues of fruits from the treatment of 250 $mg{\cdot}L^{-1}$ imazalil at 50°C for 3 min.

Endophytic fungi live within plants without causing disease symptoms (Freeman, 1904). These fungi can contribute to plant adaption and survival in unfavourable conditions. They can also produce secondary metabolites, trigger biochemical pathways or interact with other microorganisms (Aly *et al.*, 2010; Kusari *et al.*, 2012). Gonzalez and Tello (2011) assessed endophytic fungi from different organs of seven grape cultivars in central Spain, and obtained more than 500 isolates belonging to 68 species. While the endophytic composition differed according to vine cultivars, *Acremonium, Alternaria, Aureobasidium, Cladosporium, Epicoccum, Fusarium, Penicillium, Phoma*, and *Trichoderma* were isolated most frequently. Some of these fungi were reported to have antagonistic ability towards plant pathogenic fungi. Eichmeier *et al.* (2018) investigated effects of hot-water treatments (at 50°C or 53°C for 30 min) on endophytic fungi from dormant canes of Garnacha Tintorera/110 Richter and Sauvignon Blanc/SO4 grapevines. Although hot-water treatments decreased the incidence of endophytic fungi depending on temperature, HWTs did not sterilize dormant canes, and these fungi were recovered from hot-water treated plants grown for 8 months in nurseries. Although several studies have investigated curative effects of hot-water treatments (conducted generally at 50-54°C for 30, 45 or 60 min) on GTD pathogens (Habib *et al.*, 2009; Serra *et al.*, 2009), limited results are available where effects have been assessed for HWTs on endophytic fungi in dormant grapevine canes.

In evaluating current literature, we hypothesized that penetration of fungicides into dormant grapevine canes could be enhanced by heating fungicide suspensions (at low temperatures combined with extended durations) at the pre-grafting host stage. The aims of the study were: (i) to determine the efficacy of heated fungicide suspensions on latent infections of *N. parvum* and their effects on endophytic fungi in dormant grapevine canes; and (ii) to assess antagonistic abilities of some endophytic isolates against *N. parvum* in *in vitro*.

MATERIALS AND METHODS

Three fungicides (Table 1) with different modes of action were used to prepare hot-water treatment suspensions at 50°C and lower temperatures (30, 35 and 40°C). Dormant grapevine canes (*Vitis vinifera* 'Prima') were used as plant material in all experiments.

Isolation and molecular identification of endophytic and pathogenic fungi

The dormant canes were taken (in December 2017) from randomly selected vineyards in three different geographic regions of Turkey (Manisa, Tokat and Tarsus Cities; in, respectively, the Aegean, Central Anatolia and Mediterranean Regions). Fifty, one-year-old dormant canes (each with three to five buds and 50 cm long) were randomly taken from each vineyards (two vineyards per region, 25 canes from each vineyard). These canes were washed under running tap water for 10 min., and 10 cm-long cuttings were used for endophyte isolations. The cuttings were dipped into ethyl-alcohol (70%) for 1 min, sodium hypochlorite solution (2.5% NaOCl) for 3 min, and ethyl-alcohol (70%) (Schultz *et al.*, 1993). Canes were then rinsed twice with sterile distilled water and bark tissues were removed with a sterile knife. Wood tissues were cut into 3 mm^2 pieces, and seven each were plated onto potato dextrose agar (PDA, Becton Dickinson) amended with streptomycin sulfate (150 $mg \cdot L^{-1}$). Petri dishes were incubated for 10 d at 24°C in the dark, and growing fungal colonies were sub-cultured to fresh PDA (without antibiotic) for further microscope examinations and molecular identification. After morphological discrimination and microscope examinations, the incidence of endophytic species was calculated by counting colonies (at genus or species level) growing around tissues and average isolation rates were determined from 25 Petri dishes for each vineyard. For each geographic region, average incidence of fungi was calculated as percent.

For molecular identification of fungi, 10-d-old mycelium mats of the isolates (growth from hyphal tip or single spore cultures) were harvested (50–100 mg), and DNA was extracted according to the protocol of Nejat *et al.* (2009). PCR amplification of the ITS1, 5.8S and ITS2 regions of rDNA was performed using ITS4 and ITS5 primers (White *et al.* 1990), and thermocycler (SimpliAmp A24811™; Applied Biosystems) conditions were adjusted as follow; 95°C for 3 min (initial denatura-

Table 1. Fungicides and application rates used in this study.

Active ingredients	Fungicide group	Application dose (product in water·L^{-1})	Trade name and formulation	Manufacturer
Cyprodinil (37.5%) + fludioxonil (25%)	Anilinopyrimidine + phenyl-pyrrole	5 g	Switch® 62.5 WG	Syngenta
Tebuconazole (250g·L^{-1})	DMI-triazole	4 mL	Orius® 20 EW	Adama
Thiophanate-methyl (70%)	Benzimidazole	5 g	Sumitop 70 WP	Sumitomo

tion), followed by 35 cycles each of denaturation at 95°C for 1 min, annealing at 52°C for 1 min, and extension at 72°C for 1 min, and a final extension at 72°C for 10 min. PCR products were purified and sequenced (Sanger dideoxy sequencing) by MedSanTek Co. (Istanbul, Turkey), and nucleotide sequences (in forward and reverse directions) of the isolates were contiged using Chromas Lite free software (Technelysium™). Consensus nucleotide sequences were compared with those deposited in the NCBI GenBank database using the BLAST program, and identification of the isolates was decided according to maximum identity results (with 99% or 100% rates). Sequences were deposited (except the well-known species, *Aspergillus niger, Aureobasidium pullulans, Epicoccum nigrum, Nigrospora oryzae, Penicillum* sp. *Rhizopus stolonifer*) to the NCBI database and accession numbers were obtained.

Inoculation and incubation of Neofusicoccum parvum

An electric drill method (Pouzoulet *et al.*, 2013) was used to inoculate *N. parvum* in all HWT experiments. The *N. parvum* isolate (MBAi27AG, GenBank accession no: KF182330), formerly shown to be highly virulent, was selected from the fungal library of Cukurova University, Department of Plant Protection. This isolate was grown on PDA at 25°C for 10 d in the dark. Dormant grapevine cuttings (each measuring 45 cm) were taken from a Prima vineyard (grafted on 1103 Paulsen, 10 years-old) in Tarsus city (Southern Turkey), and were transferred to the laboratory in December 2017. To prevent dehydration, they were immersed in clean tap water at ambient temperature (20°C) overnight. The cuttings were then trimmed to 35 cm, and the superficially sterilized in 2.5% NaOCl solution for 3 min and rinsed twice with sterile distilled water. Cane internodes were laterally drilled with a 2 mm-diam. drill bit and 2 mm-mycelium plugs were inserted into the drill holes which were then sealed with Parafilm®. In control plants, sterile agar pieces were inoculated into the wounds and the holes were sealed with Parafilm®. The inoculated canes were planted into plastic bags containing growth mixture (fine sawdust, perlite, sand, soil and peat, in equal volumes), and were then watered and maintained in a growth room (28°C. 90% relative humidity, 12 h dark/12 h light) for 2 months. For further experiments, each of the cuttings was numbered to compare initial and final endophytic flora after all treatments. The endophytic fungi were isolated from these cuttings before inoculation with *N. parvum*, and their isolation frequencies were recorded for further comparisons.

Hot-water treatment effects on Neofusicoccum parvum *infections and endophytic fungi*

The curative effects of hot-water treatments were investigated with and without fungicides. All HWT experiments were conducted in an adjustable 10 L capacity water bath device (Memmert WB 10). Air was pumped through the device using an aquarium pump to provide homogenity of fungicide suspensions. In fungicide-HWT treatment combinations, the fungicide suspensions were prepared at the application rates specified in Table 1, and at temperature/time treatments of 30°C for 12 h, 35°C for 6 h, 40°C for 2 h or 50°C for 30 min. The plants that were previously inoculated with *N. parvum* and grown for 2 months were uprooted from pots, and their roots and shoots were removed before HWTs. The cuttings (30 cm) were firstly washed and immersed in clean tap water at ambient temperature for 1 h. Bundles of cuttings, each comprising 12 cuttings, were then dipped into heated fungicide suspensions at the temperatures and for the durations specified above. The treated cuttings were cooled in clean tap water (at 18°C*) for 1 h*. Pathogen isolations were then carried out from the wood tissues from around inoculation points (4 cm from both sides of the inoculation holes), to investigate *N. parvum* survival, and from 10 cm from inoculation points to assess presence of endophytic fungi. Inner wood tissues were cut into 3 mm pieces with sterile secateurs and placed onto PDA amended with streptomycin-sulfate in Petri plates. The plates were then incubated at 25°C for 7 d in the dark. Control canes (inoculated but not treated with HWT-fungicide combinations) were subjected to the same isolation process. All the wood chips from a cane were placed on one plate, so 12 plates were used for each treatment (completely randomized design, four replicates and three plates from each replicate). Fungal colonies were counted and incidence was calculated as percentages.

The inoculated cuttings were also subjected to hot-water treatments (without fungicides) at temperature-time combinations of: 30°C for 12 h, 35°C for 6 h, 40°C for 2 h, or 50°C, 51°C, 52°C, or 53°C, each for 30 min. After HWTs, the cooling, isolation and incidence calculation procedures described above were carried out. This experiment was repeated once and average incidences were calculated.

In vitro *antagonistic effects of endophytic fungi against* Neofusicoccum parvum

Frequently isolated endophytic species (one isolate each) were selected to check antagonistic ability against *N. parvum*. Mycelial agar discs from 10-day-old cultures of endophytic isolates were placed on one side of PDA plates. They were allowed to grow at 24°C for 72 h (because of slow growth), and *N. parvum* discs (from 10-d-old cultures) were placed on the other side of plates. These dual culture plates were incubated at 24°C for 10 d, with *N. parvum* growth observed daily and colony diameters (length-width) were measured to determine average colony diameters. Experimental control plate contained sterile agar discs opposite the *N. parvum* discs. Inhibition rates (%) were calculated according to Reyes-Chilpa *et al.* (1997) as cited by Felber *et al.* (2016). Interactions between endophytic isolates and *N. parvum* were grouped using the scale of Badalyan *et al.* (2002); where A = deadlock with mycelial contact, B = deadlock at a distance, and C = replacement, overgrowth with and without initial deadlock. Dual culture tests were arranged in a completely randomized design with five replications, and were repeated once.

Statistical analyses

Variance analyses were performed on data to reveal differences between means of the treatments, and the statistically similar groups were determined by Fisher's Least Significant Difference (LSD) test $P \leq 0.05$) in all experiments (Gomez and Gomez, 1984). The data of isolation percentages and inhibition rates were converted by Arc Sin transformations before variance analyses.

RESULTS

Endophytic and pathogenic fungi from apparently healthy cuttings

Twenty-three operational taxonomic units (OTUs) of fungi belonging to three phyla were obtained from grapevine cuttings from three regions in Turkey (Table 2). Wood pieces plated on PDA yielded more than one colony (average two) after 7-8 days of incubation. While most of the fungi (19 OTUs) were Ascomycota, the others were Basidiomycota (*Cerrena unicolor, Quambalaria cyanescens* and *Schizophyllum commune*) and Zygomycota (*Rhizopus stolonifer*). All of the OTUs were not regularly isolated from all the regions, but *Acremonium* sp. *Aureobasidium pullulans, Alternaria alternata, Aspergillus niger, Cladosporium cladosporioides, Penicil-*

Table 2. Endophytic and pathogenic fungi isolated from dormant grapevine canes in the study.

Order	Genus	Species	Isolate	GenBank Accession Number
Ascomycota	*Acremonium*	sp.	CUZFVG42	MK120286
Ascomycota	*Alternaria*	*alternata*	CUZFVG12	MK120281
Ascomycota	*Alternaria*	*tenuissima*	CUZFVG276	MK120296
Ascomycota	*Aspergillus*	*niger*	N/A	N/A
Ascomycota	*Aureobasidium*	*pullulans*	N/A	N/A
Basidiomycota	*Cerrena*	*unicolor*	CUZFVG176	MK120292
Ascomycota	*Chaetomium*	*globosum*	CUZFVG10	MK120280
Ascomycota	*Cladosporium*	*cladosporioides*	CUZFVG38	MK120283
Ascomycota	*Diaporthe*	*foeniculina*	CUZFVG125	MK120290
Ascomycota	*Diplodia*	*seriata*	CUZFVG4	MK120279
Ascomycota	*Epicoccum*	*nigrum*	N/A	N/A
Ascomycota	*Fusarium*	*equiseti*	CUZFVG87	MK120289
Ascomycota	*Geosmithia*	sp.	CUZFVG40	MK120285
Ascomycota	*Gnomonia*	*idaeicola*	CUZFVG36	MK120282
Ascomycota	*Lasiodiplodia*	sp.	CUZFVG250	MK120294
Ascomycota	*Neofusicoccum*	*parvum*	CUZFVG2	MK120278
Ascomycota	*Nigrospora*	sp.	N/A	N/A
Ascomycota	*Penicillium*	sp.	N/A	N/A
Ascomycota	*Phoma*	*glomerata*	CUZFVG212	MK120298
Zygomycota	*Rhizopus*	*stolonifer*	N/A	N/A
Basidiomycota	*Quambalaria*	*cyanescens*	CUZFVG39	MK120284
Basidiomycota	*Schizophyllum*	*commune*	CUZFVG173	MK120291
Ascomycota	*Trichoderma*	*atroviride*	CUZFVG243	MK120297

Table 3. Mean incidences of frequently isolated endophytic fungi in grapevine cuttings from three regions of Turkey.

Endophytic fungi	Incidence (%)		
	Manisa	Tarsus	Tokat
Acremonium sp.	2.0	3.6	17.1[a]
Aureobasidium pullulans	22.3	1.2	88.0
Alternaria spp.	14.3	16.3	12.6
Aspergillus niger	0.9	1.2	0.6
Cladosporium sp.	51.4	32.8	16.6
Penicillium sp.	0.9	5.2	0.8
Quambalaria cyanescens	0.1	9.9	0.1

[a]Means were for two vineyards for each location, and 25 grapevine cuttings from each vineyard were used for the isolations.

lium sp. and *Quambalaria cyanescens* were endophytic fungi isolated from all the regions and vineyards. The isolation percentages of endophytic fungi showed differences by geographical region. When compared to mean incidence rates, *Cladosporium* sp. was predominant in Manisa (51% recovery) and Tarsus (33%), but *A. pullulans* was predominant in Tokat (88%) and Manisa (22%). *Alternaria* sp. was the second most frequently isolated species in Manisa (14%) and Tarsus (16%). *Quambalaria cyanescens* was the third most common species in Tarsus (10%), but incidence of this species was very low (0.1%) in the other regions (Table 3). In apparently healthy canes, some pathogenic species (Botryosphaeriaceae,) were also isolated at low incidence rates, and only one isolate each of *S. commune* and *D. foeniculina* were obtained (Table 2).

Curative effects of hot-water treatments on Neofusicoccum parvum *infections*

No fungicide was added in the dipping tank for temperatures above 50°C due to risks of chemical degradation. The hot-water treatments below 50°C with extended durations (without fungicide), and 30 min durations at 50 or 51°C), had no curative effects on *N. parvum* infections. The 2 month incubation period ensured good establishment of the pathogen in plants, and the pathogen was re-isolated (100%) from all inoculated plants (hot-water treated and untreated). When water temperature was 52°C, the pathogen re-isolation frequency was 97%, or at 53°C was 74 % (Table 4).

Some of the hot-water treatments combined with fungicides significantly decreased fungus re-isolation rates. When comparing all treatments, the most effective combinations were tebuconazole suspensions at 40°C for 2 h and 50°C for 30 min, giving pathogen re-isolation of 66% from the cutting inoculation points. The curative effects of these combinations was less (34% recovery) than for HWT alone (26% recovery) at 53°C for 30 min, though these recovery rates were not significantly different. The efficacy of other combinations ranged from 3 to 34% (Table 4). Tebuconazole and thiophanate-methyl suspensions at 40°C reduced the incidence of *N. parvum* to some extent, but the efficacy of cyprodinil + fludioxonil decreased above 30°C (Figure 1). The 50°C for 30 min treatment did not reduce curative performance of tebuconazole but reduced the activity of thiophanate-methyl (Figure 1).

Table 4. Mean incidences and eradication rates (%) of *Neofusicoccum parvum* after different hot-water treatments and fungicide combinations.

Hot-water treatments (with and without fungicides)	Incidence of *N. parvum* (%) ± SEs	Eradication of *N. parvum* (%) ± SEs
40°C for 2 h, tebuconazole	65.7 ± 3.5 a[a]	34.3 ± 3.8 a
50°C for 30 min, tebuconazole	65.7 ± 3.5 a	34.3 ± 3.2 a
53°C for 30 min, water	74.3 ± 5.4 ab	25.7 ± 4.8 ab
40°C for 2 h, thiophanate-methyl	88.6 ± 2.9 bc	11.4 ± 2.5 bc
35°C for 6 h, tebuconazole	88.6 ± 5.4 bc	11.4 ± 2.9 bc
30°C for 12 h, cyprodinil + fludioxonil	88.6 ± 5.4 bc	11.4 ± 4.1 bc
35°C for 6 h, thiophanate-methyl	91.4 ± 3.5 cd	8.6 ± 3.1 cd
50°C for 30 min, thiophanate-methyl	94.3 ± 3.5 cde	5.7 ± 2.9 cde
35°C for 6 h, cyprodinil,+,fludioxonil	94.3 ± 5.7 cde	5.7 ± 4.2 cde
30°C for 12 h, tebuconazole	97.1 ± 2.9 de	2.9 ± 3.1 de
40°C for 2 h, cyprodinil + fludioxonil	97.1 ± 3.5 de	2.9 ± 3.0 de
52°C for 30 min, water	97.1 ± 2.9 de	2.9 ± 2.7 de
30°C for 12 h, thiophanate-methyl	100 ± 0.0 e	0 ± 0.0 e
51°C for 30 min, water	100 ± 0.0 e	0 ± 0.0 e
50°C for 30 min, water	100 ± 0.0 e	0 ± 0.0 e
50°C for 30 min, cyprodinil + fludioxonil	100 ± 0.0 e	0 ± 0.0 e
40°C for 2 h, water	100 ± 0.0 e	0 ± 0.0 e
35°C for 6 h, water	100 ± 0.0 e	0 ± 0.0 e
30°C for 12 h, water	100 ± 0.0 e	0 ± 0.0 e
Untreated control	100 ± 0.0 e	0 ± 0.0 e

[a] Mean values within each column are significantly different ($P \leq 0.05$; LSD tests).
LSD (incidence) = 14.1, F Value: 4.53. LSD (eradication) = 10.7, F value: 6.39.
SEs: Standard error values.
The experiment was repeated once and the results were calculated by averaging incidence and eradication rates obtained from both experiments.

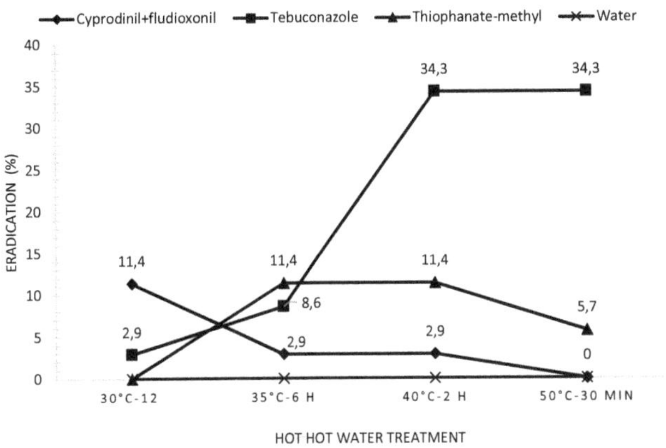

Figure 1. Mean eradication proportions (%) for *Neofusicoccum parvum* from grapevine cuttings by different fungicide suspensions and different hot-water treatments.

Effects of hot-water treatments on endophytic fungi

In the untreated control, most of the species had increased incidences during plant growth (Figure 2 a). However, with increased temperatures in the HWTs, there was progressive reductions in re-isolation rates of endophytic fungi in dormant canes (Figure 3 b-g). The 53°C for 30 min HWT almost eradicated all entophytic fungi except *Acremonium* sp. and *Alternaria alternata* (Figure 2h). *Acremonium* sp. was very tolerant to hot-water treatments, with increased re-isolation rates at from 30 to 52°C. In general, the HWTs decreased the other fungi. *Cladosporium* sp. also showed tolerance to HWTs. Compared with the initial isolation rates, final incidence of these fungi decreased for most of the treatments, but these fungi could not be fully eradicated, except at 53°C for 30 min. *Quambalaria cyanescens* was the most susceptible species, because this species was completely eradicated from dormant cuttings by temperatures of 50 to 53°C.

On the other hand, HWTs at 50°C and below, combined with fungicides gave greater eradication when compared to HWTs alone. Following these combined treatments, re-isolation rates of endophytic fungi (except *Acremonium* sp.) decreased, for almost in all of the combinations tested. Among the fungicides, tebuconazole was the most eradicative, at temperatures of 30°C to 50°C. At 30°C for 12 h dipping, most of the endophytic fungi were not be isolated after 10 d on PDA. All the tebuconazole suspension treatments completely suppressed endophytic fungi from wood pieces on isolation plates, however *Acremonium* sp. was observed at with low rates after 10 d (3 to 14%; Figure 3 b, e, h, k). For thiophanate-methyl, only *Penicillium* was eradicated at 40°C for 2h. Although incidences of *Alternaria* and

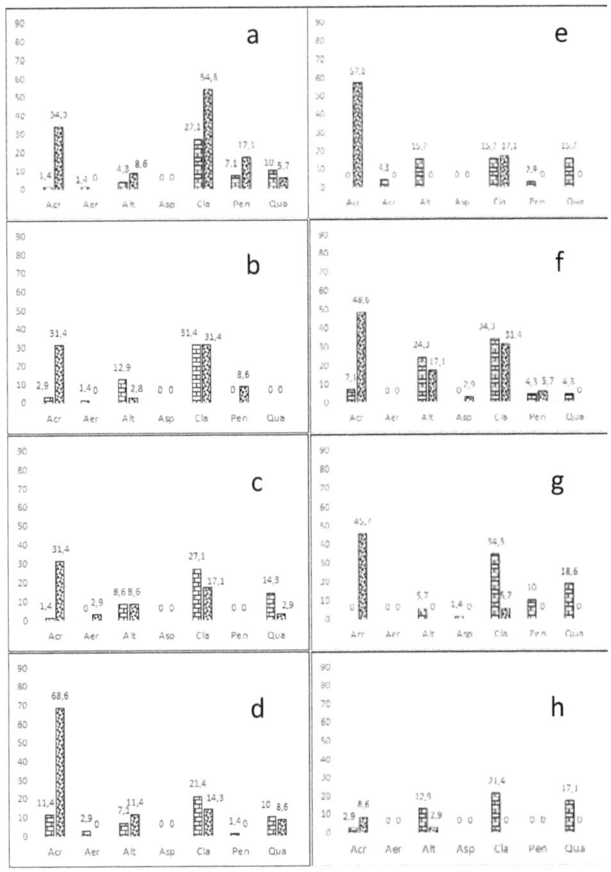

Figure 2. Mean incidence (%) of endophytic fungi in dormant grapevine cuttings after different hot-water treatments (without fungicides). The first and second columns in each histogram indicate, respectively, the re-isolation rates of endophytic fungi before and after the treatments. a) un-treated control, b) 30°C for 12 h, c) 35°C for 6 h, d) 40°C for 2 h, e) 50°C for 30 min, f) 51°C for 30 min, g) 52°C for 30 min, and h) 53°C for30 min.

Cladosporium were substantially decreased, they were not be eradicated with this treatment (Figure 3 i). The other combinations of thiophanate-methyl (30°C for 12h or 35°C for 6h) did not eradicate the other endophytic fungi (Figure 3 c, f). Cyprodinil + fludioxonil markedly suppressed occurrence of *Cladosporium* but did not affect *Acremonium* and *Alternaria* development (Figure 3 a, d, g, j).

Antagonistic effects of endophytes against Neofusicoccum parvum

The endophytic isolates gave inhibitive effects in the PDA dual culture experiment, and the inhibition interactions were of different types. *Trichoderma atroviride* gave the greatest inhibition of *N. parvum.*

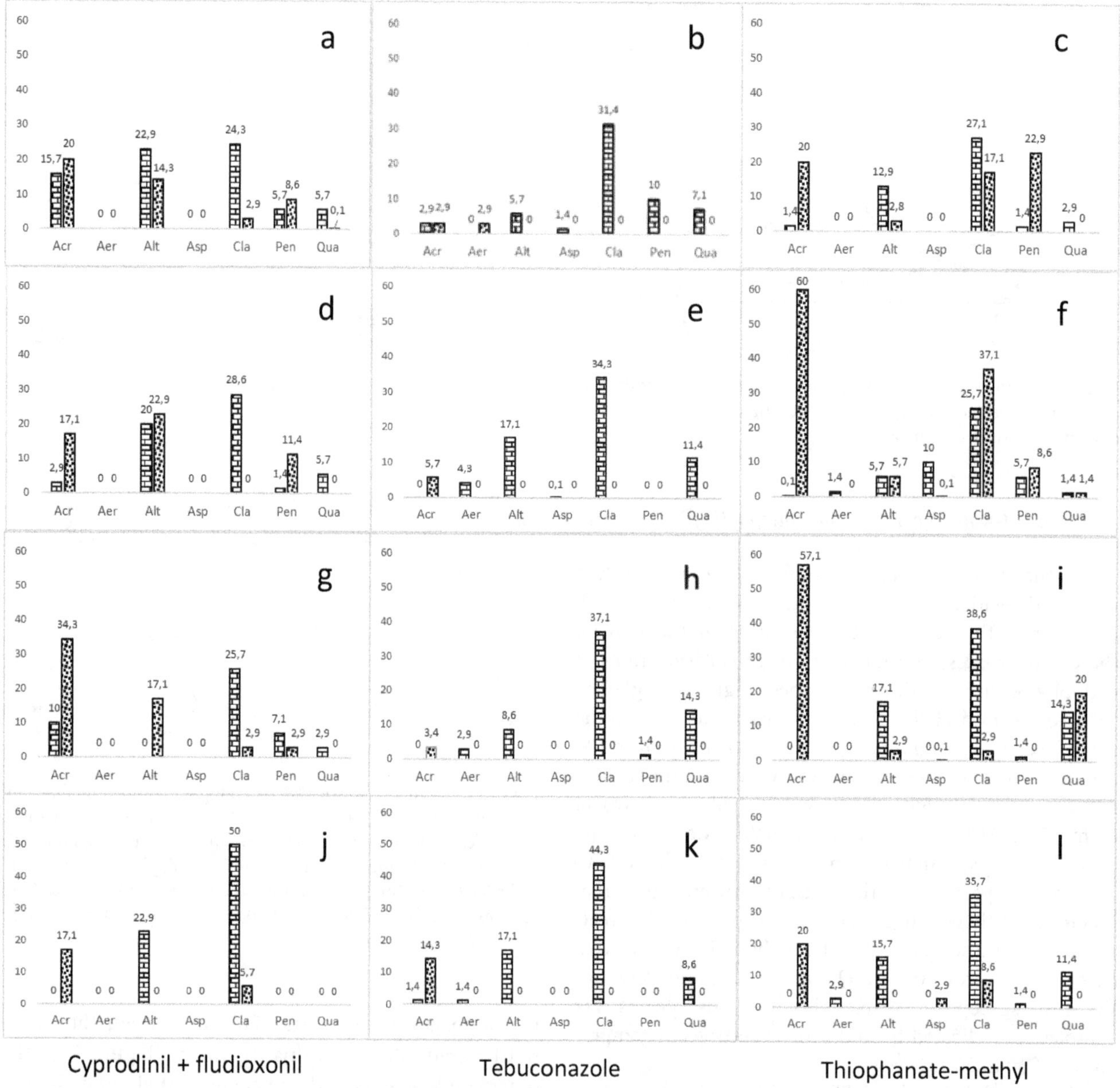

Figure 3. Mean incidence (%) of endophytic fungi in dormant grapevine cuttings after different hot-water treatments (combined with fungicides). The first and second columns in each histogram indicate, respectively, the re-isolation rates of endophytic species before or after the treatments were applied. a) b) c) 30°C for 12 h, d) e) f) 35°C for 6 h, g) h) i) 40°C for 2 h, j) k) l) 50°C for 30 min.

While *T. atroviride* (CUZFVG243) inhibited *N. parvum* growth by 69% (interaction class C), the inhibition percentages of *Acremonium* sp. (CUZFVG42), *E. nigrum* (CUZFVG88), *Geosmithia* sp. (CUZFVG40), and *Q. cyanescens* (CUZFVG39) were from 9 to 23% (class A). *Alternaria alternata* and *C. cladosporioides* moderately inhibited the pathogen, by, respectively, 31% and 30% (class B) (Table 5).

DISCUSSION

In this study, a variety of endophytic and pathogenic fungi were isolated from dormant and apparently healthy grapevine cuttings, taken from different geographical regions in Turkey. *Acremonium* sp., *Alternaria* spp., *Aureobasidium pullulans*, *Cladosporium* spp., and *Quambalaria cyanescens* were the most frequently isolated endophytic fungi, and *Diplodia seriata* was the

Table 5. Mean inhibition percentages, and competitive interactions with *N. parvum* in dual culture tests, for different isolates of grapevine endophytic fungi.

Endophytic isolates	Inhibition (%)	Interaction Class
Alternaria alternata (CUZFVG12)	31.1 ± 0.9 d [a]	B [b]
Acremonium sp. (CUZFVG42)	9.6 ± 1.0 a	A
Cladosporium cladosporioides (CUZFVG38)	29.5 ± 0.3 d	B
Epicoccum nigrum (CUZFVG88)	22.6 ± 1.1 c	A
Geosmithia sp. (CUZFVG40)	9.0 ± 1.4 a	A
Quambalaria cyanescens (CUZFVG39)	14.5 ± 1.0 b	A
Trichoderma atroviride (CUZFVG243)	69.0 ± 1.0 e	C

[a] Means values within the column are significantly different ($P \leq 0.05$), based on LSD tests.
LSD = 2.37, F value: 277.6, SEs: Standard error values
[b] Classification using the Badalyan scale (Badalyan *et al.*, 2002). A: deadlock with mycelial contact, B: deadlock at a distance, C: replacement, overgrowth with and without initial deadlock.

predominant pathogenic species from all of the regions. Halleen *et al.* (2003), screened fungi associated with healthy grapevine plants from nurseries in South Africa, and they obtained endophytic and pathogenic fungi from roots, and rootstock, grafting union and scion tissues. *Acremonium* spp. *Phoma* spp. *Alternaria* spp. *Aspergillus* spp. *Clonostachys* spp., and *Cladosporium* sp. were considered to be endophytic, but *Phaeomoniella chlamydospora, Phaeoacremonium* spp. *Botryosphaeria* spp. *Cylindrocarpon* spp. and *Phomopsis viticola* were found to be pathogenic. Pancher *et al.* (2012) investigated endophytic fungal communities from organic and Integrated Pest Management (IPM) vineyards at seven locations in Italy. Dormant canes of 'Cabernet Sauvignon' and 'Merlot' cultivars were examined for fungi using classical and molecular (ITS sequencing). Their results showed that organic vineyards had richer endophyte communities than IPM vineyards, and that *Alternaria* sp., *Epicoccum nigrum, Aureobasidium pullulans* and *Cladosporium* sp. were the most frequently isolated fungi from both types of vineyards. Kraus *et al.* (2019) investigated early development of endophytic fungi in healthy grapevines (from 2 months to 8 years old), and determined genera and species of fungi. Fast-growing fungi, such as *Alternaria* spp., *Aureobasidium pullulans, Cladosporium* spp. and *Epicoccum nigrum*, were isolated from vines that were less than 1-year-old, while many grapevine trunk diseases associated pathogens such as *D. seriata* or *Eutypa lata*, along with endophytic fungi, were obtained from perennial branches. *Alternaria* spp., *Aureobasidium pullulans, Cladosporium* spp. and *Epicoccum nigrum* made up 81% of the total fungal flora. Casieri *et al.* (2009) assessed vine wood samples taken from different regions and cultivars in Switzerland for endophytic and pathogenic fungi. They found that the composition of endophytes was different for each grape cultivar, and suggested that biochemical composition of the grape cultivars may have played an important role in fungal endophyte diversity. Endophytic and pathogenic fungi may vary according to plant age, geography, cultivar, type and age of tissues (Sieber *et al.*, 1991; Rodrigues, 1994). The results of these studies corroborate our isolation results, from 1-year-old dormant grapevine canes taken from three regions in Turkey.

The hot water treatments (below 50°C with extended duration) were not effective for eradication of *N. parvum* from inoculated dormant grapevine cuttings, but these treatments reduced re-isolation rates of the pathogen, when they were combined with fungicides. This indicates that increased water temperatures probably enhanced penetration of the fungicides into the inner wood tissues. We have been unable to find similar reported results for grapevine, but there are supportive studies conducted on postharvest disease management for fruits. Positive synergistic effects of hot-water and fungicide combinations have been reported Barkai-Golan and Appelbaum (1991), McGuire and Campbell (1993) and Smilanick *et al.* (1995). Cabras *et al.* (1999) have also demonstrated that the cuticle barrier of fruits was weakened with hot water, thus increasing the diffusion of fungicides into fruit rinds. Schirra *et al.* (1996) investigated curative effects of imazalil against blue mould of lemon (caused by *Penicillium italicum*) at the rates from 250 mg·L^{-1} to 1500 mg·L^{-1}, and with water temperatures of 20°C to 50°C for 3 min. Imazalil at 250 mg·L^{-1} and 50°C gave the same curative effect as 1500 mg·L^{-1} at 20°C, Imazalil residue in fruit rinds was 4-5 time greater for the low rate high temperature treatment than for high rate low temperature treatment. Waite *et al.* (2018) reported that dipping grapevine propagation material in hydration tanks for more than 30 min had caused softening of bark tissues which became prone to increased fungal infections. These results indicate that plant tissue softening caused by heated water may also enhance fungicide penetration. It is possible that fungicides may transport from the tips of grapevine cuttings to further parts, by passive diffusion, but this would only be possible if there was one-way flow of water during hydration. There would likely be very little water influx from the opposite ends of cuttings allowing passive infusion of fungicides in hydration tanks.

The heated suspensions of tebuconazole were more effective for eradication of *N. parvum* from dormant grapevine cuttings than those of cyprodinil + fludioxonil or thiophanate-methyl. Similar effects were also observed on endophytic fungi, so that most of the endophytes could not be isolated from the cuttings treated with hot-water and tebuconazole combinations. The fungicides assessed in the present study have been reported to be effective for preventing wound infections of vines in field conditions, or contamination of cuttings by many GTD pathogens in hydration tanks (Rolshausen *et al.*, 2010; Amponsah *et al.*, 2012; Pitt *et al.*, 2012). However, in our experiments only tebuconazole showed high efficacy for reducing of colonization by *N. parvum*. We suggest that the differences in fungicide formulations may be responsible for these results. While tebuconazole was applied as an emulsion oil formulation in water, cyprodinil + fludioxonil and thiophanate-methyl were, respectively, wettable granule and wettable powder formulations. Active ingredients in powder or granule formulations may not have passed through grapevine barriers to reach inner tissues. Reports in PubChem indicate that these fungicides are stable in water, soil and air for long periods, so maximum duration in hot-water treatments (12 h) is unlikely to degrade the fungicides (Anonymous, 2019).

The present study investigated whether some endophytic fungi obtained from dormant grapevine cuttings had antagonistic effects against *N. parvum* in dual culture tests. All the endophyte isolates used in these tests were antagonistic to the pathogen. It has been previously shown that secondary metabolites secreted by endophytic fungi affect host plant physiology and other fungi (Schulz and Boyle, 2005). In an *in vitro* study by Wang *et al.* (2013), *Cladosporium cladosporioides* produced four different secondary metabolites (cladosporin, isocladosporin, 3,5'hydroxasperentin and cladosporin-8-methyl ether) in potato dextrose broth. Of these, cladosporin was the most inhibitory to mycelium growth of *Colletotrichum acutatum, Co. fragariae, Co. gloeosporioides* and *Phomopsis viticola* in a micro dilution broth assay reducing growth of these fungi by 80 to 93%. Springer *et al.* (1981), reported cladosporin to be a plant growth regulator, suppressing growth of etiolated coleoptiles of wheat. In our results, *Cladosporium* was a predominant endophytic genus isolated from vineyards in Manisa and Tarsus cities (located in warm climates), but antagonism by *Cladosporium* was moderate to *N. parvum* mycelium growth in dual culture tests. Musetti *et al.* (2007) extracted diketopiperazine compounds from the culture filtrates of *Alternaria alternata*, and tested inhibitory effects of these against downy mildew of grapes in greenhouse conditions. Spray applications of these secondary metabolites reduced (by up to 100%) disease severity, when vine leaves were treated with the compounds 2 or 24 h after inoculation with *Plasmopara viticola*. *Acremonium* has also been reported to produce biologically active metabolites, including β-lactam and cephalosporin antibiotics, and tremorgenic indole-diterpenoids. *Acremonium* also triggers some physiological pathways related to resistance of *Gramineae* plants (Gatenbay *et al.*, 1999; Moussaif *et al.*, 1997; Lindsey *et al.*, 2002; Adinaryana *et al.*, 2003). Assante *et al.* (2005) determined secondary metabolite profiles of endophytic *Acremonium* isolates (from grapevine leaves, cv. Reginabianca), and detected four different Acremines (A, B, C, D) in an *in vitro* study. These compounds have been found to inhibit sporangium germination of *P. viticola*, and maximum inhibition (99.8%) was obtained with "Acremin C" at 1 mM concentration. *Trichoderma atroviride* is a well-known endophytic and soil-borne fungus with antagonistic ability to many fungal pathogens, and some of the *Trichoderma* isolates are widely used in the control of GTD pathogens in Europe. Kotze *et al.* (2011) reported that *T. atroviride* was effective as a pruning wound protectant against GTD pathogens, including *N. parvum*, in South Africa. Their isolates provided satisfactory protection of grapevine pruning wounds against *Phompsis viticola, E. lata, P. chlamydospora, N. australe, N. parvum, D. seriata* and *L. theobromae*, reducing disease incidence by 69-92%. In Portugal, another strain of *T. atroviride* was reported to be highly suppressive to incidence and severity of disease caused by *P. chlamydospora* and *N. parvum* (Reis et al., 2017).

In the present study, the *T. atroviride* isolate (CUZ-FVG243) exhibited antagonistic performance (69% reduction) against *N. parvum* in PDA dual cultures competing for space and hyperparasitism. This isolate needs to be tested under field conditions for useful biocontrol effects. The endophytic *Q. cyanescens* was frequently isolated as an endophyte in Tarsus vineyards, but did not show strong antagonism to *N. parvum* (14.5%). This endophyte has been reported to affect grapevine physiology by producing secondary metabolites such as resveratrol (Srivastava, 2015).

CONCLUSIONS

Hot-water treatment is an effective strategy which should be used in grapevine nurseries to produce healthy plants. In the present study, hot-water treatments using fungicide suspensions, at 40°C and for 2 h durations, facilitated fungicide penetration into inner tissues of dormant grapevine canes. Although maximum reduction of *N. parvum* infections was 34%, this reduction could be practically significant because the propagation material used was severely colonized by the pathogen. Reduction rates could be greater where infections were less severe. Hot-water treatment also eradicates endophytic fungi, and this effect could be negative for plant health. However, due to rapid proliferation and recovery by endophytes, these negative effect may be compensated by additional measures such as using antagonistic *Trichoderma* biocontrol products in nurseries. Additionally, fungicides need to be further evaluated for curative effects on latent infections by *N. parvum* in plant material, when applied in hydration and hot-water treatment systems.

ACKNOWLEDGEMENTS

This study was supported by the Cukurova University Scientific Research Projects Department, in Project FYL-2018-10507, and was a MSc Thesis Project. The authors thank The Rectorate of Cukurova University, and also Emre Yıldırım, Murat Yıldız and Nesrin Dal for laboratory assistance.

LITERATURE CITED

Adinaryana K., Prabhakar T., Srinivasulu V., Rao M.A., Lakshmi P., Ellaiah P., 2003. Optimization of process parameters for cephalosporin C production under solid state fermentation from *Acremonium chrysogenum*. *Process Biochemistry* 39: 171–177.

Aly A.H., Debbab A., Kjer J., Proksch P., 2010. Fungal endophytes from higher plants: a prolific source of phytochemicals and other bioactive natural products. *Fungal Diversity* 41: 1–16.

Amponsah N.T., Jones E., Ridgway H.J., Jaspers M.J., 2012. Evaluation of fungicides for the management of Botryosphaeria dieback diseases of grapevines. *Pest Management Science* 68: 676–683.

Anonymous, 2019. The PubChem compound database, national center for biotechnology information, https://pubchem.ncbi.nlm.nih.gov/compound/ Accession date: 11.02.2019.

Assante G., Dallavalle S., Malpezzi L., Nasini G., Burruano S., Torta L., 2005. Acremines A–F, novel secondary metabolites produced by a strain of an endophytic *Acremonium*, isolated from sporangiophores of *Plasmopara viticola* in grapevine leaves. *Tetrahedron* 61: 7686–7692.

Badalyan S.M., Innocenti G., Garibyan N.G., 2002. Antagonistic activity of xylotrophic mushrooms against pathogenic fungi of cereals in dual culture. *Phytopathologia Mediterranea* 41: 200–225.

Barkai-Golan R., Apelbaum A., 1991. Synergistic effects of heat and sodium o-phenylphenate treatments to inactivate *Penicillium* spores and suppress decay in citrus fruits. *Tropical Science* 31: 229–230.

Billones-Baaijens R., Jaspers M., Allard A., Hong Y., Ridgway H., Jones E., 2015. Management of Botryosphaeriaceae species infection in grapevine propagation materials. *Phytopathologia Mediterranea* 54: 355–367.

Cabras P., Schirra M., Pirisi F.M., Garau V.L. Angioni A., 1999. Factors affecting imazalil and thiabendazole uptake and persistence in citrus fruits following dip treatments. *Journal of Agriculture, Food and Chemistry* 47: 3352–3354.

Casieri L., Hofstetter V., Viret O., Gindro K., 2009. Fungal communities living in the wood of different cultivars of young *Vitis vinifera* plants. *Phytopathologia Mediterranea* 48: 73–83.

Caudwell A., Larrue J., Boudon-Padieu E., McLean G.D., 1997. Flavescence dorèe elimination from dormant wood of grapevines by hot-water treatment. *Australian Journal of Grape and Wine Research* 3: 21–25.

Crous P.W., Swart L., Coertze S., 2001. The effect of hot-water treatment on fungi occurring in apparently healthy grapevine cuttings. *Phytopathologia Mediterranea* 40: 464–466.

Eichmeier A., Pecenka J., Penazova E., Baranek M., Catala-Garcia S., ... Gramaje D., 2018. High-throughput amplicon sequencing-based analysis of active fungal communities inhabiting grapevine after hot-water treatments reveals unexpectedly high fungal diversity. *Fungal Ecology* 36: 26–38.

Elena G., Di Bella V., Armengol J., Luque J., 2015. Viability of Botryosphaeriaceae species pathogenic to grapevine after hot water treatment. *Phytopathologia Mediterranea* 54: 325–334.

Felber A.C., Orlandelli R.C., Rhoden S.A., Garcia A., Costa A.T., Azavedo J.L., 2016. Bioprospecting foliar endophytic fungi of *Vitis labrusca* L., Bordô and Concord cv. *Annual Microbiology* 66: 765–775.

Fourie P.H., Halleen F., 2004. Proactive control of Petri Disease of grapevine through treatment of propagation material. *Plant Disease* 88: 1241–1245.

Freeman E.M, 1904. The seed-fungus of *Lolium temulentum*, L., the Darnel. *Philosophical Transactions Royal Society B* 196: 1-27.

Gatenbay W.A., Munday-Finch S.C., Wilkins A.L., Miles C.O.J., 1999. Terpendole M, a novel indole–diterpenoid isolated from *Lolium perenne* infected with the endophytic fungus *Neotyphodium lolii*. *Agriculture Food and Chemistry* 47: 1092–1097.

Gomez K.A., Gomez A.A., 1984. *Statistical Procedures for Agricultural Research*. 2nd ed. Wiley, New York, USA,

680 pp.

Gonzalez V., Tello M.L., 2011. The endophytic mycota associated with *Vitis vinifera* in central Spain. *Fungal Diversity* 47: 29–42.

Gramaje D., Alaniz S., Abad-Campos P., García-Jiménez J., Armengol J., 2010. Effect of hot-water treatments *in vitro* on conidial germination and mycelial growth of grapevine trunk pathogens. *Annals of Applied Biology* 156: 231–241.

Habib W., Pichierri A., Masiello N., Pollastro S., Faretra F., 2009. Application of hot water treatment to control *Phaeomoniella chlamydospora* in grapevine propagation materials. *Phytopathologia Mediterranea* 48: 186.

Halleen F., Fourie P.H., 2016. An integrated strategy for the proactive management of grapevine trunk disease pathogen infections in grapevine nurseries. *South African Journal of Enology and Viticulture* 37: 104–114.

Halleen F., Fourie P.H., Crous P.W., 2007. Control of black foot disease in grapevine nurseries. *Plant Pathology* 56: 637–645.

Halleen F., Crous P.W., Petrini O., 2003. Fungi associated with healthy grapevine cuttings in nurseries, with special reference to pathogens involved in the decline of young vines. *Australasian Plant Pathology* 32: 47–52.

Kotze C., Van-Niekerk J., Mostert L., Halleen F., Fourie P., 2011. Evaluation of biocontrol agents for grapevine pruning wound protection against trunk pathogen infection. *Phytopathologia Mediterranea* 50: 247–263.

Kraus C., Voegele R.T., Fischer M., 2019. Temporal development of the culturable, endophytic fungal community in healthy grapevine branches and occurrence of GTD-associated fungi. *Microbial Ecology* 77: 866–876.

Kusari S., Hertweck C., Spiteller M., 2012. Chemical ecology of endophytic fungi: origins of secondary metabolites. *Chemistry & Biology* 19: 792–798.

Lindsey C.C., Gomes-Diaz C., Villalba J.M., Pettus T.R.R., 2002. Synthesis of the F11334's from *o*-prenylated phenols: µM inhibitors of neutral sphingomyelinase (N-SMase) *Tetrahedron* 58: 4559–4565.

McGuire R.G., Campbell C.A., 1993. Imazalil for postharvest control of anthracnose on mango fruits. *Acta Horticulture* 341: 371–376.

Moussaif M., Jacques P., Schaarwachter P., Budzikiewicz H., Thonart P., 1997. Cyclosporin C is the main antifungal compound produced by *Acremonium luzulae*. *Applied Environmental Microbiology* 63: 1739–1743.

Musetti R., Polizzotto R., Vecchione A., Borselli S., Zulini L., … Pertot I., 2007. Antifungal activity of diketopiperazines extracted from *Alternaria alternata* against *Plasmopara viticola*: An ultrastructural study. *Micron* 38: 643–650.

Nejat N., Sijam K., Abdullah S.N.A., Vadamalai G., Dickinson M., 2009. Molecular characterization of a phytoplasma associated with coconut yellow decline (CYD) in Malaysia. *American Journal of Applied Sciences* 6: 1331–1340.

Ophel K., Nicholas P.R., Magarey P.A., Bass A.W., 1990. Hot water treatment of dormant grape cuttings reduces crown gall incidence in a field nursery. *American Journal of Enology and Viticulture* 41: 325–329.

Pancher M., Ceol M., Corneo P.E., Longa C.M.O., Yousaf S., … Campisano A., 2012. Fungal endophytic communities in grapevines (*Vitis vinifera* L.) respond to crop management. *Applied and Environmental Microbiology* 78: 4308–4317.

Pitt W.M., Sosnowski M.R., Huang R., Qui Y., Steel C.C., Savocchia S., 2012. Evaluation of fungicides for the management of Botryosphaeria canker of grapevines. *Plant Disease* 96: 1303–1308.

Pouzoulet J., Mailhac N., Couderc C., Besson X., Dayde J., … Jacques A., 2013. A method to detect and quantify *Phaeomoniella chlamydospora* and *Phaeoacremonium aleophilum* DNA in grapevine-wood samples. *Applied Microbiology and Biotechnology* 97: 10163–10175.

Reis P., Letousey P., Rego C., 2017. *Trichoderma atroviride* strain I-1237 protects pruning wounds against grapevine wood pathogens. *Phytopathologia Mediterranea* 56: 580.

Rego C., Nascimento T., Cabral A., Silva M.J., Oliveira H., 2009. Control of grapevine wood fungi in commercial nurseries. *Phytopathologia Mediterranea* 48: 128–135.

Reyes-Chilpa R., Quiroz Vásquez R.I., Jiménez Estrada M., Navarro-Ocaña A., Cassini Hernández J., 1997. Antifungal activity of selected plant secondary metabolites against *Coriolus versicolor*. *Journal of Tropical Forest Production* 3: 110–113.

Rodrigues K.F., 1994. The foliar fungal endophytes of the Amazonian palm *Euterpe oleracea*. *Mycologia* 86: 376–385.

Rolshausen P.E., Urbez-Torres J.R., Rooney-Latham S., Eskalen A., Smith J.R., Gubler W.D., 2010. Evaluation of pruning wound susceptibility and protection against fungi associated with grapevine trunk diseases. *American Journal of Enology and Viticulture* 61: 113–119.

Rooney S.N., Gubler W.D., 2001. Effect of hot water treatments on eradication of *Phaeomoniella chlamydos-*

pora and *Phaeoacremonium inflatipes* from dormant grapevine wood. *Phytopathologia Mediterranea* 40: 467–472.

Schirra M., Cabras P., Angioni A., Melis M., 1996. Residue level of imazalil fungicide in lemons following pre-storage dip treatment at 20 and 50°C. *Journal of Agriculture Food and Chemistry* 44: 2865–2869.

Schultz B.U., Wanke S., Draeger H., Aust H.J., 1993. Endophytes from herbaceous plants and shrubs: effectiveness of surface sterilization methods. *Mycological Research* 97: 1447–1450.

Schulz B., Boyle C., 2005. The endophytic continuum. *Mycological Research* 109: 661–686.

Serra S., Mannoni M.A., Ligios V., Demontis A., 2009. Occurrence of *Phaeomoniella chlamydospora* in grapevine planting material. *Phytopathologia Mediterranea* 48: 177.

Sieber T.N., Sieber-Canavesi F., Dorworth C.E., 1991. Endophytic fungi of red alder (*Alnus rubra* Bong.) leaves and twigs in British Columbia. *Canadian Journal of Botany* 69: 407–411.

Slippers B., Wingfield M.J., 2007. Botryosphaeriaceae as endophytes and latent pathogens of woody plants: diversity, ecology and impact. *Fungal Biology Reviews* 21: 90–106.

Smilanick J.L., Margosan D.A., Jenson D.J., 1995. Evaluation of heated solutions of sulfur dioxide, ethanol, and hydrogen peroxide to control postharvest green mold of lemons. *Plant Disease* 79: 742–747.

Springer J.P., Cutler H.G., Crumley F.G., Cox R.H., Davis E.E., Thean J.E., 1981. Plant growth regulatory effects and stereochemistry of cladosporin. *Journal of Agriculture, Food and Chemistry* 29: 853–855.

Srivastava A., 2015. *Screening and purification of resveratrol from endophytic fungi.* MsC Thesis, Thapar University Department of Biotechnology, Thapar, Pakistan, 48 pp.

Urbez-Torres J.R., 2011. The status of Botryosphaeriaceae species infecting grapevines. *Phytopathologia Mediterranea* 50: 5-45.

Waite H., May P., 2005. The effects of hot water treatment, hydration and order of nursery operations on cuttings of *Vitis vinifera* cultivars. *Phytopathologia Mediterranea* 44: 144–152.

Waite H., Armengol J., Billones-Baaijens R., Gramaje D., Halleen F., … Smart R., 2018. A protocol for the management of grapevine rootstock mother vines to reduce latent infections by grapevine trunk pathogens in cuttings. *Phytopathologia Mediterranea* 57: 384–398.

Wang X., Radvan M.M., Tarawneh A.H., Gao J., Wedge D.E., … Cutler S.J., 2013. Antifungal activity against plant pathogens of metabolites from the endophytic fungus *Cladosporium cladosporioides. Journal of Agriculture, Food and Chemistry* 61: 4551–4555.

Wells J.M., Harvey J.M., 1970. Combination heat and 2,6-dichloro-4-nitroaniline treatments for control of *Rhizopus* and brown rot of peaches, plums, and nectarines. *Phytopathology* 60: 116–120.

White T.J., Bruns T., Lee S., Taylor J., 1990. Amplification and direct sequencing of fungal ribosomal RNA genes for phylogenetics. In: *PCR protocols: a Guide to Methods and Applications* (Innis M.A., Gelfand D.H., Snisky J.J., White T.J., ed.) Academic Press, San Diego, USA, 315–322.

13

Transmission of 16SrIII-J phytoplasmas by the leafhoppers *Paratanus exitiousus* and *Bergallia valdiviana*

Nicolás QUIROGA[1], Valeria LONGONE[2], Ximena GONZÁLEZ[1], Alan ZAMORANO[1], Ana María PINO[1], Luca PICCIAU[3], Alberto ALMA[3], Samanta PALTRINIERI[4], Nicoletta CONTALDO[4], Assunta BERTACCINI[4], Nicola FIORE[1,*]

[1] *University of Chile, Facultad de Ciencias Agronómicas, Departamento de Sanidad Vegetal, Santiago, Chile*
[2] *Instituto Nacional de Tecnología Agropecuaria, Estación Experimental Agropecuaria Mendoza, Laboratorio de Fitopatología, San Martin 3853, M5534 Luján de Cuyo, Mendoza, Argentina*
[3] *Dipartimento di Scienze Agrarie, Forestali e Alimentari (DISAFA), Università di Torino, Largo P. Braccini 2, 10095 Grugliasco, Torino, Italy*
[4] *Department of Agricultural and Food Sciences (DISTAL), Plant Pathology and Entomology, Alma Mater Studiorum, Università di Bologna, Viale G. Fanin 42, 40127 Bologna, Italy*
*Corresponding author: nfiore@uchile.cl

Competing Interests: The Author(s) declare(s) no conflict of interest.

Editor: Nihal Buzkan, Kahramanmaraş Sütçü Imam University, Turkey.

Summary. Two of the most common leafhoppers present in Chile are the Cicadellidae *Paratanus exitiosus* and *Bergallia valdiviana*. They commonly occur in vineyards of central Chile, including some vineyards infected by phytoplasmas. The present study demonstrates that *P. exitiosus* and *B. valdiviana* can transmit 16SrIII-J phytoplasmas to grapevine and periwinkle plants. This provides improved understanding of the 16SrIII-J phytoplasma epidemiology in Chilean vineyards.

Keywords. Auchenorrhyncha, transmission trials, nested-PCR, *tuf* gene, 16S rRNA gene, RFLP.

INTRODUCTION

In Chile, grapevine yellows (GY) is associated with phytoplasmas belonging to diverse ribosomal subgroups, including 16SrI-B and 16SrI-C ('*Candidatus* Phytoplasma asteris'-related), 16SrIII-J ('*Ca.* P. pruni'-related), 16SrV-A ('*Ca.* P. ulmi'), 16SrVII-A ('*Ca.* P. fraxini'), 16SrXII-A ('*Ca.* P. solani' or "stolbur") (Gajardo *et al.*, 2009; Fiore *et al.*, 2015). However, the phytoplasmas in the 16SrIII-J group were prevalent in the vineyards of the central zone of the country. This phytoplasma has been reported to infect various crops and spontaneous plant species, and a draft sequence of the genome of a Chilean strain was obtained (Gonzalez *et al.*, 2011; Zamorano and Fiore, 2016; Quiroga *et al.*, 2017a).

Phytoplasma dissemination in the field occurs with the use of infected plant materials and by insect vector transmission. The insects generally feed on weeds and only occasionally on grapevine, allowing the transmission of phytoplasmas to this species. In order to determine which insects are involved in transmission of the 16SrIII-J group-phytoplasma in Chile, an epidemiological study was carried out in symptomatic phytoplasma-infected vineyards. During dedicated surveys, several leafhoppers species (Hemiptera, Auchenorryncha, Cicadellidae) were found to be positive for the 16SrIII-J phytoplasma (N. Fiore, unpublished data), and among them, two of the most common insect species were *Paratanus exitiousus* (Beamer) and *Bergallia valdiviana* Berg 1881. This paper reports transmission of 16SrIII-J phytoplasma to periwinkle (*Catharanthus roseus* L.) and grapevine plants by both leafhoppers.

Figure 1. The two insect species used in transmission trials: A) *Paratanus exitiosus*; B) *Bergallia valdiviana*.

MATERIALS AND METHODS

In 2011 (September–December) and 2012 (January–May) surveys for *P. exitiosus* presence were carried out in three selected 16SrIII-J phytoplasma infected Chilean vineyards, two in the Metropolitana region (Buin and Pirque), and one in the Valparaiso region (Casablanca). The capture of *B. valdiviana* adults was carried out in 2012 (October–December) and 2013 (January–June) in a vineyard of cv. Pinot noir infected with 16SrIII-J phytoplasmas, in the Valparaiso region (Casablanca). *Paratanus exitiosus* and *B. valdiviana* (Figure 1) were captured using an entomological sweeping net and 150 sweepings directed to weeds present inside the vineyards. At the end of each capture, the adults of both insects were separated by species and divided in two batches. All the individuals belonging to one species were released into two entomological cages to let them feed on three grapevine plants of cv. Cabernet Sauvignon (first cage), and three periwinkle plants (second cage). The number of individuals released each month in each cage is reported in Tables 1 and 2.

Periwinkle and grapevine plants used in the transmission trials were from seeds, and tested by nested PCR (described below) to verify the absence of phytoplasmas before the transmission trials. Eighty-one periwinkle and 81 grapevine plants were used for the trials with *P. exitiosus,* while the phytoplasma transmission by *B. valdiviana* was performed to 27 periwinkle and 21 grapevine plants. All the plants were kept in a conditioned

Table 1. Numbers of *Paratanus exitiosus* individuals captured in three vineyards. Insects were released each month in cages to feed on periwinkle and grapevine plants for phytoplasma transmission trials.

Month	Number of individuals used in transmission trials to grapevine plants			Number of individuals used in transmission trials to periwinkle plants		
	Vineyard			Vineyard		
	Buin	Pirque	Casablanca	Buin	Pirque	Casablanca
September 2011	33	6**	2	32	7	2
October 2011	25**	17	20	24	18	20
November 2011	34	8	22	33**	7	22
December 2011	36	18**	9	35**	17	8**
January 2012	33	20	21	32	20	20
February 2012	31	14	30	30	14	30
March 2012	34	10	24	33	10	23
April 2012	18	5	21**	17	4	20
May 2012	33**	3	23	32	3	22

** Successful transmission trials for phytoplasma 16SrIII-J.

Table 2. Numbers of *Bergallia valdiviana* individuals captured in the Casablanca vineyard. Insects were released each month in cages to feed on periwinkle and grapevine plants for phytoplasma transmission trials.

Month	Number of individuals used in transmission trials to grapevine plants	Number of individuals used in transmission trials to periwinkle plants
October 2012	0	4
November 2012	0	16
December 2012	17	22
January 2013	20**	25
February 2013	28**	33
March 2013	24	30
April 2013	26	21**
May 2013	26	19
June 2013	0	23

** Successful transmission trials for phytoplasma 16SrIII-J.

incubator at 25°C with 16 h of photoperiod. Each feeding period lasted until the death of all insects in the cages (four to six leafhoppers per test).

The cages were inspected on a daily basis to collect the dead insects and store them in 70% ethanol for further analyses. The cages without insects were then moved into a screen house for symptom onset. Phytoplasma detection in plants was carried out every three months. All insects from each cage were also tested in order to verify the phytoplasma presence.

Total nucleic acids were extracted from plants using a chloroform/phenol method (Prince *et al.*, 1993), while insect nucleic acid was extracted with a CTAB method (Angelini *et al.*, 2001). Nucleic acid was dissolved in Tris-EDTA pH 8.0 buffer and maintained at 4°C. PCR amplification was carried out using 20 ng μL^{-1} of nucleic acid. Direct and nested PCR on the *tuf* gene were performed according to the protocol of Makarova *et al.* (2012). Further direct PCR with primer pair P1/P7 (Deng and Hiruki, 1991; Schneider *et al.*, 1995) and nested PCR with R16F2n/R2 primers on the *16S rRNA* gene (Gundersen and Lee, 1996), were performed as described by Schaff *et al.* (1992). Amplicons from nested PCRs for both genes were purified using the EZNA gel extraction kit (OMEGA Bio-tek). DNA fragments were ligated into the pGEMT-Easy Cloning Kit (Promega). Putative recombinant clones were analyzed by colony PCR, and selected fragments were sequenced in both directions by Macrogen USA Corp. The sequences were then aligned with those of classified strains deposited in GenBank using BLAST engine for local alignment (version Blast N 2.2.12). The phytoplasma identification was carried out using *in silico* restriction fragment length polymorphism (RFLP) analysis with *Hha*I, *Bst*UI, and *Rsa*I restriction enzymes, in the *i*PhyClassifier online tool (https://plant-pathology.ba.ars.usda.gov/cgi-bin/resource/iphyclassifier.cgi) (Zhao *et al.*, 2009).

RESULTS

Leafhopper surveys

In the vineyard located in Buin, the *P. exitiosus* capture rate remained constant during all months, except for decreases in October 2011 and April 2012 (Table 1). In the Pirque vineyard, the greatest numbers of individuals were collected in October and December 2011, and January 2012. In the Casablanca vineyard, *P. exitiosus* was less abundant in September 2011, remained constant during the other months and increased in February 2012. These results indicated that in the surveyed areas *P. exitiosus* tended to be more abundant during summer.

Adult individuals of *B. valdiviana* were captured in both years in greatest abundance through the summer and remained constant in autumn (Table 2). This insect was of greatest abundance in the coastal valleys, where Casablanca is located. For this reason, only the insects captured in Casablanca vineyards were used for the transmission trials.

Phytoplasma identification

Phytoplasma 16SrIII-J was detected in some grapevine and periwinkle plants used in the transmission trials, and in *P. exitiosus* and *B. valdiviana* specimens that were fed on these plants. Sequences of cloned PCR fragments from both amplified genes were the same from the positive plants and insects in the *tuf* gene sequence (438 bp). On the other hand, sequence identity of the *16S rRNA* gene clones was 99.9 to 100% (1,230 bp) in the same samples. Two sequences of the *tuf* gene were deposited in GenBank under the accession numbers (Acc. No.) MH743135, MH743136, and two of the *16S rRNA* gene and were deposited under Acc. Nos MH743137, MH743138. For the *16S rRNA* gene, the nucleotide similarity percentages of the detected phytoplasmas showed a close correlation (99.8%) with those of the strain Ch10 (GenBank Acc. No, AF147706), corresponding to the chayote witches' broom phytoplasma (16SrIII-J) from Brazil. For the *tuf* gene, the greatest nucleotide identity was 100%, with the strain Hort72, belonging to a 16SrIII-J phytoplasma from sugarbeet from Chile (GenBank Acc. No. KM658259). The 16S rRNA amplicons were also subjected to *in silico* RFLP analysis that assigned the detected phytoplasmas to the ribosomal subgroup 16SrIII-J (Figure 2).

Transmission trials with Paratanus exitiosus *to grapevine plants*

Paratanus exitiosus survived for 3 to 4 days on the grapevine plants. In the twelve months post-transmission trials, the phytoplasma presence was detected by nested PCR with primers for *tuf* and *16S rRNA* genes in five out of the 81 grapevine plants. Two plants (V47 and V78A) were from the transmission trials carried out with insects captured in Buin in two different months (respectively October 2011 and May 2012). Two plants (V43 and V61) were infected from insects captured in Pirque, during, respectively, September and December 2011. The plant V76B was infected by individuals of *P. exitiosus* captured in Casablanca, during April 2012 (Table 1). The transmission rates were 7.5% for the insects captured in Buin and Pirque, and 3.7% for those from Casablanca. At 24 months after the transmission trials were set up, two out of five grapevine plants infected with 16SrIII-J were asymptomatic (V43 and V61), while the other three plants showed short internodes, and leaves with downward curling and deformation (Figure 3).

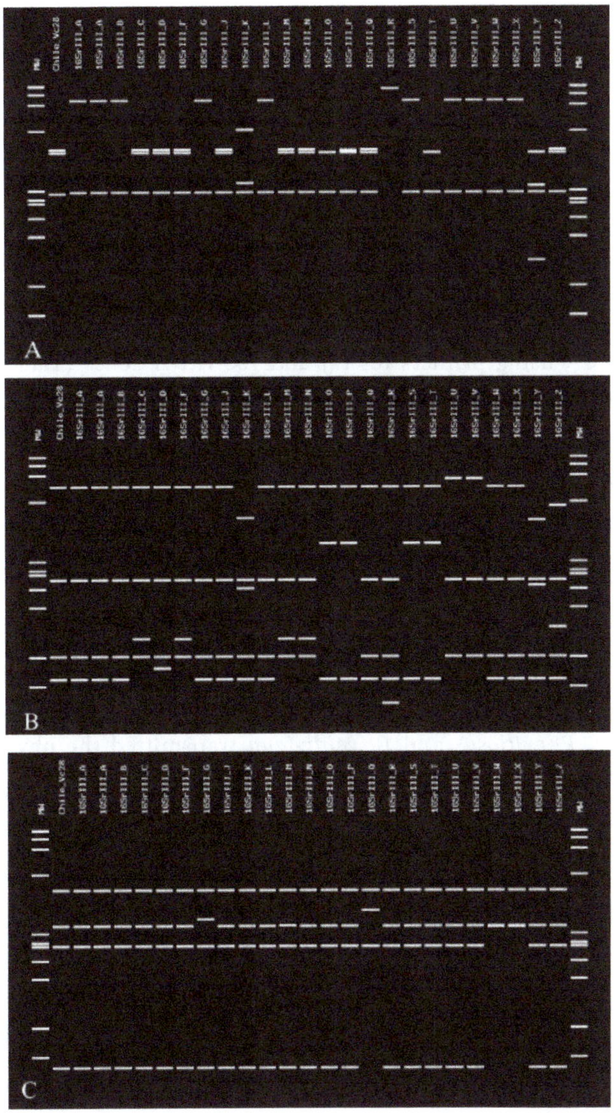

Figure 2. Virtual RFLP patterns obtained from the restriction with *HhaI* (A); *BstUI* (B) and *RsaI* (C), of a 1,200 bp *16S rRNA* gene fragment obtained from sample VC28 (phytoplasma detected in a periwinkle plant used in transmission trials with *Paratanus exitiosus*). The phytoplasmas transmitted are classified in subgroup 16SrIII-J. The strains representative of the diverse subgroups are those enclosed in the *i*PhyClassifier (Zhao *et al.*, 2009).

Transmission trials with Paratanus exitiosus *to periwinkle plants*

Paratanus exitiosus survived for 4 to 5 days on periwinkle plants. Three months after the transmission trials, in three out of the 81 periwinkle plants, the phytoplasma presence was detected by nested PCR with the primers for *tuf* and *16S rRNA* genes (Table 1). Two of the positive plants (VC28C and VC31C) were from the transmission trials carried out with insects captured in the Buin vineyard in two different months (November and December 2011, respectively). The third plant (VC33A) was infected by individuals captured in the Casablanca vineyard during December 2011. The three periwinkle plants infected with 16SrIII-J phytoplasmas showed symptoms of virescence, phyllody and witches' broom five months after the start of the transmission trials (Figure 4).

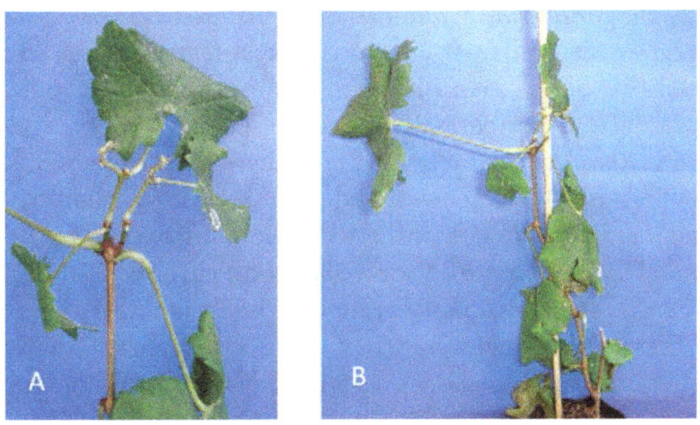

Figure 3. Symptoms associated with phytoplasma presence in grapevine after *Paratanus exitiosus* transmission. A) V43, short internodes; B) V76, leaves with downward rolling, deformations and necrosis.

Figure 4. Symptoms associated with phytoplasma presence in periwinkle after *Paratanus exitiosus* transmission. A) VC28C, flowers with virescence; B) VC31C, flowers with virescence and phyllody; C) VC33A, plant with witches' broom, small and yellow leaves.

Figure 5. Symptoms associated with phytoplasma presence in grapevine after *Bergallia valdiviana* transmission. Short internodes, and leaves with downward rolling and deformation: A) VC85A; B) V86C.

Transmission trials with Bergallia valdiviana *to grapevine plants*

Bergallia valdiviana survived for 4 to 5 days on grapevine plants. At 24 months post-transmission trials, two of the 21 plants were positive for phytoplasma presence in nested PCR tests using primers for *tuf* and *16S rRNA* genes (Table 2). The positive grapevine plants (V85A and V86C) corresponded to the transmission trials carried out with insects captured in two different months (January 2013 for V85A and February 2013 for V86C). The transmission rate was 9.5%. The two grapevines infected with 16SrIII-J were symptomatic, with short internodes, and leaves with downward rolling and deformation (Figure 5).

Figure 6. Symptoms in periwinkle after *Bergallia valdiviana* transmission; A), B) VC60B, leaf deformation and severe yellowing; C) VC62B, flowers showing virescence.

Transmission trials with Bergallia valdiviana *to periwinkle plants*

Bergallia valdiviana survived for 6 to 7 days on periwinkle plants. Two out of the 27 plants were positive for phytoplasma presence, as determined with nested PCR using primers for *tuf* and *16S rRNA* genes (Table 2) 12 months after the transmission trials. These plants (VC 60B and V62B) corresponded to the transmission trials performed with insects captured in April 2013. The transmission rate was 3.7%. The 16SrIII-J infected periwinkle plants showed symptoms of virescence, leaf deformation and severe yellowing, 15 months after the transmission trials (Figure 6).

DISCUSSION

P. exitiosus and *B. valdiviana* transmitted the phytoplasma 16SrIII-J to grapevine and periwinkle plants. Both insects live on weeds and only occasionally feed on grapevine or other crops. *P. exitiosus* transmitted the phytoplasmas at a higher rate to grapevine than to periwinkle plants, while *B. valdiviana* had the same percentage of transmission to grapevine and periwinkle. Survival times of both insects on the grapevine plants was less than on periwinkle plants. The detection of 16SrIII-J phytoplasmas and the appearance of symptoms in grapevine occurred later than in the periwinkle plants. In the plants used for the transmission assays with *B. valdiviana*, phytoplasma detection occurred later in comparison with *P. exitiosus* assays and, as expected, the symptoms took longer to appear. This could indicate that the amount of inoculum of the pathogen in the *B. valdiviana* individuals used for transmission trials was less than in *P. exitiosus*, or that its transmission efficiency was less. The main differences from the transmission trials were that higher transmission resulted from the *P. exitiosus* captured in the spring and summer periods, while *B. valdiviana* was only able to transmit the phytoplasma when collected at the end of summer and autumn. This vector difference could play a fundamental role in maintaining the phytoplasma population in host weeds during the periods of grapevine vegetative recess.

The phytoplasma 16SrIII-J and its newly identified insect vectors are widely distributed in Chile, infecting different weed species and crops of agronomic importance (Castro *et al.*, 2000; Hepp and Vargas, 2002; González *et al.*, 2010; 2011; Longone *et al.*, 2011). Taking the observed *P. exitiosus* and *B. valdiviana* phytoplasma-transmission rates into account, if environmental conditions are favorable, there is high possibility that grapevine yellows outbreaks, associated with the presence of 16SrIII-J phytoplasmas, will occur in the central zone of Chile. Recent studies have indicated that climate change could modify the habitat of these species of insects, and increase their reproduction rates in the central zone of Chile (Quiroga *et al.*, 2017b).

ACKNOWLEDGMENT

This research was supported by the National Fund for Scientific and Technological Development (FONDECYT), Chile, in Project No. 1140883. The project was carried out in the frame of the European Union Horizon 2020 Research and Innovation Programme, under grant agreement No. 727459.

LITERATURE CITED

Angelini E., Clair D., Borgo M., Bertaccini A., Boudon-Padieu E., 2001. "Flavescence dorée" in France and Italy - Occurrence of closely related phytoplasma isolates and their near relationships to Palatinate grapevine yellows and an alder yellows phytoplasma. *Vitis* 40: 79–86.

Castro S., Hepp R., Romero J., 2000. La marchitez amarilla de la remolacha azucarera de Chile es producida por un fitoplasma del grupo 16SrIII. In: *Proceedings of the Tenth Congress of the Sociedad Española de Fitopatología*, 3-6 October 2000, Valencia, Spain, (Abstract), 258 pp. (in Spanish).

Deng S.J., Hiruki C., 1991. Amplification of 16S ribosomal-RNA genes from culturable and nonculturable mollicutes. *Journal of Microbiological Methods* 14: 53–61.

Fiore N., Zamorano A., Pino A.M., 2015. Identification of phytoplasmas belonging to the ribosomal groups 16SrIII and 16SrV in Chilean grapevines. *Phytopathogenic Mollicutes* 5(1): 32–36.

Gajardo A., Fiore N., Prodan S., Paltrinieri S., Botti S., ... Bertaccini A., 2009. Phytoplasmas associated with grapevine yellows disease in Chile. *Plant Disease* 93: 789–796.

González F., Zamorano A., Pino A.M., Fiore N., 2010. Caracterización molecular de nuevos fitoplasmas en la vid en Chile. In: *Proceedings of the Nineteenth Congress of the Sociedad Chilena de Fitopatología*, 9-12 November 2010, Pucón, Chile, (Abstract), 50 pp. (in Spanish).

González F., Zamorano A., Pino A.M., Paltrinieri S., Bertaccini A., Fiore N., 2011. Identification of phytoplasma belonging to X-disease group in cherry in Chile. *Bulletin of Insectology* 64(S): S235–S236.

Gundersen D.E., Lee I-M., 1996. Ultrasensitive detection of phytoplasmas by nested-PCR assays using two universal primer pairs. *Phytopathologia Mediterranea* 35: 144–151.

Hepp R., Vargas, M., 2002. Detection by PCR of the causal agent of yellow wilt of sugar beet in leafhoppers associated with sugar beet crop. *Fitopatología* 37: 73.

Longone V., González F., Zamorano A., Pino A.M., Araya J., ... Fiore N., 2011. Epidemiological aspects of phytoplasmas in Chilean grapevines. *Bulletin of Insectology* 64(S): S91-S92.

Makarova O.V., Contaldo N., Paltrinieri S., Kawube G., Bertaccini A., Nicolaisen M., 2012. DNA barcoding for universal identification of '*Candidatus* Phytoplasmas' using a fragment of the elongation factor Tu gene. *PLoS ONE* 7(12): e52092.

Prince J.P., Davis R.E., Wolf T.K., Lee I-M., Mogen B.D., ... Barba M., 1993. Molecular detection of diverse mycoplasmalike organisms (MLOs) associated with grapevine yellows and their classification with aster yellows, X-disease, and elm yellows MLOs. *Phytopathology* 83: 1130–1137.

Quiroga N., Bustamante M., Gamboa C., Molina J., Zamorano A., Fiore N., 2017a. 16SrIII-J phytoplasmas infecting lettuce and Swiss chard crops in Chile. *Phytopathogenic Mollicutes* 7(2): 91–94.

Quiroga N., Ivulic D., Lagos J., Saavedra M., Sandoval-Rodríguez A., ... Fiore N., 2017b. Risk analysis of the establishment of *Scaphoideus titanus*, vector of "flavescence dorée" phytoplasma in grapevine, under current and estimated climate change conditions in Chile. *Phytopathogenic Mollicutes* 7 (1): 39–44.

Schaff D., Lee I-M., Davis R.E., 1992. Sensitive detection and identification of mycoplasmalike organisms by polymerase chain reactions. *Biochemistry Biophysics Research Communications* 186: 1503–1509.

Schneider B., Seemüller E., Smart C.D., Kirkpatrick B.C., 1995. Phylogenetic classification of plant pathogenic mycoplasmalike organisms or phytoplasmas. In: *Molecular and Diagnostic Procedures in Mycoplasmology* (S. Razin, J.G. Tully, ed.), Academic press, San Diego, CA, USA, 369–380.

Zamorano A., Fiore N., 2016. Draft genome sequence of 16SrIII-J phytoplasma, a plant pathogenic bacterium with a broad spectrum of hosts. *Genome Announcements* 4(3): e00602–16.

Zhao Y., Wei W., Lee I., Shao J., Suo X., Davis R.E., 2009. Construction of an interactive online phytoplasma classification tool, *i*PhyClassifier, and its application in analysis of the peach X-disease phytoplasma group (16SrIII). *International Journal of Systematic and Evolutionary Microbiology* 59(10): 2582–2593.

14

Characterization and pathogenicity of *Colletotrichum* spp. causing citrus anthracnose in Tunisia

Haïfa BEN HADJ DAOUD[1], Elena BARALDI[2], Mirco IOTTI[3], Pamela LEONARDI[2], Naima BOUGHALLEB-M'HAMDI[1,*]

[1] *Department of Biological Sciences and Plant Protection, High Institute of Agronomy of Chott Mariem, 4042 Sousse, UR13AGR03, University of Sousse, Tunisia*
[2] *Department of Agricultural and Food Sciences, DISTAL, University of Bologna, Via le Fanin 46, 40127, Bologna, Italy*
[3] *Department of Life, Health and Environmental Sciences, University of L'Aquila, via Vetoio, Coppito 1, 67100 L'Aquila, Italy*
*Corresponding author: n.boughalleb2017@gmail.com

Abstract. In the winter and spring of 2014–2015, typical anthracnose symptoms were detected on different citrus varieties in Cap-Bon and Morneg-Tunis regions of Northern Tunisia. Surveys were conducted to determine the casual agents of these symptoms. A total of seven monosporic isolates were obtained from dark lesions on fruits, flowers, leaves and twigs of citrus from six orchards. One *Colletotrichum karstii* (from the *C. boninense* species complex) and six *C. gloeosporioides* isolates were identified through morphological analysis and sequencing of their ITS rDNA sequences. Pathogenicity tests with the seven isolates were performed on symptomless, detached citrus fruits (Valencia orange and Eureka lemon). All tested isolates caused anthracnose lesions after 1 week of incubation. Koch's postulates were fulfilled by re-isolation of pathogens from the inoculated fruits. This report elucidates the diversity of anthracnose pathogens in Tunisia. This is the first report of *C. karstii* from citrus in Tunisia.

Keywords. Citrus anthracnose, *Colletotrichum karstii*, *Colletotrichum gloeosporioides*, PCR.

Competing Interests: The Author(s) declare(s) no conflict of interest.

Editor: Anna Maria D'Onghia, CIHEAM/Mediterranean Agronomic Institute of Bari, Italy.

INTRODUCTION

The citrus industry is one of the most important fruit industries worldwide. Mediterranean countries are second only to China for citrus production, and are the largest exporter after South Africa (FAO, 2016). In Tunisia, citrus orchards cover 27,000 ha and include approx. 7 million trees. During the last decade, citrus production has steadily increased from 300,000 MT in 2007/08 as new orchards have begun production, reaching 560,000 MT in 2016/17. The most important crops are Maltese and Navel oranges [*Citrus sinensis* (L.) Osbeck], clementines (*Citrus* x *Clementina*), lemons (*C. limon* L.) and mandarins (*C. reticulata* Blanco) (Haas, 2017).

Knowledge of the pathogens affecting citrus crops is important. Previous investigations have demonstrated that these crops are affected by many fungal diseases such as Alternaria brown spot (Haddad *et al.*, 2013), gummosis (Boughalleb-M'Hamdi *et al.*, 2018) and anthracnose (Ben Hadj Daoud, 2013; Rhaiem and Taylor, 2016).

In the present study, we focus on citrus anthracnose. *Colletotrichum* is among the most economically important genera of plant pathogenic fungi worldwide (Sutton, 1992; Cai *et al.*, 2009; Phoulivong, 2011). This genus has been designated the eighth most important group of plant pathogenic fungi, based on perceived scientific and economic importance (Dean *et al.*, 2012). Several species of *Colletotrichum* cause diseases (anthracnoses) of a wide range of important crops (Sutton, 1992; Hyde *et al.*, 2009a). Two species complexes, *C. acutatum* and *C. gloeosporioides* encompass the main causal agents of citrus diseases. Three important diseases of citrus caused by *Colletotrichum* spp. are anthracnose, post-bloom fruit drop and key lime anthracnose (Timmer *et al.*, 2000; Lima *et al.*, 2011; McGovern *et al.*, 2012).

Colletotrichum gloeosporioides is reported as pathogen of the main cultivated citrus species worldwide (Huang *et al.*, 2013). *Colletotrichum gloeosporioides sensu stricto* occurs more frequently in the Old World, contrasting with the greater importance of *C. acutatum* in the Americas (Lima *et al.*, 2011). Before multi-gene analysis, *C. acutatum* was identified as the only species responsible for post-bloom fruit drop (PFD) (Peres *et al.*, 2008) and key lime anthracnose (KLA) (Brown *et al.*, 1996). Similarly, *C. gloeosporioides* was reported as the only *Colletotrichum* species to cause citrus fruit anthracnose (Timmer *et al.*, 2000). PFD of citrus was first described in Belize (formerly British Honduras) in 1979, caused by *C. gloeosporioides* (Penz.) Sacc. (Denham, 1979; Fagan, 1979). This disease has been detected in Argentina, Brazil, Mexico and the United States (Schwarz *et al.*, 1978; Orozco Santos and Gonzalez Garza, 1986; McMillan and Timmer, 1989; Kuramae-Izioka *et al.*, 1997). PFD was widespread in the western hemisphere in the 1990s, including in Belize, Bermuda, Brazil, Costa Rica, Mexico and the United States (Florida), and has become less of a problem in Florida in recent years. However, PFD could re-emerge as a severe citrus disease (McGovern and Rouse, 1994; Timmer and Brown, 2000). In Florida, the causal agents of PFD and KLA were initially identified as strains of *C. gloeosporioides* (Agostini et *al.*, 1992). Genetic characterization of *Colletotrichum* isolates from Brazil suggested that PFD can be caused by both *C. acutatum* and *C. gloeosporioides* (Kuramae-Izioka *et al.*, 1997). A report from south Brazil identified *C. acutatum* as the cause of PFD (Theodoro *et al.*, 2003). *Colletotrichum gloeosporioides* was reported as a causal agent of PFD on sweet orange (Lima *et al.*, 2011; McGovern *et al.*, 2012). An extensive investigation in citrus orchards of São Paulo state (Brazil) revealed only *C. abscissum* and *C. gloeosporioides sensu stricto* associated with PFD disease (Da Silva *et al.*, 2016). Key lime anthracnose, a disease complex on leaves, flowers and fruits of Key lime, was initially reported to be caused by *C. acutatum* (Brown *et al.*, 1996; Peres *et al.*, 2008; MacKenzie *et al.*, 2009), but later classified as *C. limetticola* (Damm *et al.*, 2012a).

Following adoption of the use of multi-gene phylogenetic analyses, the polyphasic protocols for studying the genus *Colletotrichum* significantly changed the classification and species concepts this genus (Cannon *et al.*, 2012; Damm *et al.*, 2012a; 2012b; 2013; 2014; Weir *et al.*, 2012). In China, at least 13 species of *Colletotrichum* have been associated with citrus (Peng *et al.*, 2012). Huang *et al.* (2013) reported the presence of four important species complexes of *Colletotrichum* in citrus in China, namely the complexes of *C. gloeosporioides*, *C. boninense* and *C. acutatum*, and a group including *C. truncatum*. The species in the *C. gloeosporioides* species complex comprised *C. gloeosporioides* and *C. fructicola*, the *C. boninense* complex included *C. karstii* and a new species *C. citricola*, and the *C. acutatum* complex included a new species, *C. citri*. The most important causal agent is *C. gloeosporioides*, and *C. gloeosporioides sensu lato* was also reported to cause pre-harvest symptoms such as wither-tip on twigs, tear-stain (Klotz and Fawcett, 1941; Benyahia *et al.*, 2003) and stem-end rot on fruit (Kaur *et al.*, 2007).

Colletotrichum acutatum sensu stricto, as member of *C. acutatum* species complex, was found on citrus for the first time in Europe by Guarnaccia *et al.* (2017). This fungus was isolated in many citrus growing areas surveyed during 2015 and 2016, including Spain, Italy, Portugal, Greece and Malta, and from several citrus species showing typical anthracnose symptoms on leaves fruits and twigs. *Colletotrichum novaezelandiae* (from the *C. boninense* species complex) was also first reported on citrus. Infections caused by *Colletotrichum* spp. Have strongly compromised citrus production in Mediterranean countries. In Portugal, since 2003, anthracnose symptoms were detected in lemon and orange orchards. Ramos *et al.* (2006) found isolates with characteristics that did not fit those commonly described for *C. acutatum* and *C. gloeosporioides*, raising the hypothesis of other species of *Colletotrichum* were involved on pre-harvest anthracnose of citrus. Ramos *et al.* (2016) demonstrated that many isolates collected from symptomatic leaves, twigs and fruits of several citrus cultivars in Portugal were *C. gloeosporioides sensu stricto* predominated in flowers and fruits, and *C. karstii* was found in leaves and branches. In Italy, pre-harvest anthracnose symptoms appeared on orange fruits, caused by *C. gloeosporioides* and *C. karstii* (Aiello *et al.*, 2015). *Colletotrichum karstii*, a member of the *C. boninense* species complex, has been reported on many host plants with wide geographical distribution (Damm *et al.*, 2012b). In Europe, this species has been detected on tropical fruits, cotton and lupin plants (Damm *et al.*, 2012b; Ismail *et al.*, 2015). *Colletotrichum karstii* was previously found to be an important pathogen on *Orchidaceae* hosts (Yang *et al.*, 2011), and has also been isolated from citrus plants in South Africa, New Zealand (Damm *et al.*, 2012b) and China (Peng *et al.*, 2012). Peng *et al.* (2012) showed that this species causes citrus leaf anthracnose. Guarnaccia *et al.* (2017) reported the presence of three major species complexes in surveyed citrus regions in European countries namely, *C. gloeosporioides sensu stricto* and two new species (*C. helleniense* and *C. hystricis*) were identified in the *C. gloeosporioides* species complex. *Colletotrichum karstii, C. novaezelandiae* and two new species (*C. catinaense* and *C. limonicola*) in the *C. boninense* species complex, and *C. acutatum sensu stricto* was also identified as member of *C. acutatum* species complex.

Anthracnose diseases of citrus in Tunisia had not been well researched. Since early 2009, wither-tip symptoms of twigs, resembling anthracnose, were observed on cultivated citrus trees in Northern Tunisia. Using morphological and molecular characteristics and pathogenicity tests on isolates collected from withered twigs, Rhaiem and Taylor (2016) reported for the first time the presence of *Colletotrichum gloeosporioides* as the causal agent of anthracnose disease of citrus in Tunisia. This fungus was also identified on olive by Rhouma *et al.* (2010). Species of *C. acutatum sensu stricto* were found by Chattaoui *et al.* (2016) on olives fruits from Northern Tunisia, showing symptoms of anthracnose since November 2011. The *C. acutatum sensu stricto* group was recorded for the first time in the country as the causal agent of olive anthracnose. Because of the commercial yield losses in citrus orchards caused by *Colletotrichum* infections in Tunisia, further research was required to identify *Colletotrichum* spp. related to citrus anthracnose. In our surveys commencing in April 2014, severe anthracnose symptoms, such as the typical dieback and wither-tip of twigs and dark lesions on fruits, leaves and flowers, were observed on different citrus varieties in orchards of Cap-Bon and Morneg-Tunis (Northern Tunisia).

The objectives of the present investigation were to: (i) evaluate symptoms of anthracnose in surveyed citrus orchards; (ii) characterize the *Colletotrichum* spp. causing anthracnose on citrus in Tunisia, using morphological and molecular techniques; and (iii) evaluate the pathogenicity of the species on Valencia orange and Eureka lemon.

MATERIALS AND METHODS

Fungus isolations

Ten citrus fruit, twig and leaf samples with typical anthracnose symptoms were collected from orange varieties Thomson, Malti and Meski and clementine variety Cassar during spring 2014 and winter 2015. Small pieces of tissue (5 mm × 5 mm) were sterilized in sodium hypochlorite solution for 5 min, followed by several rinses with sterile distilled water (SDW), and then placed on Potato Dextrose Agar (PDA, BD DIFCO) in Petri dishes for one week at 27°C.

Monoconidial cultures were generated from each sample and stored on dry filter paper at -20°C, and were later identified morphologically. The conidia were taken from actively growing colonies and suspended in sterile water. Lengths and widths of 50 conidia from each isolate were measured under an ECLIPSE TE 2000-E microscope (×1,000 magnification) (Nikon) using differential interference contrast optics. Measurements were taken using a Nis-Elements AR (v 3.10) software (Zeiss) from images captured with a DXM1200F digital camera (Nikon). Q-value of conidia (the ratio of length to width in side view) was also calculated. The growth rate of all isolates was determined on PDA in 90 mm diam. Petri dishes, using three plates for each isolate, incubated at 25°C. Colony diameters were measured for the first 7 d after inoculation, and mean daily colony growth rates were calculated as described by Huang *et al.* (2013).

Molecular characterization and phylogeny of isolates

The identity of the isolates were examined by amplification and sequencing of ITS rDNA regions by a direct PCR approach, without DNA isolation (Iotti and Zambonelli, 2006; Mari *et al.*, 2012). Direct PCR was applied to cultures grown on PDA for 7 d at 25°C. Under a dissecting microscope, a few hyphae were taken from the aerial mycelium of each culture using a sterile needle, and were placed in a 0.2 mL capacity PCR tube containing 20 μL of sterile water. Tubes with hyphae were always maintained on ice until PCR amplification. ITS regions were amplified as described by Iotti and Zambonelli (2006), using the universal fungal primers ITS5 and ITS4 (White *et al.*, 1990). The amplifications were carried out with an initial denaturation at 94°C for 6 min, followed by 35 cycles of denaturation at 94°C for 30 s, annealing at 55°C for 30 s, and extension at 72°C for 1 min, with a final extension at 72°C for 6 min. PCR products were purified using Nucleospin PCR cleanup kit (Macherey Nagel) and sequenced using both the amplification primers by the Eurofin Genomics sequencing service (https://www.eurofinsgenomics.eu/en/custom-dna-sequencing.aspx). The sequences were analysed using the BLASTn program with default parameters (http://blast.ncbi.nlm.nih.gov/Blast.cgi) (Altschul *et al.*, 1990), and they were aligned using the ClustalW software (http://www.ebi.ac.uk/Tools/msa/clustalo/).

Bayesian inference and maximum likelihood were applied to confirm the phylogenetic affiliation of isolates from this study. Two separate phylogenies were constructed for *C. boninense* and *C. gloeosporioides* species complexes. Sequences generated in this study were used in conjunction with those of Damm *et al.* (2012b) and Weir *et al.* (2012). One or more sequences for each *Colletotrichum* sp. were selected for tree construction, by giving priority to those from type specimens (NR GenBank accession prefix) when available. The best fit model of nucleotide substitution (TrNef + I) was selected by jModelTest 2.1.4 (Darriba *et al.*, 2012). Bayesian analysis was performed by two independent runs, with four chains each, and were carried out for 5 million generations, sampling every 100 generations. A 50% majority-rule consensus tree was generated after exclusion of the first 250 trees. Maximum likelihood (ML) analysis was performed in Raxml GUI 1.1.3 (Silvestro *et al.*, 2012). The ML search included 100 random addition replicates and 1,000 nonparametric bootstrap replicates.

Pathogenicity tests

Seven *Colletotrichum* isolates were grown on PDA for 1 week at 27°C, and were used in pathogenicity tests to compare their virulence and evaluate the susceptibility of citrus varieties Valencia orange and Eureka lemon. Ten fruits of each variety were inoculated with each isolate. The fruits were washed, disinfected by immersion for 10 min in sodium hypochlorite solution , rinsed twice in SDW, and were each inoculated by placing a 10 μL droplet of spore suspension (10^5 conidia mL^{-1}) on the surface. Inocula were obtained from cultures grown on PDA for 10 d at 25°C. Conidia were washed from the dishes, passed through four layers of cheesecloth, and diluted in sterile distilled water to 10^5 conidia mL^{-1}. Fruits were previously injured using a sterile tip (wounds of 2 mm diam.). Control fruits were inoculated with SDW. Fruits were each labelled with a permanent marker, and the inoculation points were circled on each fruit. The fruits were then incubated in plastic containers at 25°C and 100% relative humidity, and were examined for lesion development 7 d after inoculation (Aiello *et al.*, 2015). After 14 d, spores from infected fruits were aseptically transferred and sub-cultured onto PDA plates, which were incubated at 25°C. The resulting cultures were checked for colony and spore morphology to confirm Koch's postulates.

Statistical analyses

Data of conidium size, colony growth rate and isolate pathogenicity were compared using one way ANOVA, and mean values among treatments were compared by the Tukey's *post hoc* test ($P < 0.05$).

RESULTS

Surveys, and Colletotrichum *species isolation*

Surveyed orchards located in major citrus production areas in Northern Tunisia showed typical symptoms of anthracnose. These consisted of typical dieback and wither-tip of twigs showing orange conidium masses, which were observed on different citrus varieties (Figure 1). On fruits, symptoms were small, light to dark irregular and sunken lesions (Figure 2). On leaves, circular, light tan, flat areas with purple margins, containing black acervuli, were observed (Figure 3). Seven *Colletotrichum* spp. isolates were obtained from surveyed orchards of citrus growing areas from Northern Tunisia. Four isolates were from orange Malti, one was from orange Meski, one was from orange Thomson and one was from clementine Cassar. Three isolates were from fruits, two were from leaves, one was from a flower and one was from a twig (Table 1).

Morphological and cultural characteristics of isolated fungi

The seven isolates were initially characterized by morphological analysis of mycelium and conidia. Colonies of most of the isolates (six) closely resembled those typically formed by *C. gloeosporioides*, with dense, raised, cottony mycelium, which, after 7 d, turned from white or pale white (for isolate C21) and pink to orange (for isolates C2, C9, C31 and C32) (Figure 4). Conidia

Figure 1. Symptoms on citrus twigs caused by *Colletotrichum* spp. A) wither-tip of twigs of clementine Cassar; B) orange conidiu masses of *Colletotrichum* spp. on twigs of clementine Cassar.

Figure 2. Symptoms on citrus fruits caused by *Colletotrichum* spp. A) Brown and dark lesions on orange, and B) on clementine. C) Stem-end rot on clementine. D) Typical anthracnose on fallen clementine fruit.

Table1. Characteristics of *Colletotrichum* species used in this study.

Isolate	Date of isolation	Location	Host citrus variety	Host tissue	GenBank[a]
C35	11/02/2015	Morneg	Malti	Fruit	MF581072
C2	14/04/2014	Hammamet	Meski	Flower	MF581073
C5	14/04/2014	Hammamet	Cassar	Leave	MF581074
C9	23/04/2014	BouArgoub	Malti	Leave	MF581075
C21	11/02/2015	Morneg	Malti	Fruit	MF581076
C31	11/02/2015	Morneg	Thomson	Twig	MF581077
C32	11/02/2015	Morneg	Malti	Fruit	MF581078

[a] Accession number of ITS sequences.

Figure 3. Anthracnose symptoms on leaves of clementine (A, B and C).

were consistently cylindrical with rounded or obtuse ends, and measured 7 to 21 × 2 to 5 µm (average = 10.9 × 3.7 µm) (Table 2). Some cultures developed black acervuli and bright orange spore masses, which were produced in the colony centres (Figure 4), as typical for *C. gloeosporioides*. In contrast, colonies of isolate C35 were similar to those produced by *C. karstii,* with moderately dense white-orange early mycelium, which, after 7 d, turned pink/orange (Figure 5). Conidia were cylindrical, hyaline and with round apices, and measured 11 to 16 × 4 to 6 µm (average = 13.1 × 4.7 µm) (Table 2). Statistical analyses of conidium dimension data revealed significant differences in length, width and Q-values. The conidia of C35 isolate were longer and their Q-value was greater than those of the other six isolates (Table 2). Statistically significant differences of conidium dimensions were also found among these six isolates.

The mean daily growth colony rate of the isolates, measured at 25°C, ranged from 4.1 to 6.9 mm (Table 2). Colonies of C35 grew more rapidly than the other isolates, and differences in colony growth rates were found between the isolates (Table 2).

PCR amplification and sequence analysis

Sequencing of the ITS regions of the seven isolates was performed in order to allow species identification. PCRs generated amplicons of 599 bp (isolates C2, C5, C9, C21, C31, and C32) and 618 bp (C35). BLASTn search against GenBank revealed that the ITS sequences of seven isolates were identical to those of *C. gloeosporioides* (for isolates C2, C5, C9, C21, C31, C32) and *C. karstii* (isolate C35). Sequence similarity ranged between 99.8 and 100% among *C. gloeosporioides* isolates (one variable position in ITS1) and 91.1 and 91.3% between *C. gloeosporioides* and *C. karstii* isolates (37 to 38 variable positions in ITS1 and six in ITS2). Phylogenetic reconstruction of the *C. boninense* and *C. gloeosporioides* species complexes placed the

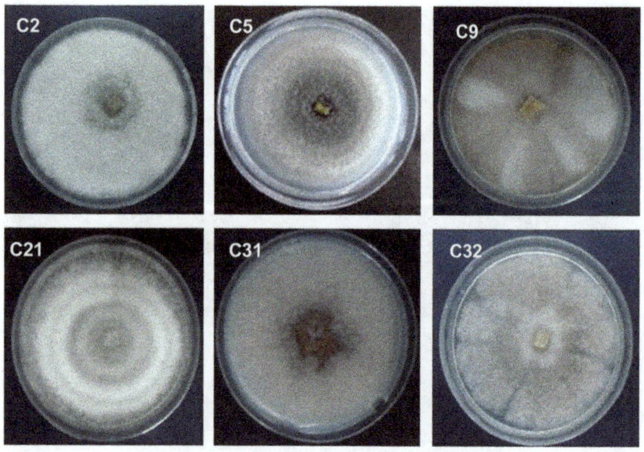

Figure 4. Morphology of *Colletotrichum gloeosporioides* isolates C2, C5, C9, C21, C31 and C32 on PDA plates incubated at 25 °C for 7 d.

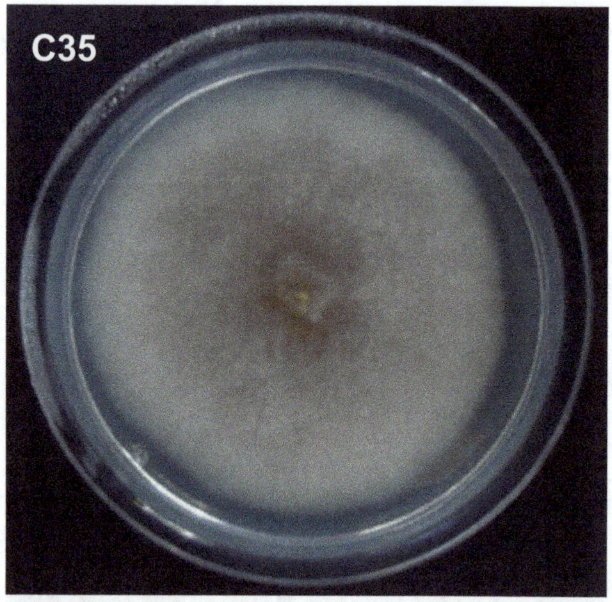

Figure 5. Morphology of *Colletotrichum karstii* (*C. boninense* complex) isolate (C35) on PDA plates incubated at 25 °C for 7 d.

isolates from this study into the clade *C. gloeosporioides sensu stricto* (isolates C2, C5, C9, C21, C31, C32) and the clade *C. karstii* (isolate C35) (Figure 6). Unlike other clades in the genus, both of these clades were well supported, by posterior probabilities and by ML bootstrap values. Two distinct subclades were also evident within the *C. gloeosporioides* clade, and respective Tunisian isolates were present in both. No correlations were found between ITS types and the metadata of the respective isolates (e.g. host tissue and variety, collection site).

Pathogenicity tests

The seven studied isolates induced symptoms on Valencia orange and Eureka lemon, which were identical to those observed in citrus orchards. One week post-inoculation, brown sunken and dark lesions appeared surrounding the inoculation points. After 14 d, the lesions contained dense mycelium, orange conidium masses and black acervuli (Figure 7). No lesions developed on fruits inoculated with water.

The data obtained from this assay were analysed to compare virulence of the studied isolates, taking into consideration the different susceptibilities of orange and lemon. Pathogen virulence and host species susceptibil-

Table 2. Conidium dimensions and growth rates of seven *Colletotrichum* isolates grown on PDA at 25°C. Isolates C2, C5, C9, C21, C31 and C32 are *C. gloeosporioides* and isolate C35 is *C. karstii*.

Character	C2	C5	C9	C21	C31	C32	C35
Length of conidia (μm)							
Mean ± SE[a]	9.6[b] ± 0.2 CD[c]	9.1 ± 0.2 D	10.9 ± 0.2 B	10.3 ± 0.2 BC	11.1 ± 0.2 B	8.9 ± 0.2 D	13.1 ± 0.2 A
Range	7.4-13.8	7.4-11.0	8.3-21.1	8.3-15.7	9.4-14.6	7.4-10.5	10.8-16.1
Width of conidia (μm)							
Mean ± SE	3.7 ± 0.1 D	3.9 ± 0.1 CD	4.1 ± 0.1 BC	4.1 ± 0.1 BC	4.2 ± 0.1 B	3.8 ± 0.1 D	4.7 ± 0.1 A
Range	2.6-5.3	2.9-5.0	2.4-4.9	2.9-4.9	3.4-5.0	2.7-4.6	4.0-5.9
Q-value of conidia							
Mean ± SE	2.6 ± 0.1A	2.4 ± 0.1B	2.7 ± 0.1A	2.6 ± 0.1AB	2.7 ± 0.1A	2.4 ± 0.1B	2.8 ± 0.1A
Growth rate (mm)							
Mean ± SE	4.6 ± 0.5B	4.3 ± 0.5 B	4.5 ± 0.5 B	4.1 ± 0.5 B	5.1 ± 0.5 B	4.3 ± 0.5 B	6.9 ± 0.5 A

[a]SE = standard errors of means.

[b]Each value represents the mean of 50 measurements.

[c]Values derived from the average of 50 measurements. Different letters accompanying means in each row indicate statistically significant differences ($P < 0.05$) according to Tukey's *post hoc* test.

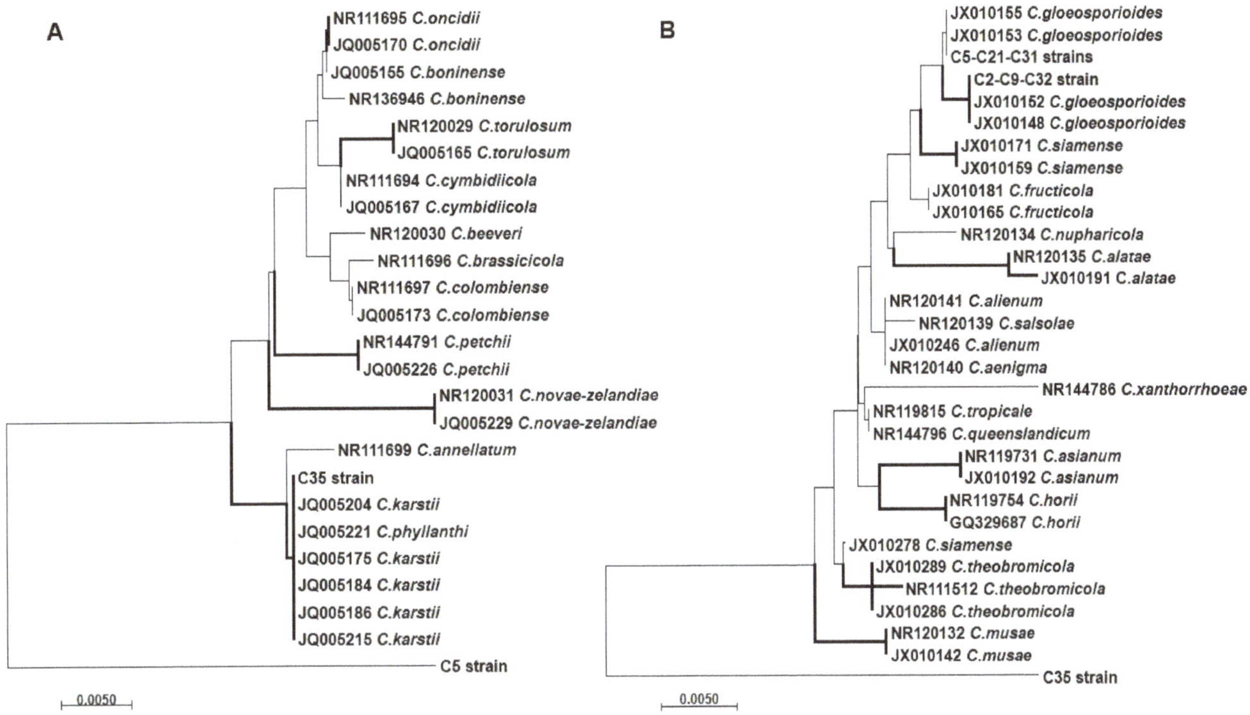

Figure 6. Maximum likelihood trees of *Boninense* (A) and *Gloeosporioides* (B) species complexes, based on the ITS1-5.8S-ITS2 region. Topologies resulting from maximum likelihood and Bayesian inference were congruent and the most likely trees are showed. Thickened branches are supported by likelihood bootstrap values of >70% and by Bayesian posterior probabilities of >95%. Taxa are labelled by their GenBank code (NR codes identify the type vouchers) or isolate code. Sequences from GenBank were selected based on results after Dumm *et al.* (2012) and Weir *et al.* (2012). Sequences of isolates C5 and C35 were differentially used as outgroups. The tree was edited using FigTree v 1.4.3.

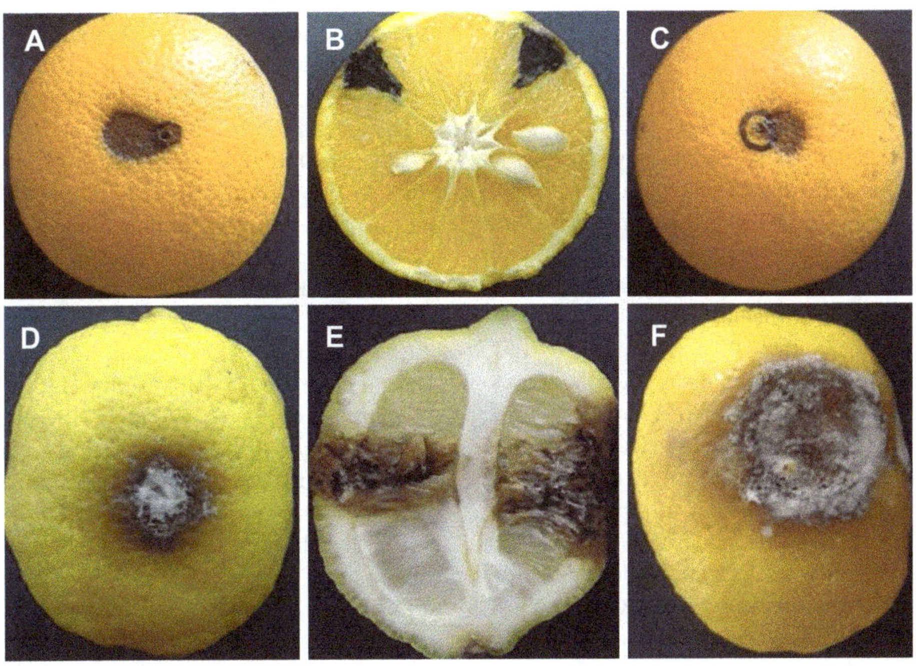

Figure 7. Lesions caused by *Colletotrichum karstii* (isolate C35) on artificially inoculated fruits of Valencia sweet orange (A and B) and on fruits of Eureka lemon (F). Lesions caused by *C. gloeosporioides* (isolate C31) on artificially inoculated fruit of Valencia sweet orange (C) and on fruits of Eureka lemon (D and E).

Table 3. Mean lesion diameter caused by *Colletotrichum* isolates on Valencia sweet orange and Eureka lemon.

Citrus variety	Isolate						
	C2	C5	C9	C21	C31	C32	C35
Valencia orange							
Inoculated fruit	5.4[a] ± 0.1[b] AB	0 ± 0.1 C	6.7 ± 0.1AB	5.3 ± 0.1 AB	8.2 ± 0.1 A	3.4 ± 0.1 BC	8.0 ± 0.1 A
Control fruit	0	0	0	0	0	0	0
Eureka lemon							
Inoculated fruit	24.0 ± 0.3 B	24.3 ± 0.3 B	22.6 ± 0.3 B	20.4 ± 0.3 B	36.8 ± 0.3 A	19.0 ± 0.3 B	13.6 ± 0.3 B
Control fruit	0	0	0	0	0	0	0

[a] Each value represents the mean of five replicates.
[b] SE = standard errors of means.
[c] Different letters accompanying means in each row indicate statistically significant differences for ($P < 0.05$), Tukey's *post hoc* test.

ity were statistically different as demonstrated in ANOVA ($P < 0.05$). For mean lesions diameter, on Valencia orange, the most aggressive isolates were C35 and C31 (Table 3), whereas no lesions were obtained with the isolate C5. On Eureka lemon, the most aggressive isolate was C31, which produced lesions three times larger than those of isolate C35. The data also showed that Eureka lemon was more susceptible to all of the *Colletotrichum* isolates compared to Valencia orange. For example, isolate C35 caused smaller lesions than the other isolates (Table 3). Isolations from lesions of inoculated fruits of the two citrus types yielded colonies identical to the original inoculated isolates, confirming Koch's postulates for these pathogenic fungi.

DISCUSSION

Symptoms of anthracnose, such as wither-tip of twigs, have been observed on cultivated citrus crops in Tunisia since 2009. *Colletotrichum gloeosporioides* was confirmed as the principal pathogen responsible for wither-tip of twigs and lesions on fruits and leaves of Encore mandarin and Minneola tangelo in Northern Tunisia (Rhaiem and Taylor, 2016). The present study has confirmed that *C. gloeosporioides* is the main causal agent of anthracnose also in clementine and sweet orange, and records for the first time in Tunisia that *C. karstii* is an anthracnose pathogen on citrus. Although ITS regions are not variable enough to separate closely related species in *Colletotrichum* (Crouch *et al.*, 2009), phylogeny obtained in the present study clearly showed that the strains were monophyletic in the *C. gloeosporioides sensu stricto* and *C. karstii* clades. Isolate C35 could also belong to *C. phyllanti*, a sister species of *C. karstii*, with identical ITS sequence, but *C. phyllanti* has only been recorded found in India on *Phyllanthus acidus*, *Bahuinia variegate* and *Bougainvillea glabra* (Damm *et al.*, 2012b; Sharma and Shenoy, 2013). *Colletotrichum karstii* could represent a species complex, although more isolates need to be analysed to validate this hypothesis (Sharma and Shenoy, 2013).

Severe symptoms of anthracnose were observed on fruits, flowers, leaves and twigs in different citrus types and varieties in Northern Tunisia. These included Malti, Meski, Thomson oranges and clementine Cassar. *Colletotrichum gloeosporioides* was isolated from citrus fruits, flowers, leaves and twigs, while the unique isolate of *C. karstii* was found on fruits.

Colletotrichum karstii is a morphologically diverse species (Alizadeh *et al.*, 2015). Colour of the upper surfaces of the colonies varies from white to grey, and pink on the reverse sides (Velho *et al.*, 2014), or also white to grey, usually with pink conidium masses, and reverse sides yellow to dark brown (Yang *et al.*, 2011). According to Taheri *et al.* (2016), the surfaces of colonies vary from white to grey with orange conidium masses and entire margins, and with pale orange reverse sides. Orange conidiomata are seen on PDA with filter paper. These variations have also been observed in other *Colletotrichum* spp. (Taheri *et al.*, 2016). In *C. karstii*, conidia also have different size ranges, including: (11.5–14.5) × (5–6.5) μm (Damm *et al.*, 2012b), (13.5–15) × (4–5) μm (Aiello *et al.*, 2015) or (10–16.5) × (4–6.5) μm (Taheri *et al.*, 2016).

During the last 2 to 3 decades, *Colletotrichum* spp. associated with citrus anthracnose have attracted considerable attention from researchers (Ramos *et al.*, 2016). Post-bloom Fruit Drop (PFD) and Key Lime Anthracnose (KLA) caused by *C. acutatum sensu lato* were known to affect citrus production in many countries (Peres *et al.*, 2008). However, these diseases were not previously reported in Tunisia. *Colletotrichum* fungi belonging to the *C. gloeosporioides* species complex include a number of economically important postharvest pathogens (Gan *et al.*, 2017). This species is very common on *Citrus* spp. in many countries, where this fungus is responsible for KLA and postharvest diseases. It also occurs on other hosts (Weir *et al.*, 2012). In Southern Italy, *C. gloeosporioides* is commonly isolated from citrus, especially from *Citrus sinensis* (Aiello *et al.*, 2015; Cannon *et al.*, 2008; Peng *et al.*, 2012). *Colletotrichum gloeosporioides* was also reported to cause preharvest symptoms in Southern Italy, including severe lesions on fruits of sweet orange (Aiello *et al.*, 2015), and wither-tip of twigs and tear stain and stem-end rot of fruits (Benyahia *et al.*, 2003; Huang *et al.*, 2013). *Colletotrichum karstii* has been isolated from citrus plants in South Africa, New Zealand (Damm *et al.*, 2012b) and China (Peng *et al.*, 2012). In Iran, five species of *Colletotrichum*, including *C. gloeosporioides sensu stricto* and *C. karstii*, were found from leaves, fruits and stems of citrus species (Taheri *et al.*, 2016). Recently, *C. karstii* was described as a causal agent of citrus anthracnose in Italy (Aiello *et al.*, 2015). In Portugal, *C. gloeosporioides* and *C. karstii* were identified from symptoms on leaves, branches, flowers and fruit, with *C. karstii* frequently occurring in branches and leaves of lemon in specific geographic locations (Ramos *et al.*, 2016).

The pathogenicity tests carried out in the present study showed that *C. karstii* and *C. gloeosporioides* isolates were virulent for the two studied citrus fruits. In Portugal, Ramos *et al.* (2016) reported that the importance of *C. karstii* was greater in lemons compared to other citrus types, although *C. karstii* can be as virulent as *C. gloeosporioides* for some citrus organs and varieties. In Italy, pathogenicity tests with isolates of *C. gloeosporioides* and *C. karstii* clearly showed their ability to cause lesions on fruits (Aiello *et al.*, 2015). However, these data demonstrated that *C. gloeosporioides* was more aggressive than *C. karstii*. Our study has shown that in Tunisia, *C. karstii* was one of the most virulent isolates on fruits of Valencia orange but it caused less severe symptoms in fruits of Eureka lemon.

Anthracnose caused by different *Colletotrichum* species could become severe threats for citrus production and for other crops when they establish in particular horticulturally important areas. Further studies on the development and spread of anthracnose caused by *C. karstii* in Tunisia are necessary, to provide knowledge of the implications of anthracnose for citrus production, and to define suitable disease management strategies.

ACKNOWLEDGEMENTS

This investigation was financed by UR13AGR03, University of Sousse, Tunisia and also supported by the Tunisian Technical Center of Citrus (CTA) for supplying the fruits used for the pathogenicity assays.

LITERATURE CITED

Aiello D., Carrieri R., Guarnaccia V., Vitale A., Lahoz E., Polizzi G., 2015. Characterization and pathogenicity of *Colletotrichum gloeosporioides* and *C. karstii* causing preharvest disease on *Citrus sinensis* in Italy. *Journal of Phytopathology* 163: 168–177.

Agostini J. P., Timmer L. W., Mitchell D. J. , 1992. Morphological and pathological characteristics of strains of *Colletotrichum gloeosporioidesf* rom citrus. *Phytopathology* 82: 1377–1382.

Alizadeh A., Javan-Nikkhah M., Zare R., Fotouhifar Kh.B., Damm U., Stukenbrock E.H., 2015. New records of *Colletotrichum* species for the mycobiota of Iran. *Mycologia Iranica* 2: 95–109.

Altschul S.F., Gish W., Miller W., Myers E.W., Lipman D.J., 1990. Basic local alignment search tool. *Journal of Molecular Biology* 215: 403–410.

Ben Hadj Daoud H., 2013. *Diagnostic, Caractérisation Morphologique, Biologique etMmoléculaire de Colletotrichum* spp.. Master, University of Sousse, Sousse, Tunisia, 77 pp.

Benyahia H., Jrifi A., Smaili C., Afellah M., Lamsetef Y., Timmer L.W., 2003. First report of *Colletotrichum gloeosporioides* causing withertip on twigs and tear stain on fruit of citrus in Morocco. *Plant Pathology* 52: 798.

Boughalleb-M'hamdi N., Benfradj N., Migliorini D., Luchi N., Santini A., 2018. *Phytophthora nicotianae* and *P. cryptogea* causing gummosis of citrus in Tunisia. *Tropical Plant Pathology* 43: 36–48.

Brown A.E., Sreenivasa Prasad S., Timmer L.W., 1996. Molecular characterization of slow-growing orange and key lime anthracnose strains of *Colletotrichum* from citrus as *C. acutatum*. *Phytopathology* 86: 523–527.

Cai L., Hyde K.D., Taylor P.W.J., Weir B.S., Waller J.M., Abang M.M., Zhang J.Z., Yang Y.L., Phoulivong S., Liu Z.Y., Prihastuti H., Shivas R.G., McKenzie E.H.C., Johnston P.R., 2009. A polyphasic approach for studying *Colletotrichum*. *Fungal Diversity* 39: 183–204.

Cannon P.F., Buddie A.G., Bridge P.D., 2008. The typification of *Colletotrichum gloeosporioides*. *Mycotaxon* 104: 189–204.

Cannon P.F., Damm U., Johnston P.R., Weir B.S., 2012. *Colletotrichum* current status and future directions. *Studies in Mycology* 73: 181–213.

Chattaoui M., Raya M.C., Bouri M., Moral J., Perez-Rodriguez M., Trapero A., Msallem M., Rhouma A.,

2016. Characterization of a *Colletotrichum* population causing anthracnose disease on Olive in northern Tunisia. *Journal of Applied Microbiology* 120: 1368–1381.

Crouch J.A., Clarke B.B., White Jr. J.F., Hillman B.I., 2009. Systematic analysis of the falcate-spored graminicolous *Colletotrichum* and a description of six new species from warm-season grasses. *Mycologia* 101: 717–732.

Damm V., Cannon P.F., Woudenberg J.H., Johnston P.R., Weir B.S., Tan Y.P., Shivas R.G., Crous P.W., 2012. The *Colletotrichum boninense* species complex. *Studies* in *Mycology* 73(1): 1–36.

Damm U., Cannon P.F., Woudenberg J.H.C., Crous P.W., 2012a. The *Colletotrichum acutatum* species complex. *Studies in Mycology* 73: 37–113.

Damm U., Cannon P.F., Woudenberg J.H.C., Johnston P.R., Weir B.S., Tan Y.P., Shivas R.G., Crous P.W., 2012b. The *Colletotrichumboninense* species complex. *Studies in Mycology* 73: 1–36.

Damm U., Cannon P.F., Liu F., Barreto R.W., Guatimosim E., Crous P.W., 2013. The *Colletotrichum orbiculare* species complex: Important pathogens of field crops and weeds. *Fungal Diversity* 61: 29–59.

Damm U., O'Connell R.J., Groenewald J.Z., Crous P.W., 2014. The *Colletotrichum destructivum* species complex – hemibiotrophic pathogens of forage and field crops. *Studies in Mycology* 79: 49–84.

DaSilva A.O., Savi D.C., Gomes F.B., Ramos Gos F.M.W., Silva G.J., Glienke C., 2016. Identification of *Colletotrichum* species associated with post-bloom fruit drop in Brazil through GAPDH sequencing analysis and multiplex PCR. *European Journal of Plant Pathology* 147: 731–748.

Darriba D., Taboada G.L., Doallo R., Posada D., 2012. jModelTest 2: more models, new heuristics and parallel computing. *Nature Methods* 9: 772.

Dean R., Van Kan J.A.L., Pretorius Z.A., Hammond-Kosack K.E., Di Pietro A., Spanu P.D., Rudd J.J., Dickman M., Kahmann R., Ellis J., Foster G.D., 2012. The top 10 fungal pathogens in molecular plant pathology. *Molecular Plant Pathology* 13: 414–430.

Denham T. G., 1979. Citrus production and premature fruit drop disease in Belize. *PANS*25: 30–36.

Fagan H.J., 1979. Post-bloom fruit drop, a new disease of citrus associated with a form of *Colletotrichum gloeosporioides*. *Annals of Applied Biology* 91: 13–20.

FAO, 2016. Agriculture Organization of the United Nations Statistics Division. Website of Food and Agriculture Organization of the United Nations. http:fao.org/.

Gan P., Nakata N., Suzuki T., Shirasu K., 2017. Markers to differentiate species of anthracnose fungi identify *Colletotrichum fructicola* as the predominant virulent species in strawberry plants in Chiba Prefecture of Japan. *Journal of General Plant Pathology* 83: 14–22.

Guarnaccia V., Groenewald J.Z., Polizzi G., Crous P.W., 2017. High species diversity in *Colletotrichum* associated with citrus diseases in Europe. *Persoonia* 39: 32–50.

Haas M., 2017. *Record Citrus Crop Highlights Market Constraints*. USDA Foreign Agricultural Service (FAS) Tunis, GAIN report-TS-17-04.

Haddad M., Boughalleb-M'hamdi N., Vicent A., Cherif M., 2013. Occurrence of Alternaria brown spot on citrus in Tunisia. Integrated Control in Citrus Fruit Crops. *IOBC-WPRS Bulletin* 95: 213–221.

Huang F., Chen G.Q., Hou X., Fu Y.S., Cai L., Hyde K.D., Li H.Y., 2013. *Colletotrichum* species associated with cultivated citrus in China. *Fungal Diversity* 61: 61–74.

Hyde K.D., Cai L., Cannon P.F., Crouch J.A., Crous P.W., Damm U., Goodwin P.H., Chen H., Johnston P.R., Jones E.B.G., Liu Z.Y., McKenzie E.H.C., Moriwaki J., Noireung P., Pennycook S.R., Fenning L.H.P., Prihastuti H., Sato T., Shivas R.G., Tan Y.P., Taylor P.W.J., Weir B.S., Yang Y.L., Zhang J.Z., 2009a. *Colletotrichum*-names in current use. *Fungal Diversity* 39: 147–182.

Iotti M., Zambonelli A., 2006. A quick and precise technique for identifying ectomycorrhizas by PCR. *Mycological Research* 110: 60–65.

Ismail A.M., Cirvilleri G., Yaseen T., Epifani F., Perrone G., Polizzi G., 2015. Characterisation of *Colletotrichum* species causing anthracnose disease of mango in Italy. *Journal of Plant Pathology* 97: 167–171.

Kaur R., Rewal H.S., Sethi A., 2007. Pre-harvest stem-end rot in citrus cultivars due to *Colletotrichumgloeosporioides*. *European Journal of Horticultural Science* 72: 20–25.

Klotz L.J. Fawcett H.S., 1941. *Color Handbook of Citrus Diseases*. University of California, Press, USA, 90 pp.

Kuramae-Izioka E.E., Lopes C.R., Souza N.L. Machado M.A., 1997. Morphological and molecular characterization of *Colletotrichum* spp. from citrus orchards affected by post-bloom fruit drop in Brazil. *European Journal of Plant Pathology* 103: 323–329.

Lima W.G., Spósito M.B., Amorim L., Golçalves F.P., de Filho P.A.M., 2011.*Colletotrichum gloeosporioides*, a new causal agent of citrus post-bloom fruit drop. *European Journal of Plant Pathology* 131: 157–165.

Mari M., M. Guidarelli, C. Martini and A. Spadoni, 2012. First report of *Colletotrichum acutatum* causing bitter rot on apple in Italy. *Plant Disease* 96: 144.

Mcgovern R.J., Rouse R.E., 1994. Grower survey of post-bloom fruit drop incidence in southwest Florida. *Pro-*

ceedings of the Florida State Horticultural Society 107: 92–94.

McGovern R.J., Seijo T.E., Hendricks K., Roberts P. D., 2012. New report of *Colletotrichum gloeosporioides* causing post-bloom fruit drop on citrus in Bermuda. *Canadian Journal of Plant Pathology* 34: 187–194.

MacKenzie S.J., Peres N.A., Barquero M.P., Arauz L.F., Timmer L.W., 2009. Host range and genetic relatedness of *Colletotrichum acutatum* isolates from fruit crops and leather leaf fern in Florida. *Phytopathology* 99: 620–631.

Darriba D., Taboada G.L., Doallo R., Posada D., 2012. jModelTest 2: more models, new heuristics and parallel computing. *Nature Methods* 9: 772.

McMillan R.T., JR., Timmer L.W., 1989. Outbreak of citrus post-bloom fruit drop caused by *Colletotrichum gloeosporioides* in Florida. *Plant Disease* 73–81.

Orozco Santos M., Gonzalez Garza R., 1986. Caída de frutopequeño y su control en naranja 'Valencia' en Veracruz. *Aric. Tec. Mex.* 12: 259–269.

Peng L.J., Yang Y.L., Hyde K.D., Bahkali A.H., Liu Z.Y., 2012. *Colletotrichum* species on citrus leaves in Guizhou and Yunnan provinces, China. *Cryptogamie Mycologie* 33: 267–283.

Peres N.A., MacKenzie S.J., Peever T.L., Timmer L.W., 2008. Post-bloom fruit drop of citrus and key lime anthracnose are caused by distinct phylogenetic lineages of *Colletotrichum acutatum*. *Phytopathology* 98: 345–352.

Phoulivong S., 2011.*Colletotrichum*, naming, control, resistance, biocontrol of weeds and current challenges. *Current Research in Environmental& Applied Mycology* 1: 53–73.

Ramos A.P., Merali Z., Talhinhas P., Sreenivasaprasad S., Oliveira H., 2006. Molecular and morphological characterisation of *Colletotrichum* species involved in citrus anthracnose in Portugal. *IOBC WPRS Bulletin* 29: 317–326.

Ramos A.P., Talhinhas P., Sreenivasaprasad S., Oliveira H., 2016. Characterization of *Colletotrichum gloeosporioides*, as the main causal agent of citrus anthracnose, and *C. karstii* as species preferentially associated with lemon twig dieback in Portugal. *Phytoparasitica* 44: 549–561.

Rhaiem A., Taylor P.W.J., 2016. *Colletotrichum gloeosporioides* associated with anthracnose symptoms on citrus, a new report for Tunisia. *European Journal of Plant Pathology* 146: 219–224.

Rhouma A., Triki M.A., Msallem M., 2010. First report of olive anthracnose caused by *Colletotrichum gloeosporioides* in Tunisia. *Phytopathologia Mediterranea* 49: 95–98.

Schwarz R.E., Klein E.H.J., Monsted P., 1978. Fungal infection of citrus flowers: probable cause of abnormal fruit drop in the Panama mist zone of Misiones, Argentina. In: *Proceedings of the Third International Congress of Plant Pathology*, 16–23 August 1978, Munich, Federal Republic of Germany, Purey, (Abstract), 130 pp.

Sharma G., Shenoy B. D., 2013. Multigene sequence-based identification of *Colletotrichum cymbidiicola*, *C. karstii* and *C. phyllanthi* from India. *Czech Mycology* 65(1): 79–88.

Silvestro D., Michalak I., 2012.raxmlGUI: a graphical front-end for RAxML. *Organisms Diversity & Evolution* 12: 335–337.

Sutton B.C., 1992. The genus *Glomerella* and its anamorph *Colletotrichum*. In: *Colletotrichum: Biology, Pathology and Control* (J.A. Bailey, M.J. Jeger, ed.), CAB international, Wallingford, UK, 1–26.

Taheri H., Javan-Nikkhah M., Elahinia S.A., Khodaparast S.A., Golmohammadi M., 2016. Species of *Colletotrichum* associated with citrus trees in Iran. *Mycologia Iranica* 3: 1–14.

Theodoro, G.F., Peres N.A.R., Verona L.A.F., 2003. Outbreak of post-bloom fruit drop of citrus, caused by *Colletotrichum acutatum*, in Santa Catarina State, southern Brazil. *New Disease Reports* 8: 4.

Timmer, L.W., Brown G.E., 2000. Biology and control of anthracnose diseases of citrus. In: *Colletotrichum: Host Specificity, Pathology, and Host-Pathogen Interaction* (D. Prusky, S. Freeman, M.B.Dickman, ed.), APS Press, St. Paul, MN, USA, 300–316.

Timmer L.W., Gamsey S.M. Graham J.H., 2000. *Compendium of Citrus diseases*. 2nd ed. APS Press, St. Paul, MN, 92 pp.

Velho A.C., Stadnik M.J., Casanova L., Mondino P., Alaniz S., 2014. First report of *Colletotrichum karstii* causing Glomerella leaf spot on apple in Santa Catarina State, Brazil. *Plant Disease* 98: 157.

Weir B.S., Johnston P.R. Damm U., 2012. The *Colletotrichum gloeosporioides* species complex. *Studies in Mycology* 73: 115–180.

White T.J., Bruns T., Lee S., Taylor J., 1990. Amplification and direct sequencing of fungal ribosomal RNA genes for phylogenetics. In: *PCR Protocols: a Guide to Methods and Applications* (M.A. Innis, D.H. Gelfand, J.J. Sninsky, T.J. White, ed.), Academic Press, Inc., New York, 315–322.

Yang Y.L., Cai L., Yu Z.N., Liu Z.Y., Hyde K.D., 2011. *Colletotrichum* species on *orchidaceae* in Southwest China. *Cryptogamie, Mycologie* 32: 229–253.

15

Activity of Italian natural chabasite-rich zeolitites against grey mould, sour rot and grapevine moth, and effects on grape and wine composition

Francesco CALZARANO[1,*], Gabriele VALENTINI[2], Giuseppe ARFELLI[1], Leonardo SEGHETTI[3], Anna Chiara MANETTA[1], Elisa Giorgia METRUCCIO[4], Stefano DI MARCO[4]

[1] *Università degli Studi di Teramo, Facoltà di BioScienze e Tecnologie Agro-Alimentari ed Ambientali, Via Renato Balzarini, 1, 64100 Teramo, Italy*
[2] *Università degli Studi di Bologna, Dipartimento di Scienze e Tecnologie Agro-Alimentari, Viale Fanin 46, 40127 Bologna, Italy*
[3] *IIS "C. Ulpiani", Via della Repubblica, 30, 63100 Ascoli Piceno, Italy*
[4] *CNR IBE, Via Gobetti 101, 40129 Bologna, Italy*
*Corresponding author: fcalzarano@unite.it

Competing Interests: The Author(s) declare(s) no conflict of interest.

Editor: Hanns-Heinz Kassemeyer, Staatliches Weinbauinstitut Freiburg, Germany.

Summary. The activity of Italian chabasite-rich zeolitites for control of grey mould, sour rot and grapevine moth was compared to that from synthetic fungicides and insecticides in three vineyards in the Abruzzo region of Italy. Persistence of zeolitites on grapevine canopies has enabled applications to be carried out before rainfall events, which are likely to predispose for infections by grey mould and sour rot pathogens. Applications for control of *Lobesia botrana* (grapevine moth) were carried out after the peak of the male flight, and the presence of eggs on grape berries was also assessed. Disease and pest control were very satisfactory and comparable to those obtained from synthetic fungicides and insecticides. In addition, there were no differences among treatments for yield, berry composition, or wine ethyl alcohol, pH and total acidity. Phenolic compounds increased in wine from zeolitite-treated plants where the treatment was carried out within 15 days from grape harvest. These promising results have strategic value, because zeolitites provided simultaneous control of grey mould, sour rot and *L. botrana*. However, since these compounds have been tested for the first time for the control of diseases and *L. botrana*, the results obtained in this study require further validation.

Keywords. Zeolites, vineyard, *Botrytis cinerea*, *Lobesia botrana*.

INTRODUCTION

Botrytis cinerea (= *Botryotinia fuckeliana;* Johnston *et al.*, 2014) and *Lobesia botrana* (Denis and Schiffermüller) (Lepidoptera: Tortricidae) are among the major diseases and pests of grapevine worldwide.

Control of grey mould, caused by *B. cinerea*, is not always effective, despite the use of specific fungicides, and incurs severe economic effects for grape production (Gullino, 1992; Elad *et al.*, 2007). During growing seasons characterized by rainy and wet conditions, up to 40–60% reduction in grape yields can be result from grey mould epidemics, leading to deterioration of organolectic characteristics of must and wine (Pearson and Goheen, 1988; Dubos, 2000; Elad *et al.*, 2007). This pathogen is widespread in all vine-growing areas and is particularly aggressive against grapevine cultivars with tight bunches (Marois *et al.*, 1986).

The pathogenic activity of *B. cinerea* is due to the ability of the fungus to synthesize lytic enzymes such as laccase and polygalacturonase (Nakajima and Akutsu, 2014), and secondary metabolites such as oxalic acid, which in turn increase the activity of polygalatturonases and laccases (Manteau *et al.*, 2003), and botrydial, a phytotoxin which causes necrosis of plant cells (Colmenares *et al.*, 2002).

Control of *B. cinerea* is based on at least two fungicide applications each growing season, the most important carried out at the pre-bunch closure growth stage, particularly for compact cluster cultivars. This is to protect micro-wounds that are very susceptible to the pathogen, and to reduce inoculum sources. The second application is usually during ripening, based on prevailing environmental conditions (Marois *et al.*, 1986).

Leaf removal at the end of flowering improves the effectiveness of the fungicide applications. The removal of the leaves may also lead to greater coverage uniformity of the fungicide on berries and provide drying conditions that reduce bunch rot development (English *et al.*, 1993; R'Houma *et al.*, 1998). Removal of leaves at the 'pea-size berries' stage can also reduce infestations of *L. botrana* (grapevine moth), which in turn may favour *B. cinerea* infections from the insect feeding wounds produced on bunches (Pavan *et al.*, 2016).

Effective *B. cinerea* control strategies are also based on the control of other biotic agents which facilitate occurrence of the pathogen (powdery mildew, grapevine moth), and on the correct adoption of cultural practices aimed at a reducing humidity inside grape bunches.

The use of anti-*Botrytis* fungicides can also cause residues in the harvested grapes, with different negative effects for end users, and possible onset of resistant pathogen strains (Fillinger and Elad, 2016).

In Italian vineyards, *L. botrana* has three generations per year (Pavan *et al.*, 2006). After emergence, the females lay eggs on flower buttons, which can be slightly damaged by the first-generation of larvae hatched from eggs. Serious damage is caused by the second and third generations of larvae that penetrate and hollow out the berries, making the bunches more susceptible to grey mould, sour rot, and powdery mildew.

Treatments against *L. botrana* are carried out based on thresholds assessed as the presence of eggs or holes on berries in vineyards normally infested by the moth, or as 5% of bunches infested by the second and third generations in vineyards normally not infested after monitoring of the male flight activity.

Restrictions on the use of synthetic products are increasing in all wine-growing areas, for environmental protection and to limit the onset of resistant pathogen and pest strains. Application of natural compounds for management of grey mould and grapevine moth would contribute to improved production health and safety and reduce potential environmental impacts of pesticides.

Zeolites are tectosilicates and consist of a mineral family with 52 mineral species (Passaglia and Sheppard, 2001; Jha and Singh, 2016). Natural zeolites have alumino-silicate frameworks whose structures contain cavities and channels filled with water and exchangeable cations. The compensating cations may move out from the crystal structures and be replaced by other cations possessing the same positive charge (Cation Exchange Capacity; CEC). In addition, water in the channels may be removed heating from 25 to 250°C. After cooling to room temperature, water can be naturally restored (reversible dehydration). Bish and Ming (2001) and Eroglu (2014) have summarized information on zeolite crystal structure, composition, occurrence, properties and applications. For each application pure zeolites are not employed, but zeolite-rich rocks ("zeolitites") are used, and these have high contents (> 50%) of pure zeolite. The most common zeolite species in zeolitic rocks are clinoptilolite, common in many parts of Europe, and chabasite and phillipsite, which are common in different regions of Italy (Eroglu, 2014).

The effectiveness of zeolitic rocks for practical applications depends both on zeolite type and concentration. Italian chabasite-rich zeolitites are well known for their high performance for several applications (Passaglia and Sheppard, 2001). The agricultural potential of zeolites, due to their CEC (ranging from 100 to 300 meq/100g), hydration properties and adsorption capacity, has been assessed for different applications. For example, these minerals are used as soil conditioners to improve physical and chemical properties of soil, and reduce leaching of NO_3^- and $(PO_4)^{3-}$ from fertilizers, as primary sources of groundwater pollution (Malekian *et al.*, 2011). Zeolite applications to soils can act as slow release fertilizers and increase water holding capacity, improving water and nutrient use efficiency (Nakhli *et al.*, 2017). These minerals can also be raw materials for plant substrates, in particular as greenhouse media for the production of vegetables or as rooting media for cuttings of fruit and ornamental plants (Pond and Mumpton, 1985).

Zeolitites have not been significantly investigated for the control of plant diseases (De Smedt *et al.*, 2015), but several studies have been carried out to investigate these materials for control of harmful insects (Kljajić *et al.*, 2010; De Smedt *et al.*, 2016; Rumbos *et al.*, 2016; Floros *et al.*, 2017). To our knowledge, there is no reported research on zeolitites for control of *B. cinerea* and *L. botrana* on grapevine.

Given the peculiar water sorbent capacity of zeolitites, due to their characteristic three-dimensional structures, the aim of the present study was to verify the effectiveness of a natural Italian chabasite-rich zeolitite, sprayed onto grapevines, for effects on grey mould and grapevine sour rot. These diseases are particularly harmful in high rainfall seasons. Sour rot is difficult to control (Hall *et al.*, 2018), and is characterized by a typical smell due to the formation of ethyl acetate and acetic acid on infected grapevine bunches. This disease is caused by a complex of filamentous fungi, saprophytic yeasts, and bacteria (Steel *et al.*, 2013). Given the positive results for zeolitites activity against insects, we also investigated their activity against grapevine moth. The study also analysed berries and wines from cv. Montepulciano vineyards to assess variations in composition, comparing the treatments of chabasite-rich zeolitites with those of synthetic fungicides and insecticides.

MATERIALS AND METHODS

Vineyard trial

Field trials were carried out in 2015 and 2016 in a vineyard located in Ari (Chieti province), and in 2017 in a vineyard located in Città Sant'Angelo (Pescara province), in the Abruzzo region of central Italy. The vineyard in Ari was established in 2001 and with the vines trained to the bilateral Guyot system, and with vine spacings of 2.5 × 1 m. The vineyard contained Montepulciano and Cococciola cultivars, each of approx. 5000 m^2, and both cultivars were on rootstock 1103 Paulsen. The vineyard in Città Sant'Angelo (CSA) was approx. 10,000 m^2 of cv. Montepulciano grafted on rootstock SO4, was established in 1989, and was trained to the Tendone system with vine spacings of 2.5 × 2.5 m. Both vineyards were located in grape growing areas usually subjected to epidemics of grey mould and sour rot and infestations of grapevine moth.

In both vineyards and in each of the 3 years of trials, the same experimental protocol was adopted. In each of the cultivars in the Ari vineyard, and in the CSA vineyard, trials were established to compare four treatments against grey mould with putative activity against sour rot, or three treatments against grapevine moth (Tables 1 and 2). Each treatment consisted of three replicates, each consisting of a plot of 162 m^2 in the Ari vineyard and 405 m^2 in CSA. Replicates were set up in randomized block designs.

In each of the three trials, an application strategy based on Italian chabasite-rich zeolitites (ICZ), (Agrisana s.r.l.) against grey mould, sour rot and grapevine moth, was compared with a strategy based on synthetic fungicides and insecticides (SYNT). A further treatment against grey mould and sour rot (SYNT/NAT) consisted of applications of the synthetic fungicides at pre-bunch closure followed by applications with zeolitites. In all the trials, untreated treatments were used as experimental controls (Tables 1 and 2). The zeolitites mineralogical composition (percent) was: Chabasite, 68.5 ± 0.9; Phillipsite, 1.8 ± 0.4; Analcime, 0.6 ± 0.3; K-feldspar, 9.7 ± 0.7; Mica, 5.3 ± 0.6; Pyroxene, 2.9 ± 0.4; and Volcanic glass, 11.2 ± 1.0. These zeolitites, collected from quarries in Sorano (GR), central Italy, were chosen for their high content in chabasite, and had high values of CEC (2.17 meq/g) and water retention (48.5 % p/p) (Malferrari *et al.*, 2013).

Applications targeting grey mould were carried out before the occurrence of conditions favorable for infection, according to weather forecasts (Tables 1, 3 and 5). Only applications at pre-bunch closure were made according to the phenological growth stage and regardless of weather conditions. The pre-infection strategy was adopted taking into account the putative adhesion, persistence and rain resistance of zeolitites. In this way the preventive activity of the synthetic products was exploited, following current control guidelines.

In the Ari vineyard, rainfall, temperature and leaf wetness duration were recorded using an agrometeorological station (DigitEco s.r.l.), which was placed at the centre of the vineyard between the two cultivars (Table 3). In the CSA vineyard, meteorological data were obtained from a control unit (Hydrographic Service - Abruzzo Region) located about 6 km from the vineyard (Table 5). The meteorological data were used to verify the amount of rainfall and leaf wetness associated with the occurrence of infections.

The applications against *L. botrana* were carried out after the insect flight peak recorded in each vineyard from four pheromone traps, and after assessment of damaged berries (Table 4).

In all vineyards, at the berry pea-size growth stage, leaf removal at the bunch zone was carried out.

Table 1. Field applications against grey mould, sour rot and grapevine moth on cv. Montepulciano and Cococciola in the Ari vineyard in 2015 and 2016.

Application date	Growth stage	Treatment	Active ingredients	Application rate (kg/L ha^{-1})
30/06/2015	73	1 - SYNT	Spinosad	0.15
		2 - ICZ	Chabasite-rich zeolitite	15
		3 - SYNT/NAT	Chabasite-rich zeolitite	15
		4 - Untreated Control	Nil	-
24/07/2015	77	1 - SYNT	Cyprodinil + Fludioxonil	0.8
		2 - ICZ	Chabasite-rich zeolitite	15
		3 - SYNT/NAT	Cyprodinil + Fludioxonil	0.8
		4 - Untreated Control	Nil	-
06/08/2015	77	1 - SYNT	Pyrimethanil	2.5
		2 - ICZ	Chabasite-rich zeolitite	15
		3 - SYNT/NAT	Chabasite-rich zeolitite	15
		4 - Untreated Control	Nil	-
14/08/2015	83	1 - SYNT	Pyrimethanil + Chlorantaniliprole	2.5 + 0.27
		2 - ICZ	Chabasite-rich zeolitite	15
		3 - SYNT/NAT	Chabasite-rich zeolitite	15
		4 - Untreated Control	Nil	-
03/09/2015	85	1 - SYNT	Cyprodinil + Fludioxonil	0.8
		2 - ICZ	Chabasite-rich zeolitite	15
		3 - SYNT/NAT	Chabasite-rich zeolitite	15
		4 - Untreated Control	Nil	-
14/07/2016	75	1 - SYNT	Pyrimethanil + Spinosad	2.5 + 0.15
		2 - ICZ	Chabasite-rich zeolitite	15
		3 - SYNT/NAT	Chabasite-rich zeolitite	15
		4 - Untreated Control	Nil	-
28/07/2016	77	1 - SYNT	Cyprodinil + Fludioxonil	0.8
		2 - ICZ	Chabasite-rich zeolitite	15
		3 - SYNT/NAT	Cyprodinil + Fludioxonil	0.8
		4 - Untreated Control	Nil	-
04/09/2016	85	1 - SYNT	Cyprodinil + Fludioxonil	0.8
		2 - ICZ	Chabasite-rich zeolitite	15
		3 - SYNT/NAT	Chabasite-rich zeolitite	15
		4 - Untreated Control	Nil	-

All pesticide applications were carried out using pneumatic sprayers, with application volume of 500 L ha^{-1}.

Disease and insect assessments

The assessments of activity against grey mould were carried out in the Ari vineyard on October 2, 2015 and October 3, 2016, close to harvesting, which was done on October 4, 2015 and October 5, 2016. In the CSA vineyard the assessments were carried out on 29 September and 11 October, 2017. For each replicate of each treatment, 60 bunches located in the central rows of the vineyard plots were considered. The percentage of infected bunches (incidence), and the percentage of infected berries (severity), were determined.

The assessments of activity against the grapevine moth were carried out on equal numbers of bunches as follows: in the Ari vineyard, on July 14, 2015 and August 4, 2016 (2nd generation infestation), and on September 10, 2015 and September 30, 2016 (3nd generation infestation); in the CSA vineyard on July 30, 2017 (2nd generation infestation) and September 29, 2017 (3nd generation infestation). The percentage of damaged bunches (incidence), and the percentage of damaged berries (severity) were determined.

The activity of the synthetic fungicides and zeolitite applications on sour rot were assessed, in Ari and CSA

Table 2. Field applications against grey mould, sour rot and grapevine moth on cv. Montepulciano in the CSA vineyard in 2017.

Application date	Growth stage	Treatment	Active ingredients	Application rate (kg/L ha^{-1})
12/07/2017	75	1 - SYNT	Pyrimethanil	2.5
		2 - ICZ	Chabasite-rich zeolitite	15
		3 - SYNT/NAT	Chabasite-rich zeolitite	15
		4 - Untreated Control	Nil	-
23/07/2017	77	1 - SYNT	Cyprodinil + Fludioxonil	0.8
		2 - ICZ	Chabasite-rich zeolitite	15
		3 - SYNT/NAT	(Cyprodinil + Fludioxonil)	0.8
		4 - Untreated Control	Nil	-
09/09/2017	83	1 - SYNT	Pyrimethanil	2.5
		2 - ICZ	Chabasite-rich zeolitite	15
		3 - SYNT/NAT	Chabasite-rich zeolitite	15
		4 - Untreated Control	Nil	-
18/09/2017	85	1 - SYNT	Cyprodinil + Fludioxonil	0.8
		2 - ICZ	Chabasite-rich zeolitite	15
		3 - SYNT/NAT	Chabasite-rich zeolitite	15
		4 - Untreated Control	Nil	-

vineyards, at the same time as the grey mould assessments, calculating the percentages of infected bunches (incidence) and infected berries (severity).

Evaluations of effects of chabasite-rich zeolitites applications on grape yields

Field applications

In the CSA vineyard in 2017, the ICZ treatment plots were first examined for the activity of the applications and then further treated with zeolitites. Each of the three plots (replicates) of the ICZ treatment was subdivided into two 202 m^2 sub-plots. The six sub-plots were treated with zeolitites on 29 September, 15 d before harvest. Three of the sub-plots were not subjected to further applications (ICZ-A treatment). The other three sub-plots were further treated with zeolitites on 11 October, 2 d before grape harvest, which was carried out on 13 October (ICZ-B treatment).

Evaluation of yield and vinification

At harvest, in each of the sub-plots treated on 29 September (IZS-A treatment), or treated on 29 September and 11 October (IZS-B treatment), and in each of the three plots of the SYNT treatment, 12 plants were selected from central rows of each plot or sub-plot (four plants per row), and the grape yields were determined. Berries (100) from each of the plots or sub-plots, alternately taken from the wings, tips and centres of clusters located at about the mid-point along the vine-shoots, were collected. These berry samples were analysed for soluble solids (Brix), pH and total acidity, according to the methods of the Official Gazette of the European Communities. Regulation (EEC) No. 2676/90 (Official Journal L 272, 3.10.1990) (Table 6). A 100 kg sample from 12 plants of each of the three plots (SYNT treatment) or sub-plots (IZS-A and IZS-B treatments), was then vinified (three repetitions per treatment).

Each sample was crushed with a stalk-remover grape crusher. Approximately 93 kg of crushed grapes were obtained from each 100 kg sample. A traditional vinification for red wine was performed, with the cap of skins punched down three times per day. Thirty mg L^{-1} of sulfur dioxide, 20 g hL^{-1} of commercial dry yeast (after rehydration) and 20 g hL^{-1} of fermentation activators (diammonium phosphate + thiamine hydrochloride) were added at the beginning and again at half of the fermentation process. The fermentation ended after 15 d, the samples reached maximum temperature of 30°C during this period. Pressings were carried out to separate the solids after about 15 d, and the wines were decanted. At the end of this process, and for each sample, approx. 70 L of wine was obtained.

Chemical analyses of the wines

The wine obtained was analyzed at 3 months from racking to determine concentrations of the parameters

Table 3. Rainfall, temperature and leaf wetness periods on rainy days at the ARI vineyard in 2015 and 2016.

Date	Maximum temperature (°C)	Minimum temperature (°C)	Rainfall (mm d^{-1})	Leaf wetness (h d^{-1})
20/06/2015	20.3	13.2	4.2	3
24/06/2015	22.1	14.1	30.5	16
07/08/2015	25.1	20.6	75.1	10
15/08/2015	24.4	19.4	1.1	2
16/08/2015	23.6	17.6	40.3	16
04/09/2015	21.1	16.3	13.2	20
08/09/2015	26.2	19.2	0.5	2
19/06/2016	17.4	12.5	15.2	10
20/06/2016	18.3	12.3	0.6	2
23/06/2016	24.7	18.8	1.4	2
06/07/2016	23.7	20.0	0.8	3
09/07/2016	27.8	22.1	1.6	1
15/07/2016	14.6	12.8	100.8	19
16/07/2016	14.1	12.8	47.6	21
01/08/2016	24.7	18.8	14.2	7
23/08/2016	20.5	16.4	0.4	4
31/08/2016	21.6	18.8	1.2	8
05/09/2016	23.7	16.3	2.6	6
06/09/2016	17.2	12.6	22.6	12
07/09/2016	16.2	14.0	10.6	22
08/09/2016	19.2	14.4	0.2	9
09/09/2016	21.2	17.7	1.0	3
10/09/2016	21.2	17.6	0.4	3
11/09/2016	19.5	16.9	6.0	10
12/09/2016	20.7	17.1	0.8	7
16/09/2016	21.3	17.8	7.2	4
18/09/2016	18.8	15.4	0.6	3
19/09/2016	16.3	13.2	8.4	6
21/09/2016	17.2	14.9	2.0	5

Table 4. Numbers of grapevine moth adults captured in traps in two grape cultivars in the Ari vineyard in 2015 and 2016.

Survey data	Cv. Montepulciano		Cv. Cococciola	
	Trap A Capture number	Trap B Capture number	Trap A Capture number	Trap B Capture number
15/06/2015	15	12	11	15
22/06/2015	137	139	146	29
30/06/2015	18	33	77	14
07/07/2015	7	15	5	5
14/07/2015	1	1	4	2
21/07/2015	2	3	2	2
29/07/2015	5	8	4	0
05/08/2015	18	11	6	3
12/08/2015	35	36	44	26
18/08/2015	18	11	27	8
26/08/2015	11	8	8	4
02/09/2015	6	4	3	2
10/09/2015	2	2	0	0
18/09/2015	0	2	2	0
25/09/2015	1	2	1	1
01/10/2015	4	5	0	1
14/06/2016	0	0	0	0
21/06/2016	0	0	0	0
28/06/2016	0	0	0	0
05/07/2016	8	7	8	6
12/07/2016	47	45	33	35
19/07/2016	11	12	15	18
26/07/2016	7	6	8	8
02/08/2016	0	0	0	0
09/08/2016	0	0	0	0
16/08/2016	0	0	0	0
24/08/2016	3	4	2	3
30/08/2016	0	0	0	0
07/09/2016	0	1	0	1
14/09/2016	1	0	0	0
21/09/2016	0	0	0	0
28/09/2016	0	0	0	0

reported in Table 6. Analyses were carried out in accordance with the methods outlined in the official Gazette of the European Community, Regulation (EEC) No. 2676/90 (Official Journal L 272, 3.10.1990).

HPLC Analyses

Determination of anthocyanins in red wine samples, filtered through 0.45 μm nylon filters, was carried out according to OIV-MA-AS 315-11 Method (2007). Thirty microliters of each sample was analysed by HPLC. The chromatographic system was an HPLC Waters Alliance equipped with a Waters 2695 separation module connected to a Waters 2996 photodiode array detector. Separation of analytes was carried out using a Supelcosil LC18 column (5 μm particle size, 250 × 4.6 mm I.D.), as described in the OIV Method. The system was controlled by Waters Empower personal computer software. Identification of the anthocyanins, detected at wavelength 520 nm, was based on retention times. For the calibration curves, i.e. peak area versus concentration, malvidin-3-glucoside chloride (Sigma-Aldrich) was used, resulting in the linear range of concentration between 10 and 100 mg L^{-1}. The calculated regression line was used to compute the amount of each analyte in the samples by interpolation, using the external standard method.

Table 5. Rainfall, temperature and leaf wetness periods on rainy days of the CSA vineyard in 2017.

Date	Maximum temperature (°C)	Minimum temperature (°C)	Rainfall (mm d^{-1})	Leaf wetness (h d^{-1})
17/06/2017	27.1	20.1	4.8	8
28/06/2017	30.3	20.2	1.8	3
14/07/2017	30.2	22.2	39.8	10
24/07/2017	35.4	22.2	2.2	2
25/07/2017	29.4	20.3	13.6	22
10/09/2017	26.1	17.4	18.6	8
11/09/2017	24.5	18.3	21.6	18
19/09/2017	23.4	15.2	12.4	11
20/09/2017	16.3	12.1	10.1	10
24/09/2017	23.1	14.3	11.2	9
06/10/2017	13.3	10.1	11.6	9
07/10/2017	14.2	10.2	10.4	8

Statistical analyses

Statistical analyses were performed with the Tukey's honest significant difference (HSD) test, (P = 0.05), comparing each treatment, in each vineyard and year of trial, for the data of incidence and severity of grey mould, sour rot and grapevine moth. Tukey's honest significant difference (HSD) test, (P = 0.05) was also used to compare yield and wine data from each treatment.

RESULTS

Vineyard trial

Grey mould

In 2015, in the untreated controls for cv. Montepulciano in the Ari vineyard, mean grey mould incidence was 18.3% of the bunches, and mean severity was 1.2%. In the treated plants, mean incidence was from 0 to 1.1%, and mean severity was from 0 to 0.05%. Mean incidence and severity in treated plants were statistically different compared to the untreated control plants, but not between the different treatments (Table 7).

In 2015, in the cv. Cococciola in the Ari vineyard, mean incidence of *B. cinerea* infections in bunches of the untreated controls was 27.2% and mean severity was 1.4%. Both of these values were statistically different compared with those for the treated plants, which showed mean incidence from 0 to 1.7%, and mean severity from 0 to 0.06% (Table 7). As for cv. Montepulciano, no statistically significant differences were detected for the ICZ, ICZ/SYNT and SYNT treated plants (Table 7).

In 2016, in the Ari vineyard, infections were detected in the untreated controls of both cultivars, with mean incidence of 16.7% and mean severity of 0.4% in cv. Montepulciano, and of 56.1 and 3.7% in cv. Cococciola. Also in this case, all the treated plants had significantly reduced infections compared to the untreated controls, without significant differences of activity among treatments (Table 7).

In 2017, in the CSA Vineyard, no infections by *B. cinerea* were detected.

Sour rot

In 2015 in the Ari vineyard, the untreated controls of cv. Montepulciano had mean incidence of sour rot of 37.2% and mean severity of 6.0% (Table 7). In cv. Montepulciano, no sour rot was detected from the SYNT treatment which was statistically less than for the other treated plants, and the untreated control (Table 7). The ICZ and ICZ/SYNT treatments gave similar mean sour rot incidence (respectively, 7.8 and 7.2%) and mean sour rot severity (both at 0.7%). These values were statistically

Table 6. Chemical analyses of cv. Montepulciano berries and wines obtained from vines treated with either Italian chabasite-rich zeolitites or synthetic fungicides and insecticides for control of grey mould, sour rot and grapevine moth.

Parameter	Unit of measurement	Sample	Method of analysis
Total polyphenols *	mg L^{-1}	Wines	Folin-Ciocalteau
Total anthocyanins	mg L^{-1}	Wines	Spectrophotometric
Total acidity	g L^{-1}	Berries and wines	Acid/base titration
Soluble solids	° Brix	Berries	Fheling
pH 20°C	-	Berries and wines	Potentiometric
Ethyl alcohol	% vol.	Wines	Distillation
O.D. 420-520-620 (IC_3)	-	Wines	Spectrophotometric
Anthocyanins	mg L^{-1}	Wines	HPLC

* Total polyphenols expressed as gallic acid equivalents.

Table 7. Mean incidence and severity of different control strategies for control of grey mould and sour rot based on applications of Italian chabasite-rich zeolitites (ICZ), synthetic fungicides (SYNT) or zeolitites replaced by synthetic fungicide at pre-bunch closure (SYNT/NAT), in the cv. Montepulciano and Cococciola in the Ari vineyard.

Survey	Treatment	Grey mould				Sour rot			
		Cv. Montepulciano		Cv. Cococciola		Cv. Montepulciano		Cv. Cococciola	
		Incidence[a] (%)	Severity[b] (%)	Incidence (%)	Severity (%)	Incidence (%)	Severity (%)	Incidence (%)	Severity (%)
02/10/2015	1 - SYNT	0.00 a[c]	0.00 a	0.00 a	0.00 a	0.00 a	0.00 a	0.00 a	0.00 a
	2 - ICZ	1.11 a	0.05 a	1.67 a	0.05 a	7.78 b	0.74 b	1.11 a	0.07 a
	3 - SYNT/NAT	0.00 a	0.00 a	1.67 a	0.06 a	7.22 b	0.65 b	0.56 a	0.05 a
	4 - untreated control	18.33 b	1.19 b	27.22 b	1.37 b	37.22 c	5.99 c	8.33 b	0.71 b
03/10/2016	1 - SYNT	0.56 a	0.01 a	18.89 a	0.76 a	18.33 a	0.34 a	1.67 a	0.05 a
	2 - ICZ	1.11 a	0.01 a	20.56 a	0.90 a	18.33 a	0.41 a	2.22 a	0.04 a
	3 - SYNT/NAT	0.56 a	0.01 a	18.89 a	0.75 a	23.33 a	0.59 a	2.22 a	0.06 a
	4 - untreated control	16.67 b	0.44 b	56.11 b	3.74 b	55.56 b	2.49 b	10.00 b	0.29 b

[a] Incidence = percentage of infected bunches on the total number of bunches.
[b] Severity = percentage of infected berries on the total number of berries.
[c] Statistical analyses were performed according to Tukey's honest significant difference (HSD) test.
Different letters indicate significant differences (P = 0.05).

different compared to those from the untreated controls (Table 7). In cv. Cococciola, all the treatments gave similar and significantly lower values than the controls, in which mean incidence of sour rot was 8.3% and mean severity was 0.7% (Table 7).

In 2016, in the cv. Montepulciano at the Ari vineyard, mean incidence of sour rot in the untreated controls was 55.6% and mean severity was 2.5%. These were significantly greater than those from the other treatments, which did not differ significantly from each other (Table 7). In the untreated controls of cv. Cococciola at the Ari vineyard, mean sour rot incidence was 10.0% and severity was 0.3%. The other treatments gave mean incidence ranging from 1.7% to 2.2% and mean severity from 0.04 to 0.06% (Table 7).

In 2017, in the CSA vineyard, no sour rot was detected.

Grapevine moth

In 2015, after the flight peak of the second generation of *L. botrana* recorded on June 22, both cv. Montepulciano and Cococciola in the Ari vineyard were treated (Tables 1 and 4). In the assessment carried out on July 14, the untreated controls of cv. Montepulciano showed mean incidence of infestation of 33.3% and mean severity of 0.7% (Table 8). In the untreated controls of the cv. Cococciola, mean incidence of infestation was 45.0% and mean severity was 1.2% (Table 8).

Spinosad applied in the SYNT treatment at both vineyards gave complete control of the infestations in cv. Montepulciano, and low levels of infestation in cv. Cococciola at mean incidence of 2.2% and mean severity of 0.03% (Table 8).

The zeolitite applications decreased the grapevine moth infestations in both cultivars, to a lesser extent than in the SYNT treatment, but statistically different compared to the untreated controls (Table 8). In the ICZ treatment for cv. Montepulciano, mean incidence of infestation was 3.9% and mean severity was 0.05%. For cv. Cococciola mean incidence of infestation was 11.1% and mean severity was 0.17% (Table 8).

In 2015, applications with chlorantaniliprole were carried out 2 d after the grapevine moth flight peak of third generation adults recorded in both cultivars on 12 August (Tables 1 and 4). The mean incidence and severity of infestations (assessed on 10 September) in the untreated controls of cv. Montepulciano were, respectively, 8.3 and 0.19%, and for cv. Cococciola were, respectively, 27.8 and 0.7% (Table 8). The treatments with chlorantaniliprole and zeolitites, carried out in the SYNT and ICZ treatments, gave similar reductions of moth infestation, which were significantly different compared to the controls (Table 8).

In 2016, single applications of Spinosad in the SYNT treatment and zeolitites in the ICZ treatment were carried out in the Ari vineyard, on 14 July. These were against the second generation of *L. botrana*, after the peak of flights registered on 12 July (Tables 1 and 4).

Table 8. Activity of Italian chabasite-rich zeolitites (ICZ) and synthetic insecticides (SYNT) on grapevine moth bunch infestations in the cv. Montepulciano and Cococciola in the Ari vineyard in the surveys carried out at the 2nd moth generation: 14/07/2015 and 04/08/2016, and at the 3nd moth generation on 10/09/2015.

Treatment	14/07/2015		10/09/2015		04/08/2016	
	Incidence [a] (%)	Severity [b] (%)	Incidence (%)	Severity (%)	Incidence (%)	Severity (%)
Cv. Montepulciano						
1 – SYNT	0.00 a [c]	0.00 a	2.22 a	0.03 a	0.00 a	0.00 a
2 - ICZ	3.89 b	0.05 b	1.11 a	0.02 a	1.67 a	0.02 a
3 - untreated control	33.33 c	0.71 c	8.33 b	0.19 b	16.67 b	0.28 b
Cv. Cococciola						
1 – SYNT	2.22 a	0.03 a	1.67 a	0.03 a	2.22 a	0.03 a
2 - ICZ	11.11 b	0.17 b	6.11 a	0.11 a	4.44 a	0.07 a
3 - untreated control	45.00 c	1.17 c	27.78 b	0.68 b	15.00 b	0.28 b

[a], [b], [c], see Table 7.

Spinosad and zeolitite applications both significantly reduced the moth infestations compared to the controls (Table 8). In the untreated control, mean incidence of infestation was 16.7% in cv. Montepulciano and 15.0% in cv. Cococciola, while mean severity was 0.3% in both cultivars (Table 8). In the SYNT and ICZ treatments the mean incidence of infestation ranged from 0 to 4.4%, and mean severity from 0 to 0.07% (Table 8).

The low number of captures of third generation moth adults did not require further treatment applications in 2016 in the Ari vineyard (Table 4). During the assessment of 30 September, there was no additional damage compared to second generation moths.

In 2017, in the CSA vineyard and throughout the growing season, no insecticide applications were carried out because no moths were captured, or because there was only sporadic *L. botrana* presence (1–2 moths per trap). During the 30 July and 29 September assessments, no grapevine moth damage was found in the grapes.

Evaluations of effects of chabasite-rich zeolitite applications on yields

At harvest, no differences among treatments were detected in grape yield per plant (Table 9). Furthermore, compared to the SYNT treatment, the treatments with zeolitites (ICZ-A, ICZ-B) had no effects on accumulation of soluble solids or acidic balance of the berries.

The chemical composition of the wines obtained after 3 months of ageing is outlined in Table 10. Although the parameters of alcohol content, pH and total acidity did not vary among the treatments, the treatments gave statistically significant differences for concentrations of polyphenols and anthocyanins, and in wine colour intensity. Polyphenol content was significantly greater from the ICZ-A treatment (mean = 3213 mg L^{-1}) compared to SYNT (mean = 2922 mg L^{-1}), whereas the ICZ-B treatment gave the least amount (mean = 2475 mg L^{-1}). Concentrations of total anthocyanins in the three wines were also statistically different. The ICZ-A treatment gave greater anthocyanin content (mean = 625 mg L^{-1}) compared to the SYNT treatment (mean = 579 mg L^{-1}) and the ICZ-B treatment (mean = 486 mg L^{-1}). The concentration of anthocyanins affected the intensity of colour (IC_3), which resulted in an increased value from the ICZ-A compared to the SYNT treatment, while the ICZ-B treatment gave greater colour discharge (Table 10).

Table 9. Mean grape yields and grape composition parameters recorded in 2017 at harvest, for cv. Montepulciano vines in the CSA vineyard, treated with Italian chabasite-rich zeolitites (ICZ-A and ICZ-B) or synthetic products (SYNT).

Treatment	Yield (Kg vine^{-1})	Soluble solids (° Brix)	pH	Total acidity (g L^{-1})
ICZ-A	10.0 a [a]	26.5 a	3.61 a	6.10 a
ICZ-B	9.8 a	26.5 a	3.61 a	6.10 a
SYNT	9.5 a	26.9 a	3.63 a	6.03 a

[a] Statistical analyses were performed according to Tukey's honest significant difference (HSD) test.
Different letters indicate significant differences (P = 0.05).

The composition of total anthocyanin fractions was determined, and the percentage of each glycosylated anthocyanin and the sum of acylated forms are outlined in Table 11. Montepulciano wines had characteristic anthocyanin profiles, with low amounts cyanidin gluco-

Table 10. Mean chemical composition values for Montepulciano d'Abruzzo wines obtained from cv. Montepulciano vines in the CSA vineyard, treated in 2017 with Italian chabasite-rich zeolitites (ICZ-A and ICZ-B) or synthetic products (SYNT), after 3 months of ageing.

Treatment	Ethyl alcohol (% vol.)	pH	Total acidity (g L^{-1})	Total polyphenols (mg $^{L-1}$)	Total anthocyanins (mg $^{L-1}$)	IC_3
ICZ-A	16.1 a[a]	3.65 a	5.93 a	3213 a	625 a	15.1 a
ICZ-B	16.1 a	3.67 a	5.93 a	2475 c	486 c	10.1 c
SYNT	16.3 a	3.67 a	6.03 a	2922 b	579 b	12.6 b

[a] See Table 9.

Table 11. Mean composition (%) of total anthocyanins in Montepulciano d'Abruzzo wines obtained from cv. Montepulciano vines in the CSA vineyard, treated in 2017 with Italian chabasite-rich zeolitites (ICZ-A and ICZ-B) or synthetic products (SYNT), after 3 months of ageing. Montepulciano d'Abruzzo total anthocyanins included glycosylate and acylated forms (sum of acetate and coumarate fractions).

Treatment	Delphinidin 3-G	Cyanidin 3-G	Petunidin 3-G	Peonidin 3-G	Malvidin 3-G	Acylated anthocyanins
ICZ-A	5.4 a[a]	0.5 a	9.3 a	9.6 a	58.0 a	17.6 a
ICZ-B	4.6 a	0.5 a	10.4 a	8.8 a	62.9 a	16.1 a
SYNT	4.6 a	0.5 a	8.8 a	9.0 a	60.2 a	17.4 a

[a] See Table 9.

side. The percentages of total glycosilated anthocyanins (delphinidin, cyanidin, petunidin, peonidin and malvidin), and of acylated anthocyanins (sum of acetates and coumarates) did not significantly differ among the different treatments. In particular, the glycosylated fractions were predominant, and malvidin 3-G was the major constituent in wine solution after 3 months of ageing.

DISCUSSION

Increasing awareness of the impacts of plant protection on the environment has led to restrictions on the use of synthetic pesticide products, with increasing research of predisposing epidemiological factors to diseases (Calzarano *et al.*, 2018), and there has been increased research on environmentally-friendly products for disease and pest management (Calzarano and Di Marco, 2018). In the present study, Italian natural chabasite-rich zeolitites were evaluated as environmentally-friendly materials for management of grapevine grey mould, sour rot and grapevine moth.

In both years of trials carried out in the Ari vineyard, applications of zeolitites gave activity for grey mould control, which was comparable to that from synthetic fungicides. This activity was verified for applications of zeolitites against *B. cinerea* at all growth stages except the pre-bunch closure growth stage (Pearson and Goheen, 1988), and in the control strategy based only on applications of zeolitites.

The control strategies for *B. cinerea* were effective also against sour rot. The side activity of the anti-*Botrytis* fungicide mixture cyprodinil + fludioxonil against the sour rot is well known (Adaskaveg *et al.*, 2011), but was greater than that from the zeolitite treatments only in 2015 in the cv. Montepulciano grapes.

However, the reductions of sour rot from the different control strategies was considerable in the presence of moderate infections, and to a lesser extent when the incidence of the disease was severe, as in 2016 in the cv. Montepulciano grapes.

The cv. Montepulciano and Cococciola have different susceptibilities to grey mould and sour rot. Severe infections of sour rot in cv. Montepulciano corresponded to minor grey mould infections. Similarly in cv. Cococciola, bunches heavily infected with *B. cinerea* were less affected by sour rot.

The activity of zeolitites against *L. botrana* was high, and was comparable to that of synthetic insecticides where mild infestations of the moth occurred. Zeolitites also gave good control of heavy infestations. This was for the second moth generation in 2015 in both cultivars in the Ari vineyard, but was less than that achieved from synthetic insecticides.

The lack of grey mould and sour rot in 2017 in the CSA vineyard could be ascribed to the dry weather conditions of that year, characterized by high temperatures and low relative humidity. Resistance to diseases in cv. Montepulciano clone could also result from the greater thickness of berry cuticles in this cultivar compared with others, despite the occurrence of rainfall predisposing to infections (Rogiers *et al.*, 2005; Mundy, 2008). The absence of grapevine moth infestations in 2017 in the CSA vineyard may have also resulted from the high temperatures that occurred (Moosavi *et al.*, 2017).

The good efficacy of the applications of zeolitites towards grey mould and sour rot could attributed to the physico-chemical properties of these materials (Bish and Ming, 2001; De Smedt *et al.* 2015). Water adsorption capacity and the resulting reduction in moisture in zeolitite-treated bunches could have caused the decreases in grey mould and sour rot (Ng and Mintova, 2008; Tatlier *et al.* 2018).

Applications of zeolitites lead to the formation of microscopic layers of mineral particles (Glenn and Puterka, 2005). High hydrophilicity of these layers of chabasite-rich zeolitites results in the absorption of condensing water and elimination of free water (Tatlier *et al.*, 2018). Direct contact of disease inoculum with host tissues could be reduced by a physical barrier developed on the treated plant surfaces. This barrier can also reduce spore germination and prevent or hinder microbial growth (Walters, 2006).

Studies on the activity of zeolitites against grapevine moth are lacking, but the insecticidal potential of zeolitite formulations have been evaluated for control of others insect pests. These include stored-grain insects (Kljajić *et al.*, 2010; Rumbos *et al.*, 2016; Kavallieratos *et al.*, 2018), *Tuta absoluta* on tomato (De Smedt *et al.*, 2016) and bean weevil (Floros *et al.*, 2017). These studies have demonstrated that zeolitites induced high adult and progeny mortality. Investigations on mechanisms of action towards different harmful insects have also been carried out for other mineral powders similar to zeolitites, such as kaolin, a clay mineral composed of aluminosilicate.

As in the case of kaolin, the epicuticles of insect larvae may be damaged by the non-sorptive particles of zeolitites for abrasion and by adsorption of epicuticular lipids to sorptive particles. Loss of water from the insect bodies leads to the death from desiccation (Ebeling, 1971; Glenn *et al.*, 1999). This specific activity has recently been observed from applying zeolitites to Pharaoh ants (Van Den Noortgate *et al.*, 2018). Plants treated with kaolin clay showed a low oviposition rates by different insects. Furthermore, hatch rate of eggs and larval development significantly decreased from these treatments, causing a high insect mortality (Puterka *et al.*, 2000; Knight *et al.*, 2000).

The larvae on which the mineral particles adhere can also die from starvation, because they are subject to tactile deterrence leading to inability to feed (Larenzaki *et al.*, 2008). Deposition of the particle films on the treated plants can also reduce the visual cues for insects, and hinder recognition of plant parts on which the adults lay their eggs (Glenn *et al.*, 1999; Mazor and Erez, 2004). The possible increases in infestations of insects not directly affected by kaolin applications (Knight *et al.*, 2001; Markò *et al.*, 2008) were not detected in the present study.

Zeolitite activity for control of grey mould, sour rot and grapevine moth has been linked with the adhesion and persistence of the mineral deposits on the treated grapes, with increased resistance to rainfall wash-off of the mineral. This characteristic is associated with the effectiveness of preventive control strategies against grey mould, based on applications carried out just before expected rainfall.

The use of zeolitites has also proved to be effective for integrated pest management for *L. botrana*. Applications were carried out immediately after the peak of adult insect flights, assessed as presence of penetration holes on grape berries (Amo-Salas *et al.*, 2011).

One of the most interesting results from the present study is the possibility of using zeolitites to simultaneously control grey mould, sour rot and grapevine moth. This could provide a clear economic advantage for viticulturists, and for the environment, in a combined disease and pest management strategy that has not been previously suggested.

Increasing of temperatures, associated with climate change, could strongly influence the production of high quality grapes, affecting the grapevine development and phenology during growing seasons. It has been established, by focusing on phenolic compounds in black berry grapevine varieties, that high temperatures negatively affect colour intensity, resulting in poor colour wines (Buttrose *et al.*, 1971; Kliewer and Torres, 1972; Mori *et al.*, 2007). Recent studies on cv. Sangiovese showed that exposure of grape bunches to high temperatures lead to reductions in synthesis of anthocyanin, while anthocyanin degradation was stimulated (Movahed *et al.*, 2016; Pastore *et al.*, 2017a; Pastore *et al.*, 2017b). Thus, in hot growing seasons as in 2017 in the CSA vineyard, the contents of phenolic compounds were low in all treatments compared to the usual amounts of these compounds in cv. Montepulciano grapes (Mattivi *et al.*, 2006).

The timing of zeolitite applications to grapevine canopies influenced the concentrations of phenolic compounds in grape berries and wine, without affecting the percentage fraction of each individual anthocyanin. No differences resulted for alcohol contents and wine total acidity. When the last treatment with zeolitites was carried out no later than 15 d from harvest (treatment ICZ-A), the polyphenol concentrations increased and the wine developed increased colour intensity. This result is similar to that from foliar applications of kaolin, which is a radiation-reflecting inert mineral able to reduce leaf surface temperatures and influence major secondary metabolism pathways leading to biosynthesis of phenolic compounds and anthocyanins (Glenn and Puterka, 2005; Song *et al.*, 2012; Conde *et al.*, 2016). Dinis *et al.* (2016) showed that kaolin-treated grape berries had enhanced total amounts of phenols, flavonoids, anthocyanins and vitamin C compared with grapes from untreated vines. The timing of zeolitite applications was shown to be a crucial factor. Application close to grape harvest (the ICZ-B treatment) gave a residue of inert dust that effectively decreased the wine phenol contents, affecting anthocyanin concentration and colour intensity. This was probably due to the hydrophilic properties of zeolitites, determining the adsorption of target molecules (Perego *et al.*, 2013; Mercurio *et al.*, 2016; Lisanti *et al.*, 2017). Recent studies have shown that natural and surfactant-modified zeolites are effective sorbents for removal of organic compounds such as phenol and 4-chlorophenol from aqueous solutions (Kuleyin, 2007; Yousef *et al.*, 2011).

The low content of cyanidin glucoside associated with the high content of malvidin glucoside was attributed to the specific anthocyanin pattern of cv. Montepulciano (Mattivi *et al.*, 2006), and to enzymatic conversion that occurred in berries during the late growing season (Versari *et al.*, 1999).

Due to lack of interference of the grape composition, particularly for phenolic compounds, the use of zeolitites could further be encouraged.

The positive results obtained in this study need to be confirmed to fully assess the potential of zeolitites as environmentally-friendly control strategies for grapevine disease and pest management. Use of these compounds could also provide clear economic advantages for grape and wine production.

LITERATURE CITED

AA.VV., 1990. Methods of community analysis to use in wine sector. In: *Official Gazette of the European Communities*, 3 October 1990. Commission Regulation (EEC). No. 2676/90.

Adaskaveg J.E., Gubler W.D., Michailides T.J., Holtz B.A., 2011. *Efficacy and timing of Fungicides, Bactericides, and Biologicals for Deciduous Tree Fruit, Nut, Strawberry, and Vine Crops 2011*. University of California – California Digital library, eScholarship.org., 49 pp.

Amo-Salas M., Ortega-López V., Harman R., Alonso-González A., 2011. A new model for predicting the flight activity of *Lobesia botrana* (Lepidoptera: Tortricidae). *Crop Protection* 30: 1586–1593.

Bish D.L., Ming D.W., 2001. *Natural Zeolites: Occurrence, Properties, Application*. M.G.S.A. Ed., 654 pp.

Buttrose M.S., Hale C.R., Kliewer W.M., 1971. Effect of temperature on the composition of 'Cabernet Sauvignon' berries. *American Journal of Enology and Viticulture* 22: 71–75.

Calzarano F., Di Marco S., 2018. Further evidence that calcium, magnesium and seaweed mixtures reduce grapevine leaf stripe symptoms and increase grape yield. *Phytopathologia Mediterranea* 57: 459–471.

Calzarano F., Osti F., Baránek M., Di Marco S., 2018. Rainfall and temperature influence expression of foliar symptoms of grapevine leaf stripe disease (esca complex) in vineyards. *Phytopathologia Mediterranea* 57: 488–505.

Colmenares A.J., Aleu J., Duran-Patron R., Collado I.G., Hernandez-Galan R., 2002. The putative role of botrydial and related metabolites in the infection mechanism of *Botrytis cinerea*. *Journal of Chemical Ecology* 28: 997–1005.

Conde A., Pimentel D., Neves A., Dinis L-T., Bernardo S., ... Moutinho-Pereira J., 2016. Kaolin Foliar Application Has a Stimulatory Effect on Phenylpropanoid and Flavonoid Pathways in Grape Berries. *Frontiers in Plant Science*, 7: 1150.

De Smedt C., Someus E., Spanoghe P., 2015. Potential and actual uses of zeolites in crop protection. *Pest Management Science* 71: 1355–1367.

De Smedt C., Van Damme V., De Clercq P., Spanoghe P., 2016. Insecticide effect of zeolites on the tomato leafminer *Tuta absoluta* (Lepidoptera: Gelechiidae). *Insects* 7: 72.

Dinis L-T., Bernardo S., Conde A., Pimentel D., Ferreira H., ... Moutinho-Pereira J., 2016. Kaolin exogenous application boosts antioxidant capacity and phenolic content in berries and leaves of grapevine under summer stress. *Journal of Plant Physiology* 191: 45–53.

Dubos B., 2000. Grey mould. In: *Parasitic Fungal Diseases in the Vine* (Féret, ed.), Bordeaux, France.

Ebeling W., 1971. Sorptive dusts for pest control. *Annual Review of Entomology* 16: 123–158.

Elad Y., Williamson B., Tudzynski P., Delen N., 2007. *Botrytis: Biology, Pathology and Control*. Springer Netherlands, Dordrecht, 403 pp.

English J.T., Kaps M.L., Moore J.F., Hill J., Nakova M., 1993. Leaf removal for control of Botrytis bunch rot of wine grapes in the midwestern United States. *Plant Disease* 77: 1224–1227.

Eroglu N., 2014. A review: Insecticidal potential of Zeolite (Clinoptilolite), toxicity ratings and general properties of Turkish Zeolites. In: *11th International Working Conference on Stored Product Protection*: 755–767.

Fillinger S., Elad Y. (Eds.), 2016. *Botrytis: The Fungus, the Pathogen and Its Management in Agricultural Systems*. New York, NY: Springer, 189–216.

Floros G.D., Kokkari A.I., Kouloussis N.A., Kantiranis N.A., Damos P., ... Koveos D.S., 2017. Evaluation of the natural zeolite lethal effects on adults of the bean weevil under different temperatures and relative humidity regimes. *Journal of Economic Entomology* 111: 482–490.

Glenn D.M., Puterka G.J., Vanderzwet T., Byers R.E., Feldhake C., 1999. Hydrophobic particle films: a new paradigm for suppression of arthropod pests and plant diseases. *Journal of Economic Entomology* 92: 759–771.

Glenn D.M., Puterka G.J., 2005. Particle films: a new technology for agriculture. *Horticultural Reviews* 31: 1–44.

Gullino M., 1992. Chemical control of *Botrytis spp*. In: *Recent Advances in Botrytis Research* (Verhoeff K., Malathrakis N.E., Williamson B., eds.), Pudoc Scientific Publishers, Wageningen, 217–222.

Hall M.E., Loeb G.M., Wilcox W.F., 2018. Control of Sour Rot via Chemical and Canopy Management Techniques. *American Journal of Enology and Viticulture* 69: 342–350.

Jha B., Singh D.N., 2016. *Fly Ash Zeolites: Innovations, Applications, and Directions* (Vol. 78). Springer, 207 pp.

Johnston P.R., Seifert K.A., Stone J.K., Rossman A.Y., Marvanova L., 2014. Recommendations on generic names competing for use in Leotiomycetes (Ascomycota). *IMA fungus* 5: 91–120.

Kavallieratos N.G., Athanassiou C.G., Boukouvala M.C., Rumbos C.I., 2018. Acaricidal effect of three zeolite formulations on different life stages of *Tyrophagus putrescentiae* (Schrank) and *Acarus siro* L. (Sarcoptiformes: Acaridae). *Journal of Stored Products Research* 78: 39–44.

Kljajić P., Andrić G., Adamović M., Bodroža-Solarov M., Marković M., Perić I., 2010. Laboratory assessment of insecticidal effectiveness of natural zeolite and diatomaceous earth formulations against three stored-product beetle pests. *Journal of Stored Products Research* 46: 1–6.

Kliewer W.M., Torres R.E., 1972. Effect of controlled day and night temperatures on grape coloration. *American Journal of Enology and Viticulture* 23: 71–77.

Knight A.L., Unruh T.R., Christianson B.A., Puterka G.J., Glenn D.M., 2000. Effects of a kaolin-based particle film on obliquebanded leafroller (Lepidoptera: Tortricidae). *Journal of Economic Entomology* 93: 744–749.

Knight A.L., Christianson B.A., Unruh T.R., Puterka G., Glenn D.M., 2001. Impacts of seasonal kaolin particle films on apple pest management. *The Canadian Entomologist* 133: 413–428.

Kuleyin A. 2007. Removal of phenol and 4-chlorophenol by surfactant-modified natural zeolite. *Journal of Hazardous Materials* 144: 307–315.

Larentzaki E., Shelton A.M., Plate J., 2008. Effect of kaolin particle film on *Thrips tabaci* (Thysanoptera: Thripidae), oviposition, feeding and development on onions: a lab and field case study. *Crop Protection* 27: 727–734.

Lisanti M.T., Gambuti A., Genovese A., Piombino P., Moio L., 2017. Treatment by fining agents of red wine affected by phenolic off-odour. *European Food Research and Technology* 243: 501–510.

Malekian R., Abedi-Koupai J., Eslamian S.S., 2011. Influences of clinoptilolite and surfactant-modified clinoptilolite zeolite on nitrate leaching and plant growth. *Journal of Hazardous Materials* 185: 970–976.

Malferrari D., Laurora A., Brigatti M.F., Coltorti M., Di Giuseppe D., ... Vezzalini M.G., 2013. Open-field experimentation of an innovative and integrated zeolitite cycle: project definition and material characterization. *Rendiconti Lincei* 24: 141–150.

Manteau S., Abouna S., Lambert B., Legendre L., 2003. Differential regulation by ambient pH of putative virulence factor secretion by the phytopathogenic fungus *Botrytis cinerea*. *FEMS Microbiology Ecology* 43: 359–366.

Markò V., Blommers L.H.M., Bogya S., Helsen H., 2008. Kaolin particle films suppress many apple pests, disrupt natural enemies and promote woolly apple aphid. *Journal of Applied Entomology* 132: 26–35.

Marois J.J., Nelson J.K., Morrison J.C., Lile L.S., Bledsoe A.M., 1986. The influence of berry contact within grape clusters on the development of *Botrytis cinerea* and epicuticolar wax. *American Journal of Enology and Viticolture* 37: 293–296.

Mattivi F., Guzzon R., Vrhvsek U., Stefanini M., Velasco R., 2006. Metabolite Profiling of Grape: Flavonols and Anthocyanins. *Journal of Agricultural and Food Chemistry* 54: 7692–7702.

Mazor M., Erez A., 2004. Processed kaolin protects fruits from Mediterranean fruit fly infestations. *Crop Protection* 23: 47–51.

Mercurio M., Bish D.L., Cappelletti P., de Gennaro B., de Gennaro M., ... Langella A., 2016. The combined use of steam-treated bentonites and natural zeolites in the oenological refining process. *Mineralogical Magazine* 80: 347–362.

Moosavi F.K., Cargnus E., Pavan F., Zandigiacomo P., 2017. Mortality of eggs and newly hatched larvae of *Lobesia botrana* (Lepidoptera: Tortricidae) exposed to high temperatures in the laboratory. *Environmental Entomology* 46: 700–707.

Mori K., Goto-Yamamoto N., Kitayama M., Hashizume K., 2007. Loss of anthocyanins in red-wine grape under high temperature. *Journal of Experimental Botany* 58: 1935–1945.

Movahed N., Pastore C., Cellini A., Allegro G., Valentini G., ... Filippetti I., 2016. The grapevine VviPrx31 peroxidase as a candidate gene involved in anthocyanin degradation in ripening berries under high temperature. *Journal of Plant Research* 129: 513–526.

Mundy D.C., 2008. A review of the direct and indirect effects of nitrogen on botrytis bunch rot in wine grapes. *New Zealand Plant Protection* 61: 306–310.

Nakhli S.A.A., Delkash M., Bakhshayesh B.E., Kazemian H., 2017. Application of zeolites for sustainable agriculture: a review on water and nutrient retention. *Water, Air, & Soil Pollution* 228: 464.

Nakajima M., Akutsu K., 2014. Virulence factors of *Botrytis cinerea*. *Journal of General Plant Pathology* 80: 15–23.

Ng E.P., Mintova S., 2008. Nanoporous materials with enhanced hydrophilicity and high water sorption capacity. *Microporous and Mesoporous Materials* 114: 1–26.

OIV, 2007. HPLC-Determination of nine major anthocyanins in red and rosé wine (Resolution Oeno 22/2003 modified by Oeno 12/2007). Compendium of international methods of analysis-OIV anthocyanins, OIV-MA-AS315-11, 1–13 pp.

Passaglia E., Sheppard R.A., 2001. The crystal chemistry of zeolites. In: Natural Zeolites: Occurrence, Properties, Applications. *Reviews in Mineralogy & Geochemistry* 45: 69–116.

Pastore C., Dal Santo S., Zenoni S., Movahed N., Allegro G., ... Tornielli G.B., 2017a. Whole plant temperature manipulation affects flavonoid metabolism and the transcriptome of grapevine berries. *Frontiers in Plant Science* 8: 929.

Pastore C., Allegro G., Valentini G., Muzzi E., Filippetti I., 2017b. Anthocyanin and flavonol composition response to veraison leaf removal on Cabernet Sauvignon, Nero d'Avola, Raboso Piave and Sangiovese *Vitis vinifera* L. cultivars. *Scientia horticulturae* 218: 147–155.

Pavan F., Zandigiacomo P., Dalla Montà L., 2006. Influence of the grape-growing area on the phenology of *Lobesia botrana* second generation. *Bulletin of Insectology* 59: 105–109.

Pavan F., Cargnus E., Kiaeianmoosavi S., Bigot G., Tacoli F., Zandigiacomo P., 2016. Bunch-zone leaf removal of grapevines to prevent damage by *Lobesia botrana* and grey mould. *Bulletin of Insectology* 69: 107–115.

Pearson R.C., Goheen A.C., 1988. *Compendium of Grape Diseases*. APS Press, St. Paul, Minnesota, USA, 93 pp.

Perego C., Bagatin R., Tagliabue M., Vignola R., 2013. Zeolites and related mesoporous materials for multi-talented environmental solutions. *Microporous and Mesoporous Materials* 166: 37–49

Pond W.G., Mumpton F.A., 1985. *Zeo-Agricolture*. Westviwe Press Ed., 296 pp.

Puterka G.J., Glenn D.M., Sekutowski D.G., Unruh T.R., Jones S.K., 2000. Progress toward liquid formulations of particle films for insect and disease control in pear. *Environmental Entomology* 29: 329–339.

R'Houma A., Chérif M., Boubaker A., 1998. Effect of nitrogen fertilization, green pruning and fungicide treatments on Botrytis bunch rot of grapes. *Journal of Plant Pathology* 80: 115–124.

Rogiers S.Y., Whitelaw-Weckert M., Radovanonic-Tesic M., Greer L.A., White R.G., Steel C.C., 2005. Effects of spray adjuvants on grape (*Vitis vinifera*) berry microflora, epicuticular wax and susceptibility to infection by *Botrytis cinerea*. *Australasian Plant Pathology* 34: 221–228.

Rumbos C.I., Sakka M., Berillis P., Athanassiou C.G., 2016. Insecticidal potential of zeolite formulations against three stored-grain insects, particle size effect, adherence to kernels and influence on test weight of grains. *Journal of Stored Products Research* 68: 93–101.

Song J., Shellie K.C., Wang H., Qian M.C., 2012. Influence of deficit irrigation and kaolin particle film on grape composition and volatile compounds in Merlot grape (*Vitis vinifera* L.). *Food Chemistry* 134: 841–850.

Steel C.C., Blackman J.W., Schmidtke L.M., 2013. Grapevine bunch rots: impacts on wine composition, quality, and potential procedures for the removal of wine faults. *Journal of Agricultural and Food Chemistry* 61: 5189–5206.

Tatlier M., Munz G., Henninger S.K., 2018. Relation of water adsorption capacities of zeolites with their structural properties. *Microporous and Mesoporous Materials* 264: 70–75.

Van Den Noortgate H., Lagrain B., Wenseleers T., Martens J., 2018. Analysis of Cuticular Lipids of the Pharaoh

Ant (*Monomorium pharaonis*) and Their Selective Adsorption on Insecticidal Zeolite Powders. *International Journal of Molecular Sciences* 19: 2797.

Versari A., Barbanti D., Galassi S., 1999. Anthocyanin composition of Montepulciano d'Abruzzo must during industrial fermentation process. *Journal of Wine Research* 10: 223–227.

Walters D.R., 2006. Disguising the leaf surface: the use of leaf coatings for plant disease control. *European Journal of Plant Pathology*, 114: 255–260.

Yousef R.I., El-Eswed B., Al-Muhtaseb A.H., 2011. Adsorption characteristics of natural zeolites as solid adsorbents for phenol removal from aqueous solutions: kinetics, mechanism, and thermodynamics studies. *Chemical Engineering Journal* 171: 1143–1149.

16

Pleurostomophora richardsiae associated with olive tree and grapevine decline in Southern Brazil

Maria Cristina CANALE[1,*], Cristiano NUNES NESI[1], Bianca Ramos FALKENBACH[2], Cesar Arthur HUNHOFF DA SILVA[2], Eduardo Cesar BRUGNARA[1]

[1] *Agricultural Research and Rural Extension Company of Santa Catarina State (Epagri), 89803-904 Chapecó, Santa Catarina, Brazil*
[2] *Agronômica, Laboratory of Phytosanitary Diagnosis and Consultancy, 91530-000 Porto Alegre, Rio Grande do Sul, Brazil*
*Corresponding author: M.C. Canale. E-mail: cristinacanale@epagri.sc.gov.br

Abstract. Olive trees showing decline symptoms were reported in an orchard in Santa Catarina, Southern Brazil. The symptoms included foliar browning and leaf drop, dieback of twigs and branches, dark streaks in internal crown wood and necrotic lesions in bark. A grapevine sample exhibiting browning streaks in the vascular tissue of branches, trunk and roots, was also analysed. A *Phialophora*-like fungus was isolated from symptomatic tissues of both hosts, and monosporic cultures were established. Based on morphology and analysis of ITS, *tub2* and *tef* sequences, the fungus was identified as *Pleurostomophora richardsiae*. Isolates inoculated onto young nursery olive trees and grapevines produced brown streaking in the wood in most of the inoculated plants. This is the first report of *P. richardsiae* occurring in olive trees and grapevines in Brazil.

Keywords. Fungal pathogenicity, molecular systematics, *Olea europaea*, *Vitis* spp. wood decay fungus.

Competing Interests: The Author(s) declare(s) no conflict of interest.

Editor: Alan J.L. Phillips, University of Lisbon, Portugal.

INTRODUCTION

Olive cultivation is emerging in Brazil as an option for farmers. The production and cultivated area is restricted to the southeast and south of the country where the established olive groves are young and have been in production for only a few decades. In 2016, national production was 647 t from 573 ha of harvested area, having increased 138% since 2014 (FAO, 2018). Government agencies estimate that approx. 110,000 L of olive oil were produced in 2017. Viticulture is mainly carried out in the south and northeast of Brazil, and approx. 0.8 million t of grapes were produced in 2014 (FAO-OIV, 2016). The total area of vineyards in Brazil is 86,000 ha, and the south region is responsible for 75.3% of the production of table grapes and wines (IBGE, 2017). Olive orchard and vineyard production supply mostly to the Brazilian domestic markets.

Studies of pathogens associated with grapevine decline in Brazil are few, and reports of pathogens in olive trees are even fewer (Correia *et al.*, 2013; Chliyeh *et al.*, 2014). During a survey of an olive orchard in Santa Catarina (SC), a state in Southern Brazil, a decline disease of olive trees was noted. At the same time, a grapevine plant brought to our laboratory for diagnosis showed brown streaking in the vascular tissues of the branches. The purpose of the study reported here was to determine the cause of these diseases.

MATERIALS AND METHODS

Sampling and fungal isolation

Symptomatic olive trees, showing foliar browning and leaf drop, wilting of apical shoots, dieback of twigs and branches, dark streaks in the internal wood, and crown and necrotic bark lesions (Figure 1) were noted during a survey in an experimental olive orchard in Itá, a municipality in SC (-27.273913, -52.339676), in 2015. The orchard was approx. 2.5 ha and contained 3-year-old olive trees of cv. Arbosana. Symptomatic trees were in a 1,000 m^2 area where the soil was poorly drained. Samples were taken from two diseased plants. At the same time (in 2016), a grapevine plant, aged approx. 3 years, with dieback symptoms, was received by our laboratory for disease diagnosis. The branches showed brown discolouration in the vascular tissue and the bark was cracking (Figure 1). The grapevine was from Riqueza municipality, also in SC (-27.064126, -53.327396). Small pieces of the necrotic and discoloured tissues were removed, surface-disinfected for 3 min in 1.5% sodium hypochlorite solution, washed twice in sterile water and then incubated on 2% potato dextrose agar (PDA, Merck) amended with 100 mg L^{-1} streptomycin sulphate (Calbiochem). One fungal isolate was obtained from each sampled plant, *i.e.*, two isolates from the two sampled olive tree and one from the diseased grapevine. After sporulation, 20 μL of a spore suspension were plated on PDA, and after 2 d of incubation individual germinated spores were transferred to fresh PDA plates. The resulting cultures produced *Phialophora*-like fungi (Vijaykrishna *et al.*, 2004). One isolate from olive (pr_OLIV) and one from the grapevine (pr_GRAP) were used for further analyses.

Morphological characterization

Morphological characterization followed the procedures of Vijaykrishna *et al.* (2004) and Carlucci *et al.* (2015) with some modifications. Colony morphology was determined after 21 d of incubation on 2% malt extract agar (MEA) at 25 ± 2°C in the dark. Dimensions and morphology of conidia, conidiophores and phialides were determined from microscopic examination of fungal tissues in lactic acid preparations. For each structure, 30 measurements were made with a ×100 objective on a Zeiss Scope A1 Axio microscope equipped with AxioCam MRC, operated through Zen Lite Software 2012 (Zeiss). Radial growth of the isolates was determined by placing individual 3 mm diam. mycelial discs on MEA in Petri dishes. Two perpendicular measurements of colony diameters were made with a digital caliper after 8 d incubation in the dark at 4°C, 25°C or 37°C, with five replicates at each temperature.

DNA isolation and sequencing

Genomic DNA of the two isolates was extracted using the Wizard Genomic DNA Purification Kit (Promega) following the manufacturer's instructions. The 5.8S rDNA gene and flanking internal transcribed spacers 1 and 2 (ITS), partial β-tubulin (*tub2*) and translation elongation factor 1-α (*tef1*) genes were sequenced to confirm fungal identification, using, respectively, the primer pairs ITS4 and ITS5 (White *et al.*, 1990), Bt2a and Bt2b (Glass and Donaldson, 1995), and EF1 and EF2 (O'Donnell *et al.*, 1998). After amplification, PCR products were purified using the QIAquick Gel Extraction Kit (Qiagen). Sequencing was performed by ATCGene (Alvorada, Rio Grande do Sul, Brazil) in forward direction. Sequences of pr_OLIV and pr_GRAP isolates were deposited in GenBank. Matches with sequences in GenBank was performed with BLAST searches.

Pathogenicity of P. richardsiae *isolates*

The two isolates obtained in this study were inoculated onto the same plant species from which they were each isolated. Isolate pr_OLIV was also inoculated onto grapevine cultivars and isolate pr_GRAP was inoculated onto an olive cultivar. One-year-old nursery plants of cv. Arbosana olive trees (*Olea europaea*) and two rootstock varieties of grapevines (*Vitis vinifera*), 'Paulsen' (*V. berlandieri* Planch. × *V. rupestris* Scheele) and 'VR04343' (*V. rotundifolia* Michx. × *V. vinifera* L.) approx. 50 cm tall, were growing in 6 L capacity plastic bags of potting mix. The plants were each wounded to about 3 mm depth with a histological needle at 7 cm above soil level. Colonized mycelial agar plugs (3 mm diam.) were placed on the wounds, and each covered with a wet

Figure 1. (a–c) Disease symptoms in olive trees. (a) Severe dieback of twigs and branches, with browning of leaves, and premature plant death. (b) Transversal cross section of an affected trunk showing dark brown to black streaking in the crown. (c) Dark streaking in the core of wood and necrotic lesions affecting the bark. (d–e) Grapevine branches showing cracks and vascular discoloration. Scale bar = 1 cm. (f–g) Longitudinal sections showing acropetal and basipetal wood discolouration from inoculation points in (e) olive tree and (f) grapevine. Scale bar = 1 cm. (h) *Pleurostomophora richardsiae* colony, on MEA after 21 d in the dark at 25°C. (i–j) Micrographs of *P. richardsiae* structures. Scale bar = 10 µm. Arrows in (i) show a phialide, a conidiophore and hyphal bundles. Arrows in (j) indicate subglobose and oblong ellipsoid conidia, and coiled hyphae.

piece of cotton and sealed with Parafilm. Mycelial plugs were taken from fungal colonies grown on PDA for 21 d at 23°C (± 2°C). Non-colonised sterile PDA plugs were used for non-inoculated controls. Five non-inoculated or inoculated plants were prepared for each isolate. Pots containing the plants were randomly arranged on a bench in a greenhouse covered with an anti-aphid net, with temperature around 25°C, and watered three times each week. The inoculation points were analysed 4 months after inoculation. A longitudinal cut was made in the wood tissue of each plant through the point of inoculation, and the length of the discoloured wood above and below the point of inoculation was measured with a digital caliper. Re-isolations were made from all of the inoculated pieces of wood to recover the isolates, in attempts to fulfil Koch's postulates.

RESULTS AND DISCUSSION

The symptoms of disease on olive trees were foliar browning and leaf drop, wilting and dieback of twigs and branches, dark streaking in the inner wood tissues and necrosis, leading to a generalized decline of the trees. The grapevine received for diagnosis showed advanced symptoms of dieback, with severe discolouration of the vascular tissues affecting the whole plant from the roots to the branches, where the bark was cracking (Figure 1). Isolates of a *Phialophora*-like fungus was isolated from the diseased plants. Detailed morphological characterization and molecular analyses identified the isolates as *Pleurostomophora richardsiae* (Nannf.) Réblová & Jaklitsch. Voucher cultures of these isolates were deposited in the culture collection of the Centre of Biological Sciences, Federal University of Pernambuco (UFPE), with access numbers 7857 (pr_OLIV) and 7858 (pr_GRAP).

Pleurostomophora richardsiae is frequently found when searching for other pathogenic fungi in necrotic lesions in the periderms of plants (Vijaykrishna *et al.*, 2004; Mostert *et al.*, 2006; Olmo *et al.*, 2015; Réblová *et al.*, 2015). This fungus is an emerging plant pathogen and is gaining recognition as an important vascular pathogen involved in Esca disease complex of grapevine, and, more recently, associated with decline of olive trees (Carlucci *et al.*, 2013; Varela *et al.*, 2016). Pathogenicity of *P. richardsiae* in grapevine was confirmed by Halleen *et al.* (2007) in South Africa, and this pathogen has been recorded in California (Rolshausen *et al.*, 2010), Italy (Carlucci *et al.*, 2015), Spain (Varela *et al.*, 2016) and Turkey (Özben *et al.*, 2017). Reports of *P. richardsiae* causing decline in olive trees are more recent. Carlucci *et al.* (2013) found *P. richardsiae* associated with olive tree decline in surveys conducted in Italy, and remarked on the aggressiveness of the fungus to this host, since it can frequently be the only pathogen isolated from affected trees.

After 21 d in the dark on MEA, colonies of pr_GRAP and pr_OLIV were cottony, light brown and paler towards the periphery, with entire margins (Figure 1), which agrees with the description provided by Carlucci *et al.* (2015). The mycelium is composed of branched septate hyphae, sometimes in bundles or forming coils (Figure 1). Neither of the isolates grew in cultures at 4°C. According to Carlucci *et al.* (2015) the minimum temperature for growth of *P. richardsiae* is 10°C. Mycelia of *P. richardsiae* developed at 25°C (mean colony diam. = 32.7 cm for pr_ OLIV and 30.2 cm for pr_GRAP), and at 37°C (mean colony diam. + 11.4 cm for pr_ OLIV and 8.5 cm for pr_GRAP). Morphology of the isolates agreed with the previous description of *P. richardsiae*, producing cylindrical and oblong ellipsoidal conidia, short, unbranched conidiophores and predominantly type II phialides. The phialides were elongated, ampulliform and constricted at the bases (Vijaykrishna *et al.*, 2004; Mostert *et al.*, 2006; Carlucci *et al.*, 2015).

ITS sequences of pr_GRAP and pr_OLIV were lodged in GenBank (NCBI), under accession numbers, respectively, MG966406.1 and MG966416.1. BLAST searches showed that these isolates had 100% identity with several *P. richardsiae* ITS sequences already deposited in GenBank, including the isolate that was associated with grapevine decline in Spain, access number JX258852.1 (Varela *et al.*, 2016). The ITS sequences also presented 99% identity with the ex-type isolate CBS 270.33 (GenBank accession number NR135933; Mostert *et al.*, 2003). Sequences of partial *tub2* and *tef1* genes were also deposited in GenBank. The accession numbers for pr_GRAP *tub2* and *tef1* sequences are MH053437 and MH053438. For pr_OLIV *tub2* and *tef1* sequences, the accession numbers are MH053439 and MH053440.

In the pathogenicity tests, both isolates produced brown streaking in the wood in most of the inoculated plants (Figure 1). Despite of the spread of the brown streaking, the isolates did not kill the inoculated nursery plants. Both isolates were re-isolated from inoculated plants.

This is the first report of *P. richardsiae* associated with diseased olive trees and grapevine in Brazil. With this report, we hope to contribute to knowledge of fungal phytopathogens that can be found in these plant species in Brazil. Further epidemiological studies are required to investigate the incidence of *P. richardsiae* in the south and other regions of Brazil, where olive trees and grapevines are cultivated, and determine the impacts of the pathogen on productivity of olive orchards and vineyards.

ACKNOWLEDGEMENTS

We thank Carmem Cella dos Santos (Epagri) for technical assistance with the microbiological and pathogenicity studies; Dr Cristiane de Lima Wesp (Epagri) for provision of grapevine and olive nursery plants; Dr Camila Andrade (Agronômica) for collaboration on gene analysis; and Prof Dr Nelson Sidnei Massola Júnior (University of São Paulo, Piracicaba, Brazil) for his valuable comments during development of this work.

LITERATURE CITED

Carlucci A., Raimondo M.L., Cibelli F., Phillips A.J.L., F. Lops, 2013. *Pleurostomophora richardsiae, Neofusicoccum parvum* and *Phaeoacremonium aleophilum* associated with a decline of olives in southern Italy. *Phytopathologia Mediterranea* 52: 517–527.

Carlucci A., Cibelli F., Lops F., Phillips A.J.L., Ciccarone C., Raimondo M.L., 2015. *Pleurostomophora richardsiae* associated with trunk diseases of grapevine in southern Italy. *Phytopathologia mediterranea* 54: 109–123.

Chliyeh M., Touati J., Selmaoui K., Touhami A.O., Filali-Maltouf A., … Douira A., 2014. Bibliographic inventory of the olive tree (*Olea europaea* L.) fungal diseases in the world. *Intenational Journal of Pure and Applied Bioscience* 2: 46–79.

Correia K.C., Câmara M.P.S., Barbosa M.A.G., Sales Jr. R., Agustí-Brisach C., … Michereff S.J., 2013. Fungal trunk pathogens associated with table grape decline in North-eastern Brazil. *Phytopathologia Mediterranea* 52: 380–387.

FAO-OIV, 2016. Table and Dried Grapes. *Food and Agriculture Organization of the United Nations and the International Organisation of Vine and Wine*, 64p.

FAO, 2018. FAOSTAT. *Food and Agriculture Organization of the United Nations*. URL http://www.fao.org/faostat/en/#data/QC.

Glass N.L., Donaldson G.C., 1995. Development of primers sets designed for use with the PCR to amplify conserved genes from filamentous Ascomycetes. *Applied and Environmental Microbiology* 61: 1323–1330.

Hallen F., Mostert L., Crous P.W., 2007. Pathogenicity testing of lesser known vascular fungi of grapevines. *Australasian Plant Pathology* 36: 277–285.

IBGE - Instituto Brasileiro de Geografia e Estatística, 2017. *Levantamento Sistemático da Produção Agrícola - 2017*. http://www.agricultura.mg.gov.br/images/Arq_Relatorios/Agricultura/2017/Fev/perfil_uva_fev_2017.pdf.

Mostert L., Crous P.W., Groenewald J.Z., Gams W., Summerbell R.C., 2003. *Togninia* (Calosphaeriales) is confirmed as teleomorph of *Phaeoacremonium* by means of morphology, sexual compatibility and DNA phylogeny. *Mycologia* 95: 646–659.

Mostert L., Groenewald J.Z., Summerbell R.C., Gams W., Crous P.W., 2006. Taxonomy and pathology of *Togninia* (*Diaporthales*) and its *Phaeoacremonium* anamorphs. *Studies in Mycology* 54: 1-113.

O'Donnell K., Kistler H.C., Cigelnik E., Ploetz R.C., 1998. Multiple evolutionary origins of the fungus causing Panama disease of banana: concordant evidence from nuclear and mitochondrial gene genealogies. *PNAS* 95: 2044–2049.

Olmo D., Armengol J., León M., Gramaje D., 2015. Pathogenicity testing of lesser-known fungal trunk pathogens associated with wood decay of almond trees. *European Journal of Plant Pathology* 143: 607–611.

Özben S., Demirci F., Değirmenci K., Uzunok S., 2017. Fisrt report of *Pleurostoma richardsiae* associated with grapevine decline disease of grapevine in Turkey. *Journal of Plant Pathology* 99: 804.

Réblová M., Jaklitsch W.M., Réblová K., Štěpánek V., 2015. Phylogenetic reconstruction of the Calosphaeriales and Togniniales using five genes and predicted RNA secondary structures of ITS, and *Flabellascus tenuirostris* gen. et sp. nov. *PLoS ONE* 10: e0144616.

Rolshausen P.E., Úrbez-Torres J.R., Rooney-Latham S., Eskalen A., Smith R.J., Gubler W.D., 2010. Evaluation of pruning wound susceptibility and protection against fungi associated with grapevine trunk diseases. *American Journal of Enology and Viticulture* 61: 113–119.

Varela C.P., Fernández R., Casal O.A., Martinez V.F., Vázquez J.P.M., 2016. First report of *Pleurostoma richardsiae* causing grapevine trunk disease in Spain. *Plant Disease* 100: 2168.

Vijaykrishna D., Mostert L., Jeewon R., Gams W., Hyde K.D., Crous P.W., 2004. *Pleurostomophora*, an anamorph of *Pleurostoma* (*Calosphaeriales*), a new anamorph genus morphologically similar to *Phialophora*. *Studies in Mycology* 50: 387–395.

White T.J., Bruns T., Lee S., Taylor J., 1990. Amplification and direct sequencing of fungal ribosomal RNA genes for phylogenetics. In: *PCR Protocols, a Guide to Methods and Applications* (Innis M.A., Gelfand D.H., Sninsky J.J., White T.J.). Academic Press Inc., New York, 315–322.

17

Race structure and distribution of *Pyrenophora tritici-repentis* in Tunisia

MARWA LARIBI[1,4,*], FERNANDA M. GAMBA[2], MARWA HASSINE[1], PAWAN K. SINGH[3], AMOR YAHYAOUI[3,4], KHALED SASSI[1]

[1] *University of Carthage, National Agronomic Institute of Tunisia, LR14AGR01 Laboratory of Genetic and Cereal Breeding, 1082, Tunis, Tunisia*
[2] *Departamento de Protección Vegetal, Facultad de Agronomía, UDELAR, Estación Experimental, Dr. M. A. Cassinoni, Ruta 3 k363, Paysandú, Uruguay*
[3] *Centro Internacional de Mejoramiento de Maíz y Trigo (CIMMYT), Km 45 Carretera México-Veracruz, El Batán, Texcoco, CP 56237, Estado de México, México*
[4] *CRP Wheat Septoria Precision Phenotyping Platform, 43 Avenue Charles Nicolle, 1082 Tunis, Tunisia*
*Corresponding author: mar.wa199@hotmail.fr

Summary. Tan spot, caused by *Pyrenophora tritici-repentis* (Ptr), is a widespread foliar disease of wheat, which is becoming important in North Africa particularly in Tunisia. To assess the pathogenic variation of Ptr in Tunisia, 84 single conidium isolates of Ptr were characterized from durum wheat cultivars, sampled during the 2017–2018 cropping season. The virulence of isolates were assessed, under controlled conditions, on a standard differential set of six wheat genotypes. Ptr races 2, 4, 5, 6, 7 and 8 were identified, the first such information available for Tunisia. Race 2, commonly found in North America, South America and Asia, was identified for the first time in North Africa, at a low frequency of 5%. Races 5 and 7 were the most frequent, representing, respectively, 39% and 43% of the isolates tested. Only 8% of the isolates were classified as race 8, while 4% were identified race 6. Race 6 was only detected at the experimental station in the North Western region of Tunisia and in a nearby farm field. Only one Ptr isolate was avirulent on all six differential genotypes, and was therefore designated race 4. The identification of six races of Ptr on durum wheat demonstrates the high diversity of the pathogen population in Tunisia.

Keywords. Physiological races, *Pyrenophora tritici-repentis*, durum wheat, tan spot.

Competing Interests: The Author(s) declare(s) no conflict of interest.

Editor: Andy Tekauz, Cereal Research Centre, Winnipeg, MB, Canada.

INTRODUCTION

The fungus *Pyrenophora tritici-repentis* (Died.) Drechs., anamorph *Drechslera tritici-repentis* (Died.) Shoemaker (synonym *Helminthosporium tritici-repentis* Died.) causes tan spot, a foliar disease that affects bread wheat (*Triticum aestivum* L.), durum wheat (*T. turgidum* L. var. *durum*) and several grass species in many areas of the world (Morrall and Howard, 1975; Hosford *et al.*, 1975; Hosford, 1982; Krupinsky, 1982; De Wolf *et al.*, 1998; Ciuffetti and Tuori, 1999; Singh *et al.*, 2010).

Tan spot was first identified in Canada in 1939 (Conners, 1939), in the United States of America in New York State in 1940 (Barrus, 1942), in Australia in 1950 (Valder and Shaw, 1952) and in Mexico in 1982 (Gilchrist *et al.*, 1984). The disease was subsequently reported as the fastest spreading disease in the Southern Cone region of South America (Kohli *et al.*, 1992), and as a damaging disease in Argentina, Brazil, Paraguay and Australia (Annone, 1998; Kohli and Diaz, 1998; Loughman *et al.*, 1998). Since the 1970s, tan spot has been considered a serious problem that has caused significant yield losses in wheat crops (Watkins *et al.*, 1978; Hosford, 1982; Rees *et al.*, 1982; 1983; Shabeer and Bockus, 1988; Sykes and Bernier, 1991). In the late 1990s, tan spot was considered to be one of the main wheat diseases in Central Asia (Postnifova and Khasanov, 1998: Lamari *et al.*, 2005), and to be increasing on wheat grown in the Mediterranean region (Nasrellah and Mergoum, 1997; Benslimane *et al.*, 2006, 2011; M.S. Gharbi, *personal communication*). The increase in tan spot severity and incidence were reported to be due to changes in pathogen virulence, wide adoption of no-till and conservation tillage practices without suitable crop rotations, and the cultivation of susceptible cultivars (Rees, 1982; Rees and Platz, 1983; Brennen and Murray, 1988; Lamari and Bernier, 1989a; Mehta and Gaudencio, 1991; Kohli *et al.*, 1992). Furthermore, increased tan spot severity is associated with the survival of the pathogen on seeds, crop residues, and grass hosts (De Wolf *et al.*, 1998; Singh *et al.*, 2010).

Typical symptoms of tan spot appear as brown necrotic lesions surrounded by chlorotic haloes, with small black points in the centre of the lesions. In some cases, extensive chlorosis occurs throughout host leaves. Lesions can coalesce, resulting in death of leaves (Lamari and Bernier, 1989a). Tan spot also affects kernels (red smudge) (Canadian Grain Commission. 1991), resulting in kernel discolouration and affecting seedling emergence, seedling vigour, yield, and grain quality. Seed infections could provide inoculum for epidemics and dispersal of pathogenic strains to new geographic areas (Vanderpool, 1963; Schilder and Bergstrom, 1995; Fernandez *et al.*, 1997; Bergstrom and Schilder, 1998; Fernandez *et al.*, 1998).

To date, eight races of *P. tritici-repentis* (Ptr) have been identified and characterized, based on their ability to induce necrosis and/or chlorosis on a wheat differential set. This includes four hexaploid wheats, 'Glenlea', 6B365, 6B662 and 'Salamouni', and two tetraploid wheats, 'Coulter' and 4B-160 (Lamari *et al.*, 1995, 1998; Strelkov *et al.*, 2002; Strelkov and Lamari, 2003; Lamari *et al.*, 2003; Singh *et al.*, 2010). Each race produces a unique necrotrophic effector (NE) (singly or in combination), which are designated Ptr ToxA, Ptr ToxB and Ptr ToxC (Lamari and Bernier, 1989c; Orolaza *et al.*, 1995; Ciuffetti *et al.*, 1998; Effertz *et al.*, 2002). The NEs are largely responsible for the necrosis and chlorosis symptoms associated with tan spot and serve as pathogenicity factors (Strelkov and Lamari, 2003). Races 2, 3 and 5 each produce a single NE, respectively Ptr ToxA, Ptr ToxC and Ptr ToxB, and are therefore considered the 'basic' races, while races 1, 6, 7 and 8 produce more than one NE each and are considered 'composite' races. Race 1 produces Ptr ToxA and Ptr ToxC, race 6 produces Ptr ToxB and Ptr ToxC, race 7 produces Ptr ToxA and Ptr ToxB, race 8 produces all three NEs (Strelkov *et al.*, 2002; Lamari *et al.*, 2003), and race 4 does not produce any active NE and is therefore avirulent.

The race structure of Ptr has been determined for several regions. In North America, 90% of the isolates have been classified as races 1 or 2 (Strelkov *et al.*, 2002; Singh *et al.*, 2007; Lamari and Strelkov, 2010; Aboukhaddour *et al.*, 2013), while races 3, 4 and 5 represent only 10% of the North American races. In the Southern Cone Region of South America, only races 1 and 2 were identified (Gamba *et al.*, 2012). Similarly, limited surveys in central Asia indicated the presence of races 1 and 2 (Lamari *et al.*, 2005). Moreover, races 1, 2, 3, 5, 7 and 8 were found in the Caucasus and the Fertile Crescent regions (Strelkov and Lamari, 2003; Lamari *et al.*, 2003, 2005; Lamari and Strelkov, 2010). In North Africa, all races except races 2 and 3 have been reported (Lamari *et al.*, 1995; Benslimane *et al.*, 2011; Gamba *et al.*, 2017). In addition to the eight well-characterized races, there have been some suggestions of the existence of other races, but no complete description has yet been published (Manning *et al.*, 2002; Meinhardt *et al.*, 2003; Andrie *et al.*, 2007; Ali *et al.*, 2010; Benslimane, 2018).

The aim of the present study was to examine the race structure and distribution of Ptr populations in the major wheat growing regions of Tunisia.

MATERIALS AND METHODS

Survey and fungal isolation

Surveys were carried out in the main wheat growing regions of Tunisia in the 2017–2018 cropping season. Each survey sample consisted of 40 leaves exhibiting typical tan spot symptoms, collected randomly from six commercial durum wheat fields (*Triticum durum*) in three regions, designated Coastal (CR), Northern (NR) and North Western (NWR), and from durum wheat growing at two experimental stations, Kodia at Bousalem and Oued Beja at Beja, at the NWR. Wheat growth

Table 1. The *Pyrenophora tritici-repentis* race structure and distribution in Tunisia.

Origin (location)	Coastal Region (CR)	Northern Region (NR)	North Western Region (NWR)	
			FF	ES
Race 2	13 %	0 %	5 %	5 %
Race 4	0 %	0 %	2 %	0 %
Race 5	38 %	34 %	46 %	29 %
Race 6	0 %	0 %	2 %	9 %
Race 7	25 %	58 %	40 %	48 %
Race 8	25 %	8 %	5 %	9 %

FF, Farmer field; ES, Experimental station.

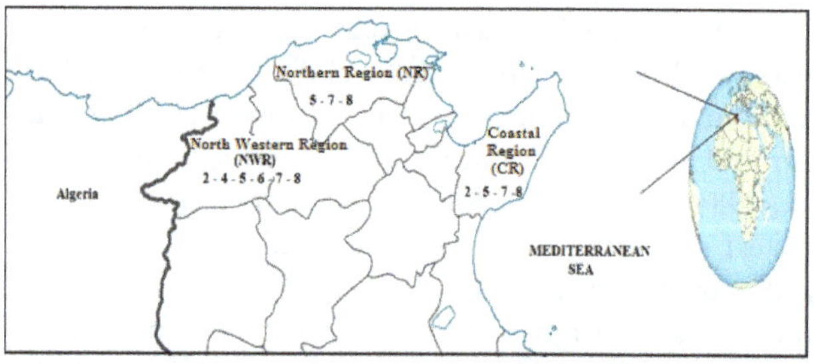

Figure 1. Races of *Pyrenophora tritici-repentis* characterized from durum wheat grown in three regions of Tunisia, North Africa.

stages at the time of the survey ranged from the beginning of stem elongation (ZGS 30) to the milk stage (ZGS 77) (Zadoks *et al.*, 1974). Leaf samples were collected and kept at room temperature overnight to dry. Fungal isolation and inoculum production were performed as described by Lamari and Bernier (1989a). Leaves were cut into 1 to 2 cm pieces, surface sterilized in 30% alcohol for 20 sec then 1% sodium hypochlorite solution for 2 min, and then washed three times, for 1 min each, with sterile distilled water. The leaf fragments were the placed in 9 cm-diam. Petri dishes each containing two layers of sterile filter paper moistened with sterile distilled water to maintain high humidity. These plates were incubated under fluorescent light for 24 h at 21°C to promote the production of conidiophores. The Petri dishes cultures were then incubated for 18 to 24 h at 15°C to induce conidial production. After incubation, leaf fragments were examined using 40× binocular magnifiers and single conidia identified as Ptr were transferred to V8-PDA medium (150 mL of V8 juice, 10 g of Potato Dextrose Agar, 3 g of $CaCO_3$, 10 g of water agar, and 850 mL of distilled water) and incubated at 20°C until the colony reached approx. 4 cm in diameter. In total, 84 single conidium isolates were obtained from the three production regions, and these isolates were subsequently phenotypically characterized on the wheat differential set (Table 1, Figure 1).

Inoculum production and inoculation

The Ptr cultures were incubated on V8-PDA medium in the dark for 7 to 8 days at 20°C, until they reached approx. 4 cm in diameter. Plates were then flooded with sterile distilled water, the mycelium in each plate was flattened with the bottom of a flamed test tube, and excess water was poured out. The cultures were subsequently incubated for 24 h under light at room temperature (20–22°C), followed by 24 h at 15°C in the dark. Conidia were then harvested by flooding the Petri dishes with sterile distilled water and dislodging the conidia with a wire loop. The inoculum concentration was adjusted to 3,000 conidia mL^{-1} using a haemocytometer (Hausser Scientific Company), and a drop of Tween 20 was added (polyoxyethylene sorbitanmonolaurate) per 100 mL to reduce surface tension in the conidium suspensions.

Wheat seedlings (see below) at the two-leaf stage were sprayed with conidium suspensions to run off, using a hand sprayer. Precautions were taken to avoid cross-infection of isolates. The inoculated seedlings were incubated in a dew chamber for 24 h (16 h light then 8 h darkness) at 20°C and 90% relative humidity (Lamari and Bernier, 1989a). All experiments were conducted at the CRP Wheat Septoria Precision Phenotyping Platform laboratory in Tunis. The seedlings were evaluated for symptom development 7 d after inoculation. Tan spot severity was assessed using the 1 to 5 scale developed by Lamari and Bernier (1989a), where: 1 = small, dark-brown to black spots, without any surrounding chloroses or tan necroses; 2 = small dark-brown to black spots, with very little chloroses or tan necroses; 3 = small, dark-brown to black spots, completely surrounded by distinct chlorotic or tan necrotic rings, not coalescing; 4 = small, dark-brown to black spots, completely surrounded by tanned chlorotic or necrotic zones, sometimes coalesced; and 5 = most lesions consisting of coalescing chlorotic or tan necrotic tissue. Scores equal to or greater than 3 indicated susceptibility, recorded as necrosis (N) and/or chlorosis (C), while those less than 3 indicate a resistant (R) reaction of the genotype to the tested isolate.

Plant material

The differential set consisted of six wheat genotypes, including the four hexaploid wheats 6B365, 'Glenlea',

Table 2. Reaction types[a] of the eight races of *Pyrenophora tritici-repentis* on six international wheat differential genotypes.

Wheat genotype	Race 1	Race 2	Race 3	Race 4	Race 5	Race 6	Race 7	Race 8
'Glenlea'	N[a]	N	R	R	R	R	N	N
6B662	R	R	R	R	C	C	C	C
6B365	C	R	C	R	R	C	R	C
'Salamouni'	R	R	R	R	R	R	R	R
'Coulter'	N	N	N	N	N	N	N	N
4B-160	C	R	N	R	N	-	-	-

[a] N, Necrosis; C, chlorosis; R, resistance. Adapted from (Lamari and Bernier, 1989a; Singh *et al.*, 2010; Lamari and Strelkov, 2010).

6B662 and 'Salamouni', and the two tetraploid wheats 4B-160 and 'Coulter'. These were proposed by Lamari *et al.* (1995) to characterize the eight known races of Ptr (Table 2). Seeds of each genotype (five per pot) were sown into 10 cm diam. pots filled with a mix of ⅔ peat moss and ⅓ soil, and then kept in a growth chamber at 21°C (day) and 19°C (night) with a 16 h photoperiod. All treatments were replicated three times.

RESULTS

Race assessments for the differential host genotypes identified Ptr races 2, 4, 5, 6, 7 and 8 occurring in Tunisia (Table 3). Race 2 induced severe necrosis in 'Glenlea' and 'Coulter', but was avirulent on 4B-160, 6B662, 6B365 and 'Salamouni'. Race 5 was recovered from all locations, and induced severe chlorosis only on 6B662, and necrosis on 'Coulter' and 4B160, but was avirulent on 6B365 'Glenlea' and 'Salamouni'. Race 6 induced extensive chlorosis on 6B365 and 6B662 and necrosis on 'Coulter,' but was avirulent on 'Glenlea' and 'Salamouni'. Race 7 was avirulent on 'Salamouni' and 6B365, but induced extensive chlorosis on 6B662 and extensive necrosis on 'Coulter'. Race 8 was avirulent on 'Salamouni', but induced extensive chlorosis on 6B662 and 6B365 and extensive necrosis on 'Glenlea' and 'Coulter'. None of the races exhibited virulence on 'Salamouni'. All the Ptr isolates produced clear symptoms on the differential set, in accordance with the designated race structure; no new virulence types were observed. The distribution of Ptr races in Tunisia is shown in Figure 1, while the different responses of the differential host lines following inoculation are illustrated in Figure 2.

The 84 Ptr isolates tested grouped into six races: (2, 4, 5, 6, 7 and 8), of which four (5%) were classified as race 2; 33 (39%) were characterized as race 5; three (4%) were race 6, 36 (43%) were race 7, and seven (8%) were classified as race 8. One isolate was avirulent on all six differential genotypes and was designated as race 4 (Table 1). Races 5, 7 and 8 were found in all the three regions (CR, NR and NWR) surveyed. Race 6 was detected only in the NWR, while race 2 was found in the CR and NWR. Only one isolate from the NWR was race 4.

Races 2, 4, 5, 6, 7 and 8 were identified from the NWR, which represents the main durum wheat growing region in Tunisia. Races 2, 5, 7 and 8 were identified from a single farm field in the CR, while races 5, 7 and 8 were detected in two farm fields in the NWR. Races 5, 7 and 8 were prevalent across the surveyed regions, with race 8 predominating in the CR, race 7 in the NR, and race 5 in the NWR particularly in the farm fields (Table 1). Races 5 and 7 were predominant in all regions surveyed; representing, respectively, 38% and 25% in the CR, 34% and 58% in the NR, 46% and 40% in farm fields in the NWR, and 29% and 48% at the experimental stations in the NWR. In the CR, race 5 was the most common; in the NWR this was race 7. Races 7 and 8 were equally prevalent (25% of isolates) in the CR. Although race 8 was found in all regions, it was most prevalent in the CR, followed by the NR and NWR. Race 2 was identified in both CR (13%) and NWR (5%), but not in the NR. Races 4 and 6 were only present in the NWR with different frequencies in farm fields compared with the experimental stations. The presence of race 4 in a farm field but not at the experimental stations could have been due to the presence of grasses around this particular farm field, as grasses are likely alternative hosts of Ptr (Ali and Francl, 2003). Race 7 was present at a slightly higher frequency at the experimental stations compared with the farm fields, possibly due to the greater diversity (including levels of Ptr resistance) of wheat germplasm growing at test sites compared to single varieties likely being grown in farm fields.

DISCUSSION

This is the first study that has characterized the race structure of Ptr in Tunisia. Race 7 was the predominant

Table 3. Reactions of six differential wheat lines to 84 isolates of *Pyrenophora tritici-repentis*, their regional origin and race designation.

Isolate	Origin	Glenlea	6B662	6B365	Salamouni	Coulter	4B-160	Race
TSPPTR1	CR[a]	R[b]	C	R	R	N	N	5
TSPPTR2	CR	R	C	R	R	N	N	5
TSPPTR3	CR	N	R	R	R	N	R	2
TSPPTR4	CR	N	C	R	R	N	-	7
TSPPTR5	CR	N	C	R	R	N	-	7
TSPPTR6	CR	R	C	R	R	N	N	5
TSPPTR7	CR	N	C	C	R	N	-	8
TSPPTR8	CR	N	C	C	R	N	-	8
TSPPTR9	NR	R	C	R	R	N	N	5
TSPPTR10	NR	N	C	R	R	N	-	7
TSPPTR11	NR	R	C	R	R	N	N	5
TSPPTR12	NR	N	C	R	R	N	-	7
TSPPTR13	NR	N	C	R	R	N	-	7
TSPPTR14	NR	N	C	R	R	N	-	7
TSPPTR15	NR	N	C	R	R	N	-	7
TSPPTR16	NR	N	C	C	R	N	-	8
TSPPTR17	NR	N	C	R	R	N	-	7
TSPPTR18	NR	N	C	R	R	N	-	7
TSPPTR19	NR	R	C	R	R	N	N	5
TSPPTR20	NR	N	C	R	R	N	-	7
TSPPTR21	NWR	R	C	R	R	N	N	5
TSPPTR22	NWR	R	C	R	R	N	N	5
TSPPTR23	NWR	R	C	C	R	N	-	6
TSPPTR24	NWR	N	C	R	R	N	-	7
TSPPTR25	NWR	N	C	C	R	N	-	8
TSPPTR26	NWR	N	C	R	R	N	-	7
TSPPTR27	NWR	N	C	C	R	N	-	8
TSPPTR28	NWR	R	C	C	R	N	-	6
TSPPTR29	NWR	N	C	R	R	N	-	7
TSPPTR30	NWR	N	C	R	R	N	-	7
TSPPTR31	NWR	N	R	R	R	N	R	2
TSPPTR32	NWR	N	C	R	R	N	-	7
TSPPTR33	NWR	R	C	R	R	N	N	5
TSPPTR34	NWR	N	C	R	R	N	-	7
TSPPTR35	NWR	R	C	R	R	N	N	5
TSPPTR36	NWR	N	C	R	R	N	-	7
TSPPTR37	NWR	N	R	R	R	N	R	2
TSPPTR38	NWR	R	C	R	R	N	N	5
TSPPTR39	NWR	N	C	R	R	N	-	7
TSPPTR40	NWR	N	R	R	R	N	R	2
TSPPTR41	NWR	R	C	C	R	N	-	6
TSPPTR42	NWR	N	C	R	R	N	-	7
TSPPTR43	NWR	N	C	R	R	N	-	7
TSPPTR44	NWR	N	C	R	R	N	-	7
TSPPTR45	NWR	N	C	R	R	N	-	7
TSPPTR46	NWR	N	C	R	R	N	-	7
TSPPTR47	NWR	N	C	R	R	N	-	7
TSPPTR48	NWR	N	C	R	R	N	N	5

(Continued)

Table 3. (Continued).

Isolate	Origin	Glenlea	6B662	6B365	Salamouni	Coulter	4B-160	Race
TSPPTR49	NWR	R	R	R	R	R	R	4
TSPPTR50	NWR	N	C	R	R	N	-	7
TSPPTR51	NWR	N	C	R	R	N	-	7
TSPPTR52	NWR	N	C	R	R	N	-	7
TSPPTR53	NWR	N	C	R	R	N	-	7
TSPPTR54	NWR	N	C	C	R	N	-	8
TSPPTR55	NWR	N	C	R	R	N	-	7
TSPPTR56	NWR	N	C	R	R	N	-	7
TSPPTR57	NWR	N	C	R	R	N	-	7
TSPPTR58	NWR	N	C	R	R	N	-	7
TSPPTR59	NWR	R	C	R	R	N	-	5
TSPPTR60	NWR	R	C	R	R	N	-	5
TSPPTR61	NWR	R	C	R	R	N	-	5
TSPPTR62	NWR	R	C	R	R	N	-	5
TSPPTR63	NWR	R	C	R	R	N	-	5
TSPPTR64	NWR	R	C	R	R	N	-	5
TSPPTR65	NWR	R	C	R	R	N	-	5
TSPPTR66	NWR	R	C	R	R	N	-	5
TSPPTR67	NWR	R	C	R	R	N	-	5
TSPPTR68	NWR	N	C	C	R	N	-	8
TSPPTR69	NWR	R	C	R	R	N	-	5
TSPPTR70	NWR	R	C	R	R	N	-	5
TSPPTR71	NWR	R	C	R	R	N	-	5
TSPPTR72	NWR	R	C	R	R	N	-	5
TSPPTR73	NWR	R	C	R	R	N	-	5
TSPPTR74	NWR	R	C	R	R	N	-	5
TSPPTR75	NWR	R	C	R	R	N	-	5
TSPPTR76	NWR	R	C	R	R	N	-	5
TSPPTR77	NWR	N	C	R	R	N	-	7
TSPPTR78	NWR	N	C	R	R	N	-	7
TSPPTR79	NWR	R	C	R	R	N	-	5
TSPPTR80	NWR	R	C	R	R	N	-	5
TSPPTR81	NWR	N	C	R	R	N	-	7
TSPPTR82	NWR	R	C	R	R	N	-	5
TSPPTR83	NWR	N	C	R	R	N	-	7
TSPPTR84	NWR	N	C	R	R	N	-	7

[a] CR, Coastal Region; NR, North Region; NWR, North West Region
[b] N, susceptible necrosis; C, susceptible chlorosis; R, resistance.

race in Tunisia, and this was expected since it is predominant in neighbouring Algeria, where 40% of isolates tested were also race 7 (Benslimane *et al.*, 2011). This race was identified in Bejaia, Algeria, the closest wheat growing region to the NWR of Tunisia, where race 7 was identified mostly from durum wheat. Isolates of race 7 of the pathogen in Syria and Azerbaijan were primarily found on tetraploid wheat hosts (Lamari *et al.*, 2005).

The presence of races 5 and 6 in Tunisia was expected, since race 5 had previously been found in eastern regions of Algeria, and in Morocco (Lamari *et al.*, 1995; Strelkov *et al.*, 2002; Benslimane *et al.*, 2011; Gamba *et al.*, 2017). Almost 95% of isolates tested from the farm fields in the NWR, located 335 km from Guelma, Algeria, were race 5, which was previously reported in Algeria by Benslimane *et al.* (2011). In addition, Lamari *et al.* (2005) identified race 5 from tetraploid wheat in the countries along the Silk Road, including Syria and Azerbaijan. Results of the present study are consistent with previous studies by Lamari and Bernier (1989b) and

Figure 2. Representative resistant (R) and susceptible (S) reactions of wheat cvs. 'Glenlea', 'Salmouni', 'Coulter' and lines 6B662, 6B365 and 4B-160, to inoculation with isolates of *Pyrenophora tritici-repentis*. 'Glenlea' developed tan necrosis (S) when inoculated with Tunisian isolates classified as races 2, 7 or 8. Line 6B662 developed chlorosis (S) in response to inoculation with isolates of races 5, 6, 7 or 8. Line 6B365 developed chlorosis (S) when inoculated with Tunisian isolates classified as races 6 or 8. 'Coulter' developed necrosis (S) when inoculated with races 2, 5, 6, 7 or 8. Line 4B-160 developed necrosis (S) when inoculated with race 5. 'Salamouni' was resistant (R) to all identified isolates.

Lamari *et al.* (1998). Since all of our isolates were from durum wheat, the current results support the hypothesis that chlorosis-inducing isolates, which lack the ability to produce the Ptr ToxA, are associated with durum wheat.

Race 6 was found only in the NWR, at the Kodia experimental station and a farm field 10 km distant from the station. It is possible that this race was spread from the experimental station to the farm field by wind, since Ptr can be dispersed by wind-blown ascospores up to 200 km (Maraite *et al.*, 1992; Francl, 1997). Race 6 was also reported to occur in the western and central regions of Algeria, and in Morocco (Benslimane *et al.*, 2011; Gamba *et al.*, 2017).

Race 8 was found only in the central areas of Algeria, but was found in all regions of Tunisia at a low frequency (8%) (Benslimane *et al.*, 2011). Unlike Algeria and Morocco, where race 1 represented, respectively, 41% and 6% of Ptr isolates, this race was not identified in the present study. This was probably because all isolates were obtained from durum wheat, while in Algeria most isolates identified as race 1 were derived from bread wheat (Benslimane *et al.*, 2011).

Race 4 was represented by only one isolate from the NWR. This low frequency was reported in other regions, such as Canada (1%), North Dakota (5%) and one single isolate from Algeria (Lamari *et al.*, 1998; Ali and Francl, 2003; Benslimane *et al.*, 2011). The occurrence of limited isolates of race 4 could be linked to the sampling protocol used in the present study, since most of the samples were collected from growing wheat plants and not from crop debris or wild grasses. In addition, avirulent isolates do not form lesions on living hosts, and are therefore unlikely to be present when fungi are isolated from leaf samples. Several studies have shown that Ptr can survive on, and be isolated from, non-cereal grasses. The race 4 isolate we detected could have originated from a grass host, as was shown in the study of Ali and Francl (2003), where 98% of the isolates obtained from non-cereal grass hosts were identified as race 4. Moreover, race 4 might be non-persistent in wheat fields since it would not compete with races that are virulent on wheat.

Race 2 has been reported from North America, the Southern Cone Region of South America, Central Asia, the Caucasus and the Fertile Crescent, Baltic states and Romania (Strelkov *et al.*, 2002; Lamari *et al.*, 2003, 2005; Singh *et al.*, 2007; Lamari and Strelkov, 2010; Lamari *et al.*, 2010; Gamba *et al.*, 2012; Aboukhaddour *et al.*, 2013; Momeni *et al.*, 2014; Abdullah *et al.*, 2017). However, the present study is the first to report the presence of Ptr race 2 in North Africa. This race was detected in the CR and NWR of Tunisia, albeit at overall low frequency (5%).

Occurrence of a greater number of Ptr races in the NWR than the other two regions could be due to the widespread use of zero tillage in the NWR. In contrast, in the NR and CR, cereal-legume crop rotations are common, likely resulting in little opportunity for the pathogen to survive in wheat stubble or on alternative hosts. Regarding races 2 and 8, their high frequency in the CR could be due to monoculture of the durum wheat cultivar 'Maali-cv' that is less cultivated in the other two regions of Tunisia. This also suggests that Ptr in the CR is very diverse. Our results are similar to those of Lamari *et al.* (2005), where five races were found within a single field in Azerbaijan.

Previous studies have reported that Ptr race 1 has been identified primarily from hexaploid wheat. In contrast, all samples in the present study were taken from tetraploid (durum) wheat. Isolates identified as race 1 originating from countries along the Silk Road likewise were mostly (70%) from hexaploid wheat, while only 22% were from tetraploid wheat (Ali and Francl, 2003; Lamari *et al.*, 2005; Benslimane *et al.*, 2011).

Leaf spot and blight diseases such as tan spot oversummer or over-winter on wheat stubble, which is retained in farm fields where zero tillage is practiced. Crop debris and stubble could act as major sources of primary inoculum of the pathogens and media allowing for sexual recombination. However, there are no reports to date that demonstrate ascospore race associations. Isolates from stubble may give different race structures following sexual reproduction during the resting stage.

Aung (2001) and Aboukhaddour *et al.* (2011) suggested that host genotypes could promote changes in pathogen population structures. In Tunisia, where cereal growers practice conservation agriculture or fallows between cereal crops, considerable straw is left on the soil, which can harbor leaf spot pathogens such as Ptr. Ascospores produced from previous crop debris can readily spread during early subsequent crop development to initiate infections (Morrall and Howard, 1975). Knowledge of when ascospores are first released would be a major asset for growers, enabling them to apply preventative measures to reduce early plant infections. Knowledge of Ptr race structure could also provide breeders with the opportunity to incorporate effective resistance genes (if available) into new varieties. Ideally, newly-developed varieties should carry several different resistance genes to maximize resistance durability.

The present study has demonstrated that the Ptr population in Tunisia is diverse. As such, and to develop effective and durable resistance, it would be prudent to test breeders' lines against all virulent races present in the region(s), to determine the resistance status of the lines, and make selections accordingly. Future studies should include Ptr isolates obtained from bread wheat and from wheat stubble and wild grasses, in order to fully characterize the races present. This would be useful in the eventual development of control measures that include choice of resistant cultivars along with appropriate fungicide disease management strategies.

ACKNOWLEDGEMENTS

This research was supported by the Department of Agronomy and Plant Biotechnology at the National Agronomic Institute of Tunisia (INAT), and the CGIAR Research Program on WHEAT through the "CRP WHEAT Tunisia Septoria Precision Phenotyping Platform". The authors thank Dr. Sarrah Ben M'Barek and Dr. Moez Fakhfakh for their assistance and support of this research.

LITERATURE CITED

Abdullah S., Sehgal S.K., Ali S., Liatukas Z., Ittu M., Kaur N., 2017. Characterization of *Pyrenophora tritici-repentis* (Tan Spot of Wheat) races in Baltic States and Romania. *Plant Pathology Journal* 33(2): 133–139. https://doi.org/10.5423/PPJ.OA.10.2016.0214

Aboukhaddour R., Cloutier S., Lamari L., Strelkov S.E., 2011. Simple sequence repeats and diversity of globally distributed populations of *Pyrenophora tritici-repentis*. *Canadian Journal of Plant Pathology* 33(3): 389–399. https://doi.org/10.1080/07060661.2011.590821

Aboukhaddour R., Turkington T.K., Strelkov S.E., 2013. Race structure of *Pyrenophora tritici-repentis* (tan spot of wheat) in Alberta, Canada. *Canadian Journal of Plant Pathology* 35(2): 256–268. https://doi.org/10.1080/07060661.2013.782470

Ali S., Francl L.J., 2003. Population race structure of *Pyrenophora tritici-repentis* prevalent on wheat and noncereal grasses in the Great Plains. *Plant Disease* 87(4): 418–422. https://doi.org/10.1094/PDIS.2003.87.4.418

Ali S., Gurung S., Adhikari T.B., 2010. Identification and characterization of novel isolates of *Pyrenophora tritici-repentis* from Arkansas. *Plant Disease* 94(2): 229–235. https://doi.org/10.1094/PDIS-94-2-0229

Andrie R.M., Pandelova I., Ciuffetti L.M., 2007. A combination of phenotypic and genotypic characterization strengthens *Pyrenophora tritici-repentis* race identification. *Phytopathology* 97(6): 69–701. https://doi.org/10.1094/PHYTO-97-6-0694

Annone J.G., 1998. Tan spot of wheat in Argentina: Importance and disease management strategies. In: *Helminthosporium Blights of Wheat: Spot Blotch and Tan Spot.* (E. Duveiller, H. J. Dubin, J. Reeves, A. McNab, ed.), February 9–14, 1997, CIMMYT, Mexico, 339-345.

Aung T.S.T., 2001. Molecular polymorphism and virulence in *Pyrenophora tritici-repentis.* MSc Thesis, University of Manitoba, Winnipeg, Manitoba, Canada, 92 pp.

Barrus M.F., 1942. Yellow spot disease of wheat in New York State. *Plant Disease Report* 26(11): 246.

Benslimane H., 2018. Virulence phenotyping and molecular characterization of a new virulence type of *Pyrenophora tritici-repentis* the causal agent of tan spot. *The Plant Pathology Journal* 34(2): 139–142. https://doi.org/10.5423/PPJ.NT.07.2017.0150

Benslimane H., Bouznad Z., Aouali S., Khalfi A., Benbelkacem A., Sayoud R., 2006. Prévalence de la tache bronzée du blé causée *Pyrenophora tritici-repentis* en Algérie. In: *6éme journées scientifiques et techniques phytosanitaires*, Alger, Algeria.

Benslimane H., Lamari L., Benbelkacem A., Sayoud R., Bouznad Z., 2011. Distribution of races of *Pyrenophora tritici-repentis* in Algeria and identification of a new virulence type. *Phytopathologia Mediterranea* 50(2): 203–211. https://doi.org/10.14601/Phytopathol_Mediterr-8746

Bergstrom G.C., Schilder A.M.C., 1998. Seed pathology of tan spot. In: *Helminthosporium Blights of Wheat: Spot Blotch and Tan Spot.* (E. Duveiller, H. J. Dubin, J. Reeves, A. McNab, ed.), CIMMYT, Mexico, 364–368.

Brennen J.P., Murray G.M., 1988. Australian wheat diseases assessing their economic importance. *Agricultural Science* 1(7): 26–35.

Canadian Grain Commission. 1991. Official Grain Grading Guide. Canadian Grain Commission, Winnipeg, Canada.

Ciuffetti L.M., Tuori R.P., 1999. Advances in the characterization of the *Pyrenophora tritici-repentis*-wheat interaction. *Phytopathology* 89(6): 444–449. https://doi.org/10.1094/PHYTO.1999.89.6.444

Ciuffetti L.M., Francl L.J., Balance G.M., Bockus W.W., Lamari L., Meinhardt S.W., Rassmussen J.B., 1998. Standardization of toxin nomenclature in the *Pyrenophora tritici-repentis*/wheat interaction. *Canadian Journal of Plant Pathology* 20(4): 421–424. https://doi.org/10.1080/07060669809500415

Conners I.L., 1939. Yellow leaf blotch. *Canadian Plant Disease Survey* 19: 12–14.

De Wolf E. D., Effertz R.J., Ali S., Francl L.J., 1998. Vistas of tan spot research. *Canadian Journal of Plant Pathology.* 20(4): 349–370.

Effertz R.J., Meinhardt S.W., Anderson J.A., Jordahl J.G., Francl L.J., 2002. Identification of a chlorosis-inducing toxin from *Pyrenophora tritici-repentis* and the chromosomal location of an insensitive locus in wheat. *Phytopathology* 92(5): 527–533. https://doi.org/10.1094/PHYTO.2002.92.5.527

Fernandez M.R., Clarke J.M., DePauw R.M., Lefkovitch L.P., 1997. Emergence and growth of durum wheat derived from red smudge infected seed. *Crop Science* 37(2): 510–514. https://doi.org/10.2135/cropsci1997.0011183X003700020033x

Fernandez M.R., DePauw R.M., Clarke J.M., Zentner R.P., McConkey B.G., 1998. Tan spot in western Canada. In: *Helminthosporium Blights of Wheat: Spot Blotch and Tan Spot* (E. Duveiller, H. J. Dubin, J. Reeves, A. McNab, ed.), CIMMYT, Mexico, 73-79.

Francl L.J., 1997. Local and mesodistance dispersal of *Pyrenophora tritici-repentis* conidia. *Canadian Journal of Plant Pathology* 19 (3): 247–255. DOI: 10.1080/07060669709500519

Gamba F.M., Strelkov S.E., Lamari L., 2012. Virulence of *Pyrenophora tritici-repentis* in the Southern Cone Region of South America. *Canadian Journal of Plant Pathology* 34(4): 545–550. https://doi.org/10.1080/07060661.2012.695750

Gamba F.M., Bassi F.M., Finckh M. R., 2017. Race structure of *Pyrenophora tritici-repentis* in Morocco. *Phytopathologia Mediterranea* 56(1): 119–126. https://doi.org/10.14601/Phytopathol_Mediterr-18830

Gilchrist, S. L., Fuentes, F. S., Bauer de la Isla, M. L., 1984. Identification of *Helminthosporium tritici-repentis (= Pyrenophora tritici-repentis),* causal agent of a leaf blight of wheat in Mexico. *Agrociencia* 56: 151–162.

Hosford R.M Jr., 1982. Tan spot developing knowledge 1902–1981, virulent races and wheat differentials, methodology, rating systems, other leaf diseases, literature. In: *Tan Spot of Wheat and Related Diseases Workshop* (R.M. Hosford Jr, ed.), July 1981, North Dakota Agricultural Experimental Station, North Dakota State University, Fargo, USA, 1–24.

Hosford R.M Jr., Morrall R. A. A., 1975. The epidemiology of leaf spot disease in a native prairie. I. The progression of disease with time. *Canadian Journal of Botany* 53(10): 1040–1050. https://doi.org/10.1139/b75-122

Kohli M.M., Diaz de Ackermann M., 1998. Evaluating southern cone wheat germplasm for spot blotch and tan spot. In: *Helminthosporium Blights of Wheat: Spot Blotch and Tan Spot* (E. Duveiller, H. J. Dubin, J. Reeves, A. McNab, ed.), CIMMYT, Mexico, 230–240.

Kohli M.M., Mehta Y.R., Diaz de Ackermann M., 1992. Spread of tan spot in the southern cone region of South America. In: *Advances in Tan Spot Research. Proceedings of International Tan Spot Workshop, 2nd* (L.J. Francl, J. M. Krupinsky, M. P. McMuller, ed.), North Dakota Agricultural Experimental Station, Fargo, USA, 86–90.

Krupinsky J.M., 1982. Observation of the host range of isolates of *Pyrenophora trichostoma*. *Canadian Journal of Plant Pathology* 4(1): 42–46. https://doi.org/10.1080/07060668209501335

Lamari L., Bernier C.C., 1989a. Evaluation of wheat lines and cultivars to tan spot *Pyrenophora tritici-repentis* based on lesion type. *Canadian Journal of Plant Pathology* 11(1): 49–56. https://doi.org/10.1080/07060668909501146

Lamari L., Bernier C.C., 1989b. Virulence of isolates of *Pyrenophora tritici-repentis* on 11 wheat cultivars and cytology of the differential host reactions. *Canadian Journal of Plant Pathology* 11(3): 284–290. https://doi.org/10.1080/07060668909501114

Lamari L., Bernier C.C., 1989c. Toxin of *Pyrenophora tritici-repentis*: Host-specificity, significance of disease, and inheritance of host reaction. *Phytopathology* 79: 740–744.

Lamari L., Strelkov S.E., 2010. The wheat/*Pyrenophora tritici-repentis* interaction: progress towards an understanding of tan spot disease. *Canadian Journal of Plant Pathology* 32(1): 4–10. https://doi.org/10.1080/07060661003594117

Lamari L., Gilbert J., Tekauz A., 1998. Race differentiation in *Pyrenophora tritici-repentis* and survey of physiologic variation in western Canada. *Canadian Journal of Plant Pathology* 20(4): 396–400. https://doi.org/10.1080/07060669809500410

Lamari L., Sayoud R., Boulif M., Bernier C.C., 1995. Identification of a new race in *Pyrenophora tritici-repentis:* implications for the current pathotype classification system. *Canadian Journal of Plant Pathology* 17(4): 312–318. https://doi.org/10.1080/07060669509500668

Lamari L., Strelkov S.E., Yahyaoui A., Orabi J., Smith R.B., 2003. The identification of two new races of *Pyrenophora tritici-repentis* from the host center of diversity confirms a one-to-one relationship in tan spot of wheat. *Phytopathology* 93(4): 391–396. https://doi.org/10.1094/PHYTO.2003.93.4.391

Lamari L., Strelkov S.E., Yahyaoui A., Amedov M., Saidov M., Djunusoba M., Koichibayev M., 2005. Virulence of *Pyrenophora tritici-repentis* in the countries of the Silk Road. *Canadian Journal of Plant Pathology* 27(3): 383–388. https://doi.org/10.1080/07060660509507236

Loughman R., Wilson R.E., Roake J.E., Platz G.J., Rees R.G., Ellison E.W., 1998. Crop management and breeding for control of *Pyrenophora tritici-repentis* causing yellow spot of wheat in Australia. In: *Helminthosporium Blights of Wheat: Spot Blotch and Tan Spot* (E. Duveiller, H. J. Dubin, J. Reeves, A. McNab, ed.), CIMMYT, Mexico, 10-17.

Manning V.A., Pandelova I., Ciuffetti L.M., 2002. Race for a novel host-selective toxin. *Phytopathology* 92 (Suppl.): S51 (Abstract).

Maraite H., Berny J.F., Goffin A., 1992. Epidemiology of tan spot in Belgium. In: *Advances in Tan Spot* (Francl, L.J., J.M. Krupinsky, M.P. McMullen, ed.), North Dakota Agricultural Experimental Station, Fargo, USA, 73–79.

Mehta Y.R., Gaudencio C.A., 1991. The effects of tillage practices and crop rotation on the epidemiology of some major wheat diseases. In: *Wheat for the Nontraditional Warm Areas* (D. F. Saunders, ed.), CIMMYT, Mexico, 266–283.

Meinhardt S., Ali S., Ling H., Francl L., 2003. A new race of *Pyrenophora tritici-repentis* that produces a putative host-selective toxin. In: *Proceedings of the Fourth International Wheat Tan Spot and Spot Blotch Workshop* (J.B. Rasmussen, T.L. Friesen, S. Ali, ed.), July 21–24, 2002, North Dakota Agricultural Experimental Station, North Dakota State University, USA, 117–119

Momeni H., Aboukhaddour R., Javan-Nikkhah M., Razavi M., Naghavi M.R., Akhavan A., Strelkov S.E., 2014. Race identification of *Pyrenophora tritici-repentis* in Iran. *Journal of Plant Pathology* 96 (2): 287–294. https://doi.org/10.4454/JPP.V96I2.036

Morrall R.A.A., Howard R.J., 1975. The epidemology of leaf spot disease in a native prairie. II. Airborne spore populations of *Pyrenophora tritici-repentis. Canadian Journal of Botany* 53(20): 2345–2353. https://doi.org/10.1139/b75-260

Nsarellah N., Mergoum M., 1997. Effect of crop rotation and straw mulch inoculation on tan spot and root rot in bread and durum wheat. *In: Heminthosporium blights of wheat: spot blotch and tan spot* (E. Duveiller, H. J. Dubin, J. Reeves, A. McNab, ed), CIMMYT, Mexico, 157–161.

Orolaza N.P., Lamari L., Ballance G.M., 1995. Evidence of a host-specific chlorosis toxin from *Pyrenophora tritici-repentis,* the causal agent of tan spot of wheat. *Phytopathology* 85(10): 1282–1287.

Postnifova E.N., Khasanov B.A., 1998. Tan spot in Central Asia. In: *Helminthosporium Blights of Wheat: Spot Blotch and Tan Spot* (E. Duveiller, H. J. Dubin, J. Reeves, A. McNab, ed.), CIMMYT, Mexico, 107–113.

Rees R.G., 1982. Yellow spot, an important problem in the North-Eastern wheat areas of Australia. In *Tan Spot of Wheat and Related Diseases*, (M. Hosford Jr, ed), North Dakota State University, USA, 68–70.

Rees R.G., Platz G.J., 1983. Effects of yellow spot on wheat: Comparison of epidemics at different stages of crop development. *Australian Journal of Agricultural Research* 34(1): 39–46. https://doi.org/10.1071/AR9830039

Rees R.G., Platz G.J., Mayer R.J., 1982. Yield losses in wheat from yellow spot: Comparison of estimates derived from single tillers and plots. *Australian Journal of Agricultural Research* 33(6): 899–908. https://doi.org/10.1071/AR9820899

Schilder A.M.C., Bergstrom G.C., 1995. Seed transmission of *Pyrenophora tritici-repentis,* causal fungus of tan spot of wheat. *The European Journal of Plant Pathology* 101(1): 81–91. https://doi.org/10.1007/BF01876096

Shabeer A., Bockus W.W., 1988. Tan spot effects on yield and yield components relative to growth stage in winter wheat. *Plant Disease* 72(7): 599–602. https://doi.org/10.1094/PD-72-0599

Singh P.K., Mergoum M., Hughes G.R., 2007. Variation in virulence to wheat in *Pyrenophora tritici-repentis* population from Saskatchewan, Canada, from 2000 to 2002. *Canadian Journal of Plant Pathology* 29(2):166–171. https://doi.org/10.1080/07060660709507453

Singh P.K., Singh R.P., Duveiller E., Mergoum M., Adhikari T.B., Elias E.M., 2010. Genetics of wheat–*Pyrenophora tritici-repentis* interactions. *Euphytica* 171(1): 1–13. https://doi.org/10.1007/s10681-009-0074-6

Strelkov S.E., Lamari L., 2003. Host-parasite interactions in tan spot (*Pyrenophora tritici-repentis*) of wheat. *Canadian Journal of Plant Pathology* 25(4): 33–349. https://doi.org/10.1080/07060660309507089

Strelkov S.E., Lamari L., Sayoud R., Smith R.B., 2002. Comparative virulence of chlorosis-inducing races of *Pyrenophora tritici-repentis. Canadian Journal of Plant Pathology* 24(1): 29–35. https://doi.org/10.1080/07060660109506967

Sykes E.E., Bernier C.C., 1991. Qualitative inheritance of tan spot resistance in hexaploid, tetraploid, and diploid wheat. *Canadian Journal of Plant Pathology* 13(1): 38–44. https://doi.org/10.1080/07060669109500963

Valder P.G., Shaw D.E, 1952.Yellow spot disease of wheat in Australia. In: *Proceedings of Linnean Society of New South Wales* 77: 323–330

Vanderpool T.C., 1963. Pink and smudge-pink discoloration of wheat seed associated with the yellow leaf disease *(Drechslera tritici-repentis). Proceedings of the Canadian Phytopathological Society* 30: 19–20.

Watkins J.E., Odvody G.N., Boosalis M.G., Partridge J.E., 1978. An epidemic of tan spot of wheat in Nebraska. *Plant Disease Report* 62(2): 132–134.

Zadoks J.C., Chang T.T., Konzak C.F., 1974. A decimal code for the growth stages of cereals. *Weed Research* 14(6): 415–421.

Chlorine and mefenoxam sensitivity of *Phytophthora nicotianae* and *Phytopthora citrophthora* from South African citrus nurseries

JAN M. VAN NIEKERK[1,2,*], ELAINE BASSON[1], CHARMAINE OLIVIER[1], GRAY-LEE CARELSE[2], VLADIMIRO GUARNACCIA[2,3]

[1] *Citrus Research International, P.O. Box 28, Nelspruit 1200, South Africa*
[2] *Department of Plant Pathology, University of Stellenbosch, Private Bag X1, Matieland 7602, South Africa*
[3] *DiSAFA, University of Torino, Largo Paolo Braccini 2, 10095, Grugliasco, TO, Italy*
*Corresponding author: janvn@cri.co.za

Competing Interests: The Author(s) declare(s) no conflict of interest.

Editor: Anna Maria D'Onghia, CIHEAM/Mediterranean Agronomic Institute of Bari, Italy.

Summary. *Phytophthora nicotianae* and *P. citrophthora* isolates were subjected to mefenoxam and chlorine sensitivity evaluations at different concentrations, and for chlorine, different exposure times. Based on mefenoxam sensitivity, the isolates of the two species were divided in six sensitivity groups with EC_{50} values ranging from sensitive (0.04 ppm mefenoxam) to highly insensitive (greater than 123.69 ppm mefenoxam), with 86% of isolates being sensitive to mefenoxam. Chlorine sensitivity testing indicated strong interactions between chlorine concentration and exposure time for both species. Increased mortality was observed with increased concentration and exposure time to chlorine. For some isolates, close to 100% mortality was only reached at 6 ppm active chlorine and at an exposure time of 60 min. Because highly mefenoxam-insensitive isolates were detected from South African citrus nurseries, this fungicide should be used with care as a curative method for management of diseases caused by *Phytophthora* spp. It is recommended that chlorination of irrigation water, at 6 ppm active chlorine and exposure of more than 60 min, is used to eliminate *P. nicotianae* and *P. citrophthora* propagules from irrigation water as a preventative measure for these diseases.

Keywords. Irrigation water, soilborne pathogens.

INTRODUCTION

The South African citrus industry annually produces approx. 1.3 million tons of citrus fruit, of which about 1.1 million tons are exported, and these fruits are produced on 77,708 ha (Edmonds, 2018). New plantings and replacement of old orchards are important to maintain this level of production. This places heavy demand on continuous production of good quality, disease- and pest-free nursery produced young trees.

Soilborne pathogens, especially *Phytophthora nicotianae* (Breda de Haan) and *P. citrophthora* (R.E. Sm. & E.H. Sm.) Leonian, have been shown to cause economic losses in citrus plantings in many countries (Meitz-Hopkins *et al.*, 2013). In established orchards, *P. nicotianae* causes collar and root rots of trees as well as infections of low hanging fruit, resulting in brown rot. This fruit rot can potentially spoil entire cartons of fruit during export if it spreads among the fruit in each carton (Graham and Feichtenberger, 2015). *Phytophthora citrophthora* can also attack aerial parts of citrus trees, such as trunks and limbs, ultimately causing tree death (Graham and Feichtenberger, 2015).

Phytophthora nicotianae has been reported to occur sporadically in citrus nurseries in many countries, including South Africa (Wehner *et al.*, 1986; Ahmed *et al.*, 2012). Diagnostic results from the Citrus Research International Diagnostic Centre (Nelspruit, South Africa) also showed that *P. citrophthora* occurs sporadically in South African citrus nurseries. These findings are important, because infected nursery trees are sources of infection of new citrus orchards (Ippolito *et al.*, 2004).

Water is a potential source of infections caused by *P. nicotianae* and *P. citrophthora* in nursery environments (Grech and Rijkenberg, 1992; Ippolito *et al.*, 2004). As a result, preventative control measures are employed to ensure that irrigation water is free from these pathogens. One of these measures is chlorination of irrigation water. This is routinely used in citrus nurseries in South Africa, and other citrus producing countries, to eradicate pathogen propagules that might be present. This practice is also employed by nurseries in other industries (Hong *et al.*, 2003; Ghimire *et al.*, 2011). In cases where nurseries are infested with one of these *Phytophthora* spp., curative fungicide treatments are applied. One such fungicide is mefenoxam, a systemic compound which has been widely studied for control of *Phytophthora* spp. in citrus, and in other tree and ornamental crops (Farih *et al.*, 1981; Davis 1982; Matheron and Matejka, 1988; Matheron *et al.*, 1997; Morales-Rodríguez *et al.*, 2014; Aiello *et al.*, 2018).

Despite the application of preventative and curative measures by South African citrus nurseries, *P. nicotianae* and *P. citrophthora* continue to occur sporadically, which indicates ineffective control measures. Hong *et al.* (2003) reported that although zoospores of *P. nicotianae* in water did not survive exposure to 2 mg kg^{-1} active chlorine, mycelium fragments could, in some cases, survive exposure of up to 8 mg kg^{-1}. This is significant, as zoospores and mycelium fragments have been isolated from irrigation water (Hwang and Benson, 2005), with current treatment of citrus nursery water in South Africa carried out with active chlorine concentrations between 3 and 6 ppm. Mefenoxam resistance, has been reported for a number of *Phytophthora* spp., including *P. nicotianae* and *P. citrophthora*, occurring on a wide variety of crops (Hwang and Benson, 2005; Hu *et al.*, 2008, 2010).

The aims of the present study were to obtain *P. nicotianae* and *P. citrophthora* isolates from South African citrus nurseries, and characterize them with regards to sensitivity to chlorine and mefenoxam. The results obtained in this study could be used to develop effective preventative and curative control measures for these pathogens in citrus nurseries.

MATERIALS AND METHODS

Collection and purification of isolates

Isolates of *P. nicotianae* and *P. citrophthora* (Table 1) were collected from South African citrus nurseries, in different provinces of South Africa, by sampling plant propagation substrate from pots of young citrus trees. These samples were placed into different compartments

Table 1. Numbers of *Phytophthora citrophthora* and *Phytophthora nicotianae* isolates from different South African provinces used for mefenoxam and chlorine sensitivity testing.

Species	Province	No. of isolates per species	No. of isolates used for mefenoxam sensitivity testing	No. of isolates used for chlorine sensitivity testing
Phytophthora citrophthora	Eastern Cape	52	48	26
	Western Cape	8	7	4
Phytophthora citrophthora	Eastern Cape	27	27	11
	Limpopo	10	10	7
	Mpumalanga	9	8	5
	North West	4	4	3
	Western Cape	11	11	6

of ice trays, with one ice tray allocated per sample. The substrate in each compartment was covered with distilled water before placing two citrus leaf discs (each 5 mm diam.) in each compartment (Grimm and Alexander, 1973). Before cutting the leaf discs from the citrus leaves, they were washed thoroughly with distilled water. The leaves had been collected from trees not subjected to any fungicide treatments. The ice trays were covered to prevent light infiltration, and were incubated at ambient room temperature on a laboratory bench for 48 h. Following incubation, the leaf discs were removed, blotted dry on absorbent paper toweling, and plated onto 90 mm Petri dishes containing PARPH medium (Jeffers and Martin, 1986). Inoculated plates were then incubated in the dark at 29°C for 48 h before being inspected for *Phytophthora* spp. colonies. Isolates were selected from the inoculated plates and transferred to water agar (WA, Biological agar, Biolab), followed by additional incubation at 29°C for 48 h. Colonies were purified from WA by hyphal tipping onto 90 mm Petri dishes containing V8 agar (Galindo and Gallegly, 1960). Purified isolates were stored in molecular grade water in 2 mL capacity micro centrifuge tubes at 25°C.

Molecular identification of isolates

Selected isolates were grown on V8 agar at 29°C for 7 d before mycelia were harvested for genomic DNA extraction, using a modified CTAB-based extraction protocol (Allen *et al.*, 2006).

PCR-RFLP analyses. The ITS region of the isolates was amplified using the primers ITS 6 (Cooke and Duncan, 1997) and ITS 4 (White *et al.*, 1990). The PCR reaction consisted of 20.0 μL of GoTaq® G2 Hot Start Green Master Mix (Promega Corporation), 1.0 μL of each primer (concentration of 10 μM), 16 μL PCR grade water and 2 μL of genomic DNA, for a total volume of 40 μL. Amplifications were conducted in a 2720 thermal cycler (Applied Biosystems). Initial denaturation was at 94°C for 5 min, followed by 32 cycles of 94°C for 30 s, annealing for 30 s at 55°C, extension at 72°C for 30 s, with a final extension step at 72°C for 5 min. PCR products were resolved in a 1% agarose gel, and DNA fragments were visualized by staining with an ethidium bromide solution. The resulting PCR products were restriction digested with enzymes *Hinf*I and *Hha*I in a single reaction, according to manufacturer's instructions (Fermentas Inc.). The PCR-RFLP products were run on a 3% agarose gel, and isolates with the same RFLP banding patterns were assigned to each RFLP group.

ITS sequencing. The ITS regions of at least two isolates of each PCR-RFLP group were sequenced, and double stranded consensus sequences were obtained. The consensus sequences were subjected to BLAST analyses in Genbank (https://blast.ncbi.nlm.nih.goc/Blast.cgi), and were identified to species level based on similarity of at least 99% to existing *P. nicotianae* or *P. citrophthora* ITS sequences on Genbank (https://www.ncbi.nlm.nih.gov/genbank/).

Mefenoxam sensitivity testing

In vitro sensitivity testing

Sensitivity testing was conducted according to a slightly amended protocol described in Timmer *et al.* (1998). A total of 60 *P. citrophthora* and 61 *P. nicotianae* isolates, from different nurseries in different citrus production areas in South Africa (Table 1), were selected and were on 90 mm Petri dishes containing corn meal agar (CMA; Sigma-Aldrich). Plates were incubated at 29°C for 5 d. After incubation, 5 mm plugs were cut from the edges of the actively growing cultures and plated onto 90 mm Petri dishes containing CMA amended with mefenoxam (Ridomil Gold® 450 EC; Syngenta) at 0, 1.0, 2.0, 5.0, 10.0, 25.0, 50.0 or 100.0 ppm. These plates were then incubated at 29°C for 2 d.

Each isolate and concentration combination was repeated using two plates, while the whole trial was repeated twice at the same time. Colony diameters of growing isolates were each measured in two directions and the average colony diameter was calculated for each isolate at each concentration. The percentage inhibition for each plate at each concentration for all the isolates was calculated, and data were subjected to statistical analyses to group isolates, and determine EC_{50}, EC_{80} and EC_{90} values for each isolate. The percentage inhibition was calculated using the following equation:
Percentage inhibition (%) = mean colony diameter of control (0 ppm) – mean diameter ÷ mean diameter of control (0 ppm) × 100.

Data analyses

The Mitscherlich function [$y = a(1\text{-}e^{-bx})$] or % Inhibition = Maximum Inhibition [$1\text{-}e^{-(Rate)(Concentration)}$] fitted the data well, and was used throughout the study. Hereafter, percentage (%) inhibition will be referred to as %Inhb, and maximum inhibition, as MaxInhb. The function was fitted for the two Petri dishes representing each isolate mefenoxam concentration combination within each of the two trials.

EC_{50}, EC_{80}, and EC_{90} values were calculated from the estimated regression parameters (MaxInhb and Rate of inhibition) for each isolate. Wherever MaxInhb < EC, these respective values could not be calculated according to the appropriate equations (EC_{50} = (-log (1 - (50 ÷ a))) ÷ b; EC_{80} = (-log (1 - (80 ÷ a))) ÷ b; EC_{90} = (-log (1 - (90 ÷ a))) ÷ b). MaxInhb did not always give realistic values, especially where the Rate of inhibition was very slow, because MaxInhb is a value at a theoretical concentration. An additional value, PInhbConc100 = $a[1\text{-}e^{-b(100)}]$, was therefore calculated. This represents the %Inhb at a fungicide concentration of 100 ppm, which gave a more realistic interpretation within the boundaries of the data than just MaxInhb. Regression parameters and EC_{50}, EC_{80} and EC_{90} values were subjected to analysis of variance (ANOVA), and cluster analysis using Ward's clustering method to cluster isolates. Principle component analysis (PCA) was also carried out for the 115 isolates and numbers of isolates in clusters as labels, to see if the grouping or clustering obtained from the cluster analyses made sense.

Chlorine sensitivity testing

Mycelium suspension preparation

Hong *et al.* (2003) found that *Phytophthora* spp. mycelium fragments were more insensitive to chlorine than zoospores. Mycelium fragment suspensions were therefore used in chlorine sensitivity tests. Ten percent V8 broth was prepared by adding 0.5 g $CaCO_3$ (Calcium carbonate; Merck) and 50 mL V8 juice (V8 Original Vegetable Juice, Campbell Soup Company) to each Schott bottle containing 450 mL filtered water, and bottles were then autoclaved at 121°C for 15 min. *Phytophthora* isolates of the two species (32 *P. nicotianae* and 30 *P. citrophthora*; Table 1), randomly selected from the populations used for mefenoxam sensitivity testing, were plated onto 90 mm Petri dishes containing CMA and incubated for 7–10 d at 29°C. After sufficient growth, the agar from each Petri dish was divided into smaller pieces using a scalpel, and placed into the prepared 10% V8 broth. The inoculated V8 broth was then placed on an orbital shaker (SHKO 20; FHM Electronics) running at 100 rpm and 29°C, for 21 d in the dark.

To prepare the mycelium broth, autoclaved filtered water was adjusted to pH 6.5, Adjustment of pH was achieved using sodium hydroxide (NaOH 40 g mol^{-1}, Merck; 2 g in 250 mL autoclaved filtered water) and hydrochloric acid (HCl 32%, Merck; 5 mL in 250 mL autoclaved filtered water). The mycelium masses harvested from the 10% V8 broth were drained using a 180-µm sieve before being each washed twice with 100 mL autoclaved filtered water. Excess water was then pressed out of the remaining fungal mycelium mass using two sterile stainless steel teaspoons. For trial purposes, a measured *Phytophthora* suspension was prepared by blending 1 g (wet mass) of mycelium in 100 mL filtered water (pH 6.5) for 30 s, followed by filtration (1,000 µm sieve) into a Schott bottle, which was then filled to 500 mL using deionized water (pH 6.5).

Chlorine sensitivity testing

Trial variables included chlorine concentration (0, 1.5, 3 or 6 ppm) and a range of chlorine exposure times (0, 5, 10, 30 or 60 min). A chlorine stock solution (SS) was prepared by adding 0.15 g chlorine granules (HTH®, South Africa) to 100 mL filtered, autoclaved water (pH 6.5). In order to achieve 0, 1.5, 3 and 6 ppm concentrations of active chlorine, 0, 0.75, 1.5 or 3.0 mL chlorine SS was added to different 500 mL Schott bottles. As a positive control, a 1.5 and 6 ppm active chlorine solution was tested using a chlorine photometer (Total Chlorine Ultra High Range Portable Photometer, HI 96771; Hanna Instruments Inc.) before the commencement of each trial set. These chlorine control solutions were also de-activated with sodium thiosulfate stock solution ($Na_2S_2O_3{*}5H_2O$; 248.21 g mol^{-1}; Merck) containing 1.47 g $Na_2S_2O_3{*}5H_2O$ in 1,000 mL of filtered, autoclaved water, and tested with Insta-Test® low range 90–10 ppm) free chlorine test strips (LaMotte) to determine whether the stock solution was still functional.

Additionally, the prepared *Phytophthora* mycelium suspension of each isolate was mixed on a magnetic stirrer plate for 10 min before being used to inoculate two PARPH plates as positive controls. Each mycelium suspension (two separate suspensions per *Phytophthora* isolate) was added (treated) to 0, 1.5, 3 or 6 ppm active chlorine and mixed for a further 30 s. Following each exposure time (0, 5, 10, 30 or 60 min), 40 mL of solution was dispensed into two containers and de-activated using the sodium thiosulfate SS. For deactivation, 0, 0.3, 0.6 or 1.2 mL of sodium thiosulfate SS was required to de-activate, respectively, 0, 1.5, 3 and 6 ppm active chlorine. De-activated solution (1 mL) from each container was used to inoculate two PARPH plates and was subsequently spread using a hockey stick and incubated for 2 d at 29°C. Free chlorine test strips were used to confirm de-activation.

Following incubation, *Phytophthora* spp. colonies were counted and percentage mortality determined using the following formula:
[(Cn-Tn)/Cn)]*100, where Cn is the number of colonies on control plates and Tn the number of colonies on treated plates.

The percentage mortality data were subjected to statistical analyses using SAS (SAS Institute Inc.). Fisher's LSD was calculated at the 5% level to compare means.

RESULTS

Molecular identification of isolates

The 121 isolates were divided into two distinct groups based on the ITS-RFLP analysis. ITS sequence analyses of representative isolates from each group identified 61 of the isolates obtained from citrus nurseries as *P. nicotianae*. The remaining 60 isolates were identified as *P. citrophthora*. The species identity was based on a 100% nucleotide homogeneity with *P. nicotianae* and *P. citrophthora* isolates lodged from previous studies on Genbank (https://www.ncbi.nlm.nih.gov/genbank/).

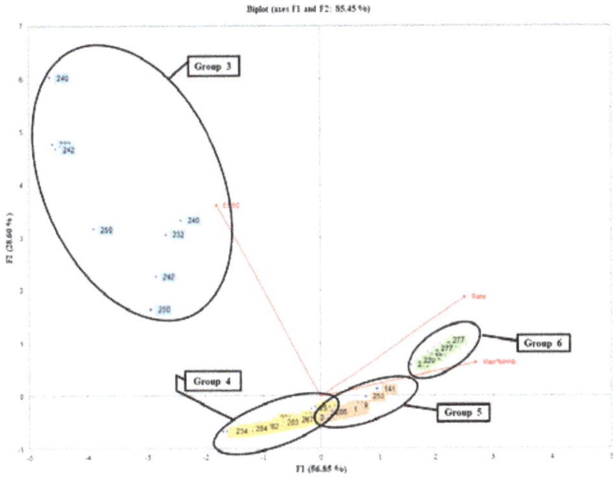

Figure 1. Principal component analysis (PCA) indicating separation of *Phytophthora nicotianae* and *Phytopthora citrophthora* isolates into four distinct mefenoxam sensitivity groups based on the regression parameters EC_{50}, rate of inhibition and Max%Inhb.

Mefenoxam sensitivity testing

Plotting of the mean EC_{50}, EC_{80} and EC_{90} values of all the isolates indicated that for a group of five isolates no EC_{50}, EC_{80} or EC_{90} values could be calculated. These isolates were grouped together in sensitivity group 1 (Table 2). A further two isolates had EC_{50} values that were greater than 100 ppm. These two isolates were placed in sensitivity group 2 (Table 2). The remaining isolates were subjected to Ward's cluster analysis and PCA to group them into sensitivity groups. Both the Ward's cluster analysis and PCA showed that the remaining *P. citrophthora* and *P. nicotianae* isolates could be aligned into four distinct sensitivity groups (Figure 1). This brought to six the total number of sensitivity groups (Table 2). The analysis of variance (ANOVA) comparing the groups using the regression parameters and EC_{50}, EC_{80} and EC_{90} data showed highly significant ($P < 0.0001$) sensitivity group effects for these three parameters. The ANOVA of the PInhbCon100 and Rate of inhibition data showed a significant sensitivity group × species interaction ($P < 0.0001$ for PInhbCon100 and $P = 0.0021$ for Rate of inhibition).

Table 2. Mean PInhbConc100 and rate of inhibition values for *Phytophthora citrophthora* and *Phytophthora nicotianae* isolates, grouped into mefenoxam sensitivity groups 1-6, following *in vitro* exposure to different mefenoxam concentrations.

Sensitivity group	Species	PInhbConc100 (%)	Rate of inhibition
1	*P. citrophthora*	39.36 h[1]	0.068 e
	P. nicotianae	30.74 i	0.036 e
2	*P. citrophthora*	44.30 g	0.002 e
	P. nicotianae	---	---
3	*P. citrophthora*	54.88 f	0.027 e
	P. nicotianae	---	---
4	*P. citrophthora*	70.29 e	1.811 d
	P. nicotianae	80.94 d	1.307 d
5	*P. citrophthora*	92.35 c	2.616 c
	P. nicotianae	96.02 b	1.858 c
6	*P. citrophthora*	100.00 a	17.752 a
	P. nicotianae	100.00 a	15.820 a
LSD		3.430	0.7200
P		< 0.0001	0.0021

[1] Means followed by the same letter are not statistically different at a 95% confidence level.

Within group 6, the isolates of *P. citrophthora* and *P. nicotianae* were 100% inhibited by a concentration of 100 ppm mefenoxam. Within group 5 the *P. nicotianae* isolates had a mean of 96.02% inhibition that was significantly more than the *P. citrophthora* isolates in the same group (92.35% inhibition) (Table 2). Both these means of group 5 were significantly less than those of group 6. Also, within group 4 the mean inhibition of *P. nicotianae* isolates was 80.94% and was statistically greater than the mean inhibition (70.29%) of the *P. citrophthora* isolates in this group (Table 2). These two means were less than that observed for the two species in groups 5 and 6. Only *P. citrophthora* isolates were grouped into group 2 and 3. In these groups the mean inhibition of isolates was 54.88% (group 3) and 44.30% (group 2), significantly less than the mean inhibition percentages for isolates in the other groups (Table 2). Isolates of both species were placed in group 1 that had the least sensitivity to mefenoxam. In this group, mean inhibition of the *P. citrophthora* isolates was 39.26% compared to 30.74% for the *P. nicotianae* isolates. The mean percentage inhibition for both species in group 1 were also the least observed in any of the groups for either pathogen species (Table 2).

The rate of inhibition results for the group × species interaction showed similar trends to those for the PInhbCon100 results. The groups with the greater PInhbCon100 means, also had the greatest rates of inhibition (Table 2). Group 6 had the greatest mean rate of inhibition of all the groups. The mean rate of inhibition of *P. citrophthora* isolates was 17.75, significantly greater than the rate for the *P. nicotianae* isolates (15.82) in this group. The mean inhibition rate of the *P. citrophthora* isolates in group 5 was 2.62 which was significantly greater again than the rate of the *P. nicotianae* isolates (1.86). This mean rate for *P. nicotianae* isolates was comparable to the rates observed for the two species in group 4 (Table 2). In this group, the mean rate of inhibition for *P. citrophthora* was 1.81 compared to 1.31 for *P. nicotianae*. In groups 1, 2 or 3, the mean rates for the two species (0.002 to 0.068) were statistically similar (Table 2).

For the isolates of both species sensitivity group 1, at 100 ppm mefenoxam, the maximum inhibition only reached a mean of 39.36% and the rate of inhibition was only 0.068 (Table 2). Consequently, no EC_{50}, EC_{80} or EC_{90} values could be determined for this group (Table 3). For group 2, the mean EC_{50} value was 123.69 ppm, which was statistically greater than the mean EC_{50} of any other group. Similarly, the mean EC_{80} (214.12 ppm) and EC_{90} (250.25 ppm) values of this group were the greatest of all the groups (Table 3). For group 3, the calculated mean EC_{50} was 76.12 ppm, which was the second greatest of all the groups. Again, no EC_{80} or EC_{90} means could be determined, possibly also due to the slow rate of inhibition of this group (Tables 2 and 3). Group 4 had mean EC_{50} (0.82 ppm) and EC_{90} (2.86 ppm) values that were statistically similar to those for groups 5 and 6. However, an EC_{90} could also not be calculated for this group (Table 2). Groups 5 and 6 had mean EC_{50} (0.45 ppm), EC_{80} (0.10 ppm) and EC_{90} (1.68 ppm) values that were statistically similar to those of group 6. For group 6, the mean EC_{50} (0.04 ppm), EC_{80} (0.10 ppm) and EC_{90} (1.15 ppm) values were the least of all the groups (Table 3). This indicated that the isolates in this group were the most sensitive to mefenoxam.

Table 3. Mean EC_{50}, EC_{80} and EC_{90} values for different mefenoxam sensitivity groups identified after *in vitro* exposure of *Phytophthora citrophthora* and *Phytophthora nicotianae* isolates to different mefenoxam concentrations.

Sensitivity group	EC_{50}	EC_{80}	EC_{90}
1	---	---	---
2	123.69 a[1]	214.12 a	250.25 a
3	76.11 b	---	---
4	0.82 c	2.86 b	---
5	0.45 c	1.11 b	1.68 b
6	0.04 c	0.10 b	0.15 b
LSD	5.101	8.545	12.608
P	< 0.0001	< 0.0001	< 0.0001

[1] Means followed by the same letter are not statistically different at a 95% confidence level.

Distribution of isolates in sensitivity groups

Within the different sensitivity groups, the number of isolates of *P. citrophthora* and *P. nicotianae* varied greatly. For *P. citrophthora*, most of the isolates (90%) were in groups 1, 2, 3 or 4, where the mean percentage inhibition at 100 ppm mefenoxam ranged from 39.36% to 70.29% (Table 4). Based on the classification of Hu *et al.* (2008), these isolates were intermediately insensitive or sensitive to mefenoxam. However, for *P. nicotianae* only 5% of isolates were in the groups 1 to 4, while 95% were in groups 5 and 6. These isolates would be classified as mefenoxam sensitive (Table 4).

Table 4. Numbers (and proportions) of *Phytophthora citrophthora* and *Phytophthora nicotianae* isolates occurring in the different mefenoxam sensitivity groups.

Species	Mefenoxam sensitivity group	No. of isolates in group
Phytophthora citrophthora (n= 60)	1	5 (8%)
	2	2 (3%)
	3	4 (7%)
	4	43 (72%)
	5	4 (7%)
	6	2 (3%)
Phytophthora nicotianae (n = 61)	1	1 (2%)
	2	0 (0%)
	3	0 (0%)
	4	2 (3%)
	5	36 (59%)
	6	22 (36%)

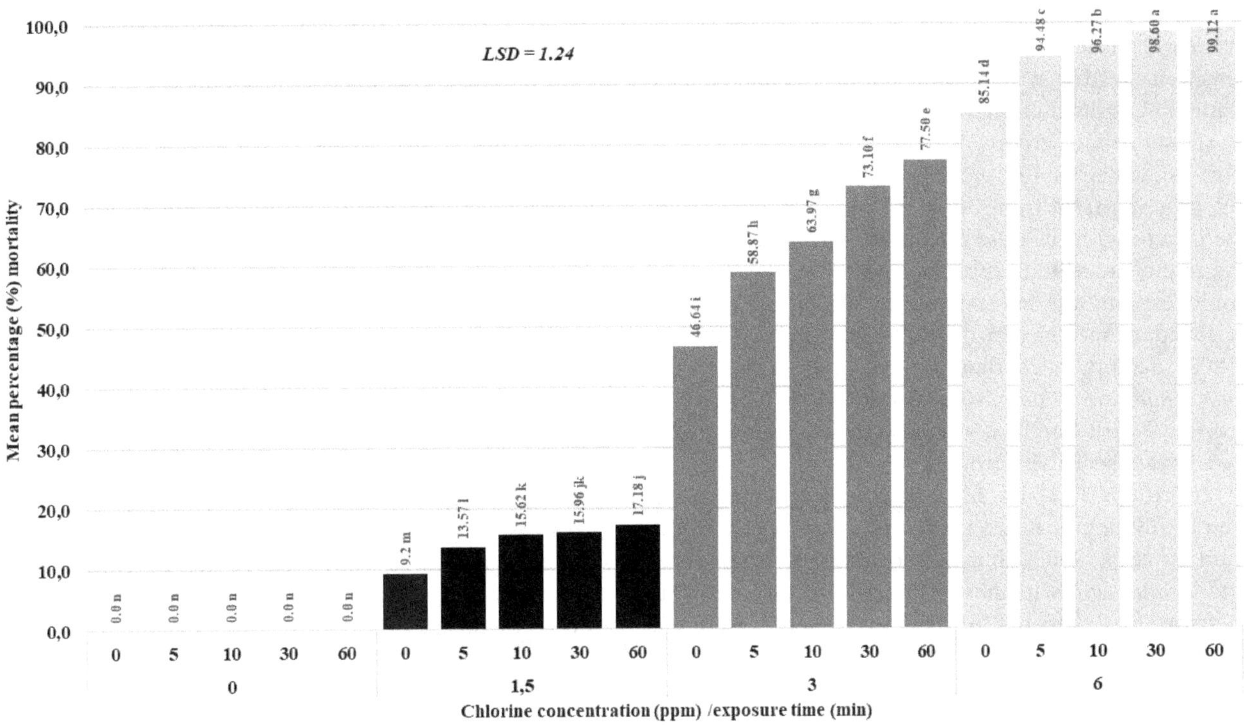

Figure 2. Mean percentage (%) mortality of *Phytophthora* spp. propagules exposed to 0, 1.5, 3 or 6 ppm active chlorine for exposure times of 0, 5, 10, 30 or 60 min.

Chlorine sensitivity testing

The analysis of variance (ANOVA) of the percentage mortality data indicated a highly significant ($P < 0.0001$) experimental repetition × species × isolate × chlorine concentration interaction, and an experimental repetition × chlorine concentration × exposure time interaction. These multifactor interactions are attributed to the significant ($P < 0.0001$) variation seen in the percentage inhibition data between the two experimental repetitions, which could be due to different mycelium suspensions used for each repetition, combined with the significant ($P < 0.0001$) variation seen in mean percentage inhibition between the different chlorine concentrations. Between isolates in the two *Phytophthora* spp., the ANOVA also indicated that there were statistical ($P < 0.0001$) differences between percentage mortalities obtained. Mean mortality of *P. citrophthora* isolates due to chlorine ranged from 27.09 to 73.47%, whereas those for *P. nicotianae* isolates were from 19.69 to 62.30%.

The results from the significant ($P < 0.0001$) chlorine concentration × exposure time interaction indicated no mortality of the *Phytophthora* spp. for 0 ppm chlorine. With chlorine concentration of 1.5 ppm, the mean percentage mortality at no exposure (chlorine deactivated immediately) was 9.20%. This then increased with each increase in exposure time to reach a maximum mean of 17.18% after 60 min exposure to 1.5 ppm active chlorine (Figure 2). When the 3 ppm chlorine treatment was deactivated immediately, the mean mortality was 46.64% and was significantly greater to be 77.50% after 60 min exposure. At 6 ppm chlorine the initial mean mortality was 85.14% with immediate deactivation. Increase in exposure time led to greater mortality at 99.12% after 60 min exposure (Figure 2).

DISCUSSION

Production of citrus trees in nurseries that are free from the soilborne pathogens *Phytophthora nicotianae* and *P. citrophthora* is regarded as essential for establishment of new orchards. This has been concluded from previous studies indicating that infected nursery trees can be sources of infection of new citrus orchards (Ippolito *et al.*, 2004). As a result, nurseries need to treat irrigation water with chlorine, because the water is a potential source of infection (Ghimire *et al.*, 2011). As a curative measure, infected plants are often treated in these nurseries with drenches of metalaxyl or mefenoxam (Hu *et al.*, 2008).

In the present study a total of 121 *P. nicotianae* and *P. citrophthora* isolates were subjected to mefenoxam sensitivity testing at eight different mefenoxam concentrations. Results indicated that the isolates were divided in six sensitivity groups. The mean percentage inhibition calculated at 100 ppm mefenoxam for the different groups varied from 30.74 to 100.00% (Table 2). In the groups with the least inhibition (groups 1, 2 and 3), the rate of increase in inhibition was also very low and could explain the low maximum inhibition achieved at 100 ppm mefenoxam (Table 2). Hwang and Benson (2005) demonstrated that isolates of *P. cryptogea, P. nicotianae* and *P. palmivora*, occurring on floriculture crops in North Carolina, were also divided into different mefenoxam sensitivity groups.

In groups 1, 2 and 3 the mean percentage inhibition at 100 ppm mefenoxam was less than 60%, indicating (Hu *et al.*, 2008), that the isolates of *P. citrophthora* and *P. nicotianae* in these groups were insensitive to mefenoxam. Compared to this, the isolates of these two species in groups 4 and 5 were intermediately sensitive, while the isolates in group 6 were sensitive to the compound. There were more *P. citrophthora* isolates than *P. nicotianae* isolates in the groups with the least sensitivity to mefenoxam. Within the groups, *P. citrophthora* isolates often had lower percentages of inhibition compared to *P. nicotianae*, indicating less mefenoxam sensitivity. This is similar to the results of Farih *et al.* (1981) and Coffey and Bower (1984), who found *P. citrophthora* isolates from citrus were less sensitive to mefenoxam than *P. nicotianae* isolates from citrus.

Mean EC_{50}, EC_{80} and EC_{90} values for the isolates in groups 1 to 4 could not be calculated (Table 3). This was probably because at 100 ppm mefenoxam, the highest concentration used in this study, the calculated percentage inhibition for *P. citrophthora* and *P. nicotianae* isolates in these groups was less than 50–90%. For these isolates, 100% inhibition would only be achieved at mefenoxam concentrations much greater than 100 ppm. This was shown by the results for the group 2 isolates, where the mean mefenoxam EC_{90} value was 250.25 ppm.

Timmer *et al.* (1998) published similar results, where they found that some isolates of *P. nicotianae* from citrus had EC_{50} values greater than 100 ppm. Similarly, Farih *et al.* (1981) reported that some isolates of *P. citrophthora* and *P. nicotianae* from citrus had 100% inhibition of mycelium growth at mefenoxam concentrations greater than 100 ppm. High levels of insensitivity to mefenoxam or metalaxyl among isolates of these two citrus pathogens are therefore not unknown. Even among *P. nicotianae* isolates from ornamental crops, Ferrin and Kabashima (1991) showed that highly insensitive isolates had EC_{50} values greater than 100 ppm. In groups 5 and 6 in the present study, the isolate EC_{50} values were well below 2 ppm, and in the case of group 6 below 0.2 ppm. These values were similar to those for isolates described in the reports cited above. These levels of mefenoxam insensitivity have also been determined for *P. infestans* from potato (Goodwin *et al.*, 1996) and *P. cinnamomi* from avocado (Coffey and Bower, 1984) and ornamental crops (Hu *et al.*, 2010).

For *P. infestans* isolates, the basis of mefenoxam or metalaxyl insensitivity or resistance, and differences in sensitivity between isolates within a species, and between species, were found to be due to genotypic differences between the isolates and species (Goodwin *et al.*, 1996). The particular level of sensitivity within a genotype was determined by insensitivity loci present (Fabritius *et al.*, 1997). Childers *et al.* (2015) also discovered that sensitive isolates of *P. infestans* can acquire mefenoxam resistance after repeated *in vitro* exposure to the fungicide. However, isolates developing resistance *in vitro* lost some resistance when they were repeatedly plated onto media not amended with mefenoxam.

It is therefore possible that sensitive isolates of *P. nicotianae* and *P. citrophthora* could acquire resistance to mefenoxam when repeatedly exposed to this fungicide in nurseries or orchards. Careful use of this fungicide in citrus nurseries is therefore important, to prevent development of highly insensitive isolates in nurseries that may reach newly established orchards. This was emphasized by the results of Timmer *et al.* (1998), who showed that insensitive isolates from citrus nurseries competed with sensitive isolates as causes of root rot. Furthermore, these insensitive isolates maintained their insensitivity after use of the fungicide ceased.

No differences in sensitivity to chlorine were observed between isolates of *P. citrophthora* and *P. nicotianae*. Increasing exposure time at particular chlorine concentrations increased percentage mortality of both pathogens. However, mean percentage mortality only came close to 100% when the isolates (32 *P. nicotianae* and 30 *P. citrophthora*) were exposed to 6 ppm chlorine for 60 min (Figure 2). Hong *et al.* (2003) tested limited numbers of isolates of *P. nicotianae, P. capsici, P. cinnamomi, P. citricola, P. citrophthora, P. cryptogea* and *P. megasperma*, from irrigation water from ornamental nurseries, for their sensitivity to chlorine. They also observed that with increasing active chlorine concentrations, mean percentage mortality of these pathogens increased. However, they did not detect a chlorine concentration and exposure time interaction. This interaction recorded in the present study could have been due to the higher numbers of isolates of the two species tested. Chlorination was introduced in South African citrus nurseries based on the study of Grech and Rijkenberg (1992). They indicated that chlorination eliminates soilborne pathogen propagules from irrigation water, consequently reducing the level of *Phytophthora* infection in roots of citrus rootstock seedlings irrigated with treated water. This treatment was also shown to not cause phytotoxic effects on the irrigated seedlings.

The present study is the first to focus on chlorine sensitivity of multiple *P. citrophthora* and *P. nicotianae* isolates subjected to a range of chlorine concentrations and exposure times. Practically, the results indicate that for complete elimination of *Phytophthora* spp. propagules from citrus nursery irrigation water, treatment with 6 ppm active chlorine for 60 min or longer, is required. Mefenoxam as a curative soil drench treatment should also be used with care, as low numbers of highly resistant *P. citrophthora* and *P. nicotianae* isolates were found in this study. For the elimination of these pesticide-resistant pathogens from nurseries, it is therefore important to make use of alternative fungicides (e.g. captan) for curative soil drenches.

ACKNOWLEDGEMENTS

The authors thank Citrus Research International and the Research for Citrus Exports Fund of the Department of Science and Technology, South African Government, for supplying facilities and funding for this study.

LITERATURE CITED

Ahmed Y., D'Onghia A.M., Ippolito A., El Shimy H., Cirvilleri G., Yaseen T., 2012. *Phytophthora nicotianae* is the predominant *Phytophthora* species in citrus nurseries in Egypt. *Phytopathologia Mediterranea* 51: 519–527.

Aiello D., Hansen Z.R., Smart C.D., Polizzi G., Guarnaccia V., 2018. Characterisation and mefenoxam sensitivity of *Phytophthora* spp. from ornamental plants in Italian nurseries. *Phytopathologia Mediterranea*, 57(2): 245–256.

Allen G.C., Flores-Vergara M.A., Krasynanski S., Kumar S., Thompson W.F., 2006. A modified protocol for rapid DNA isolation from plant tissues using cetyltrimethylammonium bromide. *Nature Protocols* 1: 2320–2325.

Childers R., Danies G., Myers K., Fei Z., Small, I.M., Fry W.E., 2015. Acquired resistance to mefenoxam in sensitive isolates of *Phytophthora infestans. Phytopathology* 105: 342–349.

Coffey M.D., Bower L.A., 1984. *In vitro* variability among isolates of six *Phytophthora* species in response to metalaxyl. *Phytopathology* 74: 502–506.

Cooke D.E.L., Duncan J.M., 1997. Phylogenetic analysis of *Phytophthora* species based on ITS 1 and ITS 2 sequences of the ribosomal RNA gene repeat. *Mycological Research* 101: 667–677.

Davis R.M., 1982. Control of Phytophthora root and foot rot of citrus with systemic fungicides metalaxyl and phosethyl aluminum. *Plant Disease* 66: 218–220.

Edmonds J., 2018. *2018 Key Industry Statistics for Citrus Growers*. Citrus Growers Association of Southern Africa, 47 pp. http://www.cga.co.za.

Fabritius A-L., Shattock R.C., Judelson H.S., 1997. Genetic analysis of metalaxyl insensitivity loci in *Phytophthora infestans* using linked DNA markers. *Phytopathology* 87: 1034–1040.

Farih A., Tsao P.H., Menge J.A., 1981. *In vitro* effects of metalaxyl on growth, sporulation and germination of *Phytophthora parasitica* and *P. citrophthora. Plant Disease* 65: 651–654.

Ferrin D.M., Kabashima J.N., 1991. *In vitro* insensitivity to metalaxyl of isolates of *Phytophthora citricola* and *P. parasitica* from ornamental hosts in Southern California. *Plant Disease* 75: 1041–1044.

Galindo A.J., Gallegly M.E, 1960. The nature of sexuality in *Phytophthora infestans. Phytopathology* 50: 123–128.

Ghimire S.R., Richardson P.A., Kong P., Hu J., Lea-Cox J.D., … Hong C., 2011. Distribution and diversity of *Phytophthora* species in nursery irrigation reservoir adopting water recycling system during winter months. *Journal of Phytopathology* 159: 713–719.

Goodwin S.B., Sujkowski L.S., Fry W.E., 1996. Widespread distribution and probable origin of resistance to metalaxyl in clonal genotypes of *Phytophthora infestans* in the United States and Western Canada. *Phytopathology* 86: 793–800.

Graham J., Feichtenberger E., 2015. Citrus Phytophthora diseases: Management challenges and successes. *Journal of Citrus Pathology* 2: 1–11

Grech N.M., Rijkenberg F.H.J., 1992. Injection of electrolytically generated chlorine into citrus microirrigation systems for the control of certain waterborne root pathogens. *Plant Disease* 76: 457–461.

Grimm G.R., Alexander A.F., 1973. Citrus leaf pieces as traps for *Phytophthora parasitica* from soil slurries. *Phytopathology* 63: 540–541.

Hong C.X., Richardson P.A., Kong P., Bush E.A., 2003. Efficacy of chlorine on multiple species of *Phytoph-*

thora in recycled nursery irrigation water. *Plant Disease* 87: 1183–1189.

Hu J.H., Hong C.X., Stromberg E.L., Moorman G.W., 2008. Mefenoxam sensitivity and fitness analysis of *Phytophthora nicotianae* isolates from nurseries in Virginia, USA. *Plant Pathology* 57: 728-736.

Hu J., Hong C., Stromberg E.L., Moorman G.W., 2010. Mefenoxam sensitivity in *Phytophthora cinnamomi* isolates. *Plant Disease* 94: 39–44.

Hwang J.H., Benson D.M., 2005. Identification, mefenoxam sensitivity and compatibility type of *Phytophthora* spp. attacking floriculture crops in North Carolina. *Plant Disease* 89: 185–190.

Ippolito A., Schena L., Nigro F., Soleti Ligoria V., Yaseen T., 2004. Real-time detection of *Phytophthora nicotianae* and *P. citrophthora* in citrus roots and soil. *European Journal of Plant Pathology* 110: 833–843.

Jeffers S.N., Martin S.B., 1986. Comparison of two media selective for *Phytophthora* and *Pythium* species. *Plant Disease* 70: 1038–1043.

Matheron M.E., Matejka J.C., 1988. Persistance of systemic activity for fungicides applied to citrus trunks to control *Phytophthora* gummosis. *Plant Disease* 72: 170–174.

Matheron M.E., Porchas M., Matejka J.C., 1997. Distribution and seasonal population dynamics of *Phytophthora citrophthora* and *P. parasitica* in Arizona citrus orchards and effect of fungicides on tree health. *Plant Disease* 81: 1384–1390.

Meitz-Hopkins J.C., Pretorius M.C., Spies C.F.J., Huisman L., Botha W.J.,... McLeod A., 2014. Phytophthora species distribution in South African citrus production regions. *European Journal of Plant Pathology* 138: 733–749.

Morales-Rodrígues C., Palo C., Palo E., Rodrígues-Molina M.C., 2014. Control of *Phytophthora nicotianae* with mefenoxam, fresh *Brassica* tissues and *Brassica* pellets. *Plant Disease* 98: 77–83.

Timmer L.W., Graham J.H., Zitko S.E., 1998. Metalaxyl-resistant isolates of *Phytophthora nicotianae*: Occurrence, sensitivity and competitive parasitic ability on citrus. *Plant Disease* 82: 254–261.

White T.J., Bruns T., Lee S., Taylor J., 1990. Amplification and direct sequencing of fungal ribosomal RNA genes for phylogenetics. In: *PCR protocols: a Guide to Methods and Applications* (M.A. Innis, D.H. Gelfand, J.J. Sninsky, T.J. White, ed.), Academic Press, New York, USA, 315-322.

Wehner F.C., Combrink H., Kotzé J.M., 1986. Root pathogens in South African citrus nurseries. *Phytophylactica* 19: 107–108.

19

Crosstalk between the cAMP-PKA pathway and the β-1,6-endoglucanase in *Verticillium dahliae*

Lugard EBOIGBE[1,3], Aliki K. TZIMA[2], Epaminondas J. PAPLOMATAS[2], Milton A. TYPAS[1,*]

[1] *Division of Genetics and Biotechnology, Department of Biology, National and Kapodistrian University Athens, Panepistemiopolis, Kouponia,15701 Athens, Greece*
[2] *Laboratory of Plant Pathology, Agricultural University of Athens, 75 Iera Odos, 11855 Athens, Greece*
[3] *Current address: Plant Biology and Biotechnology Department, Faculty of Life Science, University of Benin, PMB 1154, Benin City, Nigeria*
*Corresponding author: matypas@biol.uoa.gr

Competing Interests: The Author(s) declare(s) no conflict of interest.

Editor: Juan A. Navas-Cortes, Spanish National Research Council (CSIC), Cordoba, Spain.

Summary. In plant pathogenic fungi, different signalling pathways operate to control responses to nutrient availability during plant infection. A candidate from the cAMP-PKA signalling pathway, the cAMP-dependent protein kinase A gene, *pka*C1, and the β-1,6-endoglucanase gene, *veg*B, involved in cell wall degradation, were studied in *V. dahliae*. Double mutants of the fungus were constructed, with insertional inactivation in the *pka*C1 and *veg*B genes. Different developmental traits and virulence towards eggplant were evaluated in single and double disruption mutants. In all media tested, double mutants showed better radial growth but less conidia and microsclerotia than the wild type. An interaction between *veg*B and *pka*C1 in controlling virulence on eggplants was recorded, as double mutants were slightly less virulent than the single mutant *veg*B$^-$, but more virulent than the single mutant *pka*C1$^-$. Concomitant or independent function of the two genes and the signaling pathways they operate in for the different growth parameters and virulence are discussed.

Keywords. Protein kinase A, cAMP-mediated PKA, cell wall degrading enzyme.

INTRODUCTION

Fungi, and especially plant pathogens, can produce many plant cell wall degrading enzymes (CWDEs) for penetration and successful infection of their hosts (Kubicek *et al.*, 2014). Involvement of these enzymes in penetration, plant defense induction, and symptom expression has been studied extensively (Cooper, 1987; Walton, 1994; Di Pietro *et al.*, 2003). However, although found to contribute to pathogenesis, in most cases no specific roles have been directly attributed to specific genes coding for CWDEs, and since the activities of these enzymes from different fungi show preferences for different types of plant biomass and adaption to pathogen lifestyles, their roles in pathogenesis remains unclear (King *et al.*, 2011; Couturier *et al.*, 2012).

Probable explanations for this are: (a) that CWDEs are indirectly involved in pathogenesis, and (b) that due to the organization in large multigene families of CWDEs in fungal pathogens, and the functional specialization that each enzyme exhibits, the unequivocal identification of a particular gene involved in pathogenicity is prevented by the masking of its function by the corresponding function of other genes in the same family (Walton, 1994; Coutinho *et al.*, 2003; Zhao *et al.*, 2013; CAZy website (http://www.cazy.org/CAZY/)). There is sufficient evidence for masking from a number of studies with mutants of inactivated CWDE genes that always retained at least some residual enzyme activity. Examples are: the xylanase genes of *Cochliobolus carbonum* (Apel-Birkhold and Walton, 1996); the polygalacturonase genes of *C. carbonum* (Scott-Craig *et al.*, 1990, 1998); the pectate lyase genes of *Nectria haematococca* (*Fusarium solani* f. sp. *pisi*), (Rogers *et al.*, 2000); and the xylanase genes of the rice pathogen *Magnaporthe grisea* (Nguyen *et al.*, 2011). In the case of *C. carbonum*, double or triple xylanase mutants were shown to retain full pathogenicity on maize (Apel-Birkhold and Walton, 1996). Similarly, in *V. dahliae*, disruption of a trypsin protease gene did not affect pathogenicity (Dobinson *et al.*, 2004), while an insertional mutant with the single copy β-1,6-endoglucanase gene showed only a minor reduction in virulence (Eboigbe *et al.*, 2014).

Beta-1,6-endoglucanase function has been little studied in plant pathogenic fungi. These enzymes are produced by several fungal species under conditions of carbon starvation, together with other hydrolytic enzymes to release carbon for survival by cell autolysis (Adams, 2004; Martin *et al.*, 2007). In mycoparasitic fungi, the role of β-1,6-endoglucanases in degradation of chitin and cell wall β-glucan has been demonstrated using gene disruption studies (Amey *et al.*, 2003). In mutualistic fungi (*Neotipodium* sp.) a β-1,6-endoglucanase protein was found to be secreted in the host plant apoplast, conferring a role in fungal nutrition. Based on the absence of β-1,6-glucosidic bonds (the main substrate of β-1,6-endoglucanases) in plants, Moy *et al.*, (2002) hypothesized that β-1,6-endoglucanases may be involved in the degradation of cell walls of other fungi to prevent them from infecting host plants. However, in plant pathogenic fungi β-1,6-endoglucanases are considered to also break β-1,3-glucosidic bonds present in plant cell wall callose during parasitic attack (Martin *et al.*, 2007). Furthermore, β-glucanases secreted by fungi or produced by hosts aid β-glucan degradation, releasing hydrolytic enzymes from fungal hyphae. In support of this hypothesis, growth of *Botrytis cinerea* on media containing β-glucanases (glucanex, β-1,3-glucanase) resulted in altered morphology of the mycelium and increased activity of hydrolytic enzymes, including peroxidases, laccases, and catalases (Gil-ad *et al.*, 2001).

To avoid the problems caused by the masking effects of gene expression in multigene families, attention was focused on the transcriptional regulation of CWDE genes expression and the signals that control whole sets of pathogenicity genes (Roncero *et al.*, 2008; Kubicek *et al.*, 2014). Tonukari *et al.* (2000) earlier reported that one way to overcome the problem of functional redundancy was to abolish expression of an entire class of CWDE enzymes by knocking out components of signal transduction pathways or transcriptional activators. The sucrose non-fermenting gene *SNF1* that regulates catabolite repression in *C. carbonum* was found necessary for the expression of several CWDEs (Tonukari *et al.*, 2000). Mutation of *SNF1* in this fungus led to varying levels of repression of CWDE genes, reduced growth on complex polymers such as xylan and pectin, and also reduced virulence on its maize host. Similarly, the disruption of *SNF1* in the vascular wilt fungus *F. oxysporum* resulted in strongly impaired pathogenicity that was attributed to the low induction of CWDE genes (Ospina-Giraldo *et al.*, 2003). Thus, the *VdSNF1* gene was essential for the induction of CWDEs and virulence in *V. dahliae*, confirming the significant role of these genes in pathogenicity of the fungus (Tzima *et al.*, 2011).

Although the implication of the *SNF1* pathway in regulation of CWDEs in plant pathogenic fungi is well documented, there is little information on whether CWDEs are also influenced by the cyclic AMP- Protein kinase A (cAMP-PKA) and/or mitogen-activated protein kinase (MAPK) signalling pathways. Both pathways are necessary for fungal pathogenesis and play important roles in the formation of appressoria (Tudzynski and Tudzynski, 2001; Lee *et al.*, 2003). The implication of the MAP kinase pathway in modulation of CWDEs was clearly indicated by reduced production of polygalacturonases and pectate lyase enzymes in the *Fusarium oxysporum* f. sp. *lycopersici* MAP kinase disruption mutant *fmk*1 (di Pietro *et al.*, 2003). In *Magnaporthe grisea*, activity of the cAMP-dependent protein kinase (PKA) increased during germination of conidia and appressorium formation on hydrophobic surfaces, whereas in mutants lacking the catalytic subunit of PKA (*cpk*a), appressorium formation was impaired (Kang *et al.*, 1999). Similarly, disruption of the catalytic subunit protein kinase A gene *VdPKAC1* (hereafter, *pka*C1) of *V. dahliae* caused reduced virulence, although mutants caused typical disease symptoms (Tzima *et al.*, 2010). Therefore, since *V. dahliae* does not possess specialized infection structures that in other fungi (including *M. grisea*) are essential for host invasion (Madhani and Fink 1998), we have tested if cAMP-dependent genes including *pka*C1 influence host penetration by *V. dahliae*, in combination with the β-1,6-endoglucanase gene (*veg*B) that was recently shown to affect the virulence of the fungus (Eboigbe *et al.*, 2014).

MATERIALS AND METHODS

Fungus isolates and culture conditions

A *V. dahliae* race 2 isolate from tomato, obtained in Greece, (isolate ID number 123wt-r2; hereafter referred to as wild type or wt), and the *veg*B⁻ disruption mutant derived from the wild type (Eboigbe *et al.*, 2014) were used in this study. For long term storage, the wild type and mutant strains were stored at -80°C, as conidial suspensions in 25% glycerol. Cultures were reactivated on freshly made potato dextrose agar (PDA).

Fungus growth experiments on different carbon sources were performed using a basal medium containing mineral salts and trace elements as in minimal medium (Puhalla and Mayfield, 1974). Carbon sources were added individually to the basal medium (at final concentrations of 2% w/v) and the amended media were adjusted to pH 6.5. Radial growth assays were performed on basal medium containing 2% agar, supplemented with 2% carboxycellulose, pectin, glucose and sucrose. Growth experiments were also performed on PDA (provided by two different suppliers; Difco and BP), that was prepared according to the manufacturers' instructions. Plates were each inoculated by placing 10 μL of conidial suspension (10^7 conidia mL^{-1}) of each strain at the centre of each inoculated plate. Fungal cultures were incubated at 23°C and the colony diameters were recorded at intervals of 2 to 3 d, while colony morphology was also observed and recorded.

For conidia production, 5 mm culture plugs from the edge of actively growing mycelia of each fungal strain were transferred to the centre of PDA plates. After 10 d of incubation, 3 mL of sterilized water was added to each plate to harvest conidia. Conidia were released by scraping off fungal cultures with a glass rod. The plates were then shaken gently and a small aliquot (10 μL) of the suspension from each plate was placed on a microscope slide haemocytometer to determine number of conidia.

Construction and verification of vegB/pkaC1 double disruptants

For the construction of *veg*B⁻/*pka*C1⁻ double disruptants, the catalytic subunit of the protein kinase A gene, *pka*C1, was disrupted in the *veg*B⁻ mutant strain by *Agrobacterium tumefaciens* mediated transformation (ATMT), as described by Tzima *et al.* (2010). Mutant *veg*B⁻, is a race 2 *V. dahliae* strain in which the β-1,6-endoglucanase gene has been disrupted by insertional inactivation, using hygromycin B as a selection marker gene (Eboigbe *et al.*, 2014). For ATMT of mutant *veg*B⁻, the *Agrobacterium* strain AGL1 was used, which carries the binary vector pGK-Gen, harbouring the geneticin resistance cassette. The binary vector pGPK-Gen was constructed in a previous study, and consists of a pGKO2 backbone (Khang *et al.*, 2005), in which the *V. dahliae pka*C1 mutant allele was cloned. The *pka*C1⁻ mutant allele was constructed from a 3 kb fragment including the *pka*C1 gene, in which the 0.73 kb *Bsr*GI/*Bbv*CI portion, located 274 nucleotides downstream from the start codon, was replaced with the 1,480 bp geneticin resistance gene cassette by blunt ending both vector and insert.

Double disruptants were constructed by ATMT of the *veg*B⁻ strain as previously described (Mullins *et al.*, 2001; Tzima *et al.*, 2010). Colonies of double mutants appeared after 3-4 d incubation on PDA supplemented with 50 μg mL^{-1} geneticin, 50 μg mL^{-1} hygromycin B and 50 μM F2dU (5-Fluoro-2-deoxyuridine; Sigma), and were subjected to single conidium isolation. The binary vector pGPK-Gen harbours the HSV*tk* gene (Herpes Simplex Virus Thymidine Kinase). The nucleoside analog F2dU is converted by the HSV*tk* gene product to a toxic compound against *V. dahliae*, selecting against ectopic transformants, thus increasing transformation efficiency (Khang *et al.*, 2005).

The confirmation of *pka*C1 disruption in the *veg*B⁻ mutant, thus creating *pka*C1⁻/*veg*B⁻ double disruptants, was achieved by PCR using primers VdPK1400bp-up (5'-AGCCCAACAGCCCCATTACCC-3') and VdPK-1400bp-dn (5'-GCCCAGGCGCTTCGTCAGA-3').

Pathogenicity assays on eggplants

Virulence of the *V. dahliae* wt, *pka*C1- and *veg*B⁻ single disruption mutants and double disruptants 123Δ*VP1*, 123Δ*VP2*, 123Δ*VP3*, 123Δ*VP4* was evaluated on eggplant ('Black Beauty'). Although the wt was a tomato strain, we carried out the pathogenicity assays on eggplant because this species is more susceptible than tomato to *V. dahliae* (e.g. more rapid disease development, and increased incidence and severity of disease), thus allowing precise evaluation of plant responses to the wt and mutant isolates.

For inocula preparation, fungal cultures grown for 5 to7 d in SSN medium (Sinha and Wood, 1968) were passed through several layers of cheesecloth (to remove mycelia). The concentration of conidia in resulting suspensions was adjusted to approx. 1 × 10^7 mL^{-1}. Ten plants at the second true leaf stage were inoculated by drenching the roots with 10 mL conidialsuspensions of each isolate. Virulence assays were performed twice.

Disease severity at each observation was expressed by the percent of leaves that showed wilting symptoms and was recorded for 39 d post inoculation. Disease ratings were plotted over time to generate the disease progress curves, and the areas under the disease progress curve (AUDPC) were calculated using the trapezoidal integration method (Campbell and Madden, 1990). Disease severity was expressed as percentage of the maximum possible AUDPC for the whole period of the experiment, and is referred to as relative AUDPC (Korolev *et al.*, 2001). Relative AUDPC values calculated for each treatment were subjected to ANOVA, and means were separated by Duncan's multiple range test.

RESULTS

Construction of double mutants pkaC1⁻/vegB⁻

Gene *pka*C1 was previously cloned from a *V. dahliae* tomato race 1 strain and found to have a coding sequence of 1,787 bp (Tzima *et al.*, 2010). Since the *vegB⁻* knockout mutant contained the hygromycin resistance gene, the binary vector pGPK was altered to carry the geneticin resistance cassette (pGPK-Gen), and was used for ATMT transformation of *vegB⁻* (Figure 1A; Tzima *et al.*, 2010). Double resistant to hygromycin and geneticin transformants were isolated on appropriate selective media and were examined for stability of the genotypes for several generations. Single colony derived stable transformants were subsequently tested by PCR to examine the sizes of amplicons for the *pka*C1 gene. Representative putatively double-disruptants producing a larger amplicon (2,183 bp) in comparison with the wild

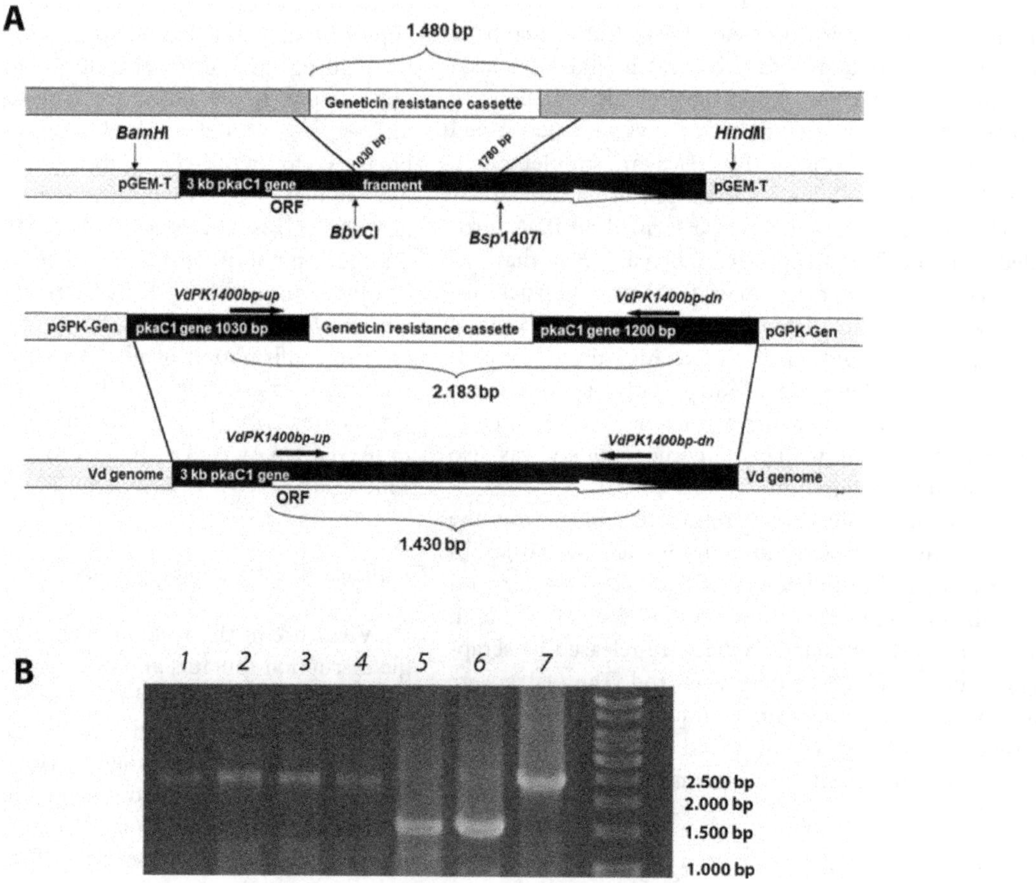

Figure 1. Generation of *Verticillium dahliae pka*C1/*veg*B double disruption mutants. A) Schematic presentation of *pka*C1 mutant allele construction and integration in the fungal genome by a double homologous recombination event. B) Disruption of *pka*C1 verified by amplification of the gene from four double disruptants 123Δ*VP1*, 123Δ*VP2*, 123Δ*VP3*, and 123Δ*VP4* (lanes 1-4; 2.183 bp), genomic DNA (lane 5) and plasmid DNA (lane 6) of clone pPKA3kb, with native fragment 1430 bp), and plasmid DNA of clone pGPK-Gen (lane 7).

type strain 1,430 bp amplicon are shown in Figure 1B. Transformants 123Δ*VP1*, 123Δ*VP2*, and 123Δ*VP4* were further examined for the possible genetic interaction between *veg*B with *pka*C1 in double knock-out *V. dahliae* mutants (*veg*B-/*pka*C1-), and for their pathogenicity in comparison with that of the single knock-out mutants and the wild type strain.

Production of conidia and utilization of carbon sources

The physiology and development of wild type *V. dahliae, veg*B$^-$ and double disruptants were examined by estimating conidium production, and their ability to degrade different carbon sources. Growth of mutant *pka*C1$^-$on glucose, sucrose and xylose was reduced compared to the wild type strain, while on pectin, carboxycellulose and Difco PDA this reduction was less apparent. Mutant *veg*B$^-$ and the double disruptants 123Δ*VP1*, 123Δ*VP2*, and 123Δ*VP4* showed slightly increased radial colony growth compared to the wild type strain and mutant *pka*C1$^-$, which was more pronounced on pectin, carboxycellulose and sucrose (Figure 2).

Colony morphology of all strains was similar in all carbon sources, except growth on Difco PDA. The *vegB*$^-$ mutant and the double disruptants formed abundant microsclerotia after 1 week of growth, while colonies of the wild type and *pka*C1$^-$ showed no microsclerotia at the same time and very few after 20 d. The use of PDA from another supplier (Scharlau) affected colony morphology of all strains with variable intensities of pigment formation. Mutant *veg*B$^-$ displayed increased and resuscitate growth compared to all other strains, while the double disruptants showed a fluffy growth phenotype with different patterns of microsclerotium production between the double disruptants. Colonies of 123Δ*VP1* were white, while 123Δ*VP2* produced microsclerotia at the colony centres and edges, and 123Δ*VP4* produced microsclerotia in internal rings within the colonies. This indicated the presence and absence of microsclerotium formation stimuli in the different PDA preparations.

In comparison with the wild type strain, production of conidia was reduced in all the other strains, at about 40-50%. The exception was *pka*C1,$^-$which showed an approx. 30% reduction.in production of conidia.

Pathogenicity assays on eggplant

Infection assays on eggplants were performed to determine the effects of single mutants *pka*C1$^-$, *veg*B$^-$, and double disruption mutants 123Δ*VP1*, 123Δ*VP2* and

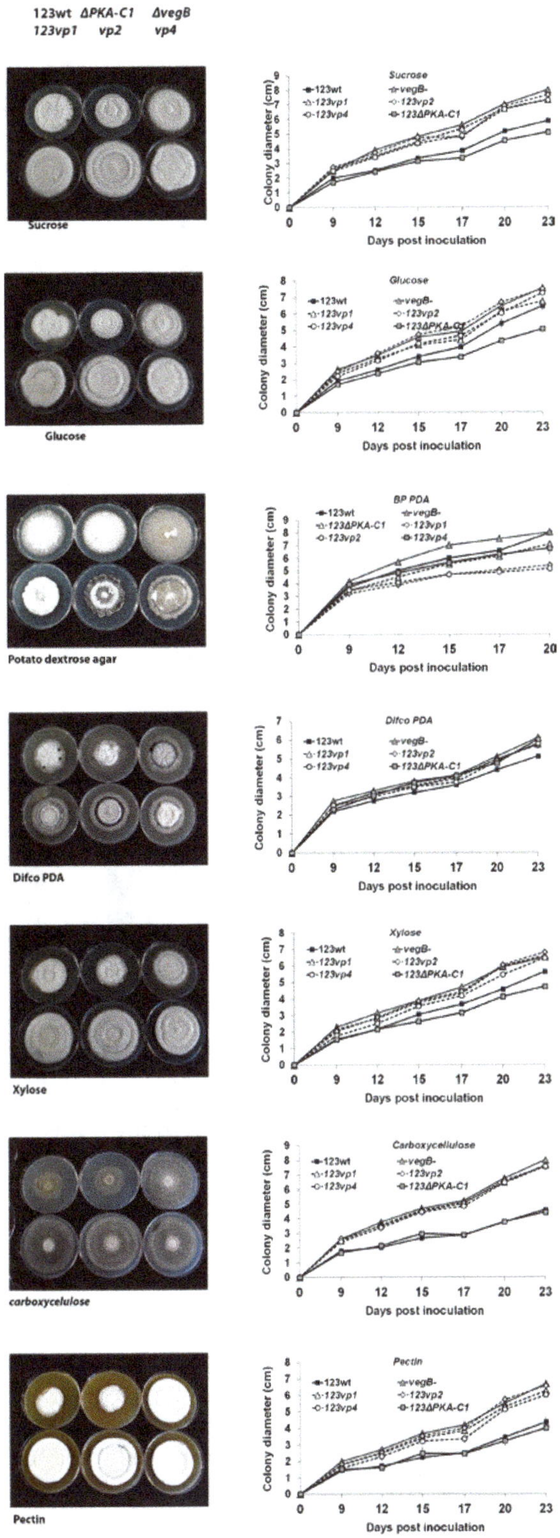

Figure 2. Radial growth of *Verticillium dahliae* wild type (123wt-r2), mutant *veg*B$^-$ and double disruptants 123Δ*VP1*, 123Δ*VP2* and 123Δ*VP4* on different carbon sources. Left: Images of fungal colonies on carbon sources after 23 d of growth. Right: Graphs presenting radial growth of the strains at specific time points.

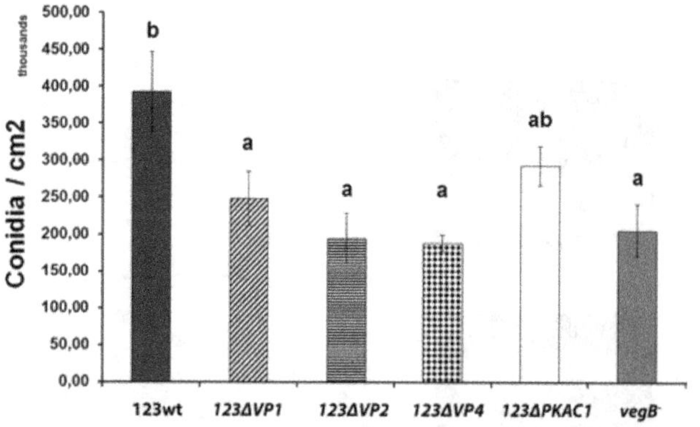

Figure 3. Production of conidia by *Verticillium dahliae* wild type (123wt-r2), mutant *veg*B- and double disruptants 123Δ*VP1*, 123Δ*VP2* and 123Δ*VP4*. Error bars indicate standard errors calculated for five replicates. Values calculated for each treatment were subjected to analysis of variance and means were separated by Fishers's least significant deifference (LSD) procedure. Columns with different letters are statistically different at $P \leq 0.05$.

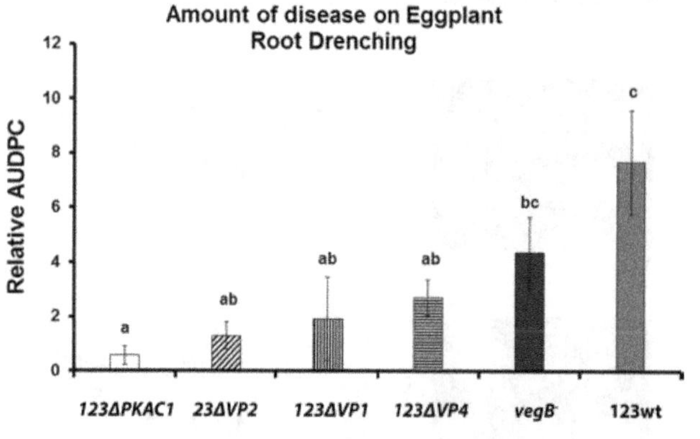

Figure 4. Disease severity caused by the *Verticillium dahliae* disruption mutants *veg*B-, *pka*C1-, double disruptants 123Δ*VP1*, 123Δ*VP2*, 123Δ*VP4* and the wild type 123wt-r2 on eggplants infected by root drenching. Disease severity was expressed as relative AUDPC. Bars indicate standard errors calculated for 20 replicates. Relative AUDPC values calculated for each treatment were subjected to analysis of variance and means were separated by Fishers's least significant deifference (LSD) procedure. Columns with different letters are statistically different at $P \leq 0.05$.

123Δ*VP4* on pathogenicity of *V. dahliae* in comparison with the wild type strain.

Mutant *pka*C1- caused significantly less disease (relative AUDPC of 0.55%) compared to the wild type strain (7.7% relative AUDPC). Disruption of *veg*B caused a smaller non-significant reduction in virulence of the *veg*B- mutant (4.4% relative AUDPC) compared to the wild type strain. Double disruptants 123Δ*VP1*, 123Δ*VP2* and 123Δ*VP4* caused intermediate amounts of disease compared to that caused by mutants *pka*C1- and *veg*B- (relative AUDPCs were 1.9% for 123Δ*VP1*, 1.3% for 123Δ*VP2* and 2.7% for 123Δ*VP4*; Figure 4).

DISCUSSION

The cAMP-PKA pathway has been extensively studied in yeast, but to a much lesser extent in plant pathogenic fungi. In plant pathogens, this signaling pathway has been implicated in morphological and physiological traits (conidiation, branching, growth), and in virulence (Yamauchi *et al.*, 2004; Tzima *et al.*, 2010; Kim *et al.*, 2011). Disruption of the catalytic subunit protein kinase A gene, *pka*C1, in a *V. dahliae* tomato race 1 strain resulted in reduced conidiation and growth, increased branching and reduced invasive growth (Tzima *et al.*, 2010).

Beta-1,6-endoglucanases, secreted by fungi, degrade mainly β-1,6-glucosidic bonds (present in fungus cell walls). However, there is accumulating evidence that these enzymes can also degrade β-1,3-glucosidic (laminarin and cell wall callose) and β-1,4-glucosidic bonds (chitin) in the environment, to achieve survival of the fungi under conditions of carbon starvation, by cell wall autolysis, as well as during plant- and myco- parasitism (Stahmann *et al.*, 1992; Amey *et al.*, 2003; Martin *et al.*, 2007). In plant pathogenic fungi, β-1,6-endoglucanases probably aid in plasticity of hyphae during plant-microbe interactions, and release of hydrolytic enzymes entrapped in the cell walls of hyphae by the presence of β-glucan (Gil-ad *et al.*, 2001). Disruption of the β-1,6-endoglucanase gene *veg*B in a *V. dahliae* race 2 isolate was reported by Eboigbe *et al.* (2014). Thus, to detect possible crosstalk between the β-1,6-endoglucanase gene and the PKA signalling pathway in *V. dahliae* race 2, double disruption mutants were created by a sequential approach.

In agreement with previous results (Tzima *et al.*, 2010; Eboigbe *et al.*, 2014), both single disruption mutants in *V. dahliae* race 2 that were examined here, *veg*B- and *pka*C1-, showed marked reduction in the production of conidia, which was more pronounced in *veg*B- than in *pka*C1-. However, no additional differences in the production of conidia were observed in the double mutants that were constructed. The *pka*C1- mutant had reduced growth on the different carbon sources tested, both in comparison with the wild type and also with the *vegB*- and the double disruption strains. VegB- and the three double mutants showed similar growth on fermentable (sucrose, glucose) and alternative (xylose, pectin, carboxycellulose) carbon sources, which was increased in comparison with the wild type strain. This indicates that the *veg*B⁻ disruption phenotype was dominant over the *pka*C1⁻ phenotype. A possible explanation for this reduced growth and conidiation phenotype of *pka*C1⁻ is that the mutant fungal strain had impaired perception of appropriate signals from the environment, thus failing to initiate conidiation and growth processes. This has been demonstrated in several other similar cases (Kronstad *et al.*, 1998; Casas-Flores *et al.*, 2006; Doehlemann *et al.*, 2006; Skamnioti and Gurr, 2007). In such mutants, the additional absence of β-1,6-endoglucanase may elevate the composition of fungal hyphae in its substrate, β-glucan (Martin *et al.*, 2007). As endoglucanases have been shown to participate in hydrolysis and re-construction of the fungal cell walls during growth and morphogenesis (Adams *et al.*, 2004), this may also further interfere with perception and response to environmental signals, thus explaining the dominant phenotype of *veg*B⁻ over *pka*C1⁻ disruption.

Glucose signalling in yeasts and filamentous fungi is controlled by different pathways including the cAMP-PKA pathway that regulates growth in response to glucose availability, whereas sucrose non-fermenting gene product (SNF1), induces transcriptional changes in the presence of low glucose concentrations or during growth on alternative carbon sources (Tonukari *et al.*, 2000; Hong *et al*, 2003; Santangelo, 2006; Zaman *et al.*, 2009). In yeasts, it has been shown that PKA activity controls the localization of the β subunit isoform (Sip1) of the Snf1 protein kinase and its complex with the Snf1 catalytic subunit in response to carbon source availability, indicating interference between the two pathways (Hedbacker *et al.*, 2004). Furthermore, microarray analysis in yeast revealed that the PKA pathway centrally controls cell growth and transcriptional changes of the majority of genes involved in response to glucose availability. SNF1 modulates the expression of a small number of genes, a significant portion of which is independent of the PKA pathway, including the majority of glucose repressed genes (Zaman *et al.*, 2009). The expression of *veg*B was reduced in *VdSNF1* mutants (Eboigbe *et al.*, 2014), indicating that *veg*B is under catabolite repression controlled by SNF1 in *V. dahliae*. Therefore, similar to its role in yeast, the SNF1 ortholog gene may also operate in *V. dahliae* and *veg*B may belong to a group of genes that function independently from the PKA pathway.

Previous results for plant–*V. dahliae* interactions (Tzima *et al.*, 2010; Eboigbe *et al.*, 2014) have shown that *pka*C1 is implicated in invasive growth of *V. dahliae* race 1, whereas *veg*B has only a minor effect on virulence in *V. dahliae* race 2. It was therefore expected that the disruption of *pka*C1 in *V. dahliae* race 2 would cause reduction in virulence. Disruption of both genes in race 2 of *V. dahliae* resulted in double mutants causing less disease in the sensitive eggplant-*V. dahliae* pathosystem than the single mutant *veg*B⁻, and more disease than the single mutant *pka*C1⁻. These results, though not strongly supported statistically, indicate a possible interaction between *pka*C1 and *veg*B in controlling virulence. Pathogenicity is complex, which includes growth on alternative carbon sources (e.g. pectin induces the production of hydrolytic enzymes), ramification into plant tissues, and confrontation with host defense mechanisms (Agrios, 2005). It is therefore difficult to speculate in which of these pathogenicity factors the functions of *pka*C1 and *veg*B may overlap, to fully explain the phenotypes observed in the double disruptant mutants of *V. dahliae* race 2. However, as disruption of *veg*B is thought to affect the activity of other hydrolytic enzymes (Gil-ad, 2001), and disruption of *pka*C1 possibly causes deregulation of the expression profiles of genes involved in similar processes, the combined action of these genes may form the basis for an explanation of the phenotypes observed. In support of this hypothesis, as well as to add to the complexity of the mechanisms involved, the disruption of a Ga subunit gene in *Botrytis cinerea* caused reduced expression of endoxylanase and endoglucanase genes, and increased expression of a pectate lyase gene, all of which aided plant cell wall degradation (Gronover *et al.*, 2001). Further investigation of the interaction of *pka*C1 and *veg*B in *V. dahliae* and the downstream targets affected is required to clarify pathogenicity mechanisms of this important plant pathogen.

LITERATURE CITED

Adams D.J., 2004. Fungal cell wall chitinases and glucanases. *Microbiology* 150: 2029–2035.

Agrios G.N., 2005. *Plant Pathology*. Elsevier Academic Press, London.

Amey R.C., Mills P.R., Bailey A., Foster G.D., 2003. Investigating the role of a *Verticillium fungicola* β-1,6-glucanase during infection of *Agaricus bisporus* using targeted gene disruption. *Fungal Genetics and Biology* 39: 264-275.

Apel-Birkhold P., Walton J., 1996. Cloning, disruption, and expression of two endo-beta 1,4-xylanase genes, *XYL2* and *XYL3*, from *Cochliobolus carbonum*. *Applied and Environmental Microbiology* 62: 4129-4135.

Campbell C.L,. Madden L.V., 1990. *Introduction to Plant Disease Epidemiology*. John Wiley & Sons, New York, NY, USA.

Casas-Flores S., Rios-Momberg M., Rosales-Saavedra T., Martinez P. -Hernandez P, ... Herrera-Estrella A., 2006. Cross talk between a fungal blue-light perception system and the cyclic AMP signaling pathway. *Eukaryotic Cell* 5: 499–506.

Cooper R.M., 1987. The use of mutants in exploring depolymerases as determinants of pathogenicity. In: *Genetics and Plant Pathogenesis* (P.R. Day, G.J. Jellis, ed.) Oxford, UK, Blackwell, 261–281.

Coutinho P.M., Stam M., Blanc E., Henrissat B., 2003. Why are there so many carbohydrate active enzyme-related genes in plants? *Trends in Plant Science* 8: 563–565.

Couturier M., Navarro D., Olive C., Chevret D., Haon M.,, ...Berrin J.G., 2012. Post-genomic analyses of fungal lignocellulosic biomass degradation reveal the unexpected potential of the plant pathogen *Ustilago maydis. BMC Genomics* 13: 57.

Di Pietro A., Madrid M.P., Caracuel Z., Delgano-Jarana J, Rocerno M.I.G., 2003. *Fusarium oxysporum*: exploring the molecular arsenal of a vascular wilt fungus. *Molecular Plant Pathology* 4: 315 - 325.

Dobinson K.F, Grant S.J., Kang S., 2004. Cloning and targeted disruption, via *Agrobacterium tumefaciens*-mediated transformation, of a trypsin protease gene from the vascular wilt fungus *Verticillium dahliae. Current Genetics* 45: 104-110.

Doehlemann G., Berndt P , Hahn M., 2006. Different signalling pathways involving a G alpha protein, cAMP and a MAP kinase control germination of *Botrytis cinerea* conidia. *Molecular Microbiology* 59: 821 - 835.

Eboigbe L., Tzima A.K., Paplomatas E.J., Typas M.A., 2014. The role of the β-1,6-endoglucanase gene *veg*B in physiology and virulence of *Verticillium dahliae. Phytopathologia Mediterranea* 53: 94-107.

Gil-ad N.L., Bar-Nun N. Meyer A., 2001. The possible function of the glucan sheath of *Botrytis cinerea* effects on the distribution of enzyme activities. *FEMS Microbiology Letters* 199: 109-113.

Gronover S.C., Kasulke D., Tudzynski P., Tudzynski B., 2001. The role of G protein alpha subunits in the infection process of the gray mold fungus *Botrytis cinerea. Molecular Plant Microbe Interactions* 14 : 1293–1302.

Hedbacker K., Townley R., Carlson M., 2004. Cyclic AMP dependent protein kinase regulates the subcellular localization of Snf1–Sip1 protein kinase. *Molecular and Cellular Biology* 24: 1836–1843.

Hong S.P., Leiper F.C., Woods A., Carling D., Carlson M., 2003 Activation of yeast Snf1 and mammalian AMP-activated protein kinase by upstream kinases. *Proceedings of the National Academy of Science USA* 100: 8839–8843.

Kang S.H., Khang C.H., Lee Y.H., 1999. Regulation of cAMP-dependent protein kinase during appressorium formation in *Magnaporthe grisea. FEMS Microbiology Letters* 170: 419-423.

Khang C.H., Park S.Y., Lee H.Y., Kang S., 2005. A dual selection based, targeted gene replacement tool for *Magnaporthe grisea* and *Fusarium oxysporum. Fungal Genetics and Biology* 35: 624 - 633.

Kim H.S., Park S.Y., Lee S., Adams E.L., Czymmek K., Kang S., 2011. Loss of cAMP-dependent protein kinase A affects multiple traits important for root pathogenesis by *Fusarium oxysporum. Molecular Plant-Microbe Interactions* 24: 719–732.

King B.C., Waxman K.D., Nenni N.V., Walker L.P., Bergstrom G.C., Gibson D.M., 2011. Arsenal of plant cell wall degrading enzymes reflects host preference among plant pathogenic fungi. *Biotechnology for Biofuels* 4: 4.

Korolev N., Perez-Artes E., Bejarano-Alcazar J., Rodriguez-Jurado D., Katan J., ..., Jimenez-Diaz R., 2001. Comparative study of genetic diversity and pathogenicity among populations of *Verticillium dahliae* from cotton in Spain and Israel. *European Journal of Plant Pathology* 107: 443-456.

Kronstad J., De Maria A., Funnell D., Laidlaw R.D., Lee N., ... Ramesh M., 1998. Signaling via cAMP in fungi: interconnections with mitogen-activated protein kinase pathways. *Archives of Microbiology* 170: 395-404.

Kubicek C.P., Starr T.L.,Glass N.L., 2014. Plant cell wall-degrading enzymes and their secretion in plant-pathogenic fungi. *Annual Review of Phytopathology* 52: 427-451.

Lee N., D'Souza C.A., Kronstad J.W., 2003. Of smuts, blasts, mildews, and blights: cAMP signaling in phytopathogenic fungi. *Annual Review of Phytopathology* 41: 399–427.

Madhani H.D., Fink G.R., 1998. The control of filamentous differentiation and virulence in fungi. *Trends in Cell Biology* 8: 348-352.

Martin K., McDougall B.M., McIlroy S., Chen J., Seviour R.J., 2007. Biochemistry and molecular biology of exocellular fungal β-(1,3)- and β-(1,6)-glucanases. *FEMS Microbiology Reviews* 31: 168–192.

Moy M.,. Li H.M., Sullivan R., , White J.F., Belanger F.C., 2002. Endophytic fungal b-1,6-glucanase expression in the infected host grass. *Plant Physiology* 130: 1298–1308.

Mullins E.D., Chen X., Romaine P., Raina R., Geiser D.M., Kang S. 2001. Agrobacterium-mediated transformation of *Fusarium oxysporum*: an efficient tool for insertional mutagenesis and gene transfer. *Phytopathology* 91: 173–180.

Nguyen Q.B., Itoh K., Vu B.V., Tosa Y., Nakayashiki H., 2011. Simultaneous silencing of endo-b-1,4 xylanase genes reveals their roles in the virulence of *Magnaporthe oryzaemmi. Molecular Microbiology* 81: 1008–1019.

Ospina-Giraldo M.D., Mullins E., Kang S., 2003. Loss of function of the *Fusarium oxysporum* SNF1 gene reduces virulence on cabbage and Arabidopsis. *Current Genetics* 44: 49-57.

Puhalla J.E., .Mayfield J.E., 1974. The mechanism of heterokaryotic growth in *Verticillium dahliae. Genetics* 76: 411-422.

Rogers L.M., Kim Y.K., Guo W., Gonzalez-Candelas L, Li D.,. Kolattukudy P.E., 2000. Requirement for either a host- or pectin-induced pectate lyase for infection of *Pisum sativum* by *Nectria hematococca. Proceedings of the National Academy of Science* 97: 9813–9818.

Roncero G.I.M., Orejas M., Di Pietro A., Hera C., Nieto F.C., 2008. Regulatory elements mediating expression of xylanase genes in *Fusarium oxysporum. Fungal Genetics and Biology* 45: 28–34.

Santangelo G.M., 2006. Glucose signaling in *Saccharomyces cerevisiae. Microbiology and Molecular Biology Reviews* 70: 253–282.

Sinha A.K., Wood R.K.S., 1968. Studies on the nature of resistance in tomato plants to *Verticillium albo-atrum. Annals of Applied Biology* 62: 319-327.

Scott-Craig J.S., Panaccione D.G., Cervone F.,Walton J.D., 1990. Endopolygalacturonase is not required for pathogenicity of *Cochliobolus carbonum* on maize. *Plant Cell* 2: 1191–1200.

Scott-Craig J.S., Cheng Y.Q., Cervone F., De Lorenzo G., Pitkin J.W., Walton J.D., 1998. Targeted mutants of *Cochliobolus carbonum* lacking the two major extracellular polygalacturonases. *Applied and Environmental Microbiology*. 64: 1497–1503.

Skamnioti P., Gurr S.G., 2007. *Magnaporthe grisea* cutinase2 mediates appressorium differentiation and host penetration and is required for full virulence. *Plant Cell* 19: 2674–2689.

Stahmann K., Pielken P., Schimz K.,Sahm H., 1992. Degradation of extracellular b-(1,3)(1,6)-D-glucan by *Botrytis cinerea. Applied and Environmental Microbiology* 58: 3347–3354.

Tonukari N.J., Scott-Craig J.S.,Walton J.D., 2000. The *Cochliobolus carbonum* SNF1 gene is required for cell wall-degrading enzyme expression and virulence in maize. *Plant Cell* 12: 237-247.

Tudzynski B., Tudzynski P., 2001. Pathogenicity factors and signal transduction in plant-pathogenic fungi. *Progress in Botany* 63: 163-188.

Tzima A.K., Paplomatas E.J., Rauyaree P., Kang S., 2010. Roles of the catalytic subunit of cAMP-dependent protein kinase A in virulence and development of the soilborne plant pathogen *Verticillium dahliae. Fungal Genetics and Biology* 47: 406-415.

Tzima A.K., Paplomatas E.J., Rauyaree P., Ospina-Giraldo M., Kang S., 2011. *VdSNF1*, the Sucrose Non-Fermenting Protein Kinase Gene of *Verticillium dahliae*, is required for virulence and expression of genes involved in cell wall degradation. *Molecular Plant-Microbe Interactions* 24: 129-142.

Walton J.D., 1994. Deconstructing the Cell Wall. *Plant Physiology* 104 :1113-1118.

Yamauchi J., Takayanagi N., Komeda K., Takano Y., Okuno T., 2004. cAMP-PKA signaling regulates multiple steps of fungal infection cooperatively with Cmk1 map kinase in *Colletotrichum lagenarium. Molecular Plant-Microbe Interactions* 7: 1355 - 1365.

Zaman S., Lippman S.I., Schneper L.,Slonim N., Broach J.R., 2009. Glucose regulates transcription in yeast through a network of signaling pathways. *Molecular Systems Biology* 5: 245.

Zhao Z., Liu H., Wang C., Xu J.R., 2013.Comparative analysis of fungal genomes reveals different plant cell wall degrading capacity in fungi. *BMC Genomics* 14: 274.

PERMISSIONS

All chapters in this book were first published in PM, by Firenze University Press; hereby published with permission under the Creative Commons Attribution License or equivalent. Every chapter published in this book has been scrutinized by our experts. Their significance has been extensively debated. The topics covered herein carry significant findings which will fuel the growth of the discipline. They may even be implemented as practical applications or may be referred to as a beginning point for another development.

The contributors of this book come from diverse backgrounds, making this book a truly international effort. This book will bring forth new frontiers with its revolutionizing research information and detailed analysis of the nascent developments around the world.

We would like to thank all the contributing authors for lending their expertise to make the book truly unique. They have played a crucial role in the development of this book. Without their invaluable contributions this book wouldn't have been possible. They have made vital efforts to compile up to date information on the varied aspects of this subject to make this book a valuable addition to the collection of many professionals and students.

This book was conceptualized with the vision of imparting up-to-date information and advanced data in this field. To ensure the same, a matchless editorial board was set up. Every individual on the board went through rigorous rounds of assessment to prove their worth. After which they invested a large part of their time researching and compiling the most relevant data for our readers.

The editorial board has been involved in producing this book since its inception. They have spent rigorous hours researching and exploring the diverse topics which have resulted in the successful publishing of this book. They have passed on their knowledge of decades through this book. To expedite this challenging task, the publisher supported the team at every step. A small team of assistant editors was also appointed to further simplify the editing procedure and attain best results for the readers.

Apart from the editorial board, the designing team has also invested a significant amount of their time in understanding the subject and creating the most relevant covers. They scrutinized every image to scout for the most suitable representation of the subject and create an appropriate cover for the book.

The publishing team has been an ardent support to the editorial, designing and production team. Their endless efforts to recruit the best for this project, has resulted in the accomplishment of this book. They are a veteran in the field of academics and their pool of knowledge is as vast as their experience in printing. Their expertise and guidance has proved useful at every step. Their uncompromising quality standards have made this book an exceptional effort. Their encouragement from time to time has been an inspiration for everyone.

The publisher and the editorial board hope that this book will prove to be a valuable piece of knowledge for researchers, students, practitioners and scholars across the globe.

LIST OF CONTRIBUTORS

Michael Fischer
JKI-Julius Kühn-Institute, Federal Research Centre for Cultivated Plants, Plant Protection in Fruit Crops and Viticulture, D-76833 Siebeldingen, Germany

Samira Peighami Ashnaei
Plant Disease Research Department, Iranian Research Institute of Plant Protection (IRIPP), Agricultural Research, Education and Extension Organization (AREEO), Tehran, Iran

Marilinda Lorenzini and Giacomo Zapparoli
Università degli Studi di Verona, Dipartimento di Biotecnologie, 37134 Verona, Italy

Maria Stella Cappello
Istituto di Scienze delle Produzioni Alimentari, CNR, 73100 Lecce, Italy

Giancarlo Perrone and Antonio Logrieco
Istituto di Scienze delle Produzioni Alimentari, CNR, 70126 Bari, Italy

Pierluigi Reveglia
National Wine and Grape Industry Centre, School of Agricultural and Wine Sciences, Charles Sturt University, Locked Bag 588 Wagga Wagga NSW 2678, Australia
University of Naples Federico II, Department of Chemical Sciences Complesso Universitario Montesant'Angelo, Via Cintia 4, 80126, Naples, Italy

Marco Masi, Alessio Cimmino and Antonio Evidente
University of Naples Federico II, Department of Chemical Sciences Complesso Universitario Montesant'Angelo, Via Cintia 4, 80126, Naples, Italy

Sami Michereff, Tamara Cinelli and Laura Mugnai
Dipartimento di Scienze delle Produzioni Agroalimentari e dell'Ambiente, Sez. Patologia vegetale ed entomologia, Università di Firenze, Piazzale delle Cascine 28, 50144 Firenze, Italy

AyşE Gulden Aday Kaya
Isparta Applied Sciences University, Yenişarbademli Vocational School, Isparta, Turkey

Şule Yeltekin and Tuğba Doğmuş Lehtijärvi
Isparta Applied Sciences Faculty of Forestry, 32260 Isparta, Turkey

Asko Lehtijärvi
Bursa Technical University, Faculty of Forestry, 16310 Bursa, Turkey

Steve Woodward
University of Aberdeen, Institute of Biological and Environmental Sciences, Aberdeen AB24 3UU, Scotland, United Kingdom

Vincenzo Mondello, Alessandro Spagnolo, Christophe Clement and Florence Fontaine
SFR Condorcet FR CNRS 3417, Université de Reims Champagne-Ardenne, Résistance Induite et Bioprotection des Plantes, RIBP EA 4707, BP 1039, Reims, Cedex 2 51687, France

Philippe Larignon
Institut Français de la Vigne et du Vin Pôle Rhône-Méditerranée, 7 Avenue Cazeaux, 30230 Rodilhan, France

Vladimiro Guarnaccia
Westerdijk Fungal Biodiversity Institute, Uppsalalaan 8, 3584 CT, Utrecht, The Netherlands
DiSAFA, University of Torino, Largo Paolo Braccini, 2, 10095 Grugliasco, TO, Italy

Pedro W. Crous and Marcelo Sandoval-Denis
Westerdijk Fungal Biodiversity Institute, Uppsalalaan 8, 3584 CT, Utrecht, The Netherlands
Faculty of Natural and Agricultural Sciences, Department of Plant Sciences, University of the Free State, Bloemfontein 9300, South Africa

Dalia Aiello and Giancarlo Polizzi
Dipartimento di Agricoltura, Alimentazione e Ambiente, Universita degli Studi di Catania, Via S. Sofia 100, 95123 Catania, Italy

Santa Olga Cacciola
Department of Agriculture, Food and Environment (Di3A), University of Catania, V. S. Sofia 100, 95123 Catania, Italy

Maria Lodovica Gullino
Centre for Innovation in the Agro-Environmental Sector, AGROINNOVA, University of Torino, Largo Braccini 2, 10095 Grugliasco (TO), Italy

Arezki Lehad, Meriem Louanchi and Mouni Aitouada
Laboratoire de Phytopathologie et Biologie Moléculaire. Ecole Nationale Supérieure d'Agronomie. Rue Hacen Badi, Belfort, El Harrach, 16000 Alger, Algeria

Ilhem Selmi and Naima Mahfoudhi
Laboratoire de Protection des Végétaux, Institut National de la Recherche Agronomique de Tunisie, Rue Hedi Karray, 2049 Ariana, Tunisia

Maria Luisa Raimondo, Antonia Carlucci, Claudio Ciccarone and Francesco Lops
Department of Science of Agriculture, Food and Environment, University of Foggia, Via Napoli, 25, 71122 Foggia, Italy

Abderraouf Sadallah
Department of Science of Agriculture, Food and Environment, University of Foggia, Via Napoli, 25, 71122 Foggia, Italy
CIHEAM- Mediterranean Agronomic Institute of Bari, Via Ceglie, 9, 70010 Valenzano (Ba), Italy
Department of Soil, Plant and Food Sciences, University of Bari Aldo Moro, Via Amendola 165/A, 70126, Bari, Italy

Tahreer Issa and Michele Tucci
Agronomic Mediterranean Institute of Bari (IAMB), Valenzano (BA), Italy

Leen Almadi, Roberto Buonaurio and Franco Famiani
Dipartimento di Scienze Agrarie Alimentari e Ambientali, Università degli Studi di Perugia, Perugia, Italy

Samer Jarrar
Canaan Center for Organic Research and Extension (CORE), Canaan, Palestine

Marco Scortichini, Luigi Zampella and Francesco Mastrobuoni
Council for Agricultural Research and the Analysis of Agricultural Economics (CREA), Research Centre for Olive, Fruit Trees and Citrus, Via Torrino 3, I-81110 Caserta, Italy

Danilo Migoni, Federica Angilè, Laura Del Coco, Chiara Roberta Girelli and Francesco Paolo Fanizzi
Department of Biological and Environmental Sciences and Technology (Di.S.Te.B.A.), University of Salento, Campus Ecotekne, Via Prov.le Lecce-Monteroni, I-73100 Lecce, Italy

Vedat Görür and Davut Soner Akgül
Department of Plant Protection, Agriculture Faculty, University of Cukurova, 01330, Balcali, Adana/Turkey

Nicolas Quiroga, Nicola Fiore, Ximena Gonzalez, Alan Zamorano and Ana Maria Pino
University of Chile, Facultad de Ciencias Agronómicas, Departamento de Sanidad Vegetal, Santiago, Chile

Valeria Longone
Instituto Nacional de Tecnología Agropecuaria, Estación Experimental Agropecuaria Mendoza, Laboratorio de Fitopatología, San Martin 3853, M5534 Luján de Cuyo, Mendoza, Argentina

Luca Picciau and Alberto Alma
Dipartimento di Scienze Agrarie, Forestali e Alimentari (DISAFA), Università di Torino, Largo P. Braccini 2, 10095 Grugliasco, Torino, Italy

Samanta Paltrinieri, Nicoletta Contaldo and Assunta Bertaccini
Department of Agricultural and Food Sciences (DISTAL), Plant Pathology and Entomology, Alma Mater Studiorum, Università di Bologna, Viale G. Fanin 42, 40127 Bologna, Italy

Haifa Ben Hadj Daoud and Naima Boughalleb-M'hamdi
Department of Biological Sciences and Plant Protection, High Institute of Agronomy of Chott Mariem, 4042 Sousse, UR13AGR03, University of Sousse, Tunisia

Elena Baraldi and Pamela Leonardi
Department of Agricultural and Food Sciences, DISTAL, University of Bologna, Via le Fanin 46, 40127, Bologna, Italy

Mirco Iotti
Department of Life, Health and Environmental Sciences, University of L'Aquila, via Vetoio, Coppito 1, 67100 L'Aquila, Italy

Francesco Calzarano, Giuseppe Arfelli and Anna Chiara Manetta
Università degli Studi di Teramo, Facoltà di BioScienze e Tecnologie Agro-Alimentari ed Ambientali, Via Renato Balzarini, 1, 64100 Teramo, Italy

Gabriele Valentini
Università degli Studi di Bologna, Dipartimento di Scienze e Tecnologie Agro-Alimentari, Viale Fanin 46, 40127 Bologna, Italy

Leonardo Seghetti
IIS "C. Ulpiani", Via della Repubblica, 30, 63100 Ascoli Piceno, Italy

Elisa Giorgia Metruccio and Stefano Di Marco
CNR IBE, Via Gobetti 101, 40129 Bologna, Italy

Maria Cristina Canale, Cristiano Nunes Nesi and Eduardo Cesar Brugnara
Agricultural Research and Rural Extension Company of Santa Catarina State (Epagri), 89803-904 Chapecó, Santa Catarina, Brazil

Bianca Ramos Falkenbach and Cesar Arthur Hunhoff Da Silva
Agronômica, Laboratory of Phytosanitary Diagnosis and Consultancy, 91530-000 Porto Alegre, Rio Grande do Sul, Brazil

Marwa Hassine and Khaled Sassi
University of Carthage, National Agronomic Institute of Tunisia, LR14AGR01 Laboratory of Genetic and Cereal Breeding, 1082, Tunis, Tunisia

Marwa Laribi
University of Carthage, National Agronomic Institute of Tunisia, LR14AGR01 Laboratoryof Genetic and Cereal Breeding, 1082, Tunis, Tunisia
CRP Wheat Septoria Precision Phenotyping Platform, 43 Avenue Charles Nicolle, 1082 Tunis, Tunisia

Fernanda M. Gamba
Departamento de Protección Vegetal, Facultad de Agronomía, UDELAR, Estación Experimental, Dr. M. A. Cassinoni, Ruta 3 k363, Paysandú, Uruguay

Pawan K. Singh
Centro Internacional de Mejoramiento de Maíz y Trigo (CIMMYT), Km 45 Carretera México-Veracruz, El Batán, Texcoco, CP 56237, Estado de México, México

Amor Yahyaoui
Centro Internacional de Mejoramiento de Maíz y Trigo (CIMMYT), Km 45 Carretera México-Veracruz, El Batán, Texcoco, CP 56237, Estado de México, México
CRP Wheat Septoria Precision Phenotyping Platform, 43 Avenue Charles Nicolle, 1082 Tunis, Tunisia

Elaine Basson and Charmaine Olivier
Citrus Research International, P.O. Box 28, Nelspruit 1200, South Africa

Jan M. Van Niekerk
Citrus Research International, P.O. Box 28, Nelspruit 1200, South Africa

Gray-Lee Carelse
Department of Plant Pathology, University of Stellenbosch, Private Bag X1, Matieland 7602, South Africa

Vladimiro Guarnaccia
Department of Plant Pathology, University of Stellenbosch, Private Bag X1, Matieland 7602, South Africa
DiSAFA, University of Torino, Largo Paolo Braccini 2, 10095, Grugliasco, TO, Italy

Milton A. Typas
Division of Genetics and Biotechnology, Department of Biology, National and Kapodistrian University Athens, Panepistemiopolis, Kouponia,15701 Athens, Greece

Lugard Eboigbe
Division of Genetics and Biotechnology, Department of Biology, National and Kapodistrian University Athens, Panepistemiopolis, Kouponia,15701 Athens, Greece
Plant Biology and Biotechnology Department, Faculty of Life Science, University of Benin, PMB 1154, Benin City, Nigeria

Aliki K. Tzima and Epaminondas J. Paplomatas
Laboratory of Plant Pathology, Agricultural University of Athens, 75 Iera Odos, 11855 Athens, Greece

Index

www.ingramcontent.com/pod-product-compliance
Lightning Source LLC
LaVergne TN
LVHW080908210926
848485LV00004B/614

* 9 7 8 1 6 3 9 8 7 1 9 0 2 *